Engineering Response to Climate Change

SECOND EDITION

Engineering Response to Climate Change

SECOND EDITION

Edited by **Robert G. Watts**

CRC Press
Taylor & Francis Group
Boca Raton London New York

CRC Press is an imprint of the
Taylor & Francis Group, an **informa** business

CRC Press
Taylor & Francis Group
6000 Broken Sound Parkway NW, Suite 300
Boca Raton, FL 33487-2742

First issued in paperback 2017

Version Date: 20130129

ISBN 13: 978-1-4398-8846-9 (hbk)
ISBN 13: 978-1-138-07411-8 (pbk)

Visit the Taylor & Francis Web site at
http://www.taylorandfrancis.com

and the CRC Press Web site at
http://www.crcpress.com

To my children

and my grandchildren,

and to yours, too.

Contents

Preface

In June 1991, we convened the Workshop on the Engineering Response to Global Climate Change: Planning a Research and Development Agenda. Approximately 70 scientists and engineers spent four days discussing what research was necessary to identify the causes and the extent of climatic change, to assess its consequences, and to prepare for mitigative and adaptive measures. Seven working groups were established: Sources and Sinks of Greenhouse Gases; Energy Demand, Energy Supply; Agricultural and Biological Systems; Water Resources; Coastal Hazards; and Geoengineering. Each working group was asked to identify a set of goals and a set of approaches to accomplishing these goals. Common goals and approaches were then identified, and small, interdisciplinary groups were convened to expand upon the various approaches. Some chapters of this book are updates of chapters from the book *Engineering Response to Global Climate Change* published in 1997. However, in a real sense, this is a new book. Several chapters introduce new material. Even so, the general idea is the same: the climate/energy problem has become largely an engineering problem. This book is more about what we as engineers can do to prevent, mitigate, or adapt to climate change.

Acknowledgments

I am especially thankful to Irma Shagla-Britton for her almost endless patience while I assembled this book and also to Kathryn Everett. It was important to each of us to have the very best scientists author the chapters. As is usually the case, those with the best credentials are also the ones with the most on their respective plates, and to them I must also offer my thanks. Editing a book is more difficult than writing one, and writing chapters is an arduous task indeed. My heartfelt thanks to each of them.

Thanks also to Jean for prodding me to prod the chapter authors when I hadn't the heart to harass my friends whose lives I knew were busy, and were doing their best to satisfy deadlines—mine and others.

I also thank Eva for protecting my door by barking especially fiercely when no one was there.

Editor

Robert G. Watts is the Cornelia and Arthur L. Jung Professor of Mechanical Engineering Emeritus at Tulane University. After receiving his undergraduate degree in mechanical engineering at Tulane University, he received an MS in nuclear engineering from the Massachusetts Institute of Technology and a PhD in mechanical engineering from Purdue University in 1965. He spent a postdoctoral year at Harvard University studying atmospheric and ocean science. During the 1977 academic year, he was at the Institute for Energy Analysis in Oak Ridge, Tennessee, and during the 1985 academic year, he was a senior scientist at the International Institute for Applied Systems Analysis in Laxenburg, Austria, where he studied the effects of climate change on agriculture. After the Tulane administration closed the department of mechanical engineering following hurricane Katrina, he taught at the United States Naval Academy in Annapolis, Maryland. He is the author of *Global Warming and the Future of the Earth* (2002), *Essentials of Applied Mathematics for Engineers and Scientists* (2009, 2012), and *Keep Your Eye on the Ball: The Science and Folklore of Baseball* (1990). He is the editor and participating author of *Engineering Response to Global Climate Change* (1997) and *Innovative Energy Strategies for CO_2 Stabilization* (2002). In 2009, he received the Dixie Lee Ray Award from the American Society of Mechanical Engineers (ASME) for his work on environmental problems and engineering solutions. He is an expert in both global warming and energy systems. He currently serves on the Environmental Protection Agency's Science Advisory Board. He is "retired" (to the extent that any teacher and scientist retires) and lives in Annapolis during hurricane season and in New Orleans when it gets cold.

Contributors

Phillip H. Burgi
Environmental and Water Resources Institute
Denver, Colorado

Darryl Calkins
U.S. Army Cold Regions Engineering Laboratory
Hanover, New Hampshire

Kate Calvin
Pacific Northwest National Laboratory
Joint Global Change Research Institute
University of Maryland
College Park, Maryland

Darienne Ciuro
Department of Atmospheric Sciences
University of Illinois
Urbana, Illinois

Leon Clarke
Pacific Northwest National Laboratory
Joint Global Change Research Institute
University of Maryland
College Park, Maryland

Matteo Convertino
Department of Agricultural and Biological Engineering
University of Florida
Gainesville, Florida

David R. Criswell
Cis-Lunar, Inc.
Houston, Texas

William R. Curtis
Engineer Research and Development Center
U.S. Army Corps of Engineers
Vicksburg, Mississippi

Robert G. Dean
College of Engineering
University of Florida
Gainesville, Florida

Jim Dooley
Pacific Northwest National Laboratory
Joint Global Change Research Institute
University of Maryland
College Park, Maryland

Hadi Dowlatabadi
Institute for Resources Environment and Sustainability
and
Liu Institute for Global Studies
The University of British Columbia
Vancouver, British Columbia, Canada

Jae Edmonds
Pacific Northwest National Laboratory
Joint Global Change Research Institute
University of Maryland
College Park, Maryland

Jiyong Eom
Pacific Northwest National Laboratory
Joint Global Change Research Institute
University of Maryland
College Park, Maryland

Richard H. French
Department of Civil and Environmental Engineering
University of Texas at San Antonio
San Antonio, Texas

Earl J. Hayter
Engineer Research and Development Center
U.S. Army Corps of Engineers
Vicksburg, Mississippi

Jeffery P. Holland
Engineer Research and Development Center
U.S. Army Corps of Engineers
Vicksburg, Mississippi

James R. Houston
Engineer Research and Development Center
U.S. Army Corps of Engineers
Vicksburg, Mississippi

Stacy E. Howington
Coastal and Hydraulics Laboratory
Engineer Research and Development Center
U.S. Army Corps of Engineers
Vicksburg, Mississippi

Bernard Hsieh
Engineer Research and Development Center
U.S. Army Corps of Engineers
Vicksburg, Mississippi

Yogesh P. Khare
College of Engineering
University of Florida
Gainesville, Florida

and

Research and Development Center
U.S. Army Corps of Engineers
Vicksburg, Mississippi

Son H. Kim
Pacific Northwest National Laboratory
Joint Global Change Research Institute
University of Maryland
College Park, Maryland

Robert E. Kopp
Department of Earth and Planetary Sciences
Rutgers University
New Brunswick, New Jersey

Page Kyle
Pacific Northwest National Laboratory
Joint Global Change Research Institute
University of Maryland
College Park, Maryland

Igor Linkov
U.S. Army Engineer Research and Development Center
Concord, Massachusetts

Patrick Luckow
Pacific Northwest National Laboratory
Joint Global Change Research Institute
University of Maryland
College Park, Maryland

Michael C. MacCracken
Climate Institute
Washington, DC

William D. Martin
Coastal and Hydraulics Laboratory
Engineer Research and Development Center
U.S. Army Corps of Engineers
Vicksburg, Mississippi

William H. McAnally
Civil and Environmental Engineering
Mississippi State University
Starkville, Mississippi

Ashish J. Mehta
College of Engineering
University of Florida
Gainesville, Florida

Barbara Miller
World Bank
Washington, DC

Arthur W. Molvik
Fusion Energy Consultant
Livermore, California

Clay L. Montague
College of Engineering
University of Florida
Gainesville, Florida

Richard Moss
Pacific Northwest National Laboratory
Joint Global Change Research Institute
University of Maryland
College Park, Maryland

Pralit Patel
Pacific Northwest National Laboratory
Joint Global Change Research Institute
University of Maryland
College Park, Maryland

Maryam Rezaei
Institute for Resources Environment and Sustainability
The University of British Columbia
Vancouver, British Columbia

Jose E. Sanchez
Coastal and Hydraulics Laboratory
Engineer Research and Development Center
U.S. Army Corps of Engineers
Vicksburg, Mississippi

Walter Short
National Renewable Energy Laboratory
Golden, Colorado

Temitope Taiwo
Nuclear Engineering Division
Argonne National Laboratory
Argonne, Illinois

Jim Thomas
U.S. Bureau of Reclamation
Denver, Colorado

Allison Thomson
Pacific Northwest National Laboratory
Joint Global Change Research Institute
University of Maryland
College Park, Maryland

James R. Tuttle
U.S. Army Corps of Engineers
Vicksburg, Mississippi

Robert G. Watts
Department of Mechanical Engineering
Tulane University
New Orleans, Louisiana

Donald J. Wuebbles
Department of Atmospheric Sciences
University of Illinois
Urbana, Illinois

Marshall Wise
Pacific Northwest National Laboratory
Joint Global Change Research Institute
University of Maryland
College Park, Maryland

Yuyu Zhou
Pacific Northwest National Laboratory
Joint Global Change Research Institute
University of Maryland
College Park, Maryland

1

The Fifth Revolution

Robert G. Watts

CONTENTS

1.1 Introduction

Certainly, one of the most important events in the long history of mankind's development was the mastery of fire. In his book, *The Next Million Years*, Charles Galton Darwin (1953) suggests that the history of humankind has seen four occasions when man took a step forward that was essentially irreversible, in the sense that the progress afforded was never lost. He refers to these as four revolutions. The first of these was the discovery of fire. It was, of course, not the discovery of fire, but its mastery, that was so important. The use of fire for cooking and warmth, and eventually for the creation of tools out of bronze and iron, changed the nature of life on earth irreversibly. The second revolution was the invention of agriculture. The people who participated in organized agricultural practices were able to provide themselves with food—both animal and vegetable—more readily than hunter–gatherers. The third of Darwin's revolutions was the urban revolution. The invention of cities brought the advantages of the division of labor and the establishment and regular practice of trade, through which people in different locations could obtain goods not available locally. According to Darwin, the fourth revolution was the scientific revolution, which he describes as "the discovery that it is possible consciously to make discoveries about the fundamental nature of the world, so that by their means man can intentionally and deliberately alter his way of life."

It is useful to contemplate what Darwin termed the fifth revolution. This will occur when humankind exhausts the store of the source of the first revolution: fire.

The sources of fire (heat) that powered the industrial revolution have been mainly coal, oil, and natural gas—the fossil fuels. Both Darwin and Haldane (1923) before him point out that resources of these fossil fuels are but centuries from exhaustion. In the long run, we shall have to find alternatives to fossil fuels to produce energy, and both Darwin and Haldane went on to suggest the usual possibilities of harnessing energy from the sun directly, or indirectly through wind, tides, or rivers. The possibility of atomic power was discussed by both Darwin and Haldane, while Darwin also discussed geothermal heat and the use of the vertical temperature gradient of the ocean, as well as the use of plants to produce alcohol as a convenient fuel.

Neither Darwin nor Haldane appears to have understood (although it had been previously suggested by other scientists) that another potential problem concerning the use of fossil fuels would soon arise. The burning of fossil fuels releases carbon dioxide (CO_2) into the atmosphere, and CO_2 strongly absorbs infrared radiation. The result is that solar radiation is allowed to penetrate the atmosphere, while infrared radiation is selectively impeded from leaving the earth's atmosphere. If the atmospheric loading of CO_2 becomes large enough, the earth's climate will change; the earth will become warmer. The consequences of a warmer earth are largely unknown, but they will surely include the shifting of climatic zones, with resulting changes in the regions where agriculture can be successfully practiced; changes in the flows of rivers and streams; and, quite possibly, changes in sea level. This means that we may be faced with Darwin's fifth revolution sooner than either Darwin or Haldane suspected.

Darwin clearly believed that the fifth revolution, brought about by the exhaustion of fossil fuels, would differ fundamentally from the first four. While the first four revolutions led to the ability to sustain large increases in population, or to improve the comfort of those populations, the fifth would have the opposite effect. Technological optimists, however, believe (Weinberg 1966) that we are far from finished with Darwin's fourth revolution. The essence of the scientific revolution lies in the discovery that nature can be understood and that humankind can, by intelligent manipulation, use this understanding to create better conditions for human life. It is in this spirit that this book has been prepared.

1.2 Nature of the Problem

The atmospheric concentrations of a number of gases that are chemically or radiatively active are increasing. Most of the increase is, by far, the result of human activities. Many of these gases absorb more strongly in the infrared part of the radiation spectrum than at the shorter wavelengths characteristic of solar radiation. The sun emits radiation principally in the wavelength band between 0.2 and 4.0 μm, while the earth and its atmosphere emit in the range of about 4 to 100 μm. The atmosphere absorbs about 23% of the incoming solar radiation, principally by ozone and water vapor. On the other hand, water vapor, CO_2, methane (CH_4), nitrous oxide (N_2O), chlorofluorocarbons (CFCs), and a number of other atmospheric gases exhibit very strong absorption of radiation in the infrared range associated with terrestrial radiation. The trapping of outgoing terrestrial radiation by the atmosphere is referred to as the greenhouse effect. Without the presence of these greenhouse gases (GHGs), the earth would be much colder than it is (see Section 1.3.1). If the concentrations of GHGs increase substantially as a result of human activities, practically all scientists now believe that the earth will become warmer. In other words, a man-made greenhouse effect will be created that enhances the natural greenhouse effect.

A number of activities that support modern societies results in the production of GHGs. For example, the best-known GHG, CO_2, is produced by the burning of fossil fuel, by deforestation, and to a lesser extent by industrial processes such as cement manufacture. Methane is produced by activities associated with energy production (e.g., coal mining and natural gas production and distribution) and with food production (ruminant animals and rice farming). These and other sources of GHGs are discussed in Chapter 2.

In light of the increasing atmospheric concentrations of GHGs, we need to speculate about what the future holds. How serious is the concern about greenhouse warming?

To estimate future GHG-induced climatic change, it is necessary to know future atmospheric concentrations of GHGs as well as the effect of these gases on the climate.

1.3 Greenhouse Effect and Climate Change

1.3.1 Natural Greenhouse

A primary reason for concern about the increasing concentrations of these gases is that they are strong absorbers of thermal radiation in the infrared wavelength range while absorbing very weakly in the short wavelength range of solar radiation (see Section 1.2). They, therefore, allow solar radiation to pass through the atmosphere, but selectively absorb terrestrial radiation emitted by the earth and its atmosphere. As noted earlier, the resulting trapping of heat in the earth's—atmospheric system is generally referred to as the greenhouse effect, and the gases as GHGs. If no such gases were present in the earth's atmosphere, the earth would be considerably colder than it is today. It would, in fact, be too cold to sustain life. Conversely, if the concentrations of GHGs were higher, the earth would be a warmer planet. Suppose, for example, that the atmosphere (including clouds) did not participate at all in the radiation balance of the earth. A global radiation balance wherein the solar radiation absorbed by the earth, $\alpha S A_c$, is set equal to the infrared radiation emitted by the earth—$\sigma \varepsilon T_{av}^4 A$—gives:

$$\alpha S A_c = \sigma \varepsilon T_{av}^4 A \tag{1.1}$$

where S is the solar constant (the radiant heat flux from the sun at the average distance of the earth from the sun, taken here as 1360 W/m^2), α the absorptivity (co-albedo) of the earth, σ the Stefan Boltzman constant, ε the emissivity of the earth, A_c the cross-sectional area of the earth, A the total surface area of the earth, and T_{av} the average surface temperature. Satellite measurements indicate that the average value of α is about 0.7, while most of the surface materials of the earth have emissivities greater than 0.95. The ratio A/A_c is 4. Using these values in the above equation gives T_{av} = 258 K (–18°C). The actual spatially and annually averaged surface temperature is approximately 33° higher than this. This difference is a manifestation of the greenhouse effect of the present atmosphere.

If one performs the same type of calculation for Mars and Venus, one finds that nature has provided us with an empirical proof of the existence of the greenhouse effect. The temperature on the surface of Mars in the absence of an atmosphere would be approximately –57°C. Its atmosphere contains CO_2, but it is so thin that there is essentially no greenhouse effect at all. On the other extreme is Venus, whose atmosphere is dense and contains large amounts of CO_2. The surface temperature of Venus without an atmosphere would be approximately –46°C. The presence of GHGs results in an actual surface temperature of 477°C.

1.3.2 Man-Made Greenhouse

There is no question that the greenhouse effect is real. Therefore, there is little doubt that increasing the concentrations of GHGs in the atmosphere will lead to more warming. There is some uncertainty, however, about the magnitude, distribution, and timing of the resulting climate change with gradually increasing GHG concentration as well as the impacts of this change on the earth and its animal and vegetable inhabitants. Nevertheless, a consensus exists among climatologists that the globally and annually averaged equilibrium temperature of the earth with a radiative force equivalent to the doubling of the CO_2 concentration would be between 2.3°C and 6.9°C warmer than the present (IPCCAR4WG1 2007), with a nonsignificant chance that it may be even greater, warmer than at any time during which humankind has been its inhabitant. Atmospheric concentrations of GHGs equivalent to doubling of CO_2 are currently predicted to occur at least by the second half of this century (IPCC 2007).

If currently prevailing human activities do not change in the next few decades, we shall be faced with a climate substantially different from that of today. In addition, the rate of change of climate will in all likelihood be unprecedented in recent history. The rate of change may well prove to be large enough to make it very difficult for many species of plant and animal life (including humans) to adapt to the new climate as they sometimes have in the past. It is therefore prudent to consider our alternatives. There are two general ways to prevent GHG-induced climate change. The first is to somehow cause the concentrations of atmospheric GHGs to reach steady-state values not too different from today's values. This can be accomplished either by decreasing the sources of these gases or by increasing the sinks. The second method is by controlling the climate itself. This may sound like science fiction, but such ideas should not be rejected out of hand. Landing a human on the moon seemed like science fiction only a few decades ago.

If climate change cannot be prevented, and it probably cannot be entirely prevented, given our current heavy commitment to burning fossil fuels and the present uncertainty as to the sources of many of the other GHGs (e.g., methane), we shall have to adapt to the changing climate. Since most adaptive methods will be more or less regionally specific, it would be most useful to know regional details of climate change and the resulting impacts. In planning adaptive strategies, we will be faced with the age-old engineering problem of designing in the face of uncertain constraints, particularly in the regional distribution of impacts and in the uncertainty of unintended consequences.

The bulk of this book is not about climate change, but about the impacts and the engineering response to climate change. Nevertheless, any response to climate change must be undertaken with an understanding of the situation we seek to avert, along with its seriousness and uncertainty. We will need to decide between adaptation and mitigation and will have to judge responses in terms of their cost, effectiveness, potential risk, and so on. Engineering responses must be envisioned within the context of the problem. We first provide a brief summary of the problem as currently conceived; more detailed discussion is available in monographs such as IPCC (2007) and Watts (2007). We begin with a brief discussion of climatic feedbacks and use this as a background for describing the results of climate change scenarios predicted by general circulation models of the earth's atmosphere and oceans as well as evidence that the earth's climate has already undergone significant change.

1.3.3 Greenhouse Effect and Evidence of Global Warming

1.3.3.1 Global Energy Balance

The earth is heated by radiation from the sun. In turn, an equal amount of radiation must leave the top of the atmosphere for the earth to maintain a balance so that it does not either heat up or cool down. The interaction of radiation with the atmosphere and the earth's surface is very complicated. However, we can get a general picture of it by summarizing the earth–atmosphere energy balance on average for the whole globe, that is, for a global average (Forster et al. 2007); see Figure 1.1. Suppose that 342 W/m^2 of radiant energy impinge on the top of the atmosphere, about the power of 3 ½ 100 W light bulbs. Not all of it reaches the surface of the earth. About 67 W/m^2 is absorbed by the atmosphere,

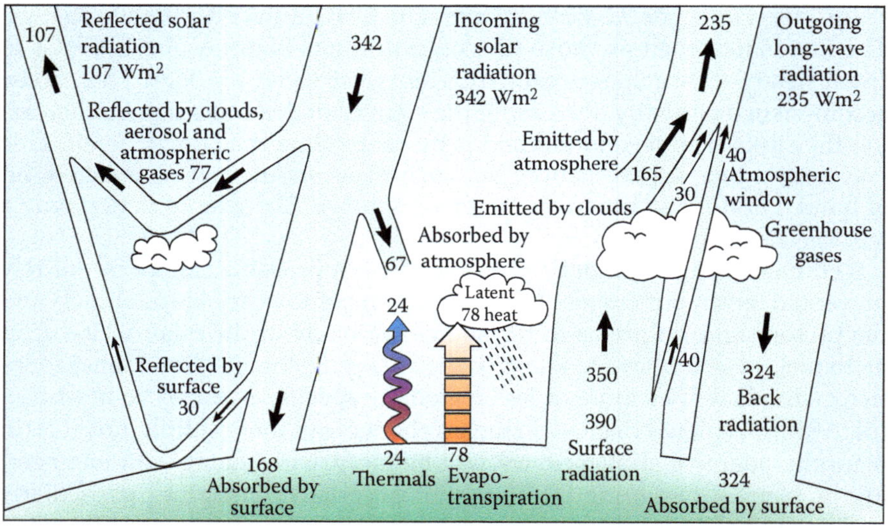

FIGURE 1.1
The earth–atmosphere global energy balance. If neither the atmosphere nor the surface of the earth (including land and sea areas) is changing in temperature, then the net amount of heat entering and leaving both the atmosphere and the land and sea areas must be zero. This figure shows the pathways of 342 W/m^2 of solar radiation bearing down on the top of the atmosphere. About 168 W/m^2 of this very high temperature (or shortwave) solar radiation is absorbed by the earth. The rest is either reflected back out of the atmosphere or absorbed by the atmosphere. In addition, about 324 W/m^2 of long-wave radiation from the atmosphere and clouds is absorbed by the surface. The total of 168 + 324 = 492 W/m^2 must now leave the surface through sensible heat (caused by the temperature difference between the earth and the atmosphere), latent heat (caused by evaporation), and long-wave radiation emitted by the surface. Most of the long-wave radiation is captured by the atmosphere, which emits about 235 W/m^2 out of the top of the atmosphere. If the amounts of water vapor and CO_2 in the atmosphere increase, more of the infrared radiation will be absorbed within the atmosphere, and both the atmosphere and the earth must compensate by emitting more radiation. To do this, they must become warmer. (From Forster, P., et al., *Climate Change 2007: The Physical Science Basis*, Contribution of Working Group I to the Fourth Assessment Report of the Intergovernmental Panel on Climate Change, Cambridge University Press, Cambridge, U.K. and New York, 2007. With permission.)

mostly by ozone, but some by water vapor and dust or other particulate pollution. Another 77 W/m^2 is reflected by clouds and aerosols (small particles in the atmosphere). The earth's surface reflects 30 W/m^2 back to space. The remaining 168 W/m^2 is absorbed by the surfaces of the land and the oceans. The surface of the earth (land and oceans) directly heats the atmosphere (sensible heat), losing about 24 W/m^2, while evaporation from the surface (latent heat) accounts for about 78 W/m^2. The earth's surface and the atmosphere emit radiation that has different characteristics from those of solar radiation. I will say more about that shortly. About 390 W/m^2 of this "long-wave" radiation are emitted by the surface. Almost all of this is absorbed by clouds, water vapor, CO_2, and several other gases. Forty watts per square meter escape through the atmosphere. About 324 W/m^2 is reemitted downward to the surface. The atmosphere, together with clouds, emits another 195 W/m^2 to space.

Notice that the sum of the energy units must balance, both at the surface and at the top of the atmosphere. The sum of the energy passing into the atmosphere from the sun must be equal to the amount of radiant energy leaving the top of the atmosphere, which is the solar radiation reflected back plus the long-wave radiation leaving from the top of the atmosphere. Otherwise the earth's atmosphere would heat up. The same is true at the surface. The 168 W/m^2 of solar energy absorbed plus the 324 W/m^2 of long-wave heating from the atmosphere must equal 390 W/m^2 of long-wave radiation emitted by the surface plus the sensible heat (24 W/m^2) plus the latent heat leaving through evaporation (78 W/m^2). Interrupting any of the processes can cause the system to be "out of balance." For example, if the radiation from the sun were to increase to, say, 350 W/m^2, the radiation escaping from the top of the atmosphere would have to increase by 8 W/m^2 to bring the system back into equilibrium. Many things could happen to cause the radiation loss from the top of the atmosphere to increase. For example, the surface and the atmosphere could become warmer, and since the radiation increases with temperature, the long-wave radiation leaving the top of the atmosphere would increase. If that happened, there would probably be less ice and snow on the warmer earth's surface. Ice and snow are excellent reflectors of solar radiation, and so less radiation would be reflected from the surface, reducing the 30 W/m^2 of solar radiation leaving the atmosphere due to reflection from the surface, and again putting the system out of balance.

1.3.3.2 Feedbacks

This means that the surface and the atmosphere would have to warm even more for the long-wave radiation to compensate for the increase in solar radiation absorbed by the surface. This is called a positive feedback (see Watts 1980 for a full discussion and an estimate of various climate feedbacks) because it reinforces the warming initiated by the increase in solar radiation. Note now that if the solar radiation increased and the atmosphere warmed, the temperature of the surface must increase. Otherwise, the heat budget at the surface would be out of balance. To compensate, in addition to less solar radiation being absorbed (the positive feedback caused by the ice/snow change mentioned earlier), the latent heat will likely increase; that is, more evaporation will occur at the surface. This means that the water vapor content of the atmosphere would increase. Water vapor is a strong GHG, and so an increase in water vapor in the atmosphere would increase the absorption of long-wave radiation and its return to the surface (the 324 W/m^2 of long-wave radiation to the surface shown in Figure 1.1). The surface must heat up even more, and so it is another positive feedback.

The water vapor feedback is illustrated in Figure 1.2. This is one of many feedbacks within the climate system. Some are definitely positive (enhancing the warming), but there are probably negative ones also. For example, in the ice/snow positive feedback mentioned earlier, the area of ice and snow will decrease when the surface warms, decreasing the 30 W/m^2 of solar radiation reflected by the surface. This means that the 168 W/m^2 of solar radiation absorbed will increase, again requiring a further increase in surface and atmospheric temperatures to bring the surface back into heat balance. But now things get just a bit more uncertain. Surely a changed surface temperature will cause a change in vegetation patterns, and this will no doubt change the way the surface (apart from snow and ice) reflects and absorbs solar radiation and emits long-wave radiation. Cloud patterns will probably also change, and herein lies one of the most important "jokers" in the deck of climate change. There are large differences in the effects of clouds on the radiation balance, depending on the locations of the clouds. Clouds reflect solar radiation back into space and also emit long-wave radiation. For low clouds, the reflected solar radiation dominates, so that an increase in low altitude clouds would lead to an increase in the 77 W/m^2 of reflected solar radiation, which would have a

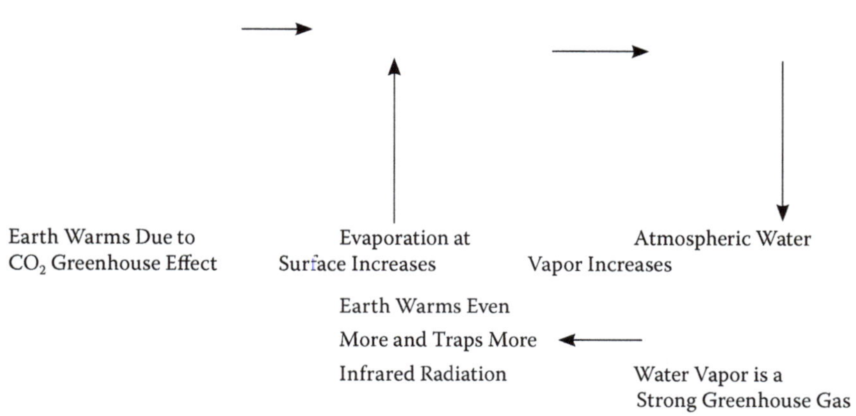

FIGURE 1.2
An illustration of the water vapor feedback. When the surface of the ocean warms due to the CO_2 greenhouse effect, more evaporation occurs. Also, a warmer atmosphere can hold more water vapor. Water vapor is itself a very strong GHG. Thus, the increase in atmospheric water vapor results in a further increase in global temperature. This effect is called a positive feedback.

cooling effect, amounting to a negative feedback and a reduction of global warming. High clouds, on the other hand, are much colder than the underlying surface, and so the long-wave radiation from the clouds is relatively small. An increase in high clouds would lead to a decrease in the 30 W/m^2 of radiation emitted into space and, therefore, would enhance warming (a positive feedback). It should also be pointed out that an increase in the water vapor content of the atmosphere need not necessarily lead to increased cloudiness. So there are, no doubt, both positive and negative feedbacks within the climate system, and they make the problem of predicting the climate challenging.

What is the bottom line? GHGs warm the atmosphere and the earth's surface. This is simple physics. We know that it is true. Positive feedbacks will make the warming effect stronger. Negative feedbacks will make it less strong. We simply do not understand some of the feedbacks very well. Paramount among these is the cloud feedback. But one thing is certain: increasing concentration of GHGs in the atmosphere has already led to global warming and will lead to further global warming in the future.

1.3.3.3 Climate Models: Predicting Global Warming

How do scientists determine how the climate will respond to increases in GHGs? In most areas of science, we rely on what we call testable hypotheses. If we want to know the drag force on a falling sphere, we hypothesize that the drag force depends on something like the square of the speed of the sphere and then we go to the laboratory and measure this drag force to test our original hypothesis. If the experimental results agree with the hypothesis, we feel that our conceptual "model" was correct. We can then write down the mathematical equation that Isaac Newton developed, that the gravitational force (the weight of the sphere) minus the drag force is equal to the mass of the sphere times its acceleration and predict the downward velocity and the acceleration of the sphere if it is dropped. We then compare the results of our experiment with our model predictions. But how do we test our hypotheses about the effect of GHGs on climate? We now have about a century of data on the current climate that we can compare to climate model predictions. And as we shall see climate models do faithfully describe the changes in climate that have occurred in the twentieth and early twenty-first centuries.

What is a climate model? Scientists have known for many years how to write the mathematical equations that govern the motions of the atmosphere and the oceans. But we also need to know the "boundary conditions"—a set of equations that govern the exchange of latent and sensible heat between the atmosphere and the oceans and land areas—and we do not know how to predict these nearly as accurately. Also, predicting the formation and motion of clouds (which, we have seen, accounts for a potentially important feedback) is difficult and uncertain. Even so, the equations describing climate are so difficult that they must be solved by programming them on very large and very fast computers.

One important test for these large, complex climate models (called general circulation models or global climate models, in either case, GCMs) is whether they can "predict" the current climate accurately.

To see how well the models perform in predicting globally averaged temperature, refer to Figure 1.3. Figure 1.3b shows (in blue) the model-predicted temperatures from 19 simulation runs using five models, with the models forced by natural forces only (including volcanic eruptions and solar changes). The average of these is shown in the solid dark blue line. The actual globally averaged temperature is shown as the black line. Figure 1.3a is produced by 58 simulations using 14 different models, with both natural and anthropogenic (man-made) forcing. The yellow lines are model predictions and the red line is

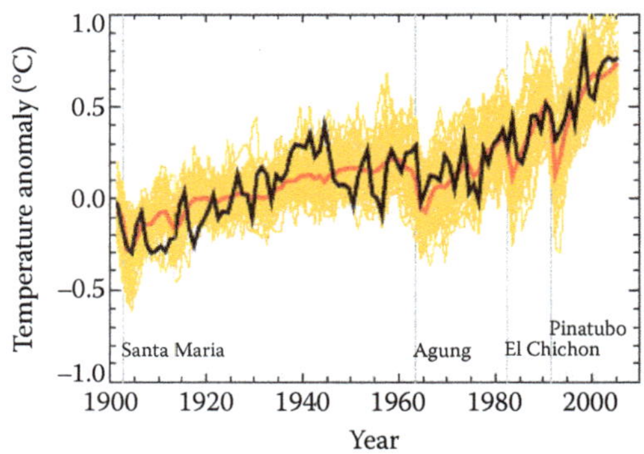

(a)

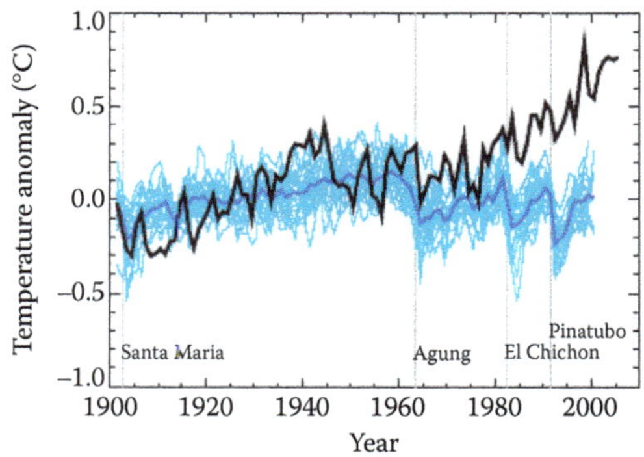

(b)

FIGURE 1.3
Global temperature anomalies (°C) from observations (black) and GCM simulations forced by (a) both anthropogenic and natural forcings and (b) natural forcings only. Data shown are global mean anomalies relative to the period 1901–1950. The yellow lines are results from 58 simulations using 14 climate models. The red line is their average. The blue lines are the results of 19 simulations from 5 models and the dark blue line is their average. In both panels, the black lines are observations. (From *Climate Change 2007: The Physical Science Basis,* Contribution of Working Group I to the Fourth Assessment Report of the Intergovernmental Panel on Climate Change, Cambridge University Press, Cambridge, U.K. and New York, 2007.)

the average of these predictions. Again, the black line represents the observed surface temperature. Note that these values are not the actual temperatures but the departure from the average temperature over the period 1910 to 1950.

Perhaps the most difficult thing for climate models to predict is precipitation. Figure 1.4 shows global precipitation distribution as observed and

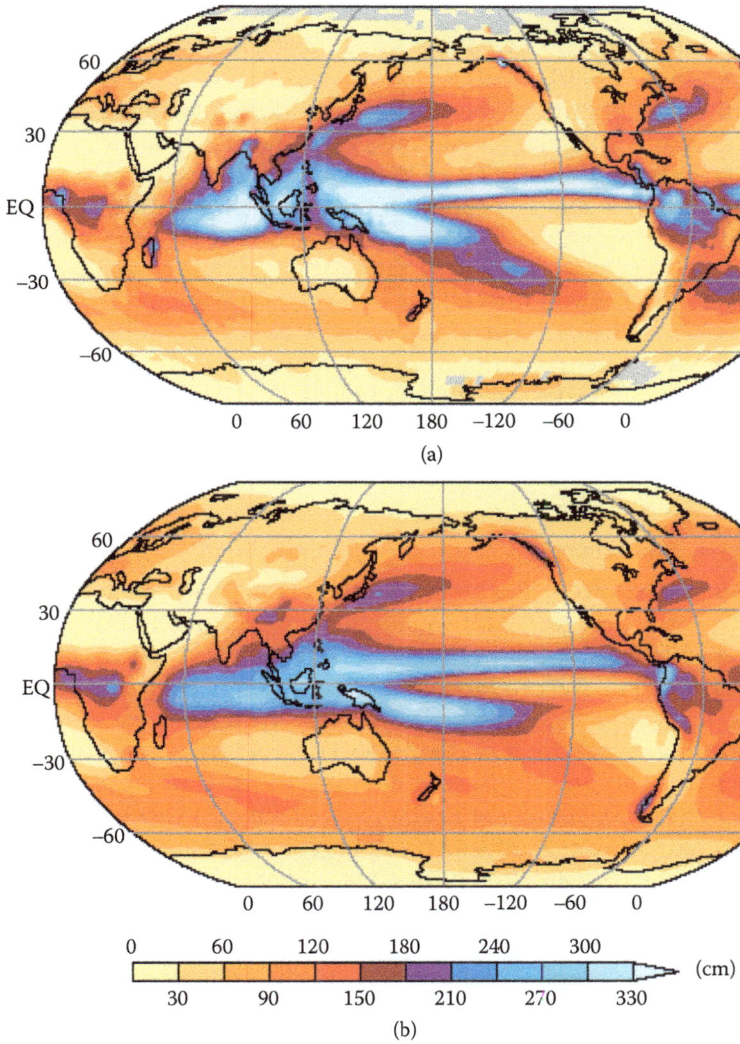

FIGURE 1.4
Annual averaged precipitation in centimeters per year, observed (a) and model predicted (b), based on a multi-model mean from the Climate Prediction Center Merged Analysis of Precipitation. Observations were not available in the small gray regions. (From *Climate Change 2007: The Physical Science Basis*, Contribution of Working Group I to the Fourth Assessment Report of the Intergovernmental Panel on Climate Change, Cambridge University Press. Cambridge, U.K. and New York, NY, 2007.)

predicted by the Climate Prediction Center merged analysis, which comes from a set of several climate models. It seems clear then that GCMs can predict the current climate quite well. They also predict the globally averaged temperature change that has occurred due to natural changes and in response to greenhouse forcing, as demonstrated in Figure 1.3a.

1.3.3.4 The Ocean

1.3.3.4.1 Temperature Delays

Of course, the temperature of the earth does not immediately respond to increased heating (Watts and Morantine 1994). The ocean has a very large heat capacity; that is, it takes a lot of heat to warm the ocean only a little. But, if the heating rate remains constant, the ocean temperature eventually responds, just as a pot of water does when you heat it on your stove. The climate responds even more slowly if the heat is applied gradually, just as the pot of water does if you turn up the heat on your stove gradually. Since CO_2 in the atmosphere is building up gradually, global warming will occur gradually. The warming resulting from GHG increases will occur decades later, and it is certainly not expected to be evenly distributed around the earth. This is why the temperature changes in Figure 1.3 respond slowly to greenhouse forcing. This will be discussed further when we look at the physical evidence of global warming.

1.3.3.4.2 Ocean Acidification

Phytoplankton, which is at the bottom of the ocean's food chain, declines when the ocean water gets warmer. Phytoplankton are the microscopic plants the zooplankton and other plant life eat. A NASA satellite tracked water temperature and the production of phytoplankton over a period from 1997 to 2006 and found that when one went up, the other went down. The ocean is warming both in high northern latitudes and in high southern latitudes. Near Antarctica, waters in summer have warmed in some places by a degree or more, threatening the krill, tiny sea creatures that many organisms depend on. In one region near the Antarctic peninsula, krill numbers have fallen by some 80% since the 1970s. This is likely to be true in the Arctic since krill feed on algae under sea ice, and the Arctic Sea ice is receding significantly.

Perhaps more significant is the potential of the changing chemistry of the ocean. More than a third of CO_2 that is emitted by the burning of fossil fuels is absorbed by the ocean. The Intergovernmental Oceanographic Commission reports that 20–25 million tons of CO_2 are being added to the ocean each day, and this can only increase as the rate of increase of CO_2 in the atmosphere increases. When CO_2 gas dissolves into the ocean, it produces carbonic acid, decreasing the pH (potential for hydrogen) of the ocean water. The pH scale goes from 0 to 14, where a lower pH means higher acidity. When CO_2 dissolves in seawater, it forms carbonic acid (H_2CO_3), which breaks down into hydrogen ions (H^+) and bicarbonate ions (HCO_3^-), and the bicarbonate ions further break down into H^+ and CO_3^{2-} ions. More H^+ ions make seawater more acidic and too much acidity can lead to the disruption of the marine environment. Increasing acidity (decreasing pH) reduces the availability of calcium carbonate (because additional H^+ ions react with [consume] carbonate ions to form bicarbonate ions) and corals as well as sea creatures with hard shells rely on calcium carbonate to form their shells. It may also affect the growth rates and reproduction rates of fish and it affects the plankton populations that the fish rely on for food, with potential disastrous results for the marine food web. The increasing ocean temperatures and the decreasing pH could impose a sort of double whammy on the ocean.

The pH of the ocean has probably been about 8.2 for at least thousands of years. Over the past few decades, it has decreased by about 0.1 unit and is projected to decrease faster in the future. The scale is not linear. A drop from pH 8.2 to 8.1 indicates a 30% increase in acidity, or concentration of hydrogen ions; a drop from 8.1 to 7.9 indicates a 150% increase in acidity. Small-sounding changes in the pH of the ocean are actually quite large and definitely in the direction of becoming less alkaline, which is the same as becoming more acidic. Some scientists believe that if it reaches 7.6, shell creatures will find it difficult to form their shells. Calcifying organisms, such as phytoplankton and zooplankton, are important organisms at the base of the ocean food chain. This is not true of all species. Ries, Cohen, and McCorkle (2009) investigated the effects of CO_2-induced ocean acidification on calcification for 69 days on 18 benthic marine organisms under controlled laboratory conditions. Under high CO_2 levels, the shells of some species, such as conchs, noticeably deteriorated. Clams, oysters, and scallops built less and less shell as CO_2 levels increased. All three crustacean species tested—the blue crab, the American lobster, and a large prawn—actually grew heavier shells at higher CO_2 levels. But most planktonic calcifying organisms tested, such as foraminifera, pteropods, and planktonic larvae of echinoderms, show a decrease in calcification in response to elevated CO_2 levels.

Professor Boris Worm, an assistant professor of marine conservation biology at Dalhousie University in Halifax, Canada, and his associates have reported an alarming increase in the number of fish species that are decreasing. Some of this is the result of overfishing, but there is little doubt that some reflect the decreasing biodiversity of the ocean and the decrease of the availability of such species as krill that are at the bottom of the food chain, and this is clearly affected by the increase in the temperature of the sea and the decrease of sea ice that contains the food source of the krill (Worm et al. 2006).

1.3.3.4.3 Coral Reefs

Coral reefs only thrive in very specific conditions because of the symbiotic relationship they have with "dinoflagellates," an algae that needs very specific conditions to live in. Less than 0.2% of the global oceans is covered in tropical reefs. Two-thirds of the coral reefs are now severely damaged, a fifth so overwhelmingly that they are unlikely to recover. As discussed above, CO_2 absorption could increase the acidity of seawater by as much as 0.5 pH units by the end of this century, from 8.2 to around 7.7. As temperatures rise and the pH decreases, corals will continue to bleach. Sea level rise can also have an effect if it rises enough that the dinoflagellates do not receive the amount of light they need.

1.3.3.5 The Hockey Stick

A group of paleoclimatologists (scientists who study ancient climates) used proxy data to see how the temperatures of the Northern Hemisphere have changed over the past 1000 years (Mann, Bradley, and Hughes 1999). The definition of the word "proxy" is "one who acts for another." Of course, we cannot directly measure the temperature of the earth hundreds or thousands of years ago. It happens, however, that there are things that we can measure today that are closely related to what the temperature was long ago. Using proxy data is common among climatologists, especially those who study ancient climates (Crowley and North 1991). This particular group examined things like tree rings in very old wood. Each year as a tree grows it makes a ring of new wood that is visually distinguishable from that of the previous year. By measuring the width and the density of each ring and counting back from the year of the most recent ring, scientists can estimate the temperature and the amount of precipitation during the growing season. The width and density of a given ring depend on both temperature and precipitation, but the relative amounts of hydrogen and deuterium seem to depend only on temperature. This allows scientists to separate the two effects. There is, of course, considerable uncertainty in getting the precise temperature. But if there is more than one proxy available, one can do "multiproxy" analysis and the estimates of temperature become more credible the more proxies we have. Other proxies are the growth and chemical analysis of corals, the abundance of sea creatures in particular locations in the sea (depending on what the sea temperature was), the abundance of certain chemical isotopes in the ocean sediments and glaciers, and the extent (growth or melting) of glaciers and ice sheets. The paleoclimatologists from the University of Massachusetts and the University of Arizona used multiproxy analysis to estimate the temperature of the Northern Hemisphere for the past 1000 years. The results were surprising. Their results are shown in the bottom panel of Figure 1.5, along with the results of many other scientists.

Two things are evident: temperatures drifted slowly downward (cooled) for 900 years until the beginning of the twentieth century. Afterward, they began to rise very rapidly. The good news is that we may have been slowly drifting into the next ice age cycle, just as we were predicted to do in

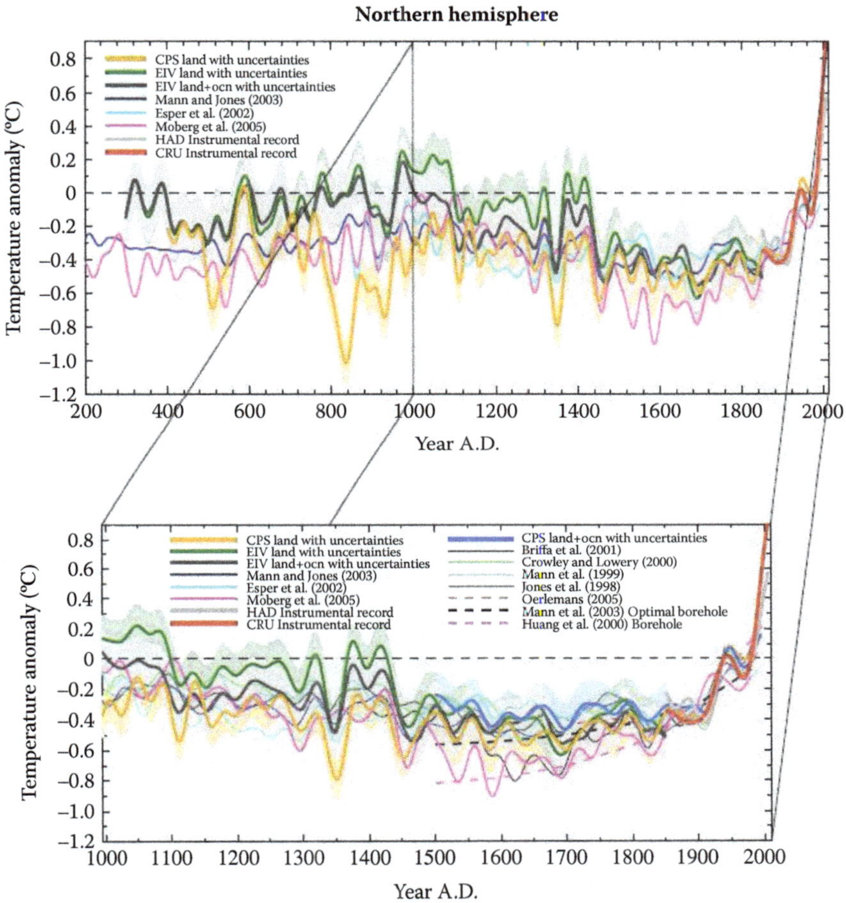

FIGURE 1.5
Globally averaged temperature for the past 1000 years, using proxy data and recent instrumental data. (From Mann, M. E., et al. *Proc. Natl. Acad. Sci. USA*, 105, 13252, 2008. With permission.)

accordance with the Milankovitch theory. The bad news is that as we pump more and more CO_2 into the atmosphere, the increase is not likely to stop. The past 20 years or so have been warmer than any such period for the past 1800 years, as shown in the top panel of Figure 1.5.

This graph, often referred to as the "hockey stick" because of its shape, has been challenged by a group of skeptics who have accused the authors of fudging the data. The claim has been investigated thoroughly by at least four groups of impartial investigators, who found the criticisms to be without basis, and an updated version (Figure 1.5, top panel) has been presented more recently and published in the proceedings of the National Academy of Sciences (Mann et al. 2008). The graph has been endorsed after a thorough review by the National Academy of Sciences.

1.3.3.6 Natural Variability

Perhaps a more detailed discussion of what is meant by natural variability is called for. Both historical data and the results of GCMs show that both the globally averaged temperatures and regional and local temperatures will have their ups and downs. Historical changes that have occurred both recently and over the past 1000 years bear out the fact that sizeable changes in the climate might occur naturally in the climate system. This is what climatologists call "natural variability." For example, there was a relatively cool period that occurred between about 4500 and 2500 years ago, ending at about the time of the dawn of the Roman Empire. The decline and fall of the empire and the beginning of the Dark Ages, about 500 to 1000 AD, saw a return to colder climates. Following this period was a relatively warm period called the Medieval Optimum that lasted from about 1100 to 1300 AD in Europe, but apparently not in the Northern Hemisphere as a whole, as can easily be seen in Figure 1.5, which shows a rather persistent cooling of the Northern Hemisphere throughout the period. It most likely occurred as a result of warming of the North Atlantic Ocean. During this period, wine grapes flourished in England and the Arctic ice pack retreated to the extent that Iceland and Greenland were settled by Europeans for the first time. About 1450, there was a return to a cooler climate. This period, which lasted through most of the nineteenth century, is known as the Little Ice Age. One can easily see why some people believe that the recent warming that has occurred is simply the result of natural climate variability as the earth emerges from the most recent cold period. But, none of the variations that occurred during the past 1000 years have been as rapid or as large as the increase that we have seen over the past 100 years. The warming of about eight-tenths of a degree Centigrade that has occurred during the past 100 years is faithfully predicted by the GCMs. It is certain that the earth's climate will warm as the atmospheric concentrations of GHGs increase. It is very likely that it will warm, on average, at about the rate predicted by the GCMs that exhibit medium sensitivity. These models predict that the global temperature will increase by 2.3 to 6.9°C by 2100, depending on how committed we are to reducing greenhouse concentrations (IPCCAR4WG1 2007). Moreover, it is important to remember that high latitudes will warm as much as two to three times as much as this globally averaged value. This we know from both models and historical data.

1.3.3.7 Fingerprints: Observations of Global Climate Change

Let us now examine some of the observational evidence of global warming so far. What should we have seen? Some evidence of what to expect comes from GCM predictions. One may perhaps be somewhat suspicious of claiming that changes in a single variable demonstrate a cause-and-effect relationship with confidence. If the earth's climate is indeed warming due to increased concentrations of GHGs, the evidence should show up in other ways besides the globally averaged temperature near the surface. Examining many such signals is often referred to as the "fingerprint method." For example, the growing season should be getting longer, and the amount of snow that falls in the mid-latitudes in the spring and fall should be decreasing because winter weather comes later and spring earlier. Examining all of these things that scientists predict will happen gives us an even better way to see whether the global greenhouse effect is beginning to take hold. It turns out that nearly all of the things that we think ought to be happening (what our models and our data and our intuition suggest) are happening.

1.3.3.7.1 Here Is What We Should Expect and What the Data Show

Obviously the earth's surface and the atmosphere should warm. The ocean surface should warm more slowly than land areas because the ocean has a larger capacity to absorb some of the heat. The data for globally averaged surface temperatures shown in Figure 1.6 are from four sources and there is very close agreement among the results. There are year-to-year changes and during the period between the mid-1940s and the mid-1970s, there was little change (this will be discussed further later), but there is a rather clear increasing trend, just as the GCMs predict (see Figure 1.3). Figures 1.7 and 1.8 show globally averaged temperatures of the atmosphere above the land and the ocean surface temperature. As we expect, the temperature of the atmosphere about the land increases faster than that of the sea surface because of the large heat capacity of the ocean.

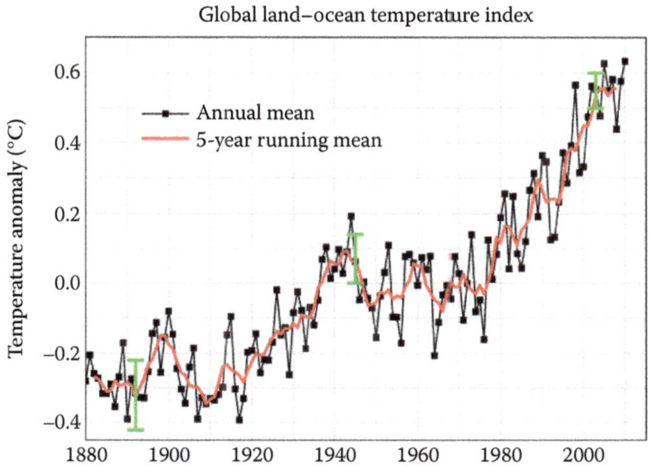

FIGURE 1.6
Line plot of global mean land–ocean temperature index, 1880 to present, with the base period 1951–1980. The black line is the annual mean and the red line is the 5-year running mean. The green bars show uncertainty estimates. (Courtesy of NASA Goddard Institute for Space Studies, http://data.giss.nasa.gov/gistemp/.)

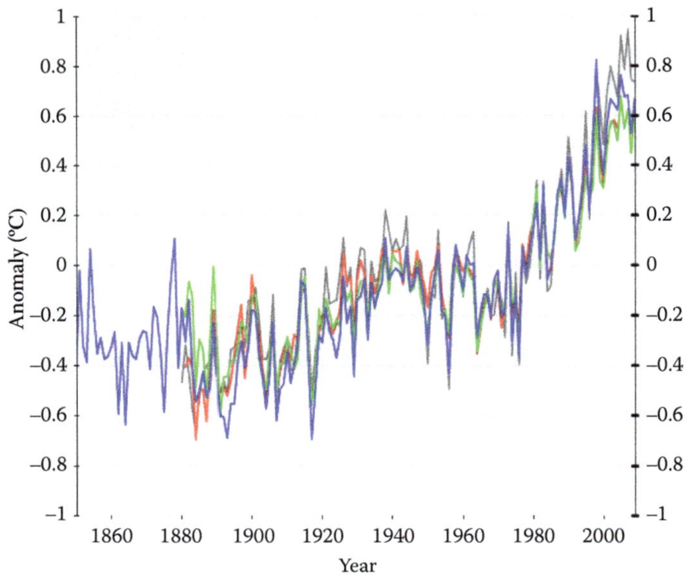

FIGURE 1.7
Land surface–air temperature departures from the 1951–1980 mean. Blue (Brohan et al. 2006), green (Hansen et al. 2001), red (Lugina et al. 2005), and gray (Smith et al. 2008). (Courtesy of National Climate Data Center, http://www.ncdc.noaa.gov/.)

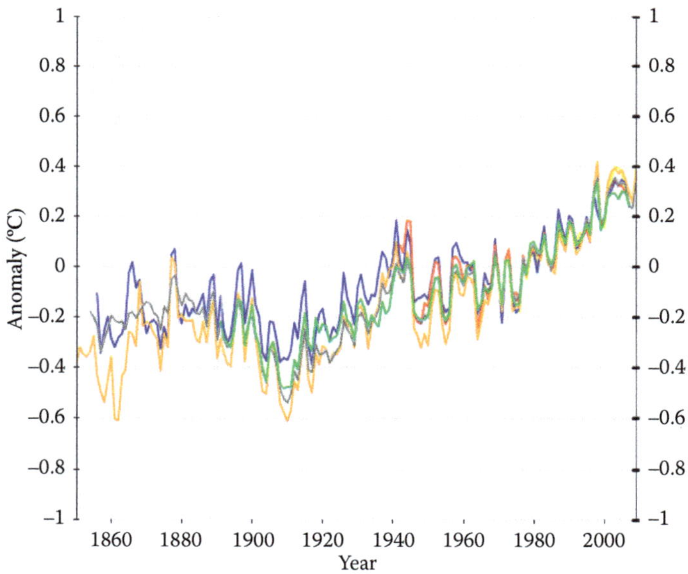

FIGURE 1.8
Sea surface temperature departures from the 1951–1980 mean. Green (Ishii et al. 2005), gray (Smith et al. 2008), orange (Rayner et al. 2006), red (Worley et al. 2005); blue (Kaplan 1998), and yellow (Berry and Kent 2009; Rajagopalan 1998). (Courtesy of National Climate Data Center, http://www.ncdc.noaa.gov/.)

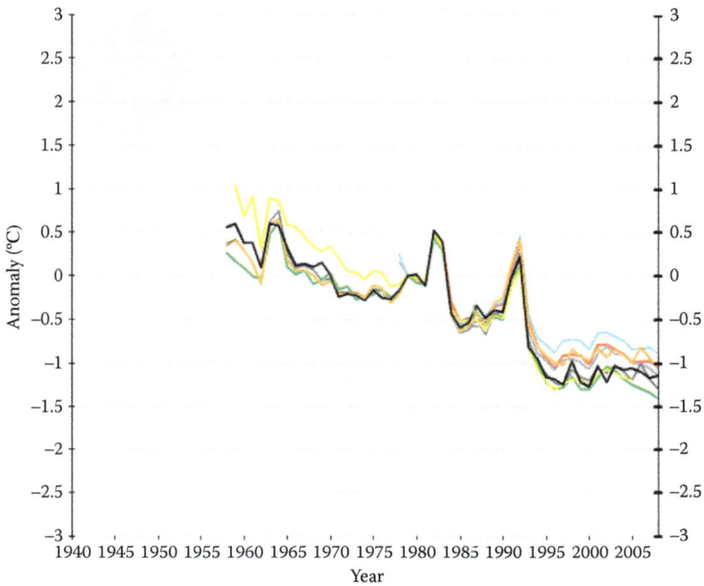

FIGURE 1.9
Globally averaged stratosphere temperature departure from the 1951–1980 mean. Black (Thorn et al. 2005), yellow (Sherwood et al. 2008), orange (Haimberger et al. 2007), green (Free et al 2005), blue (Mears et al. 2009), gray (Haimberger et al. 2008), red (Christy et al. 2003), and light gray (Zou et al. 2002). (Courtesy of National Climate Center, http://www.ncdc.noaa.gov/.)

Global surface air temperatures in 2010 tied with those in 2005 as the warmest on record, according to an analysis by researchers at NASA's Goddard Institute for Space Studies (GISS) in New York. The two years differed by less than 0.018°F. The difference is smaller than the uncertainty in comparing the temperatures of recent years, putting them into a statistical tie. In the new analysis, the next warmest years are 1998, 2002, 2003, 2006, 2007, and 2009, which are statistically tied for the third warmest year. The GISS records start in 1880.

The stratosphere should cool if the greenhouse effect is responsible for the warming, and it should warm if solar variations are the culprit. This is shown by simple radiation models and is very well understood. Figure 1.9 shows that the stratosphere has indeed cooled, confirming our expectations once again.

Higher latitudes should warm more than lower latitudes. Models predict this. More importantly, evidence from past warmer climates shows this to be true. Figure 1.10 shows a global map of temperature departures from 1951 to 1980 mean values during the Northern Hemisphere fall and winter of 2008. Figure 1.11 shows similar information, as indicated in the figure caption. Both figures show that as the earth warms, the regions at higher latitudes have indeed warmed significantly more than the regions at lower latitudes.

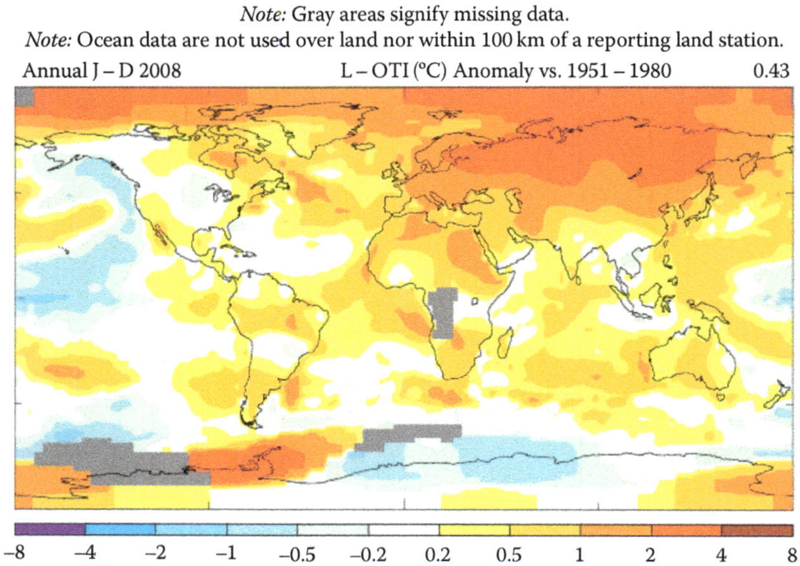

FIGURE 1.10
GISS 2008 Temperature Anomaly—C (Baseline 1951–1980); compares anomalies with locations. The temperature departure scale is shown in the band beneath the figure. (Courtesy of National Climate Data Center, http://www.ncdc.noaa.gov/.)

Simple physics suggests that the warmer air can hold more moisture, so the atmospheric water vapor content should increase. Once again, Figure 1.12 indicates that data from the real world confirm that this has occurred. The specific humidity, which is a measure of the water vapor content of the atmosphere, has increased steadily as the atmospheric temperature increased.

This leads to more precipitation. However, the increase in precipitation will not occur everywhere, and because of increased evaporation brought on by the higher temperatures, some land areas will become dryer. Thus, more intense and longer droughts are predicted in some areas, particularly in those areas where it is already dry. Data from warmer climates indicate that some regions will become wetter while, despite the increased atmospheric water vapor, some regions will become dryer due to increased evaporation from the warmer surface (Figure 1.13). Rainfall will tend to occur more in the form of very intense events. There is considerable anecdotal evidence that this is occurring both in the American continents and in Europe and Asia. Even in regions shown in Figure 1.13 that indicate a moderate short-term drought, there have been recent devastating floods because of intense precipitation events leading to river floods. Significantly increased precipitation has been observed in the eastern parts of North and South America, northern Europe, and northern and central Asia. Drying has been observed in the Sahel, the Mediterranean, southern Africa, parts of southern Asia, and Australia. In agreement with climate model predictions, more intense and

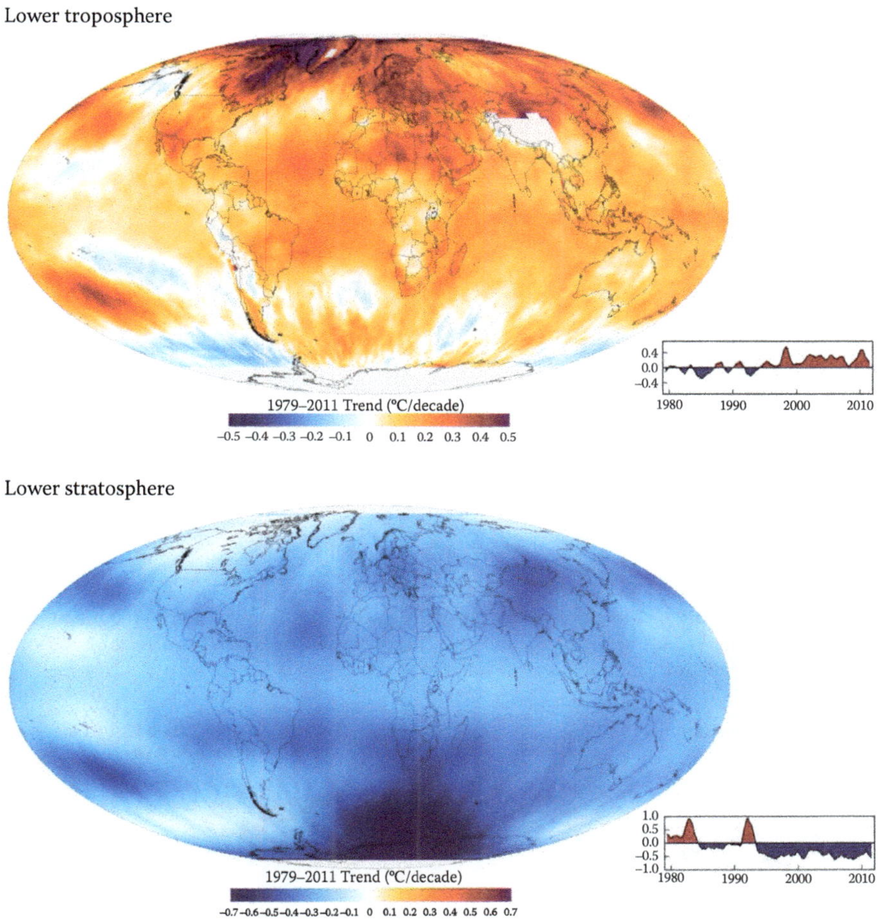

FIGURE 1.11
The map shows the 10-year average (2000–2009) global mean temperature anomaly relative to the 1951–1980 mean. The largest temperature increases are in the Arctic and the Antarctic Peninsulae. (Courtesy of NASA Earth Observatory, http://www.earthobservatory.nasa.gov.)

longer droughts have been observed over wide areas since the 1970s, particularly in the tropics and subtropics. The frequency of heavy precipitation events has increased over most land areas, consistent with warming and increases in atmospheric water vapor. Figure 1.13 shows conditions in the United States. Note that the region around Texas and Oklahoma is currently experiencing extreme to severe drought, while during June and July 2007, the region suffered severe flooding. In China, there are similarities. At least 18 million people have been affected by the worst drought in 50 years. The southwestern region of Chongqing has been worst hit, but areas of Sichuan and Liaoning are also affected. In Chongqing, there has been no rain in several months, and two-thirds of the rivers have dried up. Yet, in 2007, parts of

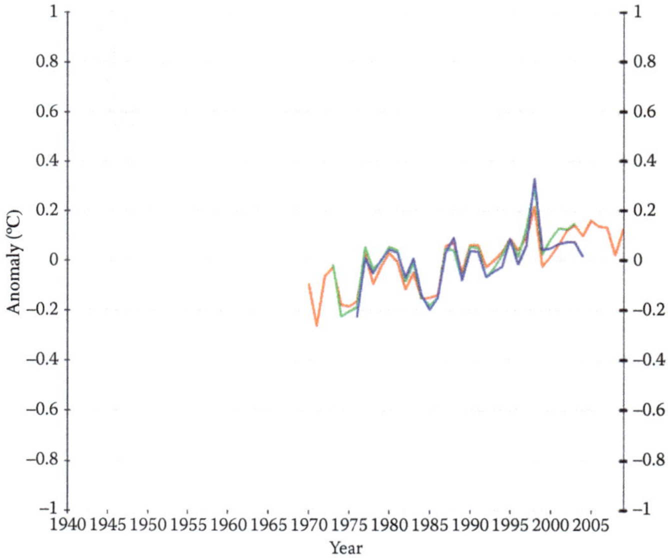

FIGURE 1.12
The specific humidity of the atmosphere has increased, relative to the 1951–1980 mean. Blue (Dai 2006), green (Willett et al. 2008), and red (Berry and Kent 2009). (Courtesy of National Climate Data Center, http://www.ncdc.noaa.gov/.)

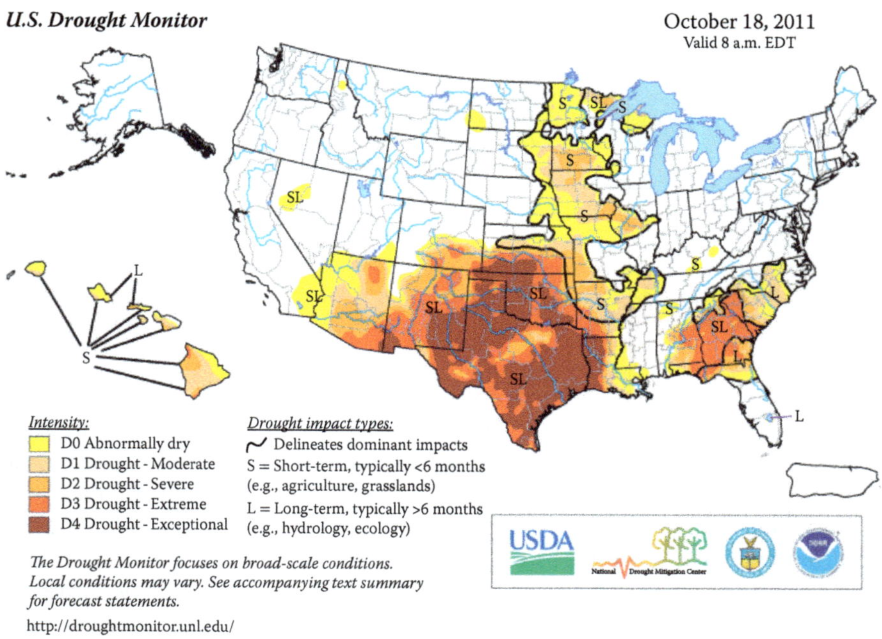

FIGURE 1.13
2011 drought conditions in the United States. (Courtesy of David Miskus, NOAA/NWS/NCEP/CPC, http://droughtmonitor.unl.edu/.)

China were hit by heavy rainfall and flooding. In spite of this, the increased evaporation has actually reduced water availability. Wetlands in the Qinghai-Tibet Plateau have shrunk by more than 10%, with those at the origin of the Yangtze River suffering the most. Water flow in the Yangtze River is down by 40% and the two largest lakes in the region, Dongting Lake and Poyang Lake, are 60% and 10% lower than their normal levels respectively. More than 2000 lakes in China have dried up. On the other hand, India is suffering from flooding from glacial lake outburst floods, which are catastrophic discharges of water due to melting of the glaciers in the Himalayas, even though large parts of the country are experiencing droughts. The further bad news is that once most of the glaciers are gone, water availability will be even more scarce. A dry spell in Morocco has slashed the country's 2007 grain crop to an estimated 2.0 million tons from 9.3 million in 2006 and the government is expected to triple soft wheat imports to 3.0 million tons. Droughts and floods are already becoming more frequent and projections are they will become more so in the future. Closer to home, the long-term drought in southern Florida has caused the water level in Lake Okeechobee, a vital source of water (as well as wildlife), to shrink to a new low (Watts 2007). The recent extended drought in Texas and the persistent drought in Georgia are examples of regions where this is occurring (Figure 1.13). Australia has suffered drought conditions for several years.

In response to increasing temperatures, the growing season should become longer in the mid- to high latitudes. The migration patterns of birds and animals will be affected. According to the most recent IPCC report, "In terrestrial ecosystems, earlier timing of spring events and poleward and upward shifts in plant and animal ranges are with *very high confidence* linked to recent warming. In some marine and freshwater systems, shifts in ranges and changes in algal, plankton and fish abundance are with *high confidence* associated with rising water temperatures, as well as related changes in ice cover, salinity, oxygen levels and circulation" (Pachauri and Reisinger 2007). Scientists have observed a wide range of the migration patterns of birds, fish, and turtles change as a result of, and apparently in response to, warming. Observations of plants, animals, and birds in the Arctic revealed that flowering and egg-laying occur some two weeks earlier in response to the early emergence of spring. Ice in northeast Greenland is melting 14 days earlier than it did in the mid-1990s. There are many studies detailing the early onset of natural events due to global warming. For example, in Scotland, butterflies arrive 8 days earlier than they did 30 years ago. Birds are nesting as much as 10 days earlier. Coral reefs are bleaching. Coral reefs get bleached when the water gets too warm. The two largest coral reefs are the Belize Barrier Reef (largest in the Western Hemisphere) and the Australian Great Barrier Reef. The Belize reef has suffered 40% damage since 1998, not because the average temperature of the water has increased very much, but because of short intervals of high water temperatures. The Great Barrier Reef suffered extensive bleaching in 2002 when the water temperature was higher than usual, and again in 2006. Scientists estimate that a quarter of the world's coral has been permanently lost, with more lost every year.

Permafrost in regions where it exists should decrease in extent. Mountain glaciers in mid- and high-latitudes should retreat. Temperatures at the top of the permafrost layer have generally increased since the 1980s in the Arctic (by up to 3°C). The maximum area covered by seasonally frozen ground, shown in Figure 1.14, has decreased by 7% in the northern hemisphere since 1900, with a decrease in spring of up to 15%. (IPCC 2007). Figure 1.14 shows how mountain glaciers worldwide have decreased in mass since 1960 (left panel) and the decrease in the area of frozen ground in the Northern Hemisphere (right panel). Note that the rate of decrease in glacier mass seems to have increased since the 1990s. In the 1850s, nearly 4500 km^2 of the Alps were glaciated. By the 1970s, the area had fallen to under 3000 km^2, a loss of about 3% per decade. From the 1970s to 2000, the rate of loss had increased to more than 8% per decade.

A couple of pictures will suffice to show how many glaciers have shrunk recently (Figures 1.15 and 1.16). Many more are available at the Byrd Polar Research Center, The Ohio State University. You can find many more of these glacier pairs by going to Glaciers Online.

Many of the glaciers in Glacier National Park are in grave danger of disappearing as did the Grinnell Glacier, shown in Figure 1.15. Particularly

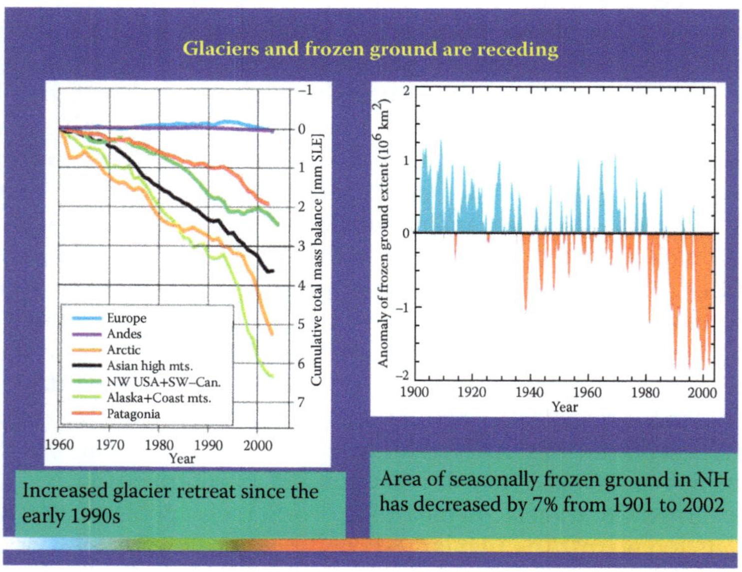

FIGURE 1.14

Glacier retreat for continental glaciers (left panel) and decrease in seasonally frozen ground. (From *Climate Change 2007: The Physical Science Basis,* Contribution of Working Group I to the Fourth Assessment Report of the Intergovernmental Panel on Climate Change, Cambridge University Press, Cambridge, U.K. and New York, 2007.)

FIGURE 1.15
The Grinnell glacier in Glacier National Park. (Courtesy of Byrd Polar Research Center, The Ohio State University.)

FIGURE 1.16
Muir glacier 1941–2003, Alaska. (Courtesy of Byrd Polar Research Center, The Ohio State University.)

striking is the Muir Glacier in Alaska, which has retreated so far since 1941 that one has to look carefully to see its tip in the background of the right photograph. The overall area of glaciers in the Alps is about half what it was in the 1850s. Additionally, tropical glaciers in the Andes, and those on the North American continent, and in Asia have been retreating. Mountain glaciers and snow cover have declined on average in both hemispheres. In the past few decades, some of the great Himalayan glaciers have either disappeared or have eroded considerably. In the Saraswati valley, north of Badrinath, the Ratakona glacier, which lies near the Mana Pass, is on the verge of disappearing. Similarly, the Pindari and Milan glaciers are also gradually receding.

Eventually, ice caps in Greenland and Antarctica should begin to melt. Greenland data up until 2006 are shown in Figure 1.17. This figure was

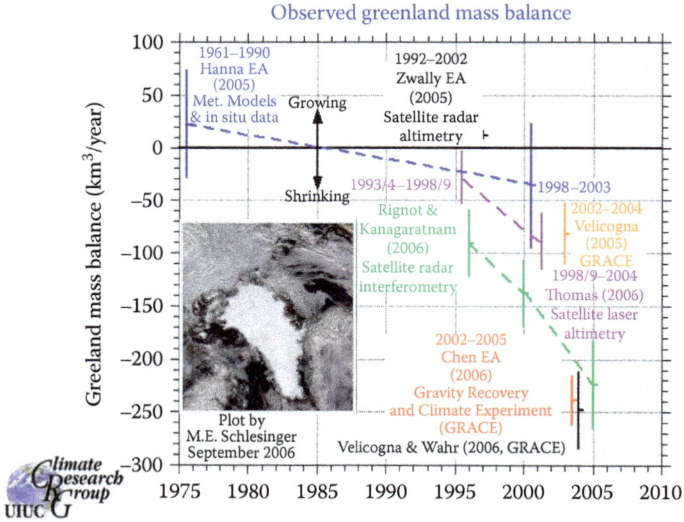

FIGURE 1.17
Greenland ice mass 1975–2006. (Courtesy of Dr. Michael Schlesinger, University of Illinois Climate Research Group, Urbana, IL.)

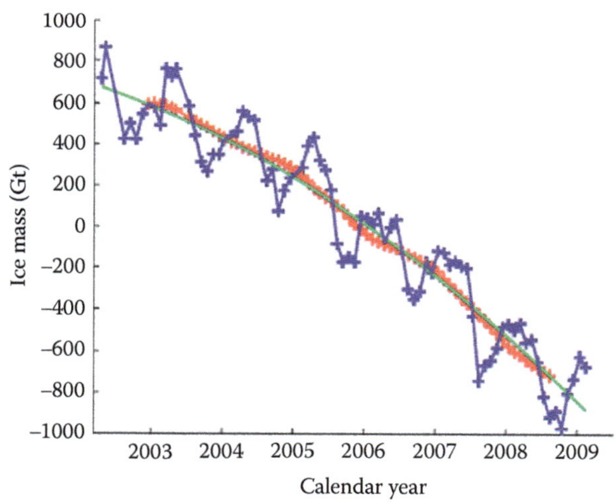

FIGURE 1.18
Decrease in the Greenland ice mass since 2002. (Courtesy of NASA Earth Observatory, http://earthobservatory.nasa.gov/Newsroom/NewImages/images.php3?img_id=17800, 2011.)

presented in a talk at the American Society of Mechanical Engineers Annual Conference in 2007 by Dr. Michael Schlesinger. Note that the rate of decrease appears to have increased over time.

This is supported by the more recent data shown in Figure 1.18, as the shape of the curved line exhibits an increasing slope.

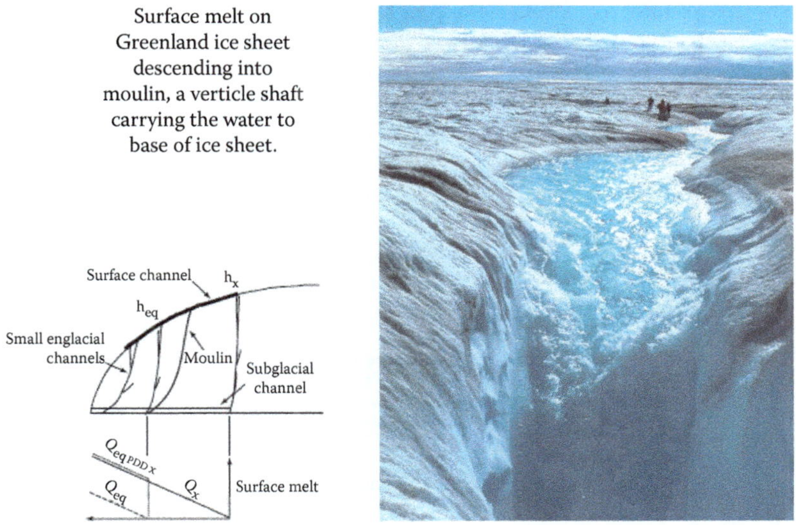

FIGURE 1.19
A moulin in Greenland. (Courtesy of Dr. Michael Schlesinger, University of Illinois, Urbana, IL.)

Flow speed has increased for some Greenland and Antarctic outlet glaciers, which drain ice from the interior of the ice sheets. Satellite measurements show that the amount of ice melt flowing into the sea from large glaciers in southern Greenland has more than doubled in the past 10 years. Glaciers normally flow very slowly. The rapid increase in the flow may come about when the bottom of the ice is lubricated by melt water cascading through crevasses in the ice called moulins, as the one shown in Figure 1.19. Climate models currently do a relatively poor job of predicting ice sheet behavior because they do not take this into account.

The Antarctic ice cap is exhibiting similar changes. As shown in Figure 1.20, the mass of the ice sheet is decreasing, and the curvature of the trend line indicates that the rate of decrease is increasing.

The area of sea ice would be expected to decrease in both hemispheres. In the Northern Hemisphere, this is true, but in the region around Antarctica, things are less certain. It appears that the area of sea ice is remaining about constant. However, in 2002, an ice shelf known as Larsen B disintegrated in just over a month (Watts 2007). Ice shelves are thick plates of ice that reach into the sea and are fed by glaciers. The Larson B sheet was about 200 m thick and the ice that was released in this short time weighed about 720 billion tons. The collapse is attributed to the strong warming, about 0.5°C per decade—a trend that has been present since the 1940s. It appears to have been related to the presence of melt water ponds on top of the ice shelf during late summer. In addition to absorbing more solar heat than does ice, the water probably enhanced fracturing of the ice by filling cracks in the ice and forcing the heavier water through to the bottom.

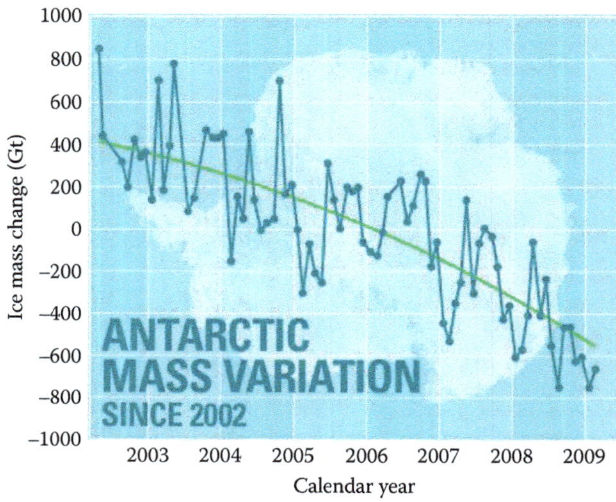

FIGURE 1.20
The recent decrease in the Antarctic ice sheet mass. (Courtesy of NASA Earth Observatory, http://www.nasa.gov/topics/earth/features/20100108.)

But in the Arctic Sea, ice is decreasing—both in area and in thickness—at an alarming rate, and at a rate that is rapidly increasing much faster than expected or predicted by models (Figure 1.21). Average Arctic temperatures increased at almost twice the global average rate in the past 100 years. In many places, it is much larger, approaching 3°C over northern Greenland. The decrease in the extent of the Northern Hemisphere sea ice is based on satellite measurements, and the decrease in thickness is based on measurements from submarines. In the 1990s, the U.S. Navy allowed scientists to go on five data-taking cruises under the Arctic Sea ice. Data from such submarines (Rothrock et al. 1999) has been classified in the past, but recently, some data became publicly available from cruises that took place between 1958 and 1976. When the two data sets were corrected for seasonal growth and compared, it was discovered that the average ice thickness had decreased by 1.3 m. Data presented at the American Geophysical Union Fall 2007 Meeting in Vienna suggests the ice is no longer showing a robust recovery from the summer melt. Most climate models simulations have predicted a loss in the September ice cover of 2.5% per decade from 1953 to 2006. But newly available data sets, blending early aircraft and ship reports with more recent satellite measurements that are considered more reliable than the earlier records, show that the September ice actually declined at a rate of about 7.8% per decade during the 1953–2006 period and the rate may be increasing (Figure 1.22). This suggests that current model projections may in fact provide a conservative estimate of future Arctic change, and that the summer Arctic Sea ice may disappear considerably earlier than IPCC projections. The 2007 IPCC reports that Arctic Sea

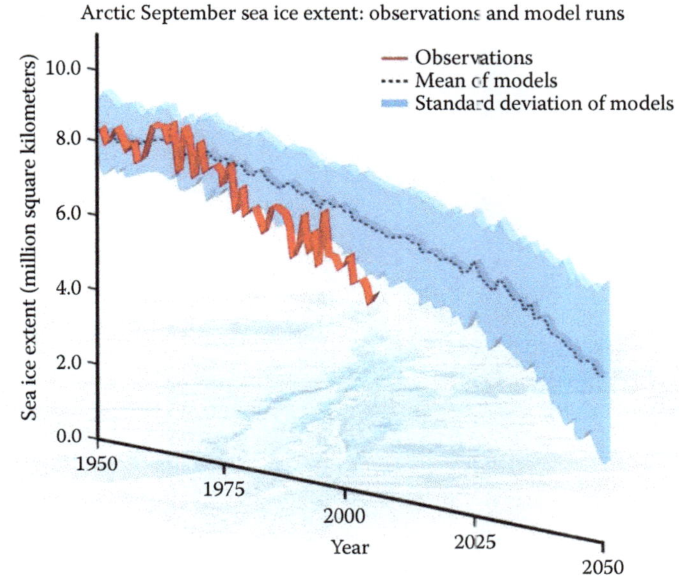

FIGURE 1.21
The rate of decrease of Arctic Sea ice exceeds that predicted by models. (Reprinted from *The Copenhagen Diagnosis: Updating the World on the Latest Climate Science*, Allison, I., et al., 60, Copyright (2009), with permission from Elsevier.)

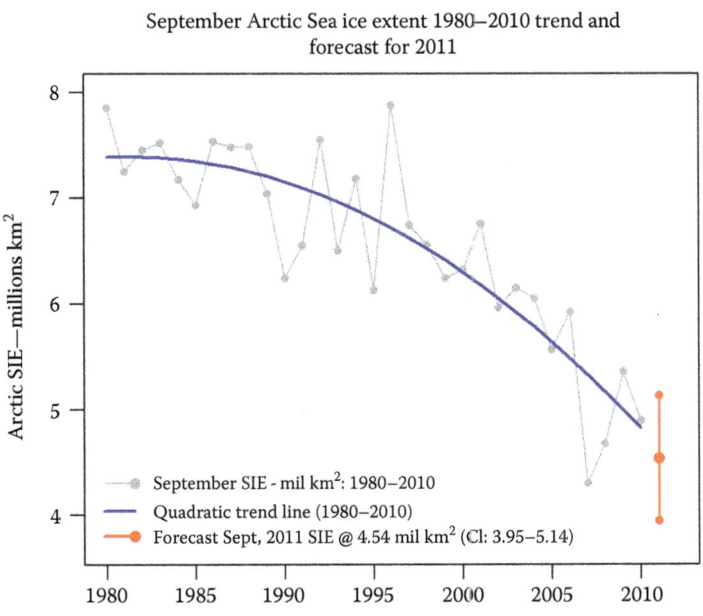

FIGURE 1.22
The observed rate of decrease of the area of Arctic Sea ice is increasing.

Sea ice extent, September 2011

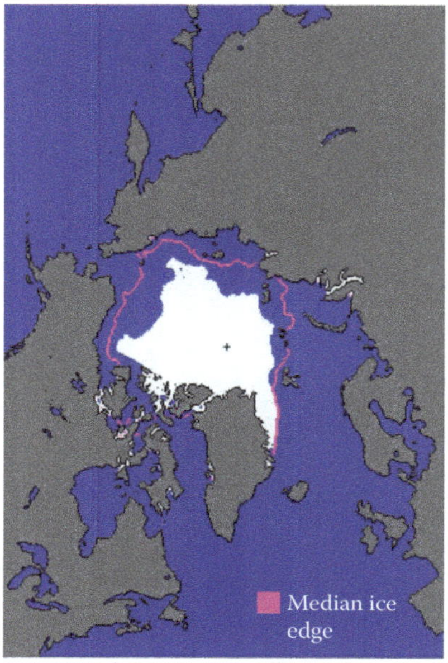

Total extent = 4.6 million sq km

FIGURE 1.23
Arctic Sea ice extent for September 2011 was 4.61 million km^2 (1.78 million square miles). The magenta line shows the 1979 to 2000 median extent for that month. The black cross indicates the geographic North Pole. (Courtesy of the National Snow and Ice Data Center, Boulder, CO.)

ice may disappear by sometime between 2050 and 2100. It may disappear much earlier.

Figure 1.23 shows data from the National Snow and Ice Data Center. Evidently, a sea route may open very soon, the long-sought northwest passage. More importantly, as the area of sea ice decreases, so does the reflectivity of the surface as the highly reflective ice is replaced by far less reflective water. This is a very important positive feedback. This will hasten global warming, and the increased high-latitude warming could hasten the melting of ice caps in Greenland.

Snowfall should decrease in all but possibly very high latitudes. At very high latitudes, it will be cold enough for snowfall, at least for a while during winter, and the increased atmospheric water vapor might even cause increased snowfall. However, the poleward extent of snow should decrease in both hemispheres. Snowfall in a limited area of east Antarctica has increased recently, as it should under global warming because the poleward transport of moisture is predicted to increase. However, the end of the snow season is occurring earlier in the Northern Hemisphere, as shown in Figure 1.24. Although snowfall can increase under warmer

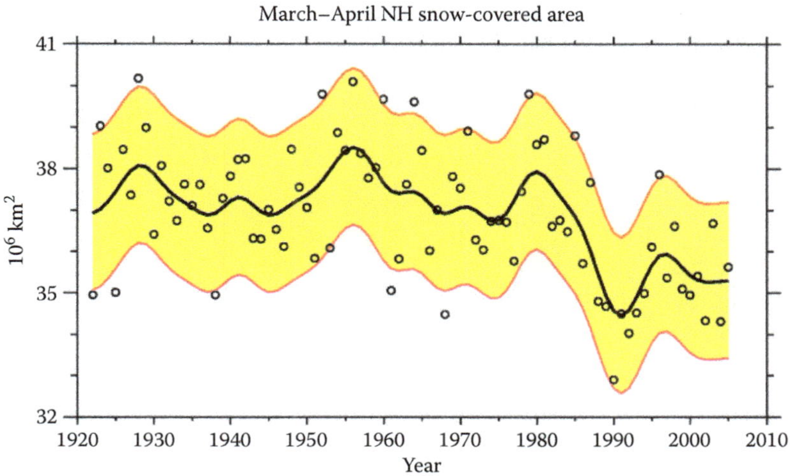

FIGURE 1.24
Northern Hemisphere snow cover for March–April. All changes are relative to corresponding averages for the period 1961–1990. Smoothed curves represent decadal average values while circles show yearly values. The shaded areas are the uncertainty intervals estimated from a comprehensive analysis of known uncertainties and from the time series. (From *Climate Change 2007: The Physical Science Basis,* Contribution of Working Group I to the Fourth Assessment Report of the Intergovernmental Panel on Climate Change, Cambridge University Press, Cambridge, U.K. and New York, 2007.)

conditions at very high latitudes, precipitation at somewhat lower latitudes can change from snow to rain if the local temperature increases, resulting in a decrease of snow cover. The extent (area) of snow cover in the Northern Hemisphere has decreased by about 10% over the past century (Solomon et al., 2007).

Sea level has been increasing over the past 150 years after remaining more or less constant for thousands of years after the end of the last glacial cycle (Figure 1.25). So far, much of the increase is attributed to the warming and thermal expansion of the ocean water. Observations since 1961 show that the average temperature of the global ocean has increased to depths of at least 3000 m and that the ocean has been absorbing more than 80% of the heat added to the climate system (Figure 1.26). Such warming causes seawater to expand, contributing to sea level rise. Note also that if so much is being added to the ocean below the surface, the rate of increase of the surface temperature would be considerably higher (Watts and Morantine 1991). The apparent increases in the rate of melting in both Greenland and Antarctica are, however, cause of concern.

Although there is little strong evidence that hurricanes are becoming more frequent, there is strong evidence that they will become more destructive. There is observational evidence for an increase of intense tropical cyclone activity in the North Atlantic since 1970. An increase in the energy of

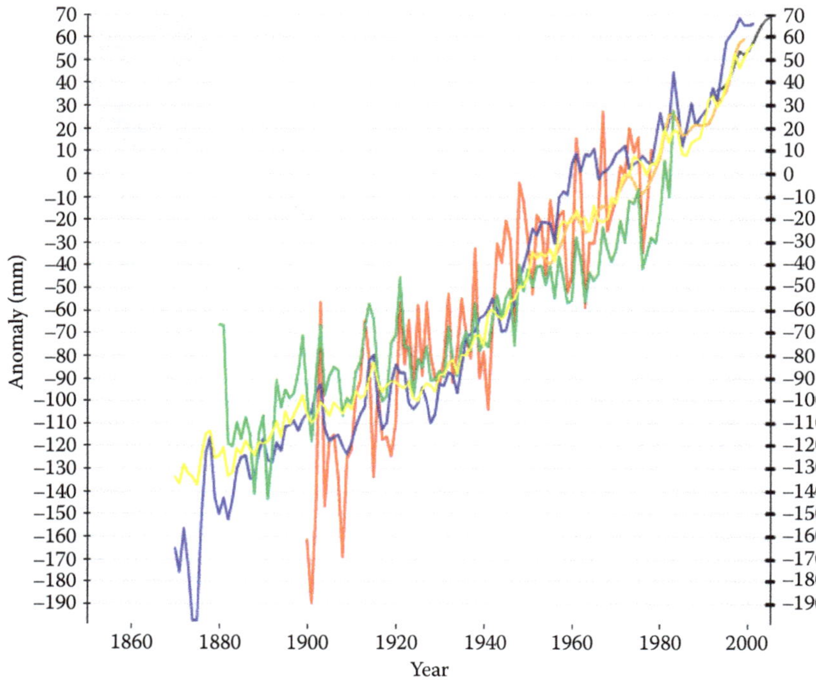

FIGURE 1.25
Sea level increase. Yellow (Church et al. 2006), green (Gornitz et al. 1987; Woodworth et al. 2008), orange (Holgate et al. 2004), blue (Jevrejeva et al. 2006), black (Leuliette et al. 2004), and red (Trupin et al. 1992). (Courtesy of National Climate Data Center, http://www.ncdc.noaa.gov/.)

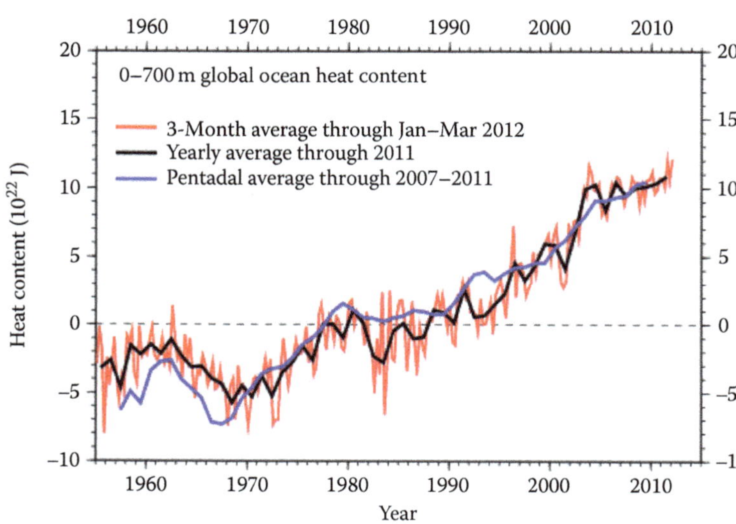

FIGURE 1.26
The increase in the heat content of the deep ocean. (Courtesy of NOAA/NESDIS/NODC Ocean Climate Laboratory; data updated from Levitus et al. [2009].)

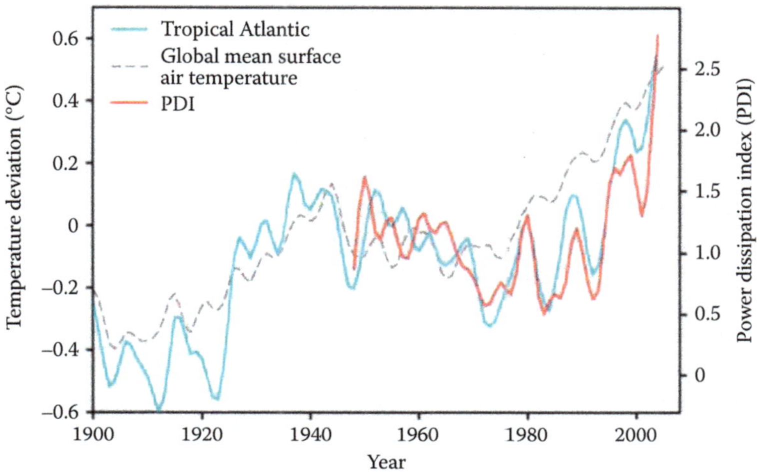

FIGURE 1.27
Temporal development of the energy of tropical storms (Power Dissipation Index–PDI, red) and the average sea-surface temperature in the tropical Atlantic from August to October (blue). For comparison, the evolution of the globally averaged near-surface air temperature is shown (dashed gray line). (After Emanuel, K., *Nature*, 436, 686, 2005.)

hurricanes is suggested in response to rising sea surface temperatures by both models and data (Figure 1.27). A number of recent studies have shown that the observed rise of sea surface temperatures in the relevant areas of the tropics is primarily due to global warming, not due to a natural cycle, and this will almost certainly affect the strength of hurricanes. This makes sense, since tropical storms are driven by evaporation from the ocean, and the tropical ocean has become warmer. Trends since the 1970s show an increasing trend in storm duration and greater storm intensity. The number of category 4 and 5 storms has increased by about 75% since 1970.

The climate change fingerprint is there. When all of the evidence is taken together, it leaves no doubt that the earth's climate has warmed during the past century.

The bottom line? The globally averaged temperature of the earth has increased (although not continuously) by about half a degree Centigrade during the past 100 years. This is entirely consistent with both estimates from the reconstruction of climates when the CO_2 content was different from the present and with the results of GCMs. One statistical study implies that the natural variability should quite possibly be cooling at present, while the presence of sulfate aerosols should almost certainly be causing a cooling, especially in the Northern Hemisphere. Yet, the long-term trend shows a definite warming. Other data provide an identifiable fingerprint of global warming. It is certain that the warming will continue as more GHGs enter the atmosphere.

1.4 Skeptics: Are Their Doubts Scientifically Valid?

An argument frequently used by congresspersons and others for delaying any action to prevent (stop) global warming is that the scientific community is divided on the issue of the science of global warming. If scientists cannot agree on such a fundamental question as "is the greenhouse effect real?" then certainly we should not base policy decisions on the prospects of global warming. The immediate response to this argument is that, by any measure, the overwhelming majority of climatologists agree on the fundamental science. Policy makers and individuals do and should, want to hear both sides of important questions, but they should be aware that while thousands of scientists believe that the greenhouse effect is real, and that warming is now occurring, and that the consequences will be very serious and harmful to the environment and to humans, only a very few are now challenging that view. Even so, the answers to scientific questions should not be based on popularity polls. When a serious scientist proposes serious questions, these must be dealt with seriously and objectively. Former vice president Al Gore has stated in his documentary, *An Inconvenient Truth* that while hundreds of scientific articles about global warming have been published in recent years, not a single article has appeared in the peer-reviewed literature that is skeptical of global warming or its impacts. This is, in fact, not true. It is true that most of the comments from the skeptics have appeared in the so-called "gray literature" (not reviewed by scientific peers), but some have appeared in peer-reviewed journals, and these claims must be taken especially seriously. Responses to many of the challenges by skeptics are fully discussed in *Global Warming and the Future of the Earth* (Watts 2008). I will discuss two more recent ones.

First consider "Climategate," the claim that scientists have been fabricating and/or fudging data and making false claims about global warming. A prominent claim is that some of the proxy data included in Figures 1.4 and 1.5 was altered by using a "trick." In fact, this and many other claims have been thoroughly investigated by at least four independent groups. The first is a report of the international panel set up by the University of East Anglia to examine the research of the Climatic Research Unit (CRU), which is available online at http://www.cru.uea.ac.uk/cru/people/briffa/yamal2009/. Then, in April 2010, the University of East Anglia set up an International Scientific Assessment Panel, in consultation with the Royal Society and chaired by Professor Ron Oxburgh. The report of the international panel assessed the integrity of the research published by the CRU and found "no evidence of any deliberate scientific malpractice in any of the work of the Climatic Research Unit." In June 2010, the Pennsylvania State University published its final investigation report, determining "there is no substance to the allegation against Dr. Michael E. Mann." In August 2011, the National Science Foundation concluded, "Finding no research misconduct or other matter raised by the various regulations and laws discussed above, this case is closed." Earlier, the U.S. National Academy of Sciences published its findings in a report, *Surface Temperature Reconstructions for the Last 2000 Years*, Committee on Surface Temperature Reconstructions for the past 2000 years, National Research Council of the U.S. National Academy of Sciences, 2006. Subsequently, in Mann et al. (2008) (Figure 1.5b), the error bars have been substantially reduced. The investigators involved have been exonerated and there seems to be little doubt of the essential accuracy of the hockey stick graph.

My second comment concerns the persistent claim that the temperature record of warming correlates with sunspot number and solar constant variations. The data shown in Figure 2.4 clearly refute this claim.

Some writers have made an issue of the fact that between about 1940 and 1975, the global temperature failed to increase, as seen clearly in Figures 1.6, 1.7, and 1.8, when atmospheric CO_2 was rapidly increasing. This is easily explained by the exchange of heat between the near-surface ocean water and the water at intermediate depths. During that period, the ocean warmed at intermediate depths (Roemmich and Wunsch 1984) sufficiently to cool the surface, completely offsetting the greenhouse warming, possible because of a decrease in the thermohaline circulation. The effect of heat exchange between the surface and the intermediate and deep ocean was first pointed out by Watts (1985) and Watts and Morantine (1991).

Finally, when questioned about the isolated spells of very warm weather or other unusual climate events, I note that scientists who are involved in climate change research generally refuse to claim that they have some "proof" of global warming. Not so of the skeptics. For example, during January 2010, an extreme cold event occurred in the eastern United States as well as most of Europe (Figure 1.28). Many skeptics were quick to claim that was an indication that global warming was a hoax. However, as Figure 1.28 shows, much of the world was undergoing extreme warming. Figure 1.29 shows the zonally averaged temperatures for various latitudes. Even at the latitudes where the temperature was cold, the average temperature over all longitudes was warm. In fact, January 2010 was globally one of the warmest on record.

Figures 1.28 and 1.29 were obtained from the National Climate Data Center website.

The claims of the skeptics are thus soundly debunked. Absent the plausibility of the other causal factors, it seems highly unlikely that global warming is caused by any other mechanism than the increase in atmospheric GHGs, the most prominent being the increase in atmospheric CO_2.

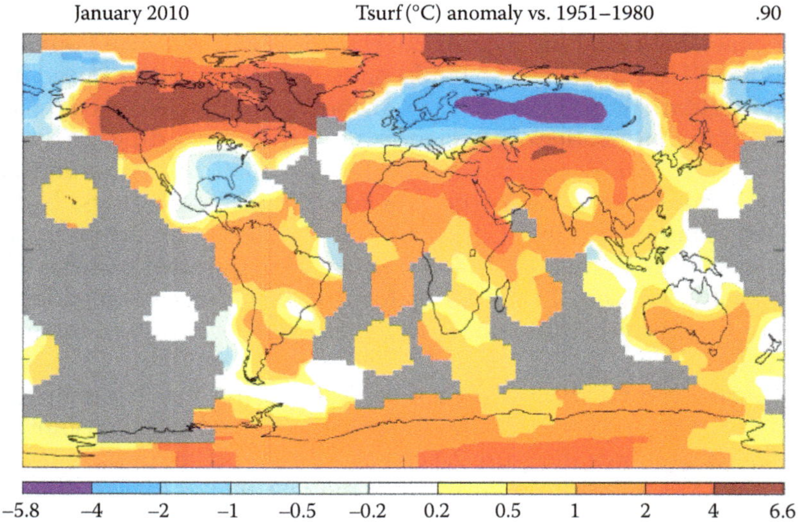

FIGURE 1.28
Surface temperature change from 1952 to 1980 mean for January 2010. (Courtesy of National Climate Data Center, http://www.ncdc.noaa.gov/.)

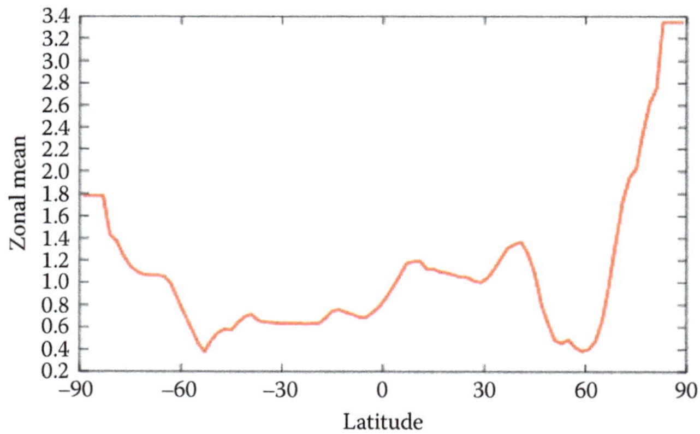

FIGURE 1.29
Zonal temperature departure from the 1951–1980 mean for January 2010. (Courtesy of National Climate Data Center, http://www.ncdc.noaa.gov/.)

1.5 About This Book

In the remainder of this book, we set the stage by discussing the separate issues of the emissions of radiatively important atmospheric constituents, energy demand, energy supply, agriculture, water resources, coastal hazards, adaption strategies, and geoengineering.

In June 1991, we convened the Workshop on the Engineering Response to Global Climate Change: Planning a Research and Development Agenda. Approximately 70 scientists and engineers spent four days discussing what research was necessary to identify the causes and the extent of climatic change, to assess its consequences, and to prepare for mitigative and adaptive measures. Seven working groups were established to discuss the following: Sources and Sinks of Greenhouse Gases; Energy Demand, Energy Supply; Agricultural and Biological Systems; Water Resources; Coastal Hazards; and Geoengineering. Each working group was asked to identify a set of goals and a set of approaches to accomplish these goals. Common goals and approaches were then identified, and small, interdisciplinary groups were convened to expand upon the various approaches. Some chapters of this book are updates of chapters from the book *Engineeering Response to Global Climate Change* that was published in 1997. Several chapters introduce new material. For example, at the time that the first book was published, the very serious problem of changing ocean chemistry was not yet recognized. The chapter on adaptation is entirely new.

In the first chapter, we have explored the nature of the problem (Section 1.2). In Section 1.3, we outlined the problem of climatic change and the concept of feedbacks. We defined climate sensitivity and reviewed the results of climate models of various complexity. We discussed the evidence that climate change has occurred and some of the important impacts that have already taken place and those that can be expected in the not so distant future. Finally, in Section 1.4, we discussed (and refuted) the various claims of the skeptics.

Chapter 2 explains the differences between the natural and human drivers of climate change and describes how humans have influenced the global climate during past decades.

The purpose of Chapter 3 is to explore the potential global scenarios of energy, economic activity, and land use in the twenty-first century. The implications of alternative economic and demographic development possibilities are discussed as well as potential stabilization regimes for energy technology choice, economic activity, land use, and land cover.

Chapter 4 provides a review of coastal hazards associated with sea level rise relative to land elevation. Changes in sea level have contributed to the remaking of the earth's surface ever since the oceans came into being. In recent decades, an additional concern has become how societies will respond to an increasing sea level. Some predictive modeling suggests that the rise will be anywhere between 0.9 m (low estimate) to 1.3 m (high estimate) by the end of this century. Increased sea levels will lead to increased coastal impacts, including, for example, shoreline recession, inundation of lowlands, storm damage to infrastructure, and ecological changes. Vulnerable countries such as the Netherlands have developed, after careful engineering analysis, responses ranging from defending the shoreline with dikes to "managed retreat" from the shoreline.

Chapter 5 presents an overview of the problems and issues likely to result from changes in the availability and distribution of water as a result of global climate change. Potential response strategies required to cope with current hydrologic variability, as well as with potential changes, are examined. Also outlined are the research and development needed to allow the flexibility, resiliency, and robustness of water resource systems necessary to deal with projected, but uncertain climatic changes and the present variability that stresses existing systems.

A review of various demand-side strategies for reducing the rate of anthropogenic production of atmospheric CO_2 through more efficient energy use is presented in Chapter 6. Although the discussion in Chapter 6 makes it clear that current industrialized societies are wasteful of energy and that a comfortable and vibrant society can be maintained with much less energy use, this chapter and Chapter 3 make it equally clear that less developed countries can only prosper if their energy use is greatly increased.

An overview of the potential of renewable electric technologies to reduce carbon emissions in the U.S. electric sector is presented in Chapter 7. It includes a description of the different renewable electric technologies, their technical potential, and their costs. It also presents the major issues associated with the introduction of large amounts of renewable electric technologies with variable resources such as wind and solar.

Chapter 8 concerns energy from nuclear fission. Nuclear energy as a source of power generation has been in existence for more than five decades and has made significant contributions to the electricity needs and GHG emissions reduction for many nations around the world. The nuclear accident in Japan, along with several prior major accidents, has increased the uncertainty of nuclear energy use as a potential response to global climate change. With the exception of a few countries, the pace of global nuclear power deployment has been more measured. More widespread global use of nuclear energy, particularly in response to climate change, will depend on several factors, including the implementation of climate mitigation policies and the successful realization of international R&D efforts to improve the cost, safety, waste management, and proliferation resistance of nuclear energy use.

The current status of magnetic and inertial fusion energy, its recent accomplishments, near-term expectations, and remaining challenges are discussed in Chapter 9. Great progress has been made by the international fusion program in its mainline efforts for both magnetic and inertial fusion energy; but because of the need for nearer term solutions, the chapter describes a possible strategy for supplementing these continuing efforts with increased efforts to develop alternative systems that offer significant advantages or simplifications in a power plant or in the development process to get there.

All commercial power needs can be obtained from solar-derived mass-less microwave photons supplied to the earth from power bases on the moon. This novel idea is discussed in Chapter 10. The sun dependably illuminates our moon with 13,000 TWs of solar power, or ~650 times the 20 TWe of commercial electric power needed by a prosperous earth. The LSP system uses solar-powered bases built on the moon to collect ≤1% of this sustainable power to dependably supply receivers on the earth with safe microwave power beams (≤ 230 We/m^2 or ≤ 20% the intensity of noon sunlight). The 20 TWe is the output on the earth from ~100,000 km^2 of receivers and then into electric power grids all about the earth without significant movement of mass within the biosphere. By 2050, a 20 TWe LSP system can enable a sustainable >$84,000 GWP/person. Additional clean LSP electric energy can be used to extract all industrial CO_2 from the earth's atmosphere. A greater than 10 T$ GL(lunar)P is possible within the twenty-first century.

Given the earlier discussion on the importance of climate change and the potential for significant changes to the earth's climate over the twenty-first century and beyond, Chapter 11 is aimed at discussing the societal basis for adaptation to climate change. Adaptation is about responding to the risks associated with climate vulnerability and building resilience in response to these risks.

Twenty years ago, when the first edition of this book was presented, fossil fuel CO_2 emissions were just over 6 billion tons of carbon per year (22,000 million metric tons of CO_2 per year in the units used in international negotiations) emissions are now up about 50%. In 1990, the world population was just over 5 billion—it recently passed 7 billion. In 1990, it had not been convincingly demonstrated that the rising CO_2 concentration was influencing the climate in significant ways—now, the human influence is clearly documented, the Arctic Sea ice is retreating, the Greenland and Antarctic ice sheets are losing mass, the rate of sea level rise has accelerated, and the ranges of many flora and fauna are shifting (IPCC 2007a, 2007b). In 1992, the nations of the world agreed on the U.N. Framework Convention on Climate Change calling for international action to limit emissions and prevent "dangerous anthropogenic interference with the climate"—now, the expected emissions of the world community are projected to lead to warming that is about double the level that world leaders suggest will cause unacceptable impacts, and about eight times the warming in 1990 just before the most serious impacts started to emerge.

With such a dire situation, it should not be surprising that perspectives on geoengineering (now more often being called climate engineering) have also changed. In the first edition of this book, an overview of geoengineering was provided, but there was no sense that it would be needed—it was merely an interesting corner of academic research. Chapter 12 updates the summary of potential options for engineering the climate. While there may be options for controlling quite local conditions, the two potential global strategies are to increase the rate of removal of CO_2 from the atmosphere, referred to as CO_2 Removal, or CDR, and to modify the energy balance at the regional and even global scale, referred to solar radiation management, or SRM.

While fossil-fuel CO_2 emissions have resulted in de facto domination of the climate by humans, climate engineering using these approaches would entail intentional human assumption of responsibility for the climate. Given the very limited international action to date to deal with climate change, the choice of how to proceed is thus really narrowing down to either not very successfully dealing with climate change and its impacts through traditional mitigation and adaptation or taking the rather audacious step of having humans take control of the global climate, seeking to keep the climate about as it is.

Although the world has so far experienced relatively limited changes in climate, it is on course, by the end of the twenty-second century, for the global average temperature to rise to levels not experienced on the earth for many tens of millions of years and to having the sea level rise at a rate of a meter or more per century. Even with increasingly aggressive actions to reduce global emissions, the world is committed to very significant changes in climate and very serious impacts, even with insightful efforts to adapt to present and projected conditions. Without climate engineering as a complement to these actions, not as a substitute for them, the earth's climate will soon move outside the ranges to which society and the environment have become adapted over the past several centuries. Chapter 12 discusses the situation the world faces and the options for taking action, even as far as moving to take over control of the global climate system, despite the very difficult governance, equity, and ethical implications involved.

1.6 Questions for Discussion

1. Is there a sixth revolution as we begin to explore the universe beyond our blue planet? What might this mean with respect to our environment?
2. What do you think about the skeptics' claims? You can go online to skepticalscience.com and find 159 claims about global warming by a variety of skeptics.
3. What impacts on society (not just the United States) will result from global warming?
4. What actions (if any) should the United States take? (You will want to return to this question after you finish this book.)
5. According to Figure 1.26, the ocean has absorbed about 1.4×10^{23} J since about 1950. The increases in global warming over those years have added an average of around 2 W/m^2 to the surface of the ocean. Calculate how much of this it would take to increase the energy of the ocean over 60 years. Refer to Watts and Morantine (1985, 1991).
6. One of the most serious impacts from increasing atmospheric CO_2 is the resulting buildup of CO_2 in the oceans with resulting acidification. A good reference to read and discuss is *Ocean Acidification: The Other CO_2 Problem*, a review paper by Doney et al. (2009). Find this paper on the internet and discuss it. The Annual Review of Marine Science is online at http://marine.annualreviews.org.

References

Allison, I., N. L. Bindoff, R. A. Bindschadler, P. M. Cox, N. de Noblet, M. H. England, J. E. Francis, N. Gruber, A. M. Haywood, D. J. Karoly, G. Kaser, C. Le Quéré, T. M. Lenton, M. E. Mann, B. I. McNeil, A. J. Pitman, S. Rahmstorf, E. Rignot, H. J. Schellnhuber, S. H. Schneider, S. C. Sherwood, R. C. J. Somerville, K. Steffen, E. J. Steig, M. Visbeck, A. J. Weaver, 2009. *The Copenhagen Diagnosis: Updating the World on the Latest Climate Science*, The University of New South Wales Climate Change Research Centre (CCRC), Sydney, Australia, p. 60.

Berry, D. I., E. C. Kent, 2009. A new air-sea interaction gridded dataset from ICOADS with uncertainty estimates. *Bull. Amer. Meteor. Soc.*, 90(5), 645–656.

Brohan, P., J. J. Kennedy, I. Harris, S. F. B. Tett, P. D. Jones, 2006. Uncertainty estimates in regional and global observed temperature changes. A new dataset from 1850. *J. Geophys. Res.*, 111, D12106.

Christy J. R., R. W. Spencer, W. B. Norris, W. D. Braswell, D. E. Parker, 2003. Error estimates of version 5.0 of MSU-AMSU bulk atmospheric temperatures. *J. Atmos. Oceanic Technol.*, 20, 613–629.

Church, J. A., N. J. White, 2006. A 20th century acceleration in global sea-level rise. *Geophys. Res. Lett.*, 33, L01602.

Crowley, T. J., G. R. North, 1991. *Paleoclimatology*, Oxford University Press, Oxford, U.K.

Dai, A., 2006: Recent climatology, variability, and trends in global surface humidity. *J. Climate*, 19, 3589–3606.

Darwin, C. G., 1953. *The Next Million Years*, Doubleday, New York.

Emanuel, K., 2005. Increasing destructiveness of tropical cyclones over the past 30 years. *Nature*, 436, 686–688.

Forster, P., V. Ramaswamy, P. Artaxo, T. Berntsen, R. Betts, D. W. Fahey, J. Haywood, J. Lean, D. C. Lowe, D. C. Myhre, J. Nganga, R. Prinn, G. Raga, M. Schulz, R. Van Doland, 2007. Changes in atmospheric constituents and in radiative forcing, Chapter 2 in *Climate Change 2007: The Physical Science Basis.* Contribution of Working Group I to the Fourth Assessment Report of the Intergovernmental Panel on Climate Change, Cambridge University Press, Cambridge, UK and New York, NY.

Free, M., D. J. Seidel, J. K. Angell, J. R. Lanzante, I. Durre, T. C. Peterson, 2005. Radiosonde atmospheric temperature products for assessing climate (RATPAC): A new dataset of large-area anomaly time series. *J. Geophys. Res.*, 110, D22101.

Gornitz, V., S. Lebedeff, 1987. Global sea-level changes during the past century. In *Sea-level Fluctuation and Coastal Evolution*, Nummedal D., Pilkey O. H., Howard J. D. (eds), The Society for Sedimentary Geology, Tulsa, Oklahoma, SEPM Special Publication No.41, p. 316.

Haimberger, L., 2007. Homogenization of radiosonde temperature time series using innovation statistics. *J. Climate*, 20, 1377–1403.

Haimberger, L., C. Tavolato, and S. Sperka, 2008. Towards elimination of the warm bias in historic radiosonde temperature records—Some new results from a comprehensive intercomparison of upper air data. *J. Climate*, 21, 4587–4606.

Haldane, J. B. S., 1923. *Daedalus or Science and the Future*, K. Paul, Trench, Trubner, London.

Hansen, J., R. Ruedy, M. Sato, M. Imhoff, W. Lawrence, D. Easterling, T. Peterson, T. Karl, 2001. A closer look at United States and global surface temperature change. *J. Geophys. Res.*, 106, 23947–23963.

Holgate, S. J., P. L. Woodworth, 2004. Evidence for enhanced coastal sea level rise during the 1990s. *Geophys. Res. Lett.*, 31, L07305.

IPCC, 2007. Climate Change 2007. Fourth Assessment Report of the Intergovernmental Panel on Climate Change. Cambridge University Press, Cambridge, UK and New York, NY.

Ishii, M., A. Shouji, S. Sugimoto, T. Matsumoto, 2005. Objective analysis of SST and marine meteorological variables for the 20th Century using ICOADS and the Kobe Collection. *Int. J. Climatol.*, 25, 865879.

Jevrejeva, S., A. Grinsted, J. C. Moore, S. J. Holgate, 2006. Nonlinear trends and multiyear cycles in sea level records. *J. Geophys. Res.*, 111, C09012.

Kaplan, A., M. Cane, Y. Kushnir, A. Clement, M. Blumenthal, B. Rajagopalan, 1998. Analyses of global sea surface temperature 1856–1991. *J. Geophys. Res.*, 103, 18567–18589.

Leuliette, E. W., R. S. Nerem, G. T. Mitchum, 2004. Calibration of TOPEX/Poseidon and Jason altimeter data to construct a continuous record of mean sea level change. *Mar. Geodesy*, 27(12), 7994.

Lugina, K. M., P. Ya. Groisman, K. Ya. Vinnikov, V. V. Koknaeva, N. A. Speranskaya, 2005. Monthly surface air temperature time series area-averaged over the 30-degree latitudinal belts of the globe, 1881–2004. In: *Trends: A Compendium of Data on Global Change*. Carbon Dioxide Information Analysis Center, Oak Ridge National Laboratory, US Department of Energy, Oak Ridge, TN.

Mann, M. E., R. S. Bradley, M. K. Hughes, 1999. Global scale temperature patterns and climate forcing over the past six centuries. *Geophys. Res. Lett.*, 26, 759–762.

Mann, M. E., et al. 2008. Proxy-based reconstructions of hemispheric and global surface temperature variations over the past two millennia. *Proc. Natl. Acad. Sci. USA*, 105, 13252–13257.

Mears, C. A., F. J. Wentz, 2009. Construction of the remote sensing systems V3.2 atmospheric temperature records from the MSU and AMSU microwave sounders. *J. Atmos. Oceanic Technol.*, 26, 1040–1056.

Pachauri, R. K., A. Reisinger (eds.), 2007. *Climate Change 2007: Synthesis Report.* Contribution of Working Groups I, II and III to the Fourth Assessment Report of the Intergovernmental Panel on Climate Change, IPCC, Geneva, Switzerland, p. 104.

Rajagopalan, B. and U. Lall, 1998. Low frequency variability in Western U.S. precipitation, *J. Hydrology*, 210, 51–67.

Rayner, N. A., P. Brohan, D. E. Parker, C. K. Folland, J. J. Kennedy, M. Vanicek, T. Ansell, S. F. B. Tett, 2006. Improved analyses of changes and uncertainties in sea surface temperature measured in situ since the mid-nineteenth century: The HadSST2 data set. *J. Climate*, 19(3), 446–469.

Ries, J. B., A. L. Cohen, D. C. McCorkle, 2009. Marine calcifiers exhibit mixed responses to CO_2-induced ocean acidification. *Geology*, 37(12), 1131–1134.

Roemmich, D., C. Wunsch, 1984. Apparent changes in the climatic state of the deep North Atlantic. *Nature*, 307, 447–450.

Rothrock, D. A., Y. Yu, G. A. Maykut, 1999. *Geophys. Res. Lett.*, 26, 3469.

Sherwood, S. C., C. L. Meyer, R. J. Allen, H. A. Titchner, 2008. Robust tropospheric warming revealed by iteratively homogenized radiosonde data. *J. Climate*, 21, 5336–5350.

Smith, T. M., R. W. Reynolds, T. C. Peterson, J. Lawrimore, 2008. Improvements to NOAA's historical merged land-ocean surface temperature analysis (1880–2006). *J. Climate*, 21, 2283–2293.

Solomon, S., D. Qin, M. Manning, Z. Chen, M. Marquis, K. B. Averyt, M. Tignor, H. L. Miller (eds.), 2007. *Climate Change 2007: The Physical Science Basis.* Contribution of Working Group I to the Fourth Assessment Report of the Intergovernmental Panel on Climate Change, Cambridge University Press, Cambridge, UK and New York, NY, p. 996.

Thorne, P. W., D. E. Parker, S. F. B. Tett, P. D. Jones, M. McCarthy, H. Coleman, P. Brohan, 2005. Revisiting radiosonde upper air temperatures from 1958 to 2002. *J. Geophys. Res.*, 110, D18105.

Trupin, A. S., J. M. Wahr, 1992. Spectroscopic analysis of global tide gauge sea level data. *Geophys. Res. Lett.*, 108, 115.

Watts, R. G., 1980. Climate models and the prediction of CO_2-induced climate change. *Clim. Change*, 2, 1980.

Watts, R. G., 1985. Global climate variations due to fluctuations in the rate of deep water formation. *J. Geoph. Res.*, 90, 8067–8070.

Watts, R. G., 2007. *Global Warming and the Future of the Earth*, Morgan and Claypool, San Rafael, CA, 113 pp.

Watts, R. G., 2008. *Global Warming and the Future of the Earth.* Morgan and Claypool Publishers, New York.

Watts, R. G., M. C. Morantine, 1991. Is the greenhouse gas-climate signal hiding in the deep ocean? *Clim. Change*, 18, 3–4.

Watts, R. G., M. C. Morantine, 1994. Time scales in energy balance climate models: Part 1. The limiting case solutions. *J. Geophys. Res*, 99, 3643–3651.

Weinberg, A. M. 1966. Can technology replace social engineering? *Bull. Atomic Sci.*, 22, 4–8.

Willett, K. W., P. D. Jones, N. P. Gillett, P. W. Thorne, 2008. Recent changes in surface humidity: Development of the HadCRUH dataset. *J. Climate*, 21, 5364–5383.

Woodworth P. L., N. J. White, S. Jevrejeva, S. J. Holgate, J. A. Church, W. R. Gehrels, 2008. Evidence for the accelerations of sea level on multi-decade and century timescales. *Int. J. Climatol.*, 29(6), 777–789.

Worley, S. J., S. D. Woodruff, R. W. Reynolds, S. J. Lubker, N. Lott, 2005. ICOADS Release 2.1 data and products. *Int. J. Climatol.* (CLIMAR-II Special Issue), 25, 823–842.

Worm, B. E. B. Barbier, N. Beaumont, J. E. Duffy, C. Folke, B. S. Halpern, J. B. C. Jackson, H. K. Lotze, F. Micheli, S. R. Palumbi, E. Sala, K. Selkoe, J. J. Stachowicz, R. Watson, 2006. Impacts of biodiversity loss on ocean ecosystem services. *Science*, 314, 787–790.

Zou, C.-Z., M. Gao, M. Goldberg, 2009. Error structure and atmospheric temperature trends in observations of the microwave sounding unit. *J. Climate*, 22, 1661–1681.

2

Radiatively Important Atmospheric Constituents

Donald J. Wuebbles and Darienne Ciuro

CONTENTS

2.1 Introduction

This chapter discusses the chemical constituents and radiative properties that characterize our atmosphere and drive our climate system. Earth's climate system is always at work; it receives, absorbs, and reflects radiative energy from the sun in higher quantities at the tropics and redistributes the heat from low to high latitudes to compensate for the lack of heat received at the poles. This transport and exchange of heat determines the global atmospheric circulation patterns and, thus, the distribution of temperature and precipitation across the globe. However, this mechanism alone cannot provide the earth with the adequate amount of heat to enable and sustain life in this planet. Instead, radiatively important gases in the earth's atmosphere, referred to as greenhouse gases, play a key role in keeping a balance between incoming short wave and outgoing long wave radiation by acting like an insulating blanket regulating the temperature of our planet. These gases, mainly water vapor (H_2O), carbon dioxide (CO_2), methane (CH_4), nitrous oxide (N_2O), and ozone (O_3), all act as effective insulators; their concentrations in our atmosphere significantly affect the overall temperature of our atmosphere. Without the greenhouse gases, our atmosphere would be unpleasantly cold and likely uninhabitable. Human-driven activities have enhanced the greenhouse effect by changing the concentrations of these mentioned gases along with introducing other chemicals like sulfates, chlorines, other halogens, and other gases and particles that chemically and physically interact in the earth's atmosphere. It is then of utmost importance to understand the structure of atmospheric gases and particles, their chemical and physical interactions, their lifetimes, and how human activities can and are altering this naturally occurring effect.

2.2 Drivers of Climate and Climate Change

The earth's climate system is constantly changing through time. These changes occur due to the system's own complex internal dynamics, which are in turn driven by direct and indirect forcings. Direct forcings include natural phenomena such as variations in the output of the sun as well as human-driven changes, while indirect forcings are those feedback mechanisms that get triggered by direct changes in the system.

In summary, there are two main direct drivers of climate and climate change:

- Natural forcings
- Human-driven forcings

Factors like the reflectivity (or albedo) to space of the earth–atmosphere system and various feedbacks within the earth's climate system are all important in determining the resulting climate. It is important to recognize the differences between climate and weather. Mark Twain is attributed to saying that climate is what we expect, weather is what we get. Climate is the long-term (often taken as 10–30 years) averages and variations in temperature, precipitation, and other weather-relevant variables.

The earth's climate history has been mainly defined by innumerable natural processes that have each shaped the system. The effect of each process varies depending on the magnitude, duration, and repetition of such change. There are several natural occurring processes that can directly impact our climate. All these processes affect in one way or another, the intensity of solar radiation received or absorbed by our atmosphere. One of these natural processes occurs in cycles and happens as a result of the interaction between the planet Earth and Sun, also known as orbital variations. Milutin Milankovitch, a Serbian mathematician, engineer, and meteorologist, proposed that the changes in the intensity of solar radiation received from the earth were primarily affected by three fundamental mechanisms, now known as the Milankovitch cycles: eccentricity, obliquity, and precession. These cycles do correlate with cyclical increases and decreases of temperature but cannot explain some shorter abrupt drops in temperature in our historical record (Figure 2.1).

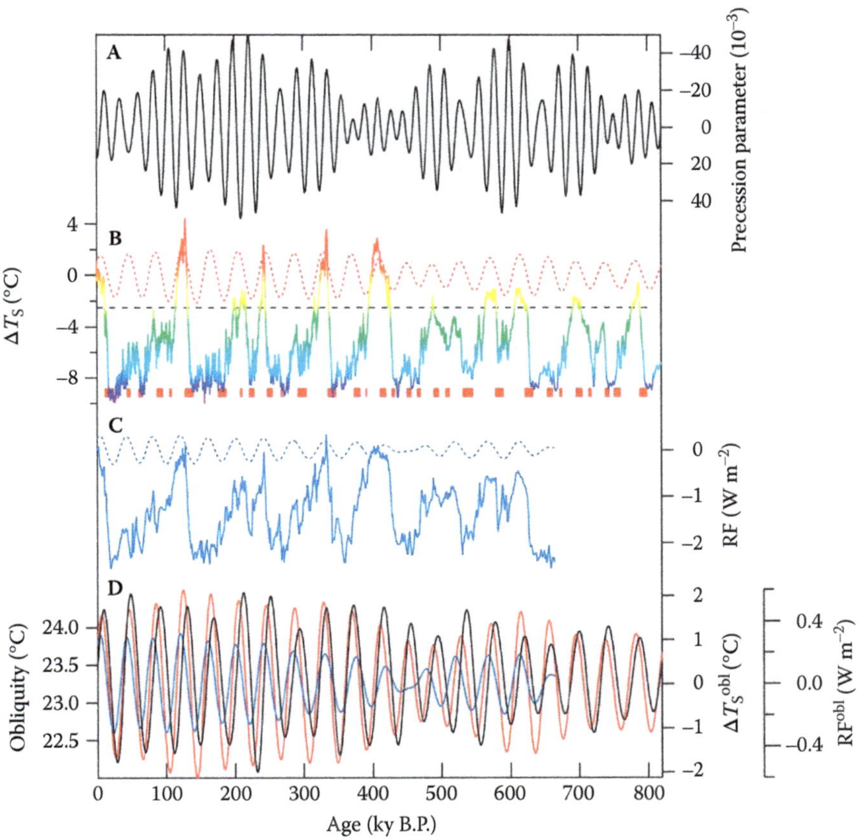

FIGURE 2.1
(A) Precession parameter displayed on an inversed vertical axis (black line). (B) EDC temperature [solid line, rainbow colors from blue (cold temperatures) to red (warm temperatures)] and its obliquity component extracted using a Gaussian filter within the frequency range 0.043 ± 0.015 ky^{-1} [dashed red line, also displayed in (D) as a solid red line on a different scaling]. Red rectangles indicate periods during which obliquity is increasing and precession parameter is decreasing. (C) Combined top-of-atmosphere radiative forcing due to CO_2 and CH_4 (solid blue) and its obliquity component [dashed blue, also displayed in (D) as a solid blue line on a different scaling]. (D) Obliquity (solid black line), obliquity component of EDC temperature (red line), and obliquity component of the top-of-atmosphere radiative forcing due to CO_2 and CH_4 (blue). (From Jouzel, J. et al., *Science, 317*, 793–796, 2007. With permission.)

Volcanic eruptions, on the other hand, are unpredictable; their effects depend on many factors and add uncertainty to the occurrence of natural fluctuations in climate. When volcanoes erupt, they release pyroclastic rocks, ash, dust, and various gases and particles into the atmosphere. Although the larger materials tend to fall out of the atmosphere quickly, others, especially sulfur dioxide (SO_2), can react to form sulfuric acid particles. These particles reflect sunlight. The radiative and chemical effects of the cloud, especially for a large explosive eruption that puts volcanic emissions high in the atmosphere, can result in an overall cooling of the troposphere by scattering some solar radiation back to space, cooling the surface but causing warming in the stratosphere owing to absorption. This cooling can last an average of a year or two depending on the magnitude of the eruption. However, in longer timescales, volcanic eruptions result in a warming by acting as a source of carbon dioxide into the atmosphere. This signal in the carbon budget is, however, very small and does not represent a significant contribution to the current increasing warming trend.

An additional natural process that we must consider is the vertical and horizontal shifting of tectonic plates. As tectonic plates move, so do continents. This redistribution of landmass has an important effect on the earth's energy balance through differences in the albedos and thermal properties of land, ocean, and ice. Additionally, tectonic processes have indirect climatic impacts through geochemical interactions and its involvement in the composition of the atmosphere and ocean. Some of these natural processes, like Milankovitch cycles and tectonic plate shifts, occur on a span of tenths of thousands to millions of years. Volcanic eruptions, however, have a fast effect and occur more often. Nonetheless, an eruption of sufficient size and of tropical location is needed to impact climate for more than a couple of years.

Although there have been many variations in the climate system due to natural forcings throughout the earth's history, the changes in climate observed over the past 4–5 decades cannot be explained by natural forcings. Human-related climatic forcings are necessary for explaining the observed global changes in temperature and other climate variables during the last half century. As an example of the many published analyses showing the human-driven attribution of recent climate changes, all of the complex numerical models of the climate system used in the last international science assessment, IPCC (2007) did analyses of the separate effects on climate from natural forcings over the last century (and beyond), and the effects also included human forcings (see Figure 2.2 for the National Center for Atmospheric Research climate model results from Meehl et al. [2004]). By separating the different signals, it is seen that natural forcing alone cannot explain the current observed warming trend over the recent decades.

Everyday human activities change climate by introducing gases into the atmosphere and changing the concentration of key constituents that alter the radiative balance. The largest known human contribution is the burning of

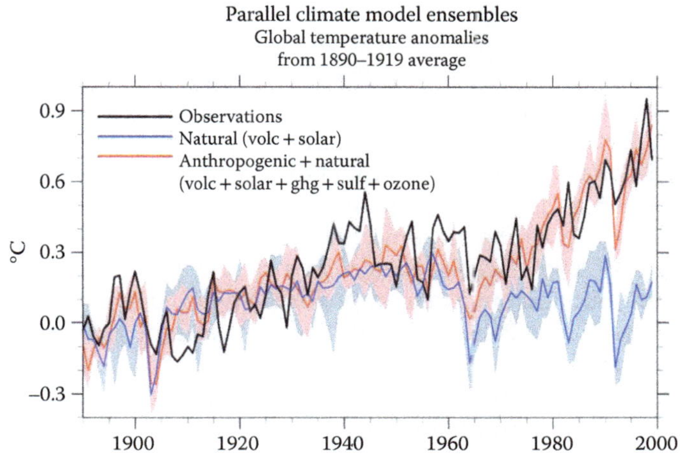

FIGURE 2.2
Global temperature anomalies from 1890 to 2000 shown for natural forcings separated from all forcings in separate climate modeling ensemble studies from the National Center for atmospheric Research climate model. Ensemble analyses used to reflect the effects of natural variability during climatic timescales. (From Meehl, G. A. et al., *J. Climate, 17*, 3721–3727, 2004. With permission.)

fossils fuels, which is the primary human-related source of carbon dioxide (CO_2). Carbon dioxide levels have increased from fossil fuel use in transportation, building temperature regulation, and manufacture of construction goods. In addition, land-use change releases CO_2 from plant matter. Especially through the removal of tropical forests, there has been a reduction in CO_2 uptake from the atmosphere, therefore, increasing the existing atmospheric concentration. Human actions also impact other important greenhouse gases like methane, nitrous oxides, and ozone. In addition, a significant amount of aerosols, specifically those associated with sulfur, black carbon, and organic compounds, are largely produced as a result of biomass and fossil fuel burning. Overall, the human influence on climate has been a warming signal since the start of the Industrial Revolution in 1750. This signal from human-driven activities exceeds that due to known natural forcings, and therefore, it is important to understand how all this natural and human forcings come together and result in a net warming or cooling of our atmosphere.

2.3 Natural Forcings

The earth travels around the sun in an elliptical orbit. At the same time, because of the existence of other planetary bodies (mostly Jupiter and Saturn), this orbit deviates from a perfect circle, which is known as eccentricity. These changes occur on a 100,000-year cycle. When the earth is closest to the sun (perihelion), more solar radiation reaches the surface of the planet, resulting in a warmer planet. When the earth's orbit is farthest from the sun (aphelion), colder temperatures are observed. However, looking at today, the orbit is closer to circular. The earth travels in a loop around the sun each year, while it additionally rotates on its tilted axis. The hemisphere tilted toward the sun experiences summer, while the hemisphere facing away experiences winter. The severity of the seasons changes when the angle of the tilt of the earth with respect to its orbit changes in a cycle called obliquity. Obliquity variations are periodic and take approximately 41,000 years to shift from approximately 22.1° to 24.5°. Obliquity at its maximum increases the radiative flux from the sun during the summers, while less is received during the winter. It is then understood that lower obliquity increases the insolation contrast between low and high latitudes and favors ice ages since it cools down summers and that in turn causes a reduction in the melting of the ice pack (Figure 2.3).

The last cycle, precession, also has to do with the tilt; however, instead of the angle changing, the direction of the tilt as the earth "wobbles" around its axis changes. The maximum precession, therefore, would result in a switch between winters and summer seasons, with winter experiencing warmer months and cooler months occurring during our current summers. This cycle occurs every 23,000 years. All these cycles contribute to changes in the amount and distribution of solar radiation that reaches the earth and the amount of energy that is transferred from the sun. These cycles correlate with glacials and interglacials; these "ice ages" occur roughly every 120,000 years. The earth came out of the last ice age about 20,000 years ago and is expected to start the long slow slide into the next ice age within the next several millennium.

Shorter term changes in climate can and have happened in the past because of variations in the output from the sun. Analyses of ice core datasets suggest that extremely small changes in surface temperature appear

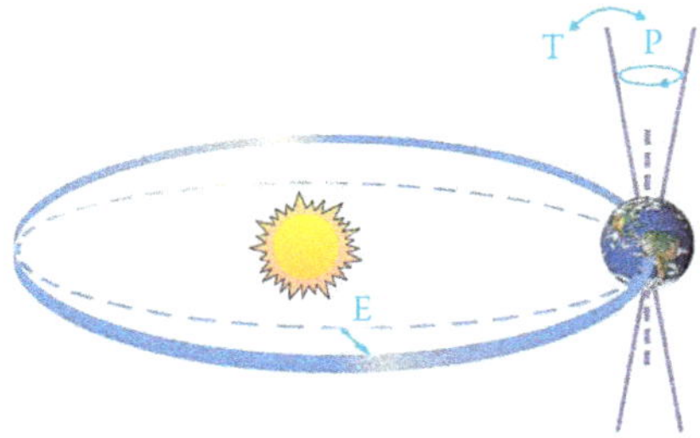

FIGURE 2.3
Milankovitch cycles of the earth. "T" is for tilt (obliquity), "P" is for precession, and "E" is for eccentricity. (From Rahmstorf, S., H.-J. Schellnhuber, *Der Klimawandel—Diagnose, Prognose, Therapie*, C. H. Beck, München, 2006. With permission.)

to result during the 11-year sunspot cycle (likely as a result of effects from atmospheric wave activity resulting from changes in ozone and temperature in the stratosphere). However, detailed measurements of the solar flux made from satellites since 1978 indicate that the net change in solar flux during the last three solar cycles has been a slight decrease (e.g., see Figure 2.4) (Gray et al. 2010, 2012). While cosmic ray penetration in the earth's atmosphere, and resulting particle production, associated with the 11-year sunspot cycle, has been proposed as a possible mechanism for effects on clouds and resulting climate changes over time, Gray et al. (2010) summarizes recent analyses indicating that this mechanism is not likely to be important.

Another natural process involved in paleoclimate change has to do with tectonic plates. However, plate tectonics plays little to no role in the changes in climate that have occurred over recent centuries. The planet Earth has not always looked the way it does now (see Figure 2.5). For ages the continents were arranged differently, all closer together, and as time passed by, tectonic plates shifted, changing the exposed surfaces, controlling land, ocean, and ice and how they interact. Although the atmosphere and clouds are primarily responsible for the planetary albedo, radiation that is not reflected or long wave radiation that is not released gets absorbed by the surface. Therefore, changes in the surfaces by tectonic shifting can also impact climate by changing the amount of surface albedo (see Table 2.1). Consequently, variations in the surface albedo of the planet have important effects on atmospheric dynamics and climate.

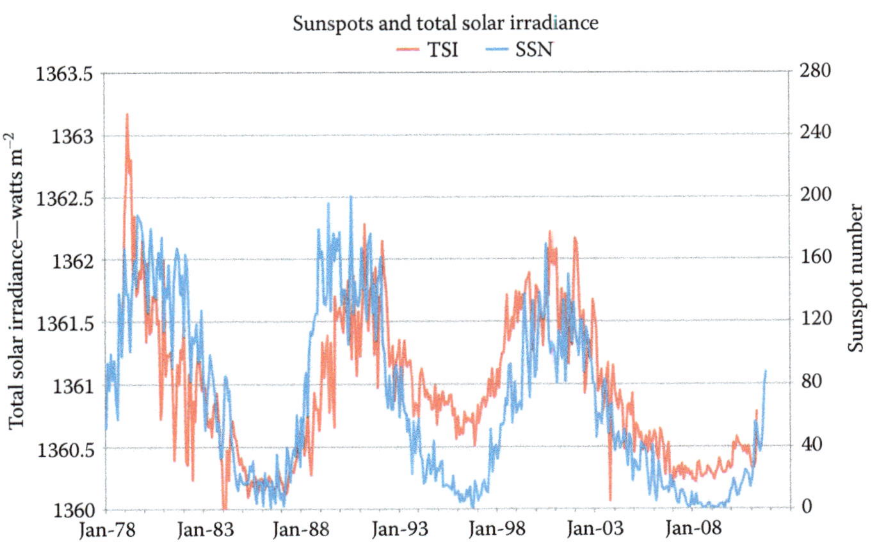

FIGURE 2.4
Satellite observed variations in sunspot number (blue) and total solar irradiance (red) since 1978. (Available at http://www.climatedata.info/Forcing/Forcing/sunspots.html.)

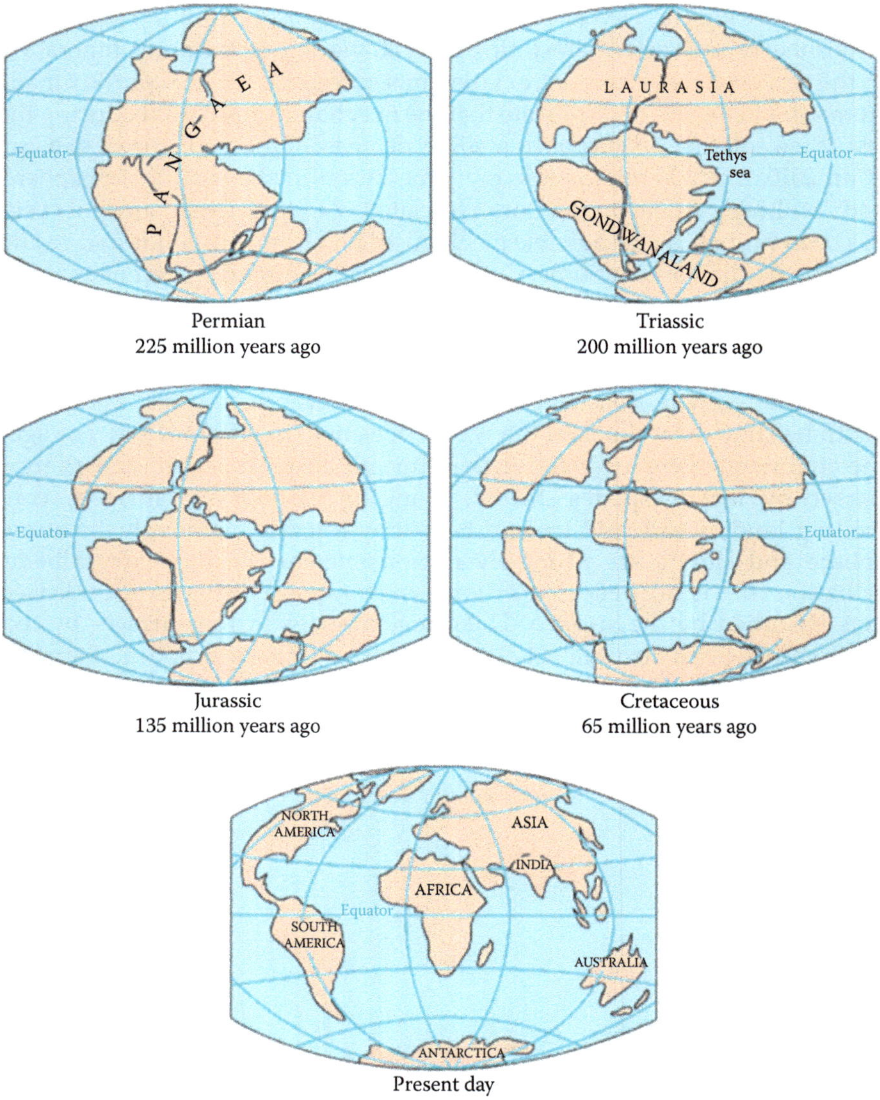

FIGURE 2.5
The earth from Pangaea to the present-day continent map. (From Kious, J., R. Tilling, *This Dynamic Earth: The Story of Plate Tectonics*, Washington, DC, USGS, 1996. With permission.)

Volcanic eruptions also contribute to the natural forcings that drive our climate. An eruption can inject into the stratosphere an enormous amount of microphysically and chemically active gases and aerosol particles. The injected ash falls rapidly from the stratosphere (several days to weeks) because of precipitation and gravity and has little impact on climate. The rest of the gaseous emissions consist primarily of water vapor, carbon dioxide,

TABLE 2.1

Average Percentage Range for Surface Albedo

Surface	Percent Reflected (%)
Water	5 to 10
Forest	5 to 20
Grasslands	18 to 25
Tundra	15 to 35
Soil	5 to 30
Desert (sand)	30 to 50
Clouds	40 to 90
Melting snow	40 to 70
Fresh snow	60 to 90

sulfur compounds (mainly SO_2), nitrogen, and halogen compounds. They are then introduced into the atmosphere with an atmospheric residence time that depends on the biogeochemical cycles of the elements (O, S, and C). Carbon dioxide and water vapor are released from a volcano when magma rises to the surface from the depths of the earth (Gerlach 2011). U.S. Geological Survey estimates an approximate contribution of 200 million tonnes of CO_2 annually from volcanic eruptions both on land and under the sea. Although this might seem like a large amount, comparing it with data from the U.S. Department of Energy's Carbon Dioxide Information Analysis Center, the carbon dioxide contribution from volcanoes is negligibly small (almost 1%) when compared with the current atmospheric reservoir size or even human emissions, and therefore, the climate impact of carbon dioxide from this source is insignificant. On longer timescales, however, this small signal contributes to the warming. Volcanic aerosols like sulfur dioxide, however, can cause an overall global cooling by scattering efficiently the visible part of the solar spectrum and increasing the optical depth of the atmosphere and thus raising the atmospheric albedo.

In addition, the stratospheric chemical composition can be disrupted because these aerosols can serve as surfaces for heterogeneous chemical reactions that act to destroy stratospheric ozone. This sort of impact in the atmosphere can only be caused by large explosive volcanic eruptions. Most tropospheric eruptions and degassing have no significant potential for climate impact beyond the short-lived effect of ash because of efficient removal of sulfates via precipitation processes. Significant climatic effects lasting for at least a few years can be caused by explosive volcanic eruptions into the stratosphere, mainly for two reasons: larger amount of SO_2 and much longer aerosol residence times of these compounds in the stratosphere before they can eventually be removed by rain out in the troposphere. It is important to note that hemispheric or global distribution of volcanic aerosols is determined by its eruptive location. Tropical eruptions are capable of global impact, since their aerosols are usually distributed globally through

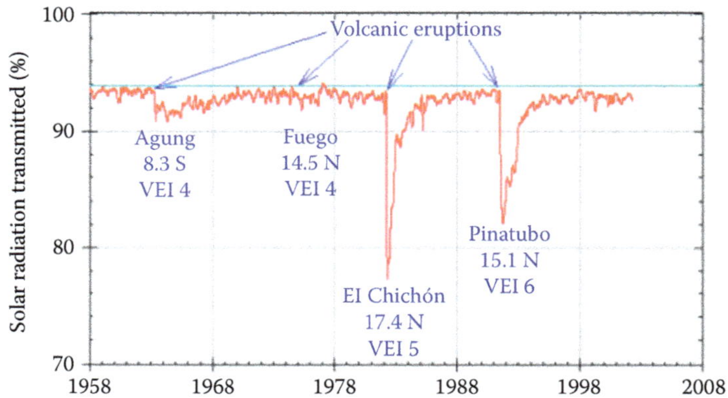

FIGURE 2.6
Reduced solar radiation due to volcanic aerosols as measured at Mauna Loa Observatory, Hawaii. (From Hofmann, D. et al., *Geophys. Monogr. Ser. 139*, 57–73, American Geophysical Union, 2003. With permission.)

poleward transport and fast horizontal circulation patterns. The earth's past climate has experienced explosive volcanoes that have resulted in temperature anomalies lasting several years to decades and longer (if the effects affect polar ice melting with resulting effects on the ocean transport of heat). In the case of the 1991 Mt. Pinatubo eruption, the solar scattering of the sulfate aerosol cloud resulted in a radiative forcing of about -0.5 W/m^2, which lasted for several months. This coincided with a global cooling on the order of a few tenths of degree Celsius, which persisted for about a year or so. Another historical volcanic eruption, the 1815 eruption of the Tambora volcano in Indonesia, had a severe impact in society at the time, by causing mass crop failure leading to starvation, civil unrest, and helping spread epidemics across country borders. The following year is referred to as the "year without a summer" in Europe and North America and the cooling effect is estimated to be on the order of 5°C. Some past eruptions are only recorded in proxy data (mainly ice cores) and in order to quantify the role of volcanism in natural climatic variations and to assess the human influence on climate; comprehensive studies on these records are needed (see Figure 2.6).

2.4 The Greenhouse Effect

The greenhouse effect is a natural process, first discovered by Fourier in 1827 and substantiated by observations made in the 1860s. The greenhouse effect is extremely important to the earth's climate. Because of the greenhouse effect, the earth has the perfect balance of temperature needed to sustain life. Solar short wave radiation passes through our atmosphere where some of it is reflected, scattered, and absorbed by the atmosphere, but where most of it reaches the earth's surface. This energy absorbed by the planet is then released from the surface as long wave (infrared) radiation, which would otherwise be transmitted to space without the existence of the greenhouse gases. Instead, this infrared radiation is intercepted, absorbed, and retransmitted by the greenhouse gases, resulting in energy being trapped in the earth–atmosphere system. Figure 2.7 summarizes the key radiative processes of the greenhouse effect. Without this effect, the average temperature of the earth would be significantly colder (approximately −18°C), rather than the present 15°C. Over 90% of the emitted long wave energy is retained and directed back to the earth's surface, where it is again absorbed and released continuing this cycle. Earth's atmosphere is mostly made of nitrogen (78.1%) and oxygen (20.9%). The rest includes the radiatively important greenhouse gases. It is important to point out that the amount of energy retained in the atmosphere by the greenhouse effect is controlled directly by that concentration of all the greenhouse gases present in the atmosphere. Each greenhouse gas has a specific heat capacity, has different lifetimes, and absorbs radiation

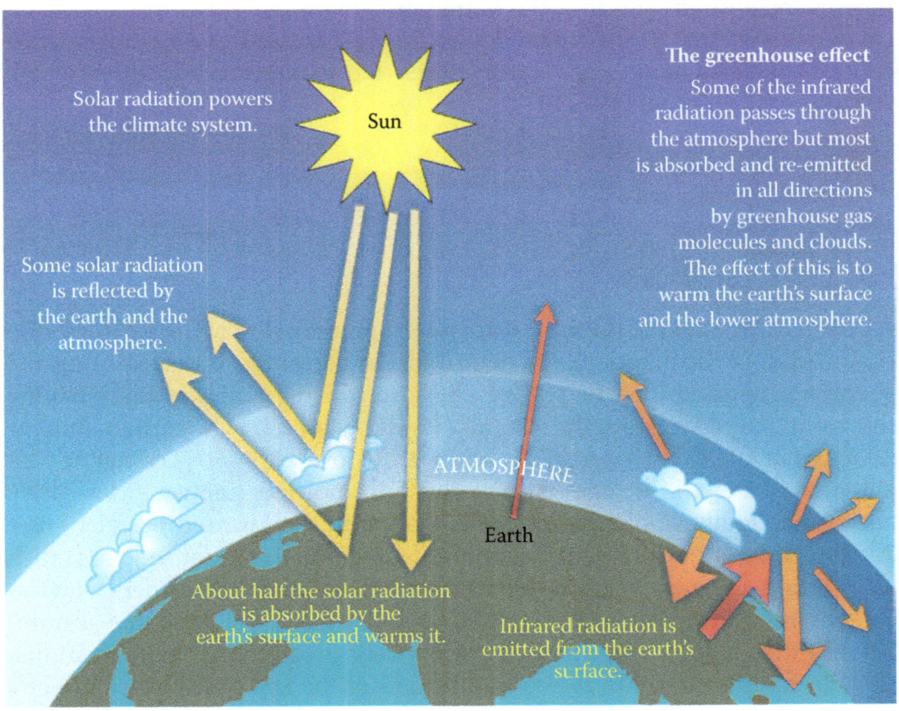

FIGURE 2.7
An idealized model of the natural greenhouse effect. (From Le Treut, H., et al., In: *Climate Change 2007: The Physical Science Basis. Contribution of Working Group I to the Fourth Assessment Report of the Intergovernmental Panel on Climate Change*, Solomon, S., et al. (eds.), Cambridge University Press, Cambridge, U.K. and New York, 2007. With permission.)

at specific wavelengths, resulting in some gases having a much stronger impact on temperature than others.

Water vapor is the single most important of the greenhouse gases in determining the greenhouse effect. Water vapor and clouds together account for 75% of the radiative forcing from infrared absorption (Schmidt et al. 2010) (these analyses assume the year 1980 atmospheric concentrations), and CO_2 accounts for 19%. However, global climate model experiments by Schmidt et al. (2010), where water vapor and clouds are removed, still leaves 33% of the absorption occurring. Because absorption line overlaps, the joint effects of removing water vapor and CO_2 or water vapor and clouds are greater than the sum of effects of removing each component individually. Accounting for uncertainties, Schmidt et al. conclude that water vapor is responsible for just over half, clouds around a quarter, and CO_2 about a fifth of the present-day total greenhouse effect. Given that the attribution is closer to 20% than 2% (as implied in some Internet blogs, perhaps resulting from an unattributed statement in a book review of the first IPCC assessment by Lindzen [1991]), it makes sense that changes in CO_2 concentrations could be important for determining changes in the earth's climate. Note that numerical climate models include changes in water vapor and clouds as part of the hydrological determination within the model, so they are treated as a feedback when considering the changes in climate either due to natural or human-related forcings on the climate system.

2.5 The Greenhouse Gases

As already mentioned, through absorption and emission of the earth's radiation, certain atmospheric gases contribute importantly to determining the average temperature of the planet. These so-called *greenhouse* gases having the right radiative properties include water vapor (H_2O), carbon dioxide (CO_2), methane (CH_4), ozone (O_3), and nitrous oxide (N_2O). (The major constituents of the air, nitrogen and oxygen, are not greenhouse gases—they absorb negligibly small amounts of radiation in the infrared wavelengths). Although the atmospheric concentrations of some greenhouse gases, such as CO_2 and CH_4, are being influenced by human activities, others, such as the chlorofluorocarbons CFC-11 ($CFCl_3$) and CFC-12 (CF_2Cl_2), and a number of other halocarbons, are almost entirely produced by human industry. Observed changes in the concentrations of the key gases being directly affected by human activities are shown in Figure 2.8. Preindustrial concentrations of these long-lived gases, as shown in Table 2.2, are largely determined from analyses of the bubbles trapped in ice cores.

The atmospheric concentrations of these important atmospheric gases are changing, largely as a result of human activities. The concentration of carbon

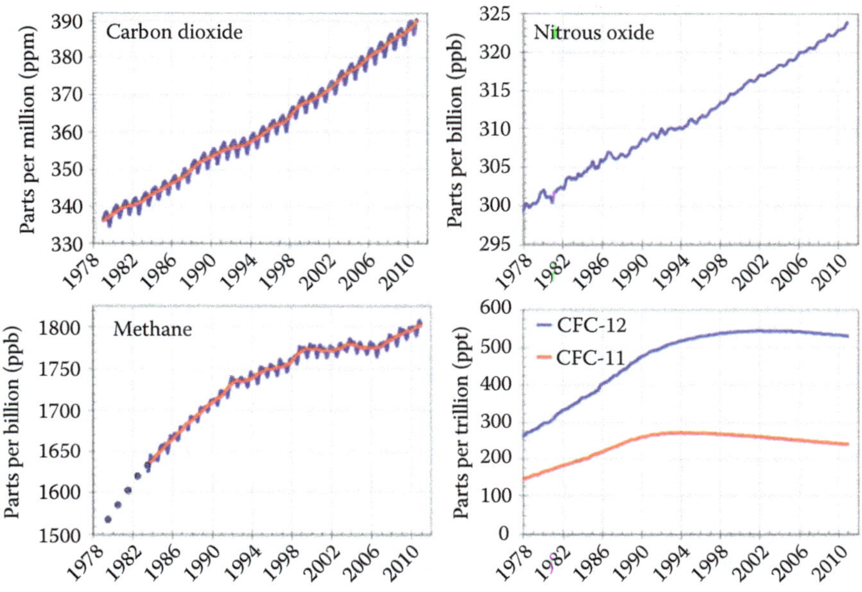

FIGURE 2.8
Global average abundances of the major, well-mixed, long-lived greenhouse gases—carbon dioxide, methane, nitrous oxide, CFC-12, and CFC-11—from the NOAA global air sampling network are plotted since the beginning of 1979. These gases account for about 96% of the direct radiative forcing by long-lived greenhouse gases since 1750. The remaining 4% is contributed by an assortment of 15 minor halogenated gases. (Available at http://www.esrl.noaa.gov/gmd/aggi/.)

dioxide is increasing at a rate of 0.5% per year; the atmospheric CO_2 concentration has increased by almost 40% since the beginning of the Industrial Revolution in the eighteenth century. Carbon cycle measurements and analyses clearly indicate that this increase is a result of fossil fuel burning and secondarily land use change (IPCC 2007). For example, as one of many different pieces of evidence, measurements of the carbon isotopes indicate that the primary source of the increasing CO_2 is from the long-buried fossil fuel carbon (IPCC 2001, 2007; Vaughn et al. 2010). Methane concentrations have more than doubled over that same time period. Nitrous oxide (N_2O) concentrations have been increasing at about 0.3% per year for a number of decades, with the increase thought to be primarily associated with the use of fertilizers leading to bacterial emissions of nitrous oxides in soils. Industrial production of chlorofluorocarbons has largely been stopped as a response to the international agreements under the Montreal Protocol, but their atmospheric concentrations increased dramatically until roughly the beginning of the twenty-first century and are now declining (for two of the important halocarbons, concentrations of $CFCl_3$ [CFC-11] have been declining since 1995, while those of CF_2Cl_2 [CFC-12] started to decline in 2004). Human-related emissions of sulfur dioxide, largely responsible for sulfuric acid particles in

TABLE 2.2

Summary of Key Greenhouse Gases by Human Activities

	CO_2	CH_4	N_2O	CFC-11	CFC-12
Atmospheric concentration Preindustrial (~1750)	280 ppm	0.8 ppm	288 ppb	0 ppt	0 ppt
Present day (2010)	390 ppm	1.8 ppm	322 ppb	240 ppt	530 ppt
Current rate of change per year	~2 ppm (~0.5%/year)	0.01 ppm (0.3%)	0.8 ppb (0.25%)	−2 ppt (−0.8%)	−2 ppt
Atmospheric lifetime (years)	100[a]	8[b]	114	55	100

Source: Based on various sources, including WMO (2010) and NOAA ESRL (http://www.esrl.noaa.gov/gmd/aggi/).

ppm, parts per million by volume; ppb, parts per billion (thousand million) by volume; ppt, parts per trillion (million million) by volume.

[a] CO_2 does not have a single atmospheric lifetime because it is removed from the atmosphere and recycled into the oceans and biosphere. Lifetime gets longer with time as CO_2 is gradually transported to the deep ocean.

[b] Response time after a perturbation is longer for methane, ~12 years, due to effects of CH4 on atmospheric hydroxyl, OH.

the atmosphere, appear to have increased substantially over the past century, but have declined in Europe and North America during the last decade while continuing to increase in China and India. Ozone concentrations in the stratosphere have decreased since the 1970s, but are now showing the beginning signs of a slow recovery. Tropospheric ozone concentrations appear to have increased over much of the twentieth century.

Water vapor has the largest greenhouse effect in the earth's atmosphere; however, its concentration in the troposphere is determined internally within the climate system, so tropospheric water vapor is not significantly affected, at least directly, by human activities. In contrast, the atmospheric concentrations of a number of other greenhouse gases are changing, primarily as a result of human activities. Table 2.2 summarizes the atmospheric concentrations and trends of some of the significant greenhouse gases directly affected by human activities; there are a number of other greenhouse gases that also need to be considered in evaluating human effects on climate forcing.

In general, carbon dioxide and other greenhouse gases are relatively inefficient absorbers of solar radiation; however, these gases are strong absorbers of infrared (IR, also called long wave) radiation (heat), the type of radiative energy emitted by the earth. The greenhouse gases reemit the absorbed long wave radiation at a rate dependent on the local atmospheric temperature, which tends to be cooler than the earth's surface temperature. Some of this radiation reaches space. Some, however, is emitted downward, leading to a net trapping of long wave radiation and resulting in a radiative (or *greenhouse*) forcing on climate. As the concentrations of greenhouse gases increase, this net trapping of infrared radiation is enhanced, leading to a tendency for warming on the earth's surface.

The contribution of a gas to the greenhouse effect depends on the wavelengths of infrared radiation that the gas absorbs, the amount of the radiative absorption per molecule of the gas, the atmospheric concentration of the gas, and whether other gases absorb infrared radiation at the same wavelengths. At a number of wavelengths in the range of the infrared, the relatively high concentrations of H_2O and CO_2 in the present-day atmosphere almost completely absorb the radiation emitted at the earth's surface before it can be lost into space. Increases in the concentrations of these gases lead primarily to increased absorption in the wings (edges) of the absorption spectrum, with the result that the net trapping of infrared radiation because of these gases increases logarithmically with concentration. Gases absorbing at infrared wavelengths similar to those of the H_2O and CO_2 spectral lines contribute little to the greenhouse effect unless their atmospheric concentrations approach those of H_2O and CO_2.

Absorption by gases in the 7- to 13-μm region of the infrared wavelengths is particularly relevant to climate forcing. This wavelength region is referred to as the window region, because there is little absorption of this radiation by H_2O and CO_2. Most gases other than CO_2, with the potential to affect climate, including CH_4, N_2O, and the CFCs, have absorption lines in the window region. Some of these gases, such as CH_4 and N_2O, have sufficient present-day atmospheric concentrations that significant absorption is occurring; the net result is that their absorption (and the radiative forcing on climate) increases approximately as the square root of their concentration. Other gases, such as the CFCs and other halocarbons, exhibit absorption that increases linearly with concentration.

The net result of these varying radiative characteristics is that comparable increases in the concentrations of the various greenhouse gases have vastly different effects on radiative forcing (see Table 2.3). For example, the addition of one molecule of CH_4 to the atmosphere has about 21 times the effect on

TABLE 2.3

Radiative Forcing (RF) on Climate Relative to CO_2 per Unit Molecule Change and per Unit Mass Change in the Atmosphere for Various Greenhouse Gases for Present-Day Concentrations

Trace Gas	RF Per Molecule Relative to CO_2	RF Per Unit Mass Relative to CO_2
CO_2	1	1
CH_4	26	72
N_2O	214	214
CFC-11	17,700	5,700
CFC-12	22,700	8,300

climate that the addition of one CO_2 molecule has; one CFC-12 molecule has about 15,800 times the effect as an additional molecule of CO_2.

Ozone (O_3) is a gas naturally formed in the atmosphere, but its concentration is also influenced by other atmospheric gases. Ozone has several important effects on climate. Although the direct radiative effects of CO_2 and the other trace gases considered above depend largely on their concentrations in the troposphere, the climatic effect of ozone depends on its distribution throughout the troposphere and stratosphere. Ozone is an important absorber of ultraviolet (UV) and visible wavelengths (sunlight) in the atmosphere, and its concentrations determine the amount of UV radiation reaching the earth's surface. It is the absorption of solar radiation by ozone that explains the increase in temperature with altitude in the stratosphere. Ozone is also a greenhouse gas, with a large infrared absorption band at 9.6 μm. The balance between these radiative processes determines the net effect of ozone on climate.

2.6 Concerns about Human Effects on Climate

Concentrations of greenhouse gases in the atmosphere have changed naturally over long time scales. For example, analysis of air bubbles trapped in ice cores from Antarctica that date back 800,000 years indicates that CO concentrations in the atmosphere have varied from as low as 180 parts (molecules) per million (molecules of air) by volume (written ppmv, or more often as just ppm) during past ice ages to as high as 280 ppm during interglacial periods. However, for a thousand years before the Industrial Revolution, the quantities and proportions of greenhouse gases remained relatively constant. During the past 200 years, as world population increased and technology and agriculture have developed, the abundances of greenhouse gases have changed markedly. For example, as seen in Table 2.2, atmospheric concentrations of CO_2 have increased by more than 30%, from 280 ppm in the 1700s to the current 390 ppm. The primary causes for this increase appear to be the combustion of fossil fuels with a smaller effect due to land use change and deforestation. Concentrations of CH_4 have more than doubled over the past two centuries. Although there remain many uncertainties about absolute magnitudes, the increase in CH_4 concentrations appears to be due to a combination of sources, including rice production, belching by farm animals like cattle and sheep, biomass burning, coal mining, and natural gas mining, production, and distribution. CFC-11, CFC-12, and a number of other halocarbons have no significant natural sources; they are produced by the chemical industry for use as refrigerants, solvents, aerosol propellants, foam-blowing agents, and other applications.

Evaluation of potential effects on climate requires estimates of future concentrations of greenhouse gases. These concentrations depend upon the magnitude of human-related emissions and on any changes occurring in biospheric processes, including those in response to climate change (e.g., changes in natural emissions of isoprene and other hydrocarbons as a result of warmer temperatures). The range of estimates for future emissions indicate that, without major human intervention, future concentrations of important greenhouse gases like CO_2, CH_4, and N_2O are likely to increase throughout this century. Ultimately, increases in greenhouse gas emissions, specifically carbon dioxide, will result in more infrared radiation trapped and held, which will in turn accelerate the increase in the observed temperature at the surface.

2.7 The Role of Particles in the Atmosphere

The air that we breathe is never pristine. It contains invisible solid and liquid bits of matter in a scale of micrometers to nanometers. These particles can be anything from dust, smoke, pollen, salt, spores, viruses, and bacteria, commonly known as aerosols. Aerosols are produced both naturally and from anthropogenic sources and can have a profound impact in our daily life by impacting global climate, local weather, visibility, and most importantly, personal health. It is important then to understand the chemical and physical characteristics of aerosols and their interaction, since the concentration, size, and chemical structure of them in the atmosphere are spatially and temporally variable, and their climatic and health impact depend greatly on these parameters. About 90% of the atmosphere's aerosols are of natural occurrence, while only 10% are human induced. Naturally produced aerosols include sulfates from volcanic eruptions, organic carbon from forest fires, sea salt from wind-blown sea spray, dust lifted by common desert storms, and some pollen and bacteria from biological processes. Anthropogenic aerosols account for the remaining 10%, and they are more common in urban areas where industrial activity is common and population is greater. Combustion of fossil fuels produces a significant amount of sulfur dioxide, which reacts with hydroxyl (OH) or on cloud droplets, resulting in the formation of sulfate particles. Another human-driven process is biomass burning to clear land and eliminate waste. This process results in smoke mainly composed of organic carbon and black carbon. It is important to notice also that human activities have enhanced the ability of natural aerosol occurrence due to land use changes and mining. These activities erode, expose, and degrade the soil, making the soil susceptible to be easily picked up and lifted into the atmosphere. We usually measure the size distribution of aerosols due to it being a very significant characteristic. Aerosols with radii smaller than 0.1 μm are referred to as Aitken nuclei. For radii between 0.1 and 1 μm and larger than 1 μm, aerosols are referred to as large nuclei and giant nuclei, respectively. Another important measurement is an aerosol's optical depth, which measures the atmospheric loading in a vertical column of air from the top of the atmosphere to the earth's surface. This measurement is specifically useful when the concentration of a uniformly distributed (well mixed) aerosol is needed at the surface and it becomes a useful quantitative measurement for the evaluation of air quality control.

Overall, aerosols affect the earth's energy budget in two ways: by direct and indirect processes. The direct process is simple; aerosols interact by scattering and absorbing incoming and outgoing radiation, causing an overall cooling in the troposphere. The indirect process is, however, more complex and involves cloud microphysics (summarized in Charlson et al. 1992). Many atmospheric aerosols can act as cloud condensation nuclei (CCN), leading to cloud formation. Aerosols act as ideal surfaces in which water vapor prefers to condense to and form cloud droplets. By increasing aerosol particle concentrations, the availability of CCN and ice nuclei also increases, leading to more cloud droplets. The decreased competition for condensation sites decreases the size of the droplets (liquid water content) and yields brighter clouds, which in turn increases cloud albedo and results in an overall cooling effect. The semidirect effect involves suspended aerosols near clouds or in clouds that contributes to the evaporation of the clouds and, thus, decreasing their lifetime (see Figure 2.9). All these cloud microphysical processes impact global precipitation patterns and cloud cover.

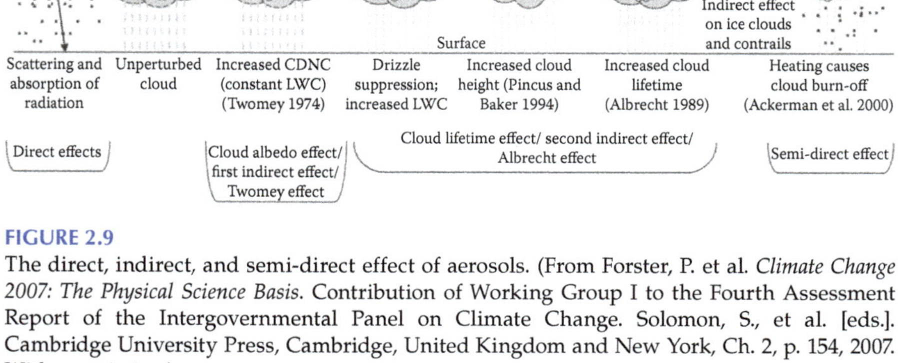

FIGURE 2.9
The direct, indirect, and semi-direct effect of aerosols. (From Forster, P. et al. *Climate Change 2007: The Physical Science Basis*. Contribution of Working Group I to the Fourth Assessment Report of the Intergovernmental Panel on Climate Change. Solomon, S., et al. [eds.]. Cambridge University Press, Cambridge, United Kingdom and New York, Ch. 2, p. 154, 2007. With permission.)

The overall radiative cooling caused by aerosols is believed to mask a percentage of the otherwise persistent warming trend by greenhouse gases, and cutting aerosol emissions would remove this cooling, further enhancing the greenhouse gas effect. Figure 2.10 shows the estimated changes in radiative forcing on climate from natural and human-related forcing from 1750 to 2005, based on the analyses in the last international science assessment (IPCC 2007). This figure shows the importance of the particles in determining the overall radiative forcing on climate over this time period. Because of the limited understanding of the underlying physical and chemical processes of aerosols, it is still unclear whether clouds provide a positive or negative feedback to an increase in atmospheric greenhouse gases. Furthermore, the uncertainties of aerosol, cloud, and precipitation interactions and feedback effects are among the main reasons for the large range of estimates on the projected increase of future global surface temperature increase (globally and annually averaged changes of 1–8°C are projected by IPCC [2007]) over the next century or so.

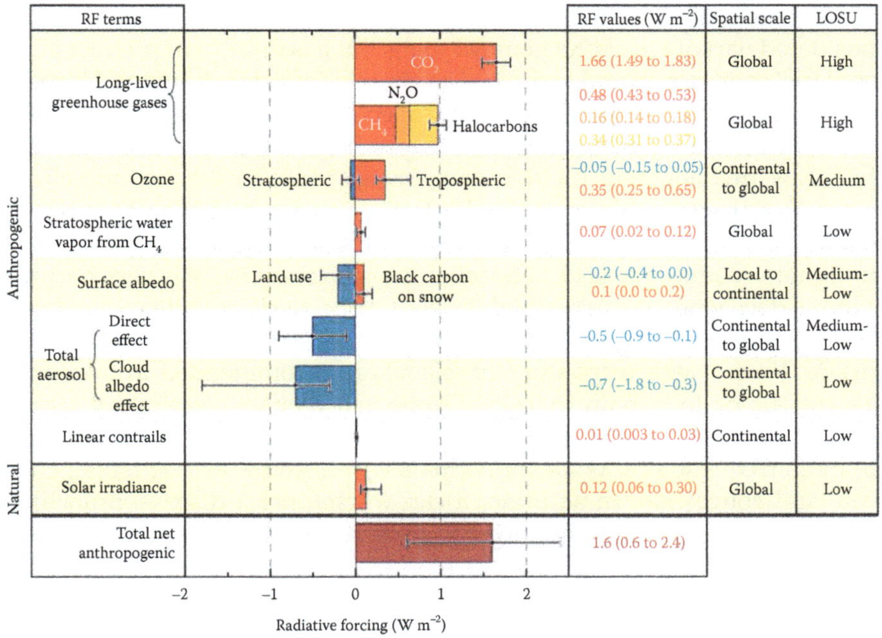

FIGURE 2.10
Factors affecting the radiative forcing of climate from 1750 to 2005, including the current estimated vales for the net forcing and analysis of associated uncertainties. (From IPCC, *Climate Change 2007: The Physical Science Basis*. Contribution of Working Group I to the Fourth Assessment Report of the Intergovernmental Panel on Climate Change. Solomon, S. et al. [eds.]. Cambridge University Press, Cambridge, United Kingdom and New York, SPM 2, p. 4, 2007. With permission.)

However, particles are not only important in terms of radiative forcing on climate, they are also directly harmful to human health and the environment. When inhaled, high concentrations of aerosols lead to respiratory diseases such as asthma. They also decrease visibility, which often leads to unsafe conditions for transportation. A significant number of epidemiological studies show that very fine particulate matter and traffic-related air pollution are correlated with severe health effects (Bernstein et al. 2004). These include but are not limited to mortality, cardiovascular, respiratory, and allergic diseases, and even cancer. We are exposed everyday to vehicle exhaust and other industrial activities, breathing sufficiently small particles able to penetrate the membranes of the respiratory tract and enter the blood circulation or into the brain. The most common form of air pollution is smog. Smog, however, can form in two different ways. The most common is from fossil fuel burning stacks that release ash, soot, carbon residue, sulfur dioxide, and traces of heavy metals. Sulfur dioxide reacts with oxygen gas to give sulfur trioxide or with water to get sulfurous acid (H_2SO_3). It can again react with water to get sulfuric acid (H_2SO_4). These acid compounds are capable of damaging buildings, soil, crops, water quality among other things through acid rain. The photochemical smog has three main compounds: nitrous oxides, ozone, and volatile organic compounds. Carbon monoxide is often considered part of the photochemical smog, but it is in fact a byproduct of the fossil fuel combustion and it also has a very toxic effect on humans. Nitrogen oxides result from oxygen (O_2) and nitrogen (N_2) interaction often found in car exhausts. Nitrogen further reacts to produce NO, NO_2, and NO_3. These chemicals are dangerous on their own, but a greater concern is their ability to produce tropospheric ozone and nitrate particles. Volatile organic compounds, on the other hand, react with nitrogen oxides in the presence of sunlight to produce the last chemical of smog: ozone. Ozone is very harmful at the surface when it reaches high levels. Ozone alerts are often issued during the summertime when sunlight promotes the production of ozone at the surface. People with respiratory problems are recommended to take extra precautions or to remain indoors. Ozone can damage respiratory tissues through inhalation by causing tissue decay and cell damage. It can also create more frequent attacks for individuals with asthma, eye irritation, chest pain, and congestion. In addition, it can worsen heart disease, bronchitis, and emphysema. In addition, organic particles produced from the organic gases can affect climate.

The main sink for particulate matter is through wet and dry deposition. Precipitation washes out the majority of airborne fine aerosols, while other aerosols can experience dry deposition, which does not involve precipitation and relies on convective transport, adhesion to existing surfaces, or diffusion. Aerosols can last from hours to weeks depending on an array of factors including meteorological conditions, lifetime or residence time, stratospheric or tropospheric location, and environmentally present constituents.

Air quality policies have heavily targeted the reduction of sulfur dioxide and the components of photochemical smog in recent years. However, policies have been slowly hindered by the acceleration of warming, because they result in the removal of the "cooling" effect of some aerosols. Despite the efforts, smog is still a common problem in big cities and policies should be further developed to reduce these pollutants and protect human health and the environment.

The increasing temperature trend and the known cooling effect that some types of aerosols have on the atmosphere have also sparked scientists and engineers to propose an array of geoengineering projects with the goal of offsetting the warming until new renewable nonfossil fuel energy is put into place. However, as discussed in Chapter 14, our climate system is very complex and experimenting without complete understanding could result in unintended and potentially dangerous side effects on our health, ecosystems, agricultural yields, and even the climate itself.

To summarize, aerosols play an important role in determining local, regional, and even global climate by both direct radiative effects and by complex and not yet fully understood indirect effects on cloud dynamics. In addition, they have a critical impact on human health and require important attention from both scientists and policymakers.

2.8 Past and Projected Changes in Climate Forcing

2.8.1 Past Climate

For years, paleoclimatologists have been constructing a blueprint of how the earth's temperature has changed over the centuries. Temperature data were unavailable before 1850. Therefore, scientists rely on tree rings, ocean sediments, ice cores, and historical documents to derive temperature estimates and fill in the gap in the temperature records. These reconstructed records show that climate on earth has changed on all timescales, including before human activity could have unintentionally played a role. Earth has experienced several extremes: both warmer and much colder periods that caused mass extinctions of some existing species. For the past millions of years or so, climate has fluctuated between ice ages and warmer interglacial periods with temperatures similar to those of the past few millennia. There is strong evidence supporting that these temperature changes are linked to the Milankovitch cycles (oscillations in the planet's orbit and inclination). However, recent studies also indicate that roughly half of the cooling during an ice age actually results from the effects of the ice formation on the carbon cycle and resulting effects on the atmospheric concentrations of CO_2 and CH_4.

Some smaller timescale changes are explained by the occurrence of explosive volcanic eruptions that injected sulfate aerosols into the stratosphere and troposphere, causing changes in the climate forcing. Although most of these effects are for only a few years, longer term effects can result. As an example, the recent study by Miller et al. (2012) indicates that the Little Ice Age during the 1600s and 1700s largely resulted from sea–ice/ocean feedback effects on ocean circulation processes in the Arctic, resulting from the insolation and cooling effects from the atmospheric sulfate aerosols injected by a series of major volcanic eruptions.

Feedback mechanisms of the system also interact within the system to keep a balance. For example, geologic evidence suggests that the earth may have experienced periods from 750 to 650 million years ago when a significant fraction of the planet was mostly frozen, with this scientifically controversial period often referred to as "snowball earth" (the global extent of the ice remains controversial but is being assessed by scientists under the auspices of the International Geoscience Programme). While this remains controversial, there is no question about the importance of the ice ages in the earth's history. Ice cover can produce further cooling by increasing the surface albedo and reflecting more of the sun's energy back into space. Covering all the land with ice blocks the chemical weathering of rocks in charge of removing CO_2 from the atmosphere, and this in turn leads to warming as concentrations of carbon dioxide rise and a reduction of ice cover results.

In addition to the ice–albedo feedback, there are many other feedbacks on climate in the earth system. Another important feedback mechanism is the changes in atmospheric water vapor concentration. A warmer atmosphere can hold more water and evapotranspiration increases as well. Since water vapor is a greenhouse gas, an increase in the concentrations results in a positive feedback to further increase the temperature. Changes in water vapor can also lead to changes in clouds, adding additional feedbacks on climate, with the result of the climate feedback depending on what types of clouds are formed. There are also feedbacks involving the carbon cycle and other biogeochemical cycles, most of which are positive feedbacks, where a change in climate can further amplify the amount of the initial change. In summary, climate system feedbacks are important in determining the state of the climate.

2.8.2 Present Climate

Present climate, however, is forced by an additional driver: human activities. The natural atmosphere contained many greenhouse gases, whose atmospheric concentrations were determined by the ongoing biogeochemical reactions that served as sources and sinks. Before the industrial era, human activities effects were limited to local impacts, primarily due to fires. The industrial revolution brought with it technological and economic progress with the development of steam-powered ships and railways. It further gained momentum with the invention of the internal combustion engine and the generation of electrical power. The use of fossil fuels became more prominent and amplified with the increasing population in developing countries, raising the concentration of carbon dioxide and other greenhouse gases in the atmosphere.

In 1896, Swedish chemist Svante Arrhenius, based on concerns about the societal use of coal, did theoretical studies assuming a doubling of atmospheric CO_2 that indicated human activities using fossil fuels could possibly lead to a significant effect on global temperatures. However, it was not until the middle of the twentieth century, when conclusive highly precise measurements of atmospheric constituents were begun, that scientists realized that human activities were having a significant effect on the atmosphere and climate. At this point, countries had developed a strong economic and social dependence on fossil fuels and used it to promote further growth. Population growth also triggered an increase in land use changes and biomass burning for food production and for increased construction, which also increased the release of carbon dioxide into the atmosphere and reduced the intake by eliminating forests and other existing biomes. Carbon dioxide and other greenhouse gases were not the only players in this, anthropogenic aerosols also increased as technology evolved, complicating the climate sensitivity and enhancing existing forcing on the climate system (as shown in Figure 2.10). Since the Industrial Revolution, concentrations of carbon dioxide have increased by almost 40% (as shown in Figure 2.8), increasing from about 280 ppm to more than 390 ppm. In addition, the concentrations of methane have more than doubled and those of nitrous oxide have risen by about 10%–15% (IPCC 2007) (Figure 2.8).

A similar trend to that of carbon dioxide is observed in globally averaged temperatures with an increase of approximately 0.74°C ± 0.18°C from 1906 to 2005 (IPCC 2007). In fact, 2000–2009 appears to have been the hottest decade in the last 2000 years (Mann et al. 2008). This increase in temperature is reflected on the records both over land and for sea surface temperatures. It is important to understand that oceans play an important role in our climate and its variability. Our planet is 71% covered in water and approximately 97% of that is saltwater. Water, because of its intrinsic characteristics, has a much greater heat capacity than that of air, and thus, the net intake of energy by the oceans is almost 20 times greater than that of our atmosphere. This absorbed heat is mainly located in the upper layers of the ocean, where it can lead to particular variations on climate on different timescales. Also, increasing sea temperatures leads to sea level rise through the thermal expansion and the melting of ice. Rising sea level is consistent with the current warming (Figure 2.11). Global average sea level has risen at an average rate of 1.8 mm/year (earlier decades) to a 3.1 mm/year in the last two decades with added contributions from thermal expansion, melting ice caps, glaciers, and polar ice sheets. Snow and ice cover, on the contrary, have decreased with the observed warming (Figure 2.11). Satellite data indicates that the annual

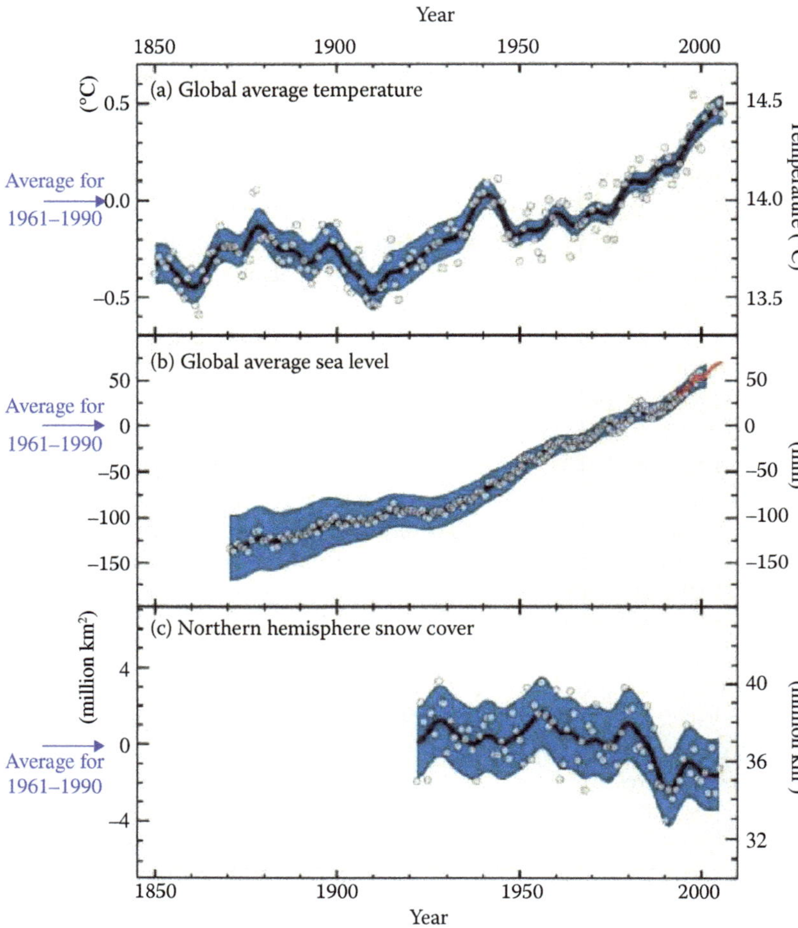

FIGURE 2.11
Observed changes in (a) global average surface temperature, (b) global average sea level from tide gauge (blue) and satellite (red) data and (c) northern hemisphere snow cover for March–April. All changes are relative to corresponding averages for the period 1961–1990. Smoothed curves represent decadal average values while circles show yearly values. The shaded areas are the uncertainty intervals estimated from a comprehensive analysis of known uncertainties (a and b) and from the time series (c). (From IPCC, *Climate Change 2007: The Physical Science Basis.* Contribution of Working Group I to the Fourth Assessment Report of the Intergovernmental Panel on Climate Change. Solomon, S. et al. [eds.]. Cambridge University Press, Cambridge, United Kingdom and New York, SPM 2, p. 4, 2007. With permission.)

average Arctic Sea ice extent has decreased extensively over the last few decades during the summer seasons.

Furthermore, increase in sea surface temperatures could produce better conditions for tropical cyclogenesis, leading potentially to an increase in number and intensity of tropical cyclones. However, there remains a lot of uncertainty about the past recognition of hurricanes, such that it is not clear if there are any changes currently occurring in trends for either number or intensity. Observations suggest that the 2005 Atlantic hurricane season was perhaps the most active on record, in part because of the high sea–surface temperatures recorded in the tropical North Atlantic. The 2010 Atlantic hurricane season follows closely behind in third place, providing ample room in the Caribbean for tropical cyclone conditions. Measurements of the ocean's heat content, often referred to "Tropical Cyclone Heat Potential," helps meteorologists have an idea of locations in the Atlantic or Pacific basin where formation or fast intensification of tropical cyclones is enhanced. Numbers above 90 kJ/cm^2 are usually considered as hotspots for hurricane intensification (see Figure 2.12).

The increase in carbon dioxide by human-driven activities also has impacts on the ocean's chemistry. Ocean pH had been stable for about 20 million years. Over the last 200 years, the oceans have absorbed more than 500 billion tons of carbon dioxide. When the ocean surface absorbs the carbon dioxide, several chemical interactions occur. The net result is an increase in hydrogen ions and a reduction in carbonate ions, calcium carbonate minerals, and seawater pH. The pH scale is a logarithmic scale that goes from 1 to 14. Numbers higher than seven are basic, while numbers lower than seven are considered acidic. Estimates show that the ocean's pH has dropped 0.1 on the scale. However, this being a logarithmic scale, small changes like a 0.1 drop represents a change of as much as 30%. This lowering of the pH will not make the ocean itself acidic, but it would shift it to a less basic state from the

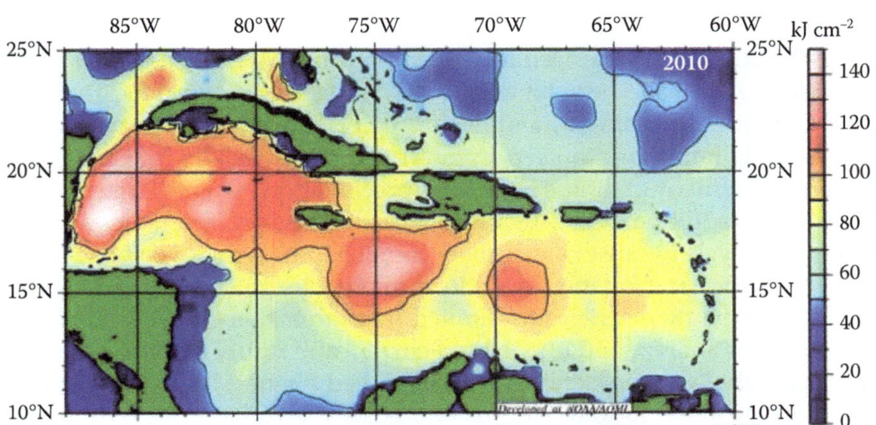

FIGURE 2.12
The image shows the tropical cyclone heat potential (TCHP) or oceanic heat content in kilojoules per square centimeter (kJ/cm^2) for September 22, 2010. TCHP above 90 kJ/cm^2 (orange colors) is commonly associated with rapid intensification of hurricanes. (From Tropical Cyclone heat potential daily maps, NOAA/AOML. http://www.aoml.noaa.gov/phod/cyclone/data/ca.html.)

preindustrial levels (pH of 8.1). To make calcium carbonate, creatures need two ingredients: calcium ions and carbonate ions. When carbon dioxide gets dissolved in water, it results in carbonic acid. This acid disassociates and reacts with carbonate ions, making them unavailable. Depleting the ocean of the important compounds that organisms need to build their exoskeletons and shells will disrupt the lifetime and even maybe existence of corals, sea urchins, sea stars, crabs, phytoplankton, and other marine creatures. This, in turn, would create an imbalance of minerals and food for fish and other predators and entirely threaten their habitat and the entire food chain. Coral reefs are more susceptible to these changes since they require more of the depleted minerals and their decay could further impact coastal cities because of their role in reducing erosion and storm surges. Coral reefs also serve as shelter and food to many sea creatures, and the impacts could result in a heavy decrease in biological diversity in our oceans. Current acidification is likely irreversible in our lifetimes. Natural processes that could help counter acidification (weathering of rocks on land) occur on a very slow timescale to counter the actions of human activities. In addition, even if CO_2 emissions were to be reduced or stopped, it would still take thousands of years for the ocean chemistry to recover and return to preindustrial levels.

Increasing global surface temperatures has led to changes in precipitation and atmospheric moisture, since it increases the water-holding capacity of the atmosphere and promotes a more active hydrological cycle. The general trend since 1950 in precipitation is an increase in higher northern latitudes (40°N–70°N). This is also true for the southern latitudes while a decrease is observed in the tropics and more profoundly in the northern subtropics (0°N–30°N). Observations show that the eastern parts of North and South America, most of northern Europe, and north/central Asia have become wetter. On the other hand, the Mediterranean, southern Africa, and some parts of southern Asia have become drier. Overall, observations show that the amount, frequency, intensity, and even type of precipitation are changing. However, it is important to keep in mind that many factors influence precipitation patterns, and for the most part, precipitation exhibits large natural variability. In addition, it is difficult to measure the specific changes in the hydrological cycle because of the many complex processes involved (evaporation, transport, precipitation, and cloud microphysics). Also, quality of the data, incomplete time series, and limited measurements can be limitations when trying to analyze and understand current and future behavior. In spite of all these challenges, scientists have observed changes over the last few decades in the amounts and patterns of precipitation. Therefore, it is important to monitor these changes in order to better predict future changes in the hydrological cycle.

Changes in some extreme events have already been observed. The frequency and intensity of heat waves, heavy precipitation, hurricanes, and other events have increased in recent years. Other extreme events, like drought, have become more common in different areas of the world. At the same time, there has been a decrease in the likelihood of cold waves in many parts of the world, although the weakening of the jet stream resulting from decreasing Arctic Sea ice may lead to more unusual midlatitude cold waves.

As discussed earlier, available evidence points to the most important factor to the changes in climate over recent decades, with the dramatic increase in global temperatures and the changes in the probability for severe weather events, being the emissions and corresponding atmospheric concentration increase in CO_2, with CH_4, N_2O, and CFCS also contributing to the positive radiative forcing on climate. Changes in stratospheric ozone, that is, the decrease in ozone due to CFCs and other halocarbons, should also have their largest effect during that time period, but because ozone absorbs both solar radiation and infrared radiation, the net effect is a decrease in radiative forcing because of this effect over that time frame (IPCC 2007).

While the growth in emissions of CO_2 declined in 2009, during the worst of the global economic crisis, by 2011 CO_2 emissions from human activities were higher than they have ever been. The current emissions are larger than even the highest Special Report on Emission Scenarios (SRES) developed by economists and other experts had projected just a decade ago (Figure 2.13). In addition, developing countries have surpassed the emissions of the developed countries and at a very accelerated pace (see Figure 2.14) (Peters et al. 2011). Unless countries accept responsibility and act together to greatly reduce future emissions, the future scenarios will not only become harder to address but also more immediate adaptation will be necessary.

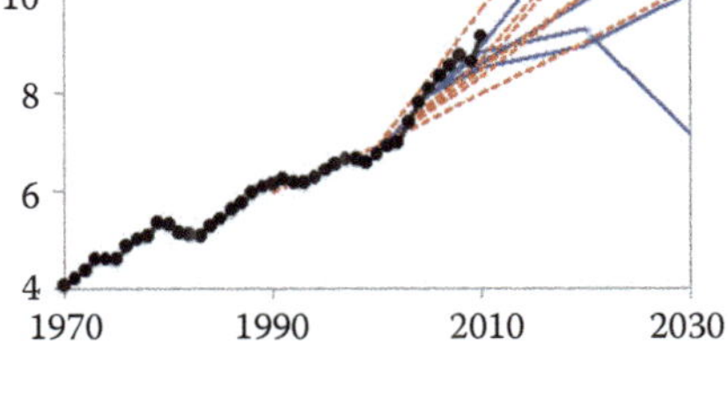

FIGURE 2.13
Global emissions of CO_2 from fossil fuel burning based on EIA analyses (http://www.eia.gov/) compared to projected emissions from the SRES and RCP scenarios. (From Personal communication, Katharine Hayhoe, 2012.)

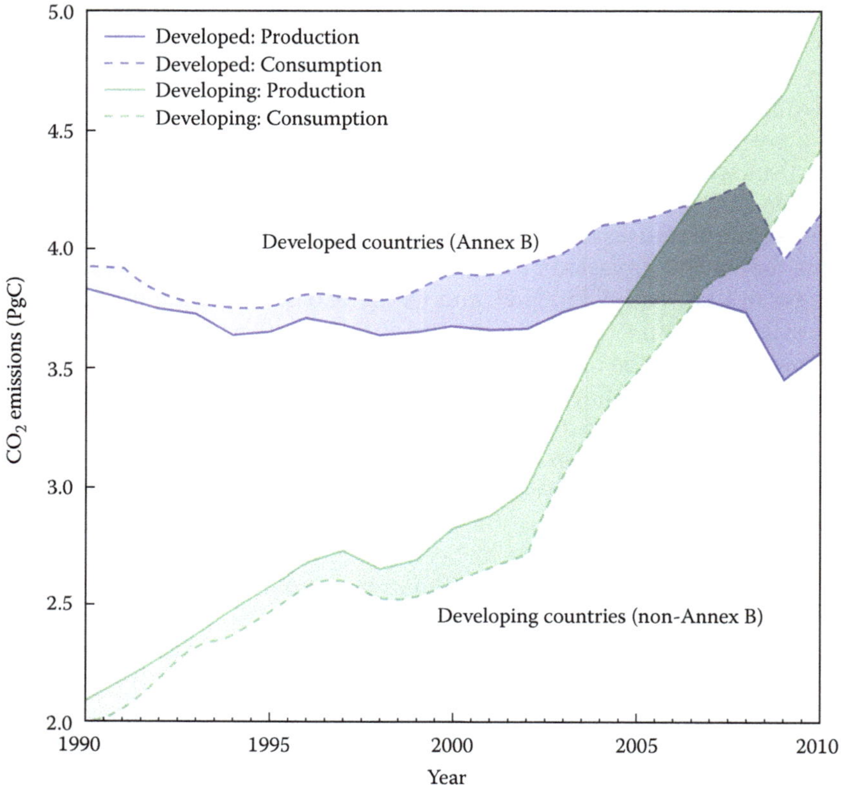

FIGURE 2.14
CO_2 emissions in petagrams of carbon (PgC) by developed (blue) and developing (green) countries in the last 20 years (1990–2010). It also shows production (solid lines) and consumption (dotted lines). (From Peters, G. P. et al., *Nature Climate Change*, 2, 2–4, 2011. With permission.)

2.8.3 Future Climate

Future changes in climate will depend extensively on human emissions of greenhouse gases and the concentration of aerosols in our atmosphere. These emissions can never be fully predicted in a deterministic sense since they are strongly dependent on population, socio-economic growth, and the development of government-enforced environmental policies. Instead, scientists use scenarios to address the future projections by establishing reasonable ranges on the magnitude of future emissions and evaluating the resulting change in climate. The IPCC SRES evaluates different emission scenarios based on a range of future assumptions about population, economics, energy choices, and lifestyle choices. On one end of the range are noninterventionist emissions scenarios, which are often referred to as "business as usual," while at the other end, we have policy intervention, stabilization, and cost-benefit scenarios. Stabilization scenarios are those with atmospheric concentration targets, usually setting long-term goals. Cost-benefit scenarios, on the other hand, exercises the idea of applying the most efficient policy intervention that still meets the most cost-efficient margin of reducing emissions. All these scenarios, although carefully thought of and evaluated, cannot anticipate certain events, like political shifts, population behavior, and advancements in technology.

The IPCC concluded that the current observed trends in temperature and its global to regional impacts in the environment will have discernible implications on some social and economic systems. Increased frequency of extreme events (floods, droughts, tropical cyclones) has put significant stress in some areas of the world. With no current or near-future intervention in carbon emissions, acceleration of climate change in the future could further affect these systems and make the impacts more profound. To better prepare, mitigate, and adapt to these coming changes, we need to improve our understanding of the likely global, regional, and local distribution of these impacts.

The continuous increase of human emissions is already affecting our current climate. These changes will only become more pronounced as they intensify and the implications on the global economy and the human race, in general, are indisputable (see WGII report of IPCC [2007]). We discussed in Section 2.8.2 the current observed changes (from sea level rise to increased extreme events), but future changes are also important to consider in order to develop effective adaptation and mitigation policy strategies and to help reduce some of the greater potential impacts.

Although they are not perfect and never will be, the ability of climate model simulations to accurately reconstruct current and past climate has increased greatly in recent years (IPCC 2007). As a result, the confidence in the predictions of future climate change has also increased. Although future climate predictions lack complete certainty because of the chaotic nature and some intrinsic characteristics of the system, there is confidence in many of the estimated ranges for temperature, sea level rise, and other projections. With increasing emissions, temperature is expected to rise. By how much will be determined by the future pathway of emissions. Model results for global temperature changes following six emission scenarios used by the IPCC in the Fourth Assessment Report are shown in Figure 2.15. These results project global surface temperature changes between 1°C and more than 6°C. This temperature increase is expected to be greater over land than the oceans and more over higher latitudes rather than at the tropics. Daily minimum temperatures are also expected to rise faster than the daily maximum temperature, reducing the diurnal temperature range and forcing a decrease in cold episodes and frost days per year.

Increasing surface temperatures will result in a higher overall rate of evaporation. Higher evaporation rates will further add more water vapor to the atmosphere, driving higher rates of precipitation. However, changes in precipitation will vary spatially and temporally. While some places have

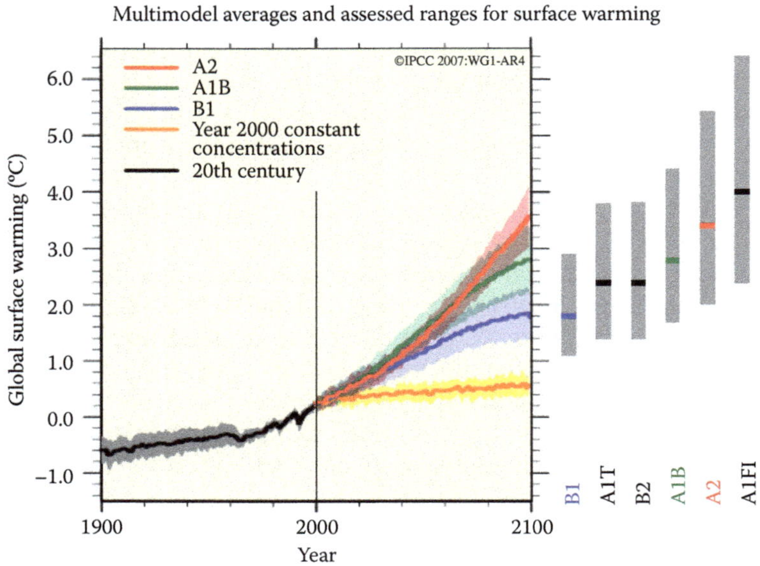

FIGURE 2.15
IPCC future emission scenarios showing future change in surface temperature in degrees Celsius as compared to the baseline temperatures (1980–1999). Solid lines represent the likeliest trend while shaded regions show the possible range. The grey bars on the right represent the temperature for the year 2100. Colored stripes represent the best estimate and shaded gray the possible range. (From IPCC, *Climate Change 2007: The Physical Science Basis.* Contribution of Working Group I to the Fourth Assessment Report of the Intergovernmental Panel on Climate Change. Solomon, S. et al. [eds.]. Cambridge University Press, Cambridge, United Kingdom and New York, Ch 10, p. 750, 2007. With permission.)

an overall increase in snow and rain, others have less precipitation overall. All model scenarios predict an overall global increase in precipitation (Figure 2.16). Three of the IPCC scenarios project changes in precipitation between 3% and 5%. Notice how some outlying models even show an increase of 7% and 8%. Spatial distribution of these model results show the increasing percent in precipitation at higher latitudes during both winter and summer months (Figure 2.17). Models also project drier midlatitudes during summer months while a larger variation spatially during the winter.

Models also suggest that more of the precipitation that does occur will occur as larger events. Short periods of intense rainfall followed by long periods without rain are expected in some regions. In the mid-continental areas, drier climate is expected during the summer, increasing the possibilities of more intense droughts and heat waves in those regions.

Another likely effect of increasing temperatures is the accelerated loss of sea ice, glaciers, ice caps, and permafrost due to melting. The continued dominance of summer melting over winter precipitation will further increase the loss of mass of these features, resulting in sea level rise and a freshening (reduction of salinity) of the oceans. Furthermore, reducing the

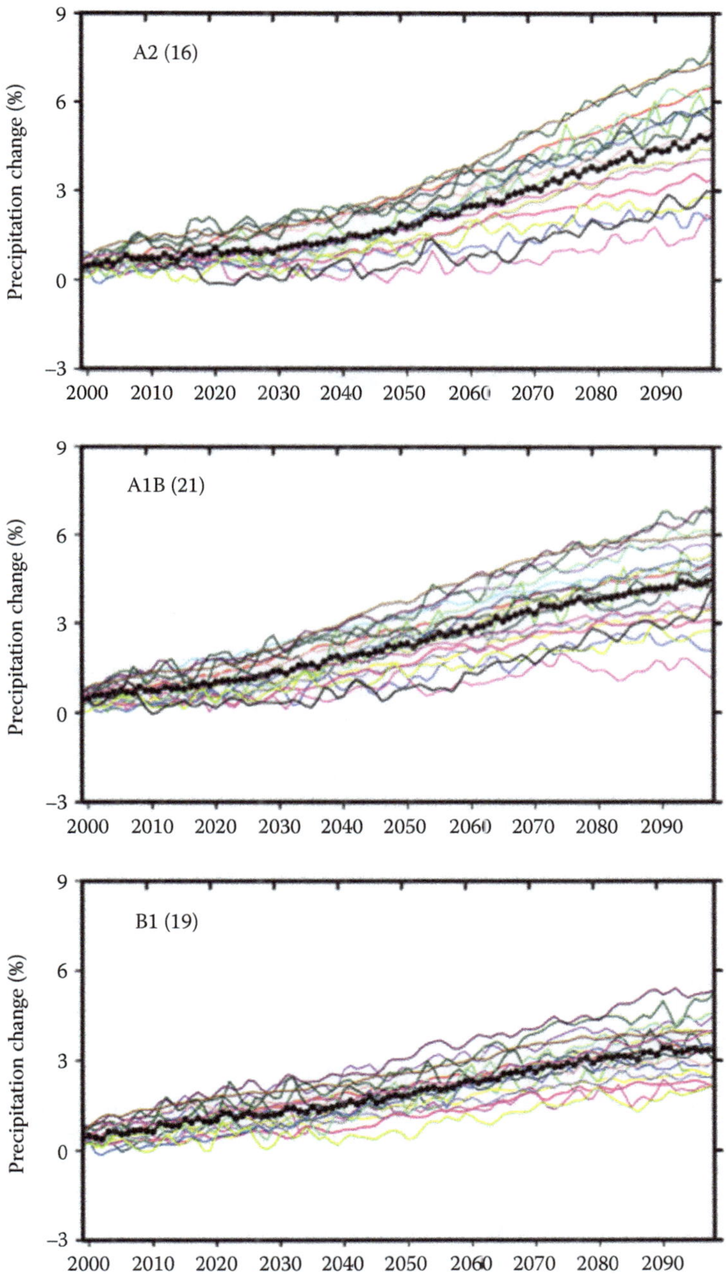

FIGURE 2.16
Projected changes in the total global precipitation from 2000 to 2100 relative to (1980–1999) average. Colored lines represent different climate models and the black line represents the average for all the models. These projections were made using three IPCC scenarios A2 (top), A1B (middle), and B1 (bottom).

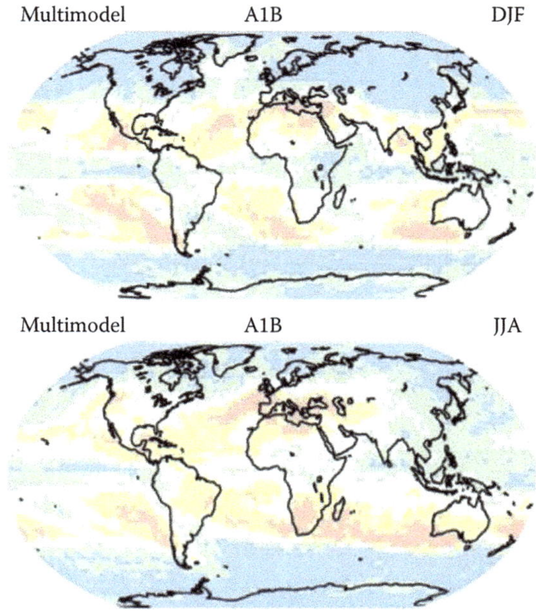

FIGURE 2.17
Spatial global distribution of precipitation changes in 2100 for SRES scenario A1B (upper middle scenario). Blue and green represent increases while red and yellow represent decreases in precipitation. Winter months (December, January, February; top) and summer months (June, July, August; bottom) are also represented. (From IPCC, *Climate Change 2007: The Physical Science Basis.* Contribution of Working Group I to the Fourth Assessment Report of the Intergovernmental Panel on Climate Change. Solomon, S. et al. [eds.]. Cambridge University Press, Cambridge, United Kingdom and New York, p. 127, 2007. With permission.)

snow and ice cover tends to accelerate warming by enhancing the ice–albedo positive feedback. Less snow cover on the ground would tend to decrease surface albedo, resulting in more incoming radiation being absorbed, warming the ground even more and melting more ice and snow. Also, melting permafrost can result in a positive feedback, potentially leading to further climate change. When permafrost melts, it can release methane that, as we discussed, is another significant greenhouse gas that produces warming. Some other implications include endangering water supplies for several regions, given that glacial meltwater is an important source for their water consumption and agriculture needs in some parts of the world. High latitude ecosystems have already experienced some of these changes. Polar bears heavily rely on ice packs and have been struggling to adapt to the current changes.

The fast loss of ice due to melting will continue to contribute to the rising sea levels currently observed on the models. As we explained earlier, sea level rises via two very basic mechanisms. Thermal expansion and glacier ice melting have so far shared the driving of this rise. The IPCC scenarios

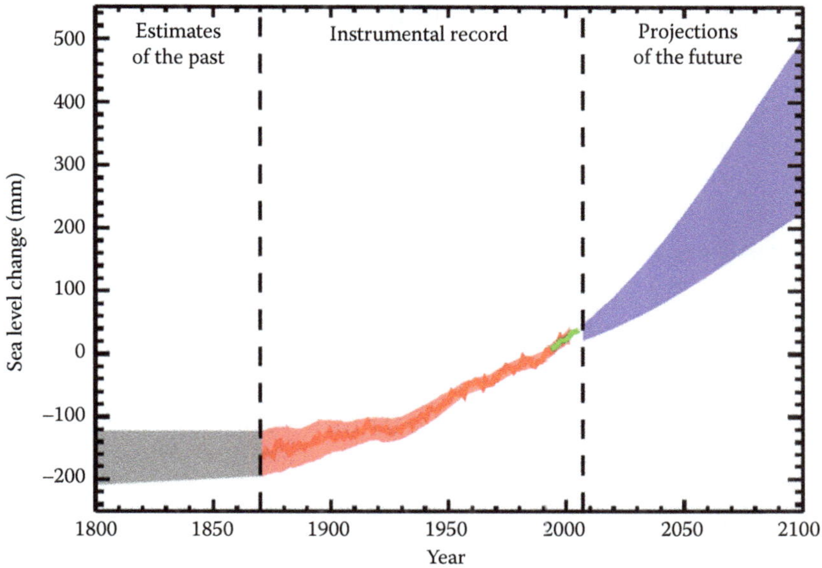

FIGURE 2.18
Past estimates of sea level rise (left), current observations on sea level rise (middle), and future IPCC projections of sea level rise (right). (From IPCC, *Climate Change 2007: The Physical Science Basis.* Contribution of Working Group I to the Fourth Assessment Report of the Intergovernmental Panel on Climate Change. Solomon, S. et al. [eds.]. Cambridge University Press, Cambridge, United Kingdom and New York, p. 111, 2007. With permission)

projected a sea level rise for the year 2100 between 20 and 50 cm above the already observed twentieth century levels based primarily on thermal expansion (Figure 2.18). However, more recent studies suggest that sea level could increase by as much as 1–2 m when accounting for the increasing likelihood of major melting from glaciers and the ice fields of Greenland.

2.9 Conclusions

Human activities are driving significant changes in the concentrations of radiatively important gases and particles in the earth's atmosphere. Available scientific evidence suggests that the changes in these concentrations are the only primary explanation for explaining the majority of the observed changes in global climate over the last 4–5 decades. While changes in temperature are part of these trends, it is the concerns about severe weather under a changing climate that are the larger concern about potential impacts on human society and ecosystems. Measurements already indicate that more precipitation is coming as larger events, and this trend is likely to continue.

Other indicators of climate change include sea level rise. Future changes in climate largely depend on human choices in future emissions of the radiatively important gases and particles, but could also be greatly influenced by natural variations in the incoming radiation from the sun and in the extent of major volcanic eruptions.

2.10 Questions for Discussion

1. Use the web to determine how much human-related emissions of CO_2, CH_4, and N_2O occur in the United States (or your country) each year. Compare those emissions to the total global emissions per year. How does your country rank with other countries in terms of total emissions?
2. Could human-driven climate change prevent the next ice age from occurring? Even if we could, would that be a good idea?
3. Consider the trade-offs to controlling the emissions of particles to prevent direct human health effects (as well as effects on other animals and plants) relative to the cooling effects of some particles on climate. Even if more global warming were to occur, would such controls still be a good idea?
4. Is climate change already affecting your region?
5. Looking at future projections of climate change on your region (use the web to get more information than what is in this chapter), what might be good consequences of climate change on your region and what challenges might be of concern?

References

Bernstein, J. A., N. Alexis, C. Barnes, I. L. Bernstein, A. Nel, D. Peden, D. Diaz-Sanchez, S. M. Tarlo, P. B. Williams, 2004. Health effects of air pollution. *J. Allergy Clin. Immunol.*, 114, 1116–1123.

Charlson, R. J., S. E. Schwartz, J. M. Hales, R. D. Cess, J. A. Coakley, Jr., J. E. Hansen, D. J. Hoffman, 1992. Climate forcing by anthropogenic aerosols. *Science*, 255, 423–430.

Forster, P., V. Ramaswamy, P. Artaxo, T. Berntsen, R. Betts, D. W. Fahey, J. Haywood, J. Lean, D. C. Lowe, G. Myhre, J. Nganga, R. Prinn, G. Raga, M. Schulz, R. Van Dorland, 2007. Changes in atmospheric constituents and in radiative forcing. In: *Climate Change 2007: The Physical Science Basis. Contribution of Working Group I to the Fourth Assessment Report of the Intergovernmental Panel on Climate*

Change. Solomon, S., D. Qin, M. Manning, Z. Chen, M. Marquis, K. B. Averyt, M. Tignor, and H. L. Miller (eds.). Cambridge University Press, Cambridge, United Kingdom and New York, NY.

Gerlach, T. M., 2011. Volcanic versus anthropogenic carbon dioxide. *Eos Trans. AGU*, 92(24), 201–202.

Gray, L. J., J. Beer, M. Geller, J. D. Haigh, M. Lockwood, K. Matthes, U. Cubasch, D. Fleitmann, G. Harrison, L. Hood, J. Luterbacher, G. A. Meehl, D. Shindell, B. van Geel, W. White, 2010. Solar influences on climate. *Rev. Geophys.*, 48, RG4001, doi:10.1029/2009RG000282.

Gray, L. J. et al., 2012. Correction to solar influences on climate. *Rev. Geophys.*, 50, RG1006, doi:10.1029/2011RG000387.

Hofmann, D., J. Barnes, E. Dutton, T. Deshler, H. Jäger, R. Keen, M. Osborn, 2003. Surface-based observations of volcanic emissions to stratosphere. *Geophys. Monogr. Ser.*, 139, 57–73. American Geophysical Union.

Intergovernmental Panel on Climate Change (IPCC), 2001. *Climate Change 2001: The Scientific Basis*. Houghton, J. T., Y. Ding, D. J. Griggs, M. Noguer, P. J. van der Linden, X. Dai, K. Maskell, and C. A. Johnson (eds.). Cambridge University Press, Cambridge, UK.

Intergovernmental Panel on Climate Change (IPCC), 2007. *Climate Change 2007: The Physical Science Basis*. Contribution of Working Group I to the Fourth Assessment Report of the Intergovernmental Panel on Climate Change (IPCC). Cambridge University Press, Cambridge, United Kingdom and New York, NY, 996 pp.

Jouzel, J. et al., 2007. Orbital and millennial Antarctic climate variability over the last 800,000 years. *Science*, 317, 793–796.

Kious, J., R. Tilling, Geological Survey (U.S.), 1996. *This Dynamic Earth: The Story of Plate Tectonics*, USGS, Reston, VA.

Lindzen, R. S., 1991. Review: Climate change: the IPCC Scientific Assessment. *Q. J. R. Meteorol. Soc.*, 117, 651–652.

Mann, M. E., Z. Zhang, M. K. Hughes, R. S. Bradley, S. Miller, S. Rutherford, F. Ni, 2008. Proxy-based reconstructions of hemispheric and global surface temperature variations over the past two millennia. *Proc. Natl. Acad. Sci. USA*, 105, 13, 252–13,257, doi:10.1073/pnas.0805721105.

Meehl, G. A., W. M. Washington, C. M. Ammann, J. M. Arblaster, T. M. L. Wigley, C. Tebaldi, 2004. Combinations of natural and anthropogenic forcings in twentieth-century climate. *J. Climate*, 17, 3721–3727, doi: http://dx.doi.org/10.1175/1520-0442(2004)017<3721:CONAAF>2.0.CO;2.

Miller, G. H. et al. 2012. Abrupt onset of the Little Ice Age triggered by volcanism and sustained by sea-ice/ocean feedbacks. *Geophys. Res. Lett.*, 39, L02708, doi:10.1029/2011GL050168.

Peters, G. P., G. Marland, C. LeQuéré, T. Boden, J. G. Canadell, M. R. Raupach, 2011. Rapid growth in CO_2 emissions after the 2008–2009 global financial crisis. *Nat Clim Chang.*, 2, 2–4, doi:10.1038/nclimate1332.

Rahmstorf, S., H.-J. Schellnhuber, 2006. *Der Klimawandel—Diagnose, Prognose, Therapie*. 144 S., 7,90 Euro, C. H. Beck, München.

Schmidt, G. A., R. Ruedy, R. L. Miller, A. A. Lacis, 2010. The attribution of the present-day total greenhouse effect. *J. Geophys. Res.*, 115, D20106, doi:10.1029/2010JD014287.

Vaughn, B. H., C. U. Evans, J. W. C. White, C. J. Still, K. A. Masarie, J. Turnbull, 2010. Global network measurements of atmospheric trace gas isotopes. *Isoscapes*, Part 1, 3–31, doi: 10.1007/978-90-481-3354-3_1.

World Meteorological Organization, 2010. *Scientific Assessment of Ozone Depletion: 2010*. WMO Global Ozone Research and Monitoring Report number 52, Geneva.

Websites

Carbon Dioxide Information Analysis Center: http://cdiac.ornl.gov/

U.S. Geological Survey: http://www.usgs.gov/

3

Scenarios of Future Socio-Economics, Energy, Land Use, and Radiative Forcing

Jiyong Eom, Richard Moss, Jae Edmonds, Kate Calvin, Leon Clarke, Jim Dooley, Son H. Kim, Robert E. Kopp, Page Kyle, Patrick Luckow, Pralit Patel, Allison Thomson, Marshall Wise, and Yuyu Zhou*

CONTENTS

* The authors are grateful for research support provided by the U.S. Department of Energy's Office of Policy and International Affairs. The authors also wish to express appreciation to the Integrated Assessment Research Program in the Office of Science of the U.S. Department of Energy under Contract No. DE-AC05-76RL01830 for long-term support that enabled the development of the Global Change Assessment Model, which was used in the conduct of this research. This research used Evergreen computing resources at the Pacific Northwest National Laboratory's Joint Global Change Research Institute at the University of Maryland in College Park, which is supported by the Integrated Assessment Research Program in the Office of Science of the U.S. Department of Energy under Contract No. DE-AC05-76RL01830. The views and opinions expressed in this paper are those of the authors alone. Authors are researchers at the Pacific Northwest National Laboratory's Joint Global Change Research Institute at the University of Maryland in College Park, with the exception of Robert Kopp of Rutgers University.

3.1 Introduction

This chapter discusses scenarios of potential future anthropogenic developments that could affect the concentrations of greenhouse gases, aerosols, and short-lived species and thereby the climate. Scenarios of this type have been published for more than a quarter of a century. The earliest work on this problem focused on fossil fuel and industrial CO_2 emissions and employed time-trend extrapolations of fossil fuel resource utilization (Rotty 1977, 1978). This work was followed by the development of models that employed economic principles and considered energy markets explicitly in the development of scenarios that explored emissions and concentrations of CO_2 in the absence of policies to limit anthropogenic emissions. See, for example, Nordhaus (1979), JASON (1979), Niehaus and Williams (1979), Häfele (1981), Nordhaus and Yohe (1983), and Edmonds and Reilly (1983, 1985). This literature evolved with other climate sciences and was extended to include land use change emissions of CO_2, non-CO_2 greenhouse gases, aerosols, and short-lived species, as well as a wide range of potential climate policy regimes. The range of global CO_2 emissions reported in the literature by Nakicenovic et al. (2000) is shown in Figure 3.1. This broad literature is summarized in a variety of reviews and assessments, including Leggett et al. (1992), Edmonds et al. (1994), Weyant et al. (1996), Special Report on Emissions Scenarios (SRES) (Nakicenovic et al. 2000), Fisher et al. (2007), van Vuuren et al. (2008), and van Vuuren and Riahi (2011).

Since the Second Assessment Report of the Intergovernmental Panel on Climate Change (IPCC 1996), the climate modeling community has employed a small number of scenarios, designed to span uncertainty in greenhouse gas emissions and concentrations in the literature, as common points of reference in the assessment process. The first set of scenarios used by the climate modeling community was the IS92 scenarios (Leggett et al. 1992). These were supplanted by the SRES (Nakicenovic et al. 2000) used in the IPCC Third and Fourth Assessment Reports (IPCC 2001, 2007). The IPCC Firth Assessment Report will use a set of four scenarios referred to as Representative Concentration Pathways (RCPs) (van Vuuren et al. 2011). Each RCP is labeled

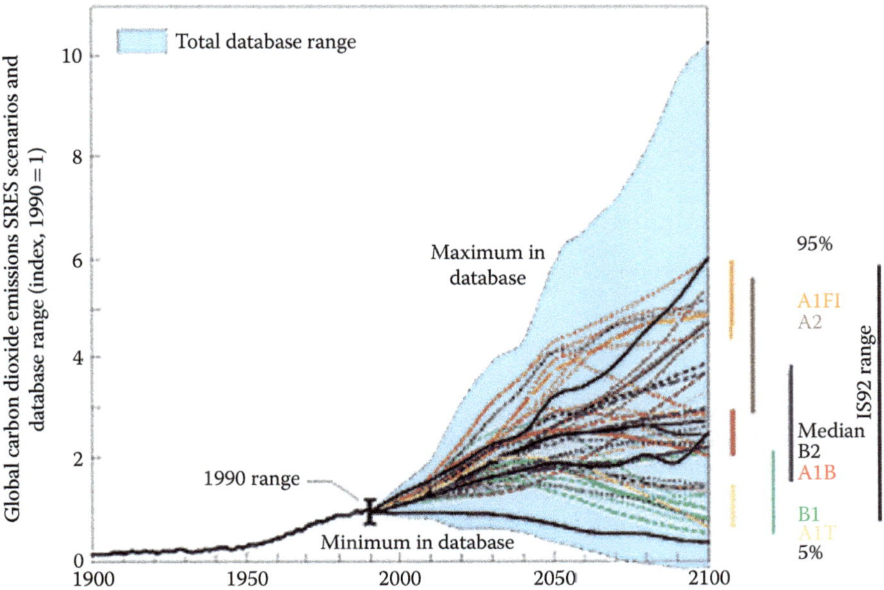

FIGURE 3.1
Range of global energy-related and industrial CO_2 emissions for the 40 SRES scenarios. The dashed time-paths depict individual SRES scenarios and the shaded area, the range of scenarios from the SRES database. The median (50th), 5th, and 95th percentiles of the frequency distribution are shown. The statistics associated with scenarios from the literature do not imply the probability of occurrence (e.g., the frequency distribution of the scenarios may be influenced by the use of IS92a as a reference for many subsequent studies). The 40 SRES scenarios are classified into groups that constitute four scenario families. Jointly, the scenarios span most of the range of the scenarios in the literature. The emissions profiles are dynamic, ranging from continuous increases to those that curve through a maximum and then decline. The colored vertical bars indicate the range of the four SRES scenario families in 2100. The black vertical bar shows the range of the IS92 scenarios. (From Nakicenovic, N. et al., *IPCC Special Report on Emissions Scenarios*, Cambridge University Press, Cambridge, U.K., 2000, Figure 1.6. With permission.)

with its associated radiative forcing in the year 2100 (in Wm^{-2}): RCP2.6, RCP4.5, RCP6.0, and RCP8.5. The four scenarios are shown in Figure 3.2.

The range of climate forcing, including scenarios that include and exclude explicit consideration of greenhouse emissions mitigation policies, is very broad, as shown in Figure 3.2. Radiative forcing in the year 2100 ranges from as little as 2.5 Wm^{-2} to approximately 9 Wm^{-2}.* The wide range of scenarios of radiative forcing in the literature, as those developed and described in

* Radiative forcing is a measure of climate forcing and is discussed in Chapter 2. By definition, pre-industrial radiative forcing is zero, and a doubling of the concentration of atmospheric CO_2 relative to pre-industrial levels (pre-industrial CO_2 concentrations were approximately 280 ppm), results in a radiative forcing of approximately 3.7 Wm^{-2}. This, in turn, implies a long-term (1,000-year equilibrium) change in the average earth surface temperature equivalent to the climate sensitivity, that is generally taken to range between 2°C and 5°C, though higher and lower values, while less likely, are also found in the literature as discussed in Chapter 2.

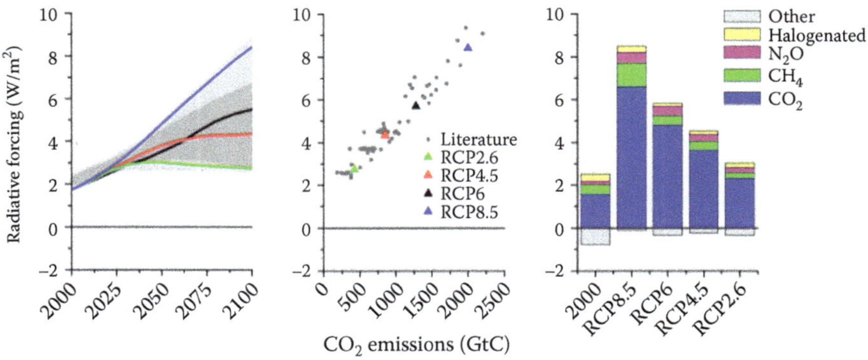

FIGURE 3.2
Trends in radiative forcing (left), cumulative twenty-first century CO_2 emissions versus 2100 radiative forcing (middle) and 2100 forcing level per category (right). Grey area indicates the 98th and 90th percentiles (light/dark grey) of the literature. (With kind permission from Springer Science+Business Media: *Climatic Change*, The Representative Concentration Pathways: An Overview, 109, 2011, 5–31, van Vuuren, D. P. et al., Copyright © 2011.)

this chapter, reflects uncertainty in both the potential future developments in key drivers such as population, economic development, technology availability, and the emissions policy environment. The range of reference scenarios, in which explicit policies or measures are undertaken to address climate change, covers a narrower range, with year 2100 values ranging from as high as approximately 9 Wm^{-2} to as low as 4.5 Wm^{-2}. Note, however, that there are no reference scenarios in the open literature that result in radiative forcing of 2.6 Wm^{-2} or less in the year 2100, a level that is consistent with a long-term, steady-state climate change of approximately 2°C, without stringent policy intervention assumptions.

The RCP scenarios (van Vuuren et al. 2011) constitute an initial element in a larger and more general assessment framework design described in Moss et al. (2010). Van Vuuren et al. (2012), Kriegler et al. (2010), and CWT (2011) build on the Moss et al. architecture. CWT (2011) proposed the development of new scenarios that would produce RCP replications, characterized by the same levels of radiative forcing in the year 2100 as the RCPs, but in addition, they would explore a variety of socioeconomic pathways that could underpin scenarios leading to specific radiative forcing levels in 2100.* Socioeconomic pathways would be developed for scenarios so as to explore uncertainty in both the challenges to mitigation and challenges to adaptation. This chapter presents a set of such scenarios, building on prior work reported in Eom et al. (2012).

In this chapter, we develop future global and regional scenarios of energy, economic activity, land use, emissions of greenhouse gases, aerosols, and short-lived species, and calculate radiative forcing for the resulting atmospheric

* Since climate models are driven by radiative forcing, Moss et al. (2011) reasoned that, to a first approximation, a wide range of underlying socioeconomic circumstances could be associated with a given anthropogenic climate change.

composition in the twenty-first century. We investigate the implications of alternative economic, demographic, and technological development possibilities, as well as potential stabilization regimes for energy technology choice, economic activity, land use, and land cover, based on the GCAM integrated assessment model (Edmonds and Reilly 1985; Kim et al. 2006; Clarke et al. 2007).

Section 3.2 describes the approach we have taken to develop alternative socioeconomic pathways. Section 3.3 provides an overview of GCAM 3.0, the integrated assessment model (IAM) employed to develop the scenario set reported in Section 3.4. Section 3.4 discusses GCAM scenario results, focusing on the implications for energy, emissions, climate, and land use change. Section 3.5 discusses issues related to the quantification of the challenges to mitigation and adaptation that the future society might face with regard to climate change. Section 3.6 concludes with a summary of insights that could be useful for future work.

3.2 Exploring Alternative Socioeconomic and Ecosystem Drivers

While climate modelers want scenarios of emissions, concentrations, and radiative forcing, other climate change researchers need scenarios with additional and/or other information. The other two climate research communities that use scenarios are the IAM and IAV (impacts, adaptation, and vulnerability) research communities. These communities are broad, with varying scenario needs. The IAV community, for example, often finds it useful to have a narrative as well as a quantitative representation of key scenario elements. IAM researchers often find it useful to have scenario elements that include quantitative descriptions of key external drivers such as population, economic activity, technology availability, and policy.

In this section, we describe the elements, forcing assumptions, and associated motivation for a set of six reference scenarios with both narrative and quantitative elements that we will use to explore the uncertainty in potential future radiative forcing, challenges to mitigation, and challenges to adaptation. In exploring the latter, we focus on factors that affect the level of vulnerability of societies. We have adopted a simple concept of vulnerability that incorporates both sensitivity and adaptive capacity (IPCC 2001). We had to address a number of key issues, including: (1) deciding what variables or dimensions of socioeconomic conditions to include; (2) establishing a geographic scale of the scenarios; and (3) developing an approach to vary the degree and type of vulnerability so that some narratives reflect relatively high vulnerability and some reflect low vulnerability.

The issue of scale is extremely important because many IAV analyses are conducted at the local scale, but it is not feasible to develop scenarios at this scale using IAMs. Thus, we focus on conditions and dynamics in the international system and among large regions, such as those defined by IPCC (Watson et al. 1997). Conditions on this scale provide an international context and boundary conditions for influences on vulnerability at national and sub-national scales. The degree of coupling between these international scenarios and locally-developed scenarios can vary from equivalency (very tight coupling) to complementary (loose coupling), and can be developed in processes that range from joint to independent (Zurek and Henrichs 2007).

3.2.1 Development of Storyline Elements

The scenario set in this chapter is designed around three suites of factors important for understanding future vulnerability:* (1) population and demographic conditions, which strongly affect demands, resource flows in the economy and the environment; (2) socioeconomic development trajectory, in this case measured by eventual progress towards the Millennium Development Goals (MDGs), which influence exposure, sensitivity, and the level of economic and social resources available for adaptation and mitigation; and (3) the technology used in the production and consumption of energy, agriculture, and other goods/services, which determine the intensity of resource use and environmental stresses.

The overall logic and a summary of basic conditions included in the scenarios are provided in Table 3.1. We will now explore each of the three areas that influence vulnerability and the capacity for adaptation and mitigation.

3.2.2 Millennium Development Goals

The MDGs play an important role in shaping the quantitative inputs to the scenarios described in this chapter and in framing the six storylines. The MDGs are a comprehensive and specific set of poverty reduction targets established in 2000 by all 193 United Nations member states and are intended to improve conditions for the bottom billion of humanity currently living in extreme poverty (U.N. Millennium Project 2005). The goals include:

- Eradicating extreme poverty and hunger
- Achieving universal primary education
- Promoting gender equality and empowering women
- Reducing child mortality rates
- Improving maternal health
- Combating HIV/AIDS, malaria, and other diseases
- Ensuring environmental sustainability
- Developing a global partnership for development

* Including vulnerability and capacity for adaptation, as well as requirements and capacity for mitigation.

TABLE 3.1

A Summary of Socioeconomic Conditions in the Scenarios

	POP6/MDG+ Sustainability and Equity	POP6/MDG– Collapse	POP9/MDG+ Consumerism	POP9/MDG– Muddling Through	POP14/MDG+ Social Conservatism	POP14/MDG– Crowded Chaos
Summary	Priority is given to sustainable economic development. The MDGs are eventually achieved in a world where population peaks and declines to 6 billion by 2100.	A shock causes economic collapse and growing inequality. Population peaks and declines to 6 billion. Local environmental problems are serious.	Robust free market growth increases prosperity and leads to convergence. Population stabilizes at 9 billion. The environment is highly managed. Infrastructure must be robust to climate extremes.	Sporadic economic growth and slow diffusion of technology reduce economic opportunities for developing countries and slow attainment of the MDGs. Population reaches 9 billion.	Women adopt traditional family roles and fertility is high, with 14 billion by 2100. High economic growth, social cohesion, and humanitarian assistance result in progress towards the MDGs.	Social strife and economic problems, low MDG progress. But, low child mortality, 14 billion population. Environment stress and resource shortages cause international conflict.

(Continued)

TABLE 3.1 *(Continued)*

A Summary of Socioeconomic Conditions in the Scenarios

	POP6/MDG+ Sustainability and Equity	POP6/MDG− Collapse	POP9/MDG+ Consumerism	POP9/MDG− Muddling Through	POP14/MDG+ Social Conservatism	POP14/MDG− Crowded Chaos
Population Growth and Characteristics	-Peaks at 8.0 billion in 2040 and declines to 5.5 billion in 2100 -Rapid progress in child/maternal health -Universal access to family planning -Advances in opportunities for women -High educational attainment -Aging population structure	-Peaks at 8.0 billion in 2040 and declines to 5.5 billion in 2100 -Low completed fertility -Low educational attainment -Women participate in the workforce	-Peaks at 9.4 billion in 2070 and stabilizes at 9 billion in 2100 -Child and maternal health improve, -High educational attainment -Aging population structure -High international mobility -Women active in workforce	-Peaks at 9.4 billion in 2070 and stabilizes at 9 billion in 2100 -Modest progress in child/maternal health -Restricted access to family planning -Limited opportunities for women -Low educational attainment -Balanced age structure	-Reaches 14.0 billion in 2100 -Rapid reduction in child mortality -High fertility -Restricted access to family planning -Women not active in workforce -Moderate educational attainment -Young population structure	-Reaches 14.0 billion in 2100 -Progress in child/maternal health increases fertility -Restricted access to family planning -Low educational attainment

Economic Growth, Productivity, Employment, Preferences, and Equity	-Rapid growth -High labor productivity -Low unemployment -High dependency ratio -Even income distribution -Non-material, low meat diets	-Medium growth then none -Low labor productivity -High unemployment and dependency -Skewed income distribution -Consumption of essential goods	-Moderate/high growth -High labor productivity -Convergence -Low unemployment and dependency -Preference for material goods, individual mobility, meat	-Slow/moderate growth -Low labor productivity -High unemployment -Moderate income distribution -Conventional preferences but limited incomes	-Rapid growth -Moderate labor productivity -Low unemployment -Improving equity -Moderate levels of material consumption and some reduction in meat consumption	-Slow growth -Low labor productivity -High unemployment -Poor income distribution -Consumption and preferences follow usual patterns for rich and poor
Energy and Agricultural Technology	-Rapid innovation, advanced technologies -Demand managed -Low fossil fuels -High renewables -Phase out of nuclear and CCS -High agricultural productivity	-Slow innovation -High energy demand -High fossil fuels -Low nuclear and CCS -Low renewables -Low agricultural productivity	-Fast innovation -High energy demand -High fossil fuels (advanced to control local pollution) -Low nuclear and CCS -Low renewables -Medium ag productivity	-Moderate innovation -High energy demand -Medium fossil fuels -Medium renewables -Medium nuclear and CCS -Medium agricultural productivity	-Rapid innovation -Moderate energy demand -Low coal/fossil fuel -High renewables -High nuclear and CCS -High agricultural productivity	-Slow innovation -High energy demand -High fossil fuels -Low renewables -Low nuclear and CCS -Low agricultural productivity

(Continued)

TABLE 3.1 *(Continued)*

A Summary of Socioeconomic Conditions in the Scenarios

	POP6/MDG+ Sustainability and Equity	POP6/MDG– Collapse	POP9/MDG+ Consumerism	POP9/MDG– Muddling Through	POP14/MDG+ Social Conservatism	POP14/ MDG– Crowded Chaos
Government and International Affairs	-Strong int'l governance (instant accession and UCT) -Effective development assistance -Low tariffs -Strong local institutions	-Weak int'l governance (delayed accession, FFICT) -Low ODA -High tariffs -Weak local and national governments	-Good international governance (delayed accession and FFICT) -Low tariffs -High ODA -Limited government upholds rule of law	-Weak int'l governance (delayed accession, FFICT) -Low ODA -High tariffs -Domestic goals and development, mixed government effectiveness	-Strong int'l governance (instant accession and UCT) -High humanitarian assistance -Low tariffs -Strong national and local government	-Weak int'l governance (delayed accession, FFICT) -Basic medical dev. asstnc -High tariffs -Police states
Environmental Conditions and Policies	-Low stresses -Strong environmental policies accepted and enforced	-Use of old, inefficient technologies causes multiple stresses -Environment a low priority	-Environmental stress is moderate -High levels of wealth support remediation -Strong local environmental policy	-High environmental stresses, poor environmental conditions -Weaker environmental policies	-Multiple environmental stresses -Highly managed systems -Strong policies to maintain basic ecosystem services	-High environmental stresses, poor environmental conditions -Weak environmental policies

Settlement Patterns, Neighborhoods, and Infrastructure	-Concentrated settlements -Modern infrastructure -Public transit available and used -High social cohesion and strong community organizations	-Mixed settlement patterns -Existing infrastructure decays -Low social cohesion and weak community organizations	-High urban density, low land fragmentation -Infrastructure highly engineered to buffer climate impacts -Medium social cohesion and strong community organizations supported by private sector	-Diffuse settlements -Poor urban conditions (poor sanitation, etc.) -Ineffective local institutions and planning -Low social cohesion and weak community organizations	-Tolerable urban conditions -Modern infrastructure -Public transit available and used -Strong planning enforced -High social cohesion and strong community organizations	-Diffuse settlements -Poor urban conditions overall, but water and sanitation ok -Low female education and workforce -Low social cohesion and weak community organizations

The intent of the MDGs is that they should have been achieved by 2015. But, progress towards reaching the goals has been uneven. Much development assistance is in the form of debt relief, military assistance, or natural disaster recovery, which do not necessarily advance socioeconomic development. Some countries have achieved many of the goals, while others are not on track to realize any. Nevertheless, these broad societal goals and their generalization to longer time frames provide a useful point of reference for crafting scenarios that explore the challenges to mitigation and adaptation as well as uncertainty in future radiative forcing.

MDGs are also useful because of their recognized importance in development circles, and because they have direct relevance for many societies in terms of their vulnerability or resilience in the face of climate variability and change. The attainment of the MDGs is closely related to population growth and technology change. However, it is also possible to envision different combinations of population dynamics and societal progress, as depicted in our scenarios.

In crafting our scenarios in Table 3.1, we identify two broad qualitative scenario groups—those that meet MDGs and those which do not. In the first scenario group, "MDG+," overall progress is accelerated, even if the objectives are not met in 2015. The notion is that over time, progress is sufficient so that development stays ahead of population increases and the number of those living in extreme poverty drops over time, even if the total global population itself continues to increase. In the second scenario group, "MDG−," progress is not sustained, and the proportion and number of those living in extreme poverty increases.

We focused on a number of factors, including:

- Population growth and characteristics such as progress in child/maternal health, access to family planning, advances in opportunities for women, educational attainment, and population–age structure
- Economic growth rates, labor productivity, employment levels, consumer preferences, and equity with countries
- Domestic governance and international relations, including accession to climate policy, the level and effectiveness of development assistance, level tariffs/access of developing countries to developed country markets for their exports, and strength/effectiveness of national to local institutions in solving domestic problems and providing resources to assist those who are vulnerable to climate variability and change, but have limited adaptive capacity
- Settlement patterns, neighborhoods, and infrastructure, representing whether development is concentrated in urban areas or sprawls, as is often currently the case, the effectiveness of local planning, the level

of social cohesion and willingness to help one's neighbors, and access to sanitation, safe drinking water, and modern forms of energy

- Environmental conditions and policies, including whether stresses on the environment are high or low, the extent of natural ecosystems, the level of engineering to replace lost ecosystem services, and the effectiveness of environmental policies

We use these factors to guide the quantification of key exogenous assumptions used to create the six quantitative scenarios in Table 3.1.

3.2.3 Population and Economic Growth

3.2.3.1 Global Populations

Population scenarios are taken from the U.N.'s World Population Prospects (The 2008 Revision—Long Range Projections) released in 2011 (U.N. Population Division 2011), which extends the 2008 Revision to provide demographic profiles beyond 2050 through the year 2300. Three demographic scenarios were chosen from the U.N.'s World Population Prospects: low, medium, and high fertility variants, with mortality and international migration trends held constant. In the medium-fertility scenario (POP9), total fertility rates in all countries are assumed to converge toward a below-replacement fertility level of 1.85 by 2050, though not all countries reach that level by 2050; thereafter, countries return to the replacement level of 2.05 by the year 2175. Similar convergence assumptions were made for high- and low-fertility variants (POP14 and POP6) with the 2050 asymptotes of 2.35 and 1.35 and the long-range asymptotes of 2.30 and 1.80, respectively. Note that, in this study, we modified the U.N.'s population prospects only for the United States, in order to reflect the trend in the 2008 U.S. Census Projection, which led to an increase in 2050 global populations by about 0.5%.*

The three population scenarios present very different trends in the global population (Figure 3.3). While the high-fertility population (POP14) continues to grow, eventually reaching 14 billion of global population by the end of

* For U.S. population in POP9, we employed the U.S. population in 2008 U.S. Census national projection (extending only to 2050), instead of the U.S. population projected by U.N. medium-fertility scenario—they are very different from the census population, which is higher than the U.N. population by 11% in 2050; Regarding the years after 2050, we assumed that the U.S. population will slowly converge to U.S. population in U.N. medium-fertility scenario, with the convergence completing by 2150. Similarly, it was assumed that new U.S. populations in POP6 and POP14 split after 2010 and gradually converge to their U.S. population counterparts in U.N. low-fertility and high-fertility scenarios. The new U.S. population scenarios have the variance of 23% in 2050, which is lower than the variance of U.N. projected U.S. population scenarios (28%), and the gap in variance diminishes over time.

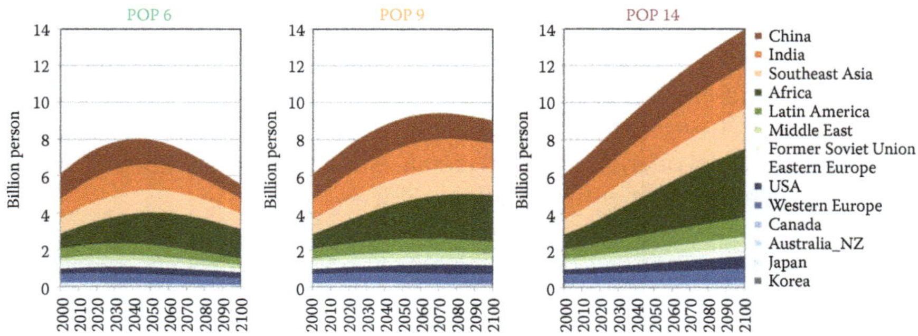

FIGURE 3.3
Global population. (From U.N. World Population Prospects, the 2008 Revision–Long Range Projections [2011] with USA population adapted to reflect U.S. Census Projection [2008].)

the century, the other scenarios all exhibit peak-and-decline behavior, with the low-fertility population (POP6) and medium-fertility population (POP9) declining from the peaks to year-2100 populations of about 6 billion and 9 billion. Note that each population scenario constitutes two of the six SSPs explored in this study.

Asia regions (China, India, Southeast Asia, Japan, and South Korea) begin the century with 55% of the world's population. However, in all of the scenarios, their dominant position is gradually undermined, largely by the more rapid population growth in other developing regions, particularly in Africa. By the end of the century, Asia regions constitute from 43% (POP6) to 47% (POP14) of the world's population. Africa's population grows remarkably, with its population being very sensitive to fertility assumptions and its global population share increasing from its current 15% to 29% (POP6) or 27% (POP14) in 2100.

3.2.3.2 Labor Participation Rates

We also specify the changes in labor participation rates coupled with the population scenarios for the 14 GCAM regions, and they have been used, in combination with the assumed growth in labor productivities, to develop the GDP pathways for the individual GCAM regions. The common understanding is that the labor participation rate is closely linked to the country's demographic structure: the lower the proportion of population in work ages (typically between ages 15 and 64), the smaller the labor force.* This is not

* In the U.N. population prospects, work-age population rate, as represented by the proportion of population in ages 15–64, fluctuates over time, generally peaking earlier in the century for developed economies and continuing to decline thereafter, with some moderation by the end of the century. In developing economies, including Africa, India, Southeast Asia, and the Middle East, work-age population rate peaks later in the century as they do not face severe populating aging in the near term.

always the case, however. The labor force participation rate could increase despite an aging population if workers are allowed to retire later (delayed retirement), if more people in work ages become economically active (more labor force participation), or if more of the potential labor force participates in productive activities (higher employment rate). For convenience, we define the rate of total labor force (including those unemployed) to total population as the gross labor participation rate. Thus, the change in gross labor participation rate may reflect not only the change in demographic composition, but also the changes in retirement ages and labor force participation.

It is assumed that—given ongoing social and institutional changes throughout the world, as well as aging populations—current variance in gross labor participation rates across countries will diminish and ultimately converge to the levels of the most advanced countries that have almost completed their demographic transition (Figure 3.4). Gross labor participation rates of the countries in Canada, Australia/New Zealand, and Latin America all converge to the current level of United States region (52%). The rates of the countries in Eastern Europe, Middle East, the former Soviet Union, and Africa, all converge to the current level of Western Europe (47%). The rates of the countries of China, India, Southeast Asia, and Korea, all converge to the current level of Japan (52%). With these assumptions, the Middle East and, to a lesser extent, Africa and India will experience rapid increases in gross labor participation rate, whereas China and Canada will have decreases in their rate over the century.

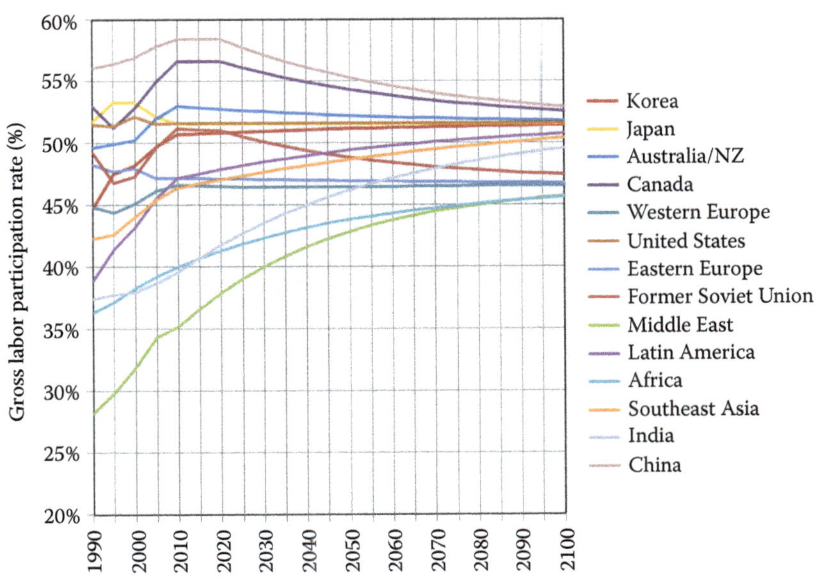

FIGURE 3.4
Gross labor participation rates (Pre-2010 data obtained from World Bank [2011]).

In this study, the same set of assumptions about gross labor participation rate pathways has been made for all of the six scenarios. By implementing the same gross labor participation pathways for different population scenarios, we implicitly assume that the variation in work–age population rate driven by different population assumptions is offset by the variation in retirement ages or labor force participation. That is, compared to a higher population world, a lower population world characterized by lower work–age population rate is likely to increase its retirement age and labor force participation in the long run, which may result in higher labor productivity due to more experienced labor staying longer and greater labor market competition. This is also consistent with our storyline development approach that lower (or higher) population assumptions are generally associated with faster (or slower) labor productivity growth assumptions.

3.2.3.3 Labor Productivity Growth

The GDP is computed as the simple product of labor force and labor productivity. The gross labor participation rate and the assumptions about total population and employment rates generate the economy's total labor force. Assuming that employment rates in all countries are held fixed to the base year levels, the percentage change in per capita GDP over time is represented as follows:

$$\begin{aligned}\%\Delta \text{ GDP per capita} = {} & \%\Delta \text{ labor productivity} \\ & + \%\Delta \text{ gross labor participation rate}\end{aligned}$$

With this relationship, an increase in the gross labor participation rate reinforces an increase in GDP per capita, whereas a decrease in the gross labor participation rate offsets it. Since we have assumed that the paths of gross labor participation rate are the same for all of the six scenarios, the variation in per capita GDP across the alternative scenarios originates only from assumed variation in labor productivity growth rates.

The structure of the labor productivity growth is, undoubtedly, a major determinant of the future GDP. It is assumed that all countries' labor productivities gradually converge into a marker productivity frontier that grows at a fixed percentage rate (the convergence completes at the end of the twenty-third century). The United States is chosen as the marker country in all of the six scenarios with varying annual labor productivity growth rate: POP6/MDG+ (Sustainability and Equity) with 1.6%, POP6/MDG– (Collapse) with 1.6% followed by 0.0%, POP9/MDG+ (Consumerism) with 1.4%, POP9/MDG– (Muddling Through) with 0.9%, POP14/MDG+ (Social Conservatism) with 1.1%, and POP14/MDG– (Crowded Chaos) with 0.9%. Note that, with some exceptions, the growth rate assumption broadly represents the inverse relationship between the fertility rate and labor productivity.* The important question is how other countries might evolve, tapering the labor productivity gap with the United States. Because individual countries are currently at different development stages of the national economy, political stability, and social institutions, they may exhibit substantial differences in the timing and magnitude of the rate of productivity growth over the course of the century. The modeling structure needs to capture these possible heterogeneities, although its implementation would require some reasoned judgment.†

A group of countries in the same GCAM regions, representing similar geopolitical and economic development status, is assumed to share similar trends in labor productivity growth. In POP9/MDG+ (Consumerism), for example, industrialized regions (Japan, Western Europe, Canada, and Australia/New Zealand) continue to have moderate productivity growth rates ranging from 1.3% to 2.0%, whereas less developed regions have varying degrees of relatively fast productivity growth over some income ranges. Africa and the Southeast Asia countries, for example, experience the peak growth rate of around 5% around the mid-century, but the former Soviet Union, Latin America, and the Middle East have the peak growth rate of around 3% around the mid-century. The two most rapidly growing economies—China and India—are assumed to have passed or nearly passed the peak growth, so that the growth rates of the two countries continue to decline over the century. All of these assumptions still give enough variation in labor productivities among the 190 countries throughout the twenty-first century (Figure 3.5).

* We have developed estimates of labor productivity growth that are linked to fertility rates: higher-fertility (i.e., higher population) scenarios exhibit slower growth in labor productivity than lower-fertility scenarios. Indeed, cross-sectional data of 189 countries in 2000 and 2005 indicates a negative relationship between the total fertility rate and calculated labor productivity, with a much tighter relationship observed for lower-income and higher-fertility observations (not shown).

† To implement the country-level differences in the growth of labor productivity while maintaining the convergence-in-productivity feature, we developed a bell-shaped Gaussian function, given by

$$P_{i,t} = \exp\left[\ln \overline{P_{i,t}} - \ln\left(\frac{\overline{P_0}}{P_{i,0}}\right) \cdot \frac{\exp\{-r(s+t)^{\sigma}\}}{\exp(-r \cdot s^{\sigma})}\right]$$

where $P_{i,t}$ is the labor productivity of country i in year t, $P_{i,0}$ is the country's productivity in base year, $\overline{P_t}$ is the productivity of the marker country in year t, and $\overline{P_0}$ is the marker country's productivity in the base year. This structure ensures that each country's productivity rises monotonically, asymptotically converging to the level of the marker country. The above equation also has three shape parameters allowing for implementing differences in the pattern of productivity growth: (1) the rate of convergence (r) influencing overall rate of productivity growth with the changed width of the bell-shaped function; (2) the peakedness of convergence (σ) influencing the maximum growth rate with changed kurtosis of the function; and (3) the time shift of convergence (s) influencing the starting point of growth path with changed initial location on the function. It is assumed that a group of countries in the same GCAM regions, broadly representing similar geopolitical and economic development status, share the same set of shape parameters.

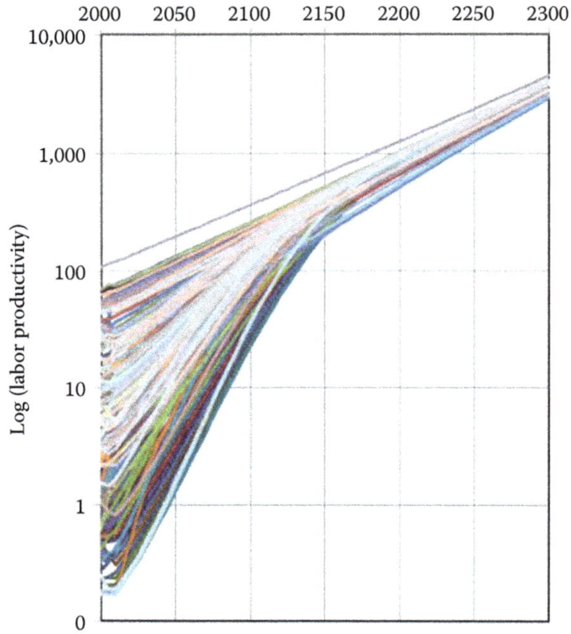

FIGURE 3.5
Labor productivity growth of 190 countries in POP9/MDG+ (Pre-2010 data calculated based on World Bank [2011]).

3.2.3.4 GDP Development

The variations in population and labor productivity assumptions in the six scenarios collectively result in wide-ranging GDP growth paths over the century (Figure 3.6). Given a population assumption, MDG–achieving scenarios have higher GDP than the other scenarios, particularly in the low population case. Compared to the current global GDP, mid-century global GDP increases by 2–3 times and the end-of-the-century global GDP increases by 2–12 times, depending on the scenarios. Perhaps the most important observation is that, by the end of the century, presently less developed economies (i.e., regions except for the United States, Western Europe, Canada, Australia/New Zealand, Japan, and Korea) account for more than 70% of the world's economic output, and the dominant position of currently industrialized economies wane to the levels of under 30%, with the exception of POP6/MDG- (Collapse) (47%).

Per capita GDP scenarios also cover a wide range of uncertainties about global and regional income growth over the century (Figure 3.7). Compared to the current global per capita GDP, the end-of-the-century global per capita GDP increases by 2–11 times, depending on the scenarios. Although developing economies generally make substantial progress in closing the income gap with the presently industrialized economies, there is varying degree of

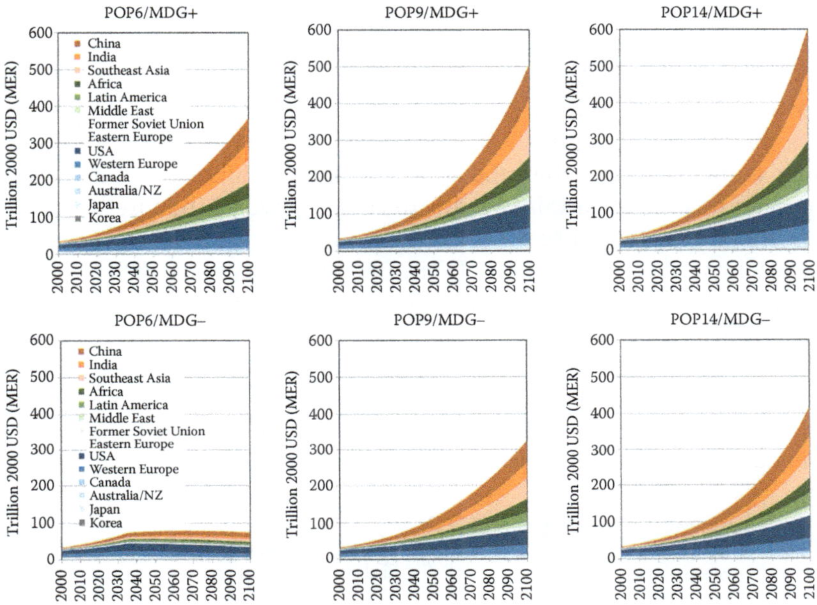

FIGURE 3.6
Total GDP by GCAM region in the six scenarios.

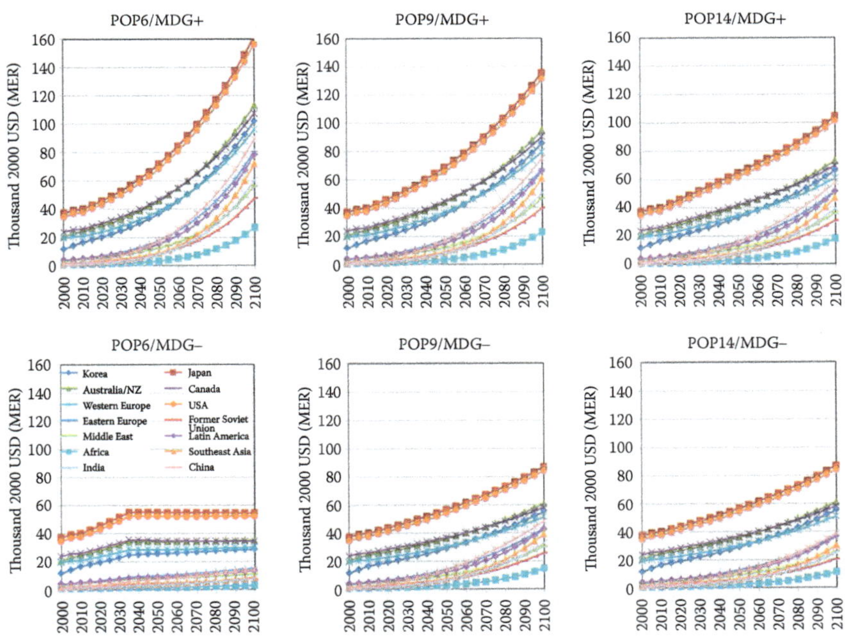

FIGURE 3.7
Per capita GDP by GCAM region in the six scenarios.

income disparity among different regions among the scenarios. Consistent with the scenario storylines, POP14/MDG– (Crowded Chaos) and POP6/MDG– (Collapse) have larger income gaps than the other scenarios. Another important point is that the regions exhibit varying degree of income spread across the scenarios (Figure 3.8). The spread in Asia and other developing regions is much greater than the spread in industrialized regions, suggesting greater income uncertainty in the developing world than in the industrialized world over the century.

In the long term, the per capita GDP in any region is highest in POP6/MDG+ (Sustainability and Equity), followed by POP9/MDG+ (Consumerism), POP14/MDG+ (Social Conservatism), POP9/MDG– (Muddling Through), POP14/MDG– (Crowded Chaos), and POP6/MDG– (Collapse). Note that POP6/MDG– follows the same path as POP6/MDG+ until 2040, and then stays nearly the same, resulting in the lowest per capita GDP among the six scenarios by the end of the century. The post-2040 increases in POP6/

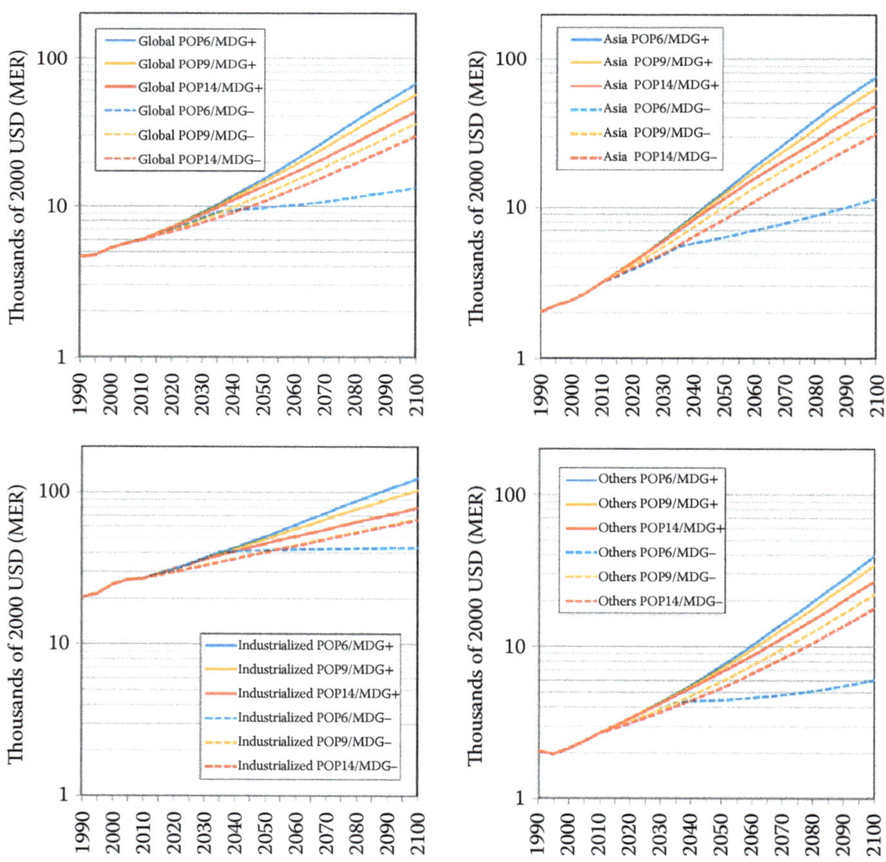

FIGURE 3.8
Per capita GDP by region (global, Asia, industrialized, and other developing regions).

MDG–, mostly occurring in developing regions, are due to the increased labor participation assumed above.

One notable interaction is the competing effect of per capita GDP and population—or, equivalently, of the assumed negative correlation between labor productivity and labor force—across the scenarios. Thus, in all regions, the variation of the total GDP between the scenarios (absent of POP6/MDG–) is much less than the variation of per capita GDP. We suppose that the above variation in per capita GDP between the scenarios is large enough as even a slightly larger variation would result in much less variation in total GDP, which would not be useful to explore a potential range of global economic development.

3.2.4 Technology Assumptions

The progress of technologies used in the production and consumption of energy, agriculture, and other goods and services also influences the challenges to mitigation and adaptation. In particular, the availability and performance of renewable and fossil-fuel based technologies, agricultural productivities, and the efficiency of end uses will have major consequences. Several distinct sets of technology assumptions have been developed, representing ranges of technological progress, so that they constitute the six consistent scenarios by mixing and matching with the assumptions about population/economic growth and development pathways (Table 3.2). These technology assumptions have been used as inputs to the GCAM integrated assessment model.

The Med Tech assumption set is characterized by modest progress in nuclear power, renewable generation technologies (solar, wind, and geothermal), crop productivity, and end use technologies (transportation, buildings, and industry), as well as greater availability of carbon capture and storage (CCS) and fossil fuel resources.

The High Tech assumption set presents a more optimistic path throughout energy and agricultural systems whereas the Low Tech assumes lower levels of improvement and reduced resource availability than the Med Tech case (Table 3.2). The case where no new construction of nuclear power plants are allowed to be built beyond those already operating and under construction and no deployment of CCS has also been taken into account (Lower Tech) because it has major consequences to the challenges to mitigation. Thus, all of the technology categories, except for nuclear power and CCS, have three variants in this specification.

The first three assumptions on nuclear power are distinguished by different capital recovery factors used when leveling the costs of building nuclear power plants. This is indicative of social environments that are more or less supportive of nuclear power deployment. Operations and maintenance (O&M) costs are also assumed to vary. The fourth possibility is representative of a world that chooses not to build new plants for reasons beyond economic concerns, for example, safety and environmental reasons.

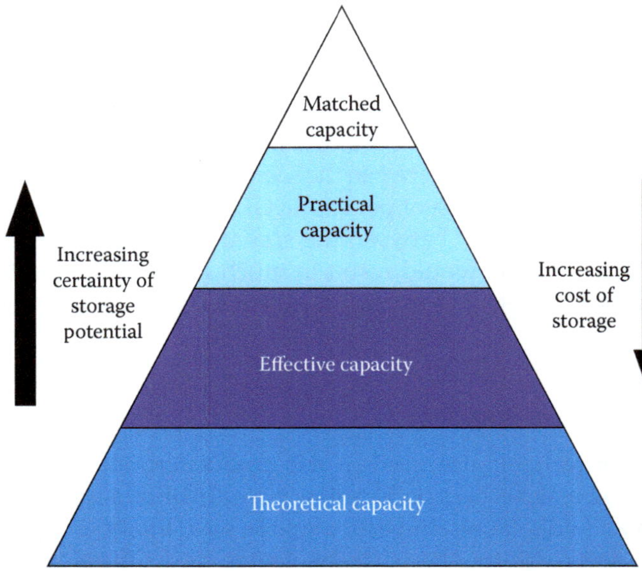

FIGURE 3.9
Standardized, community-wide CO_2 storage resource grades. (From IEA GHG, CO_2 Storage in Depleted Gas Fields. 2009/01, IEA Greenhouse Gas R&D Programme, Cheltenham, U.K., 2009.)

The availability of carbon capture and storage (CCS) is explored through the region-specific availability of storage resources. The Med Tech scenario is developed based on updated values of Dooley et al. (2005), regionally disaggregated to the 14 GCAM energy regions. The updated supply curves used in this study have a total of 27,900 $GtCO_2$ of effective storage capacity, and 7,200 $GtCO_2$ of practical capacity (Figure 3.9).* We use the effective capacity as the medium technology assumption as it is likely to be a reasonable estimate of what is actually available to use. The resource is split into 4 distinct grades; the lowest cost grade costs \$0.25/tCO_2 (in 2005 \$ rates) and encompasses 0.5% of the resource in each region. The cost rises from the level, with 70% of the land-based storage resource being available at costs under \$10/tCO_2 (Dahowski et al. 2011). The remainder is available at a cost of \$75/tCO_2.

The regional, land-based storage is not treated in this study. We assign a higher cost for each ton of CO_2 stored off-shore, \$96/tCO_2. This cost is conservative, several times the \$32/tCO_2 estimate in Decarre et al. (2010). This offshore resource is not characterized in detail and, instead, is assumed to be vast to the point where cost is likely to be more limiting than the available

* There is no justifiable reason to compute a "theoretical capacity" for geologic storage as it is simply impossible to come anywhere close to being able to meet the definition of theoretical capacity in practice, which is "the physical limit of what the geological system can accept. It assumes that the system's entire capacity to store CO_2 in pore space, or dissolved at maximum saturation in formation fluids, or adsorbed at 100% saturation in the entire coal mass, is accessible and utilized to its full capacity" (IEA GHG, 2011).

TABLE 3.2

Specifications of Technology Assumptions

	High Tech	Med Tech	Low Tech	Lower Tech
Nuclear power	Lower capital recovery factor with capital and O&M costs declining at 0.3% per year	Base capital recovery factor with capital and O&M costs declining at 0.1% per year	Higher capital recovery factor with fixed capital and O&M costs	No new nuclear power plant
Carbon capture & storage (CCS)	Lower cost non tradable regional land-based storage with larger capacity, expensive global-access offshore storage	Non tradable regional land-based storage combined with expensive global-access offshore storage	Total available resource to 5% of the medium case. Cost scales up rapidly without offshore storage	No deployment
Fossil fuel extraction	Extraction costs of coal, oil, and gas resource drop by 0.75% per year	Extraction costs of coal, oil, and gas resource drop by 0.5% per year	Extraction costs of coal, oil, and gas resource drop by 0.25% per year	NA
Advanced grid for renewable tech	1:1 backup required when renewables supply 50% of capacity	1:1 backup required when renewables (central PV, CSP, rooftop PV, wind) supply 25% of capacity	1:1 backup required when renewables supply 15% of capacity	NA
Solar tech	Capital and O&M costs decline at a faster rate (double)	Capital and O&M costs decline	Capital and O&M costs decline at a slower rate (50%)	NA
Wind tech	Capital and O&M costs drop at 0.5% per year	Capital and O&M costs drop at 0.25% per year	Capital and O&M costs do not drop	NA

(Continued)

TABLE 3.2 *(Continued)*

Specifications of Technology Assumptions

	High tech	Med Tech	Low Tech	Lower Tech
Geothermal tech	Faster improvement in hydrothermal/EGS available with the improvement rate of 0.5% per year or more	Base improvement in hydrothermal/EGS available only after the exhaustion of hydrothermal resource/EGS improves at 0.25% per year or more	No improvement in hydrothermal/EGS not available	NA
Building tech	Faster improvements in end use efficiencies	Base improvements in end use efficiencies	Slower improvements in end use efficiencies	NA
Transportation tech	Faster declines in fuel intensities in all modes	Base declines in fuel intensities in all modes	Slower declines in fuel intensities in all modes	NA
Industry tech	Faster improvements in end use efficiencies	Base improvements in end use efficiencies	Slower improvements in end use efficiencies	NA
Crop production	Crop yield improvements converging to 0.5% per year by 2050	Crop yield improvements converging to 0.25% per year by 2050	Crop yield improvements converging to 0% per year by 2050	NA

resource. Our Low Tech case assumes that only the least expensive (and most well understood) resources are available, making up about 5% of the resource in the Med Tech case. The High Tech case has modest reductions in storage cost, and also allows for some nearby regions to share and trade storage resources. Similar to the nuclear power case, the fourth case of no CCS deployment is taken into account again for reasons beyond economics.

The ability of power grids to accommodate large-scale deployment of intermittent renewable energy technologies is explored through varying the amount of backup power required. The level of backup required increases with increasing renewable penetration along an s-curve, starting at zero and reaching a peak at renewables levels of 15%, 25%, and 50%, reflecting realistic penetration estimates (NREL 2008; NREL 2010).

The improvements in renewable generation are technology-dependent. For solar, wind, and geothermal, the High Tech case doubles the rate of technology improvement assumed in the Med Tech case. Even in the Low Tech case, solar continues to improve, reflecting the early state of the development of this technology, albeit at a slower rate. Wind is assumed to be fully mature in the Low Tech case, and no further improvements are seen beyond today's technology. Geothermal is characterized largely by the availability of enhanced geothermal systems (EGS), substantially deeper resources with a much larger supply than conventional geothermal, which will be exhausted before the century end. These EGS resources are not available in the Low Tech case.

The energy intensities of end use services in the transportation, buildings, and industrial sectors are broadly classified with low, medium, and high improvement rates. The assumptions used in this study are largely consistent with those already in GCAM publications (Kyle et al. 2009; Kyle et al. 2010).

Assumptions about the improvements in crop production affect both the availability of food to feed global populations as well as the availability of land for bioenergy use. Agricultural productivity change is aggregated by GCAM region and commodity from an internal FAO database used in support of Bruinsma (2009), which has 108 countries and 34 commodities. We aggregated the data for irrigated, rainfed, and total agricultural production, harvested area, and yields. The projection years are 2005 (base year), 2030, and 2050, which allows annual productivity change rates to be calculated for each region and crop. The same rate is applied to biomass as to other crops. The High Tech case assumes a substantially improved rate of crop productivity improvement, while the Low Tech case assumes no improvements beyond today's levels.

3.2.5 Summary of GCAM Input Assumptions in Six Alternative Scenarios

The combinations of population and economic growth, proxies of development trajectories, and energy and agricultural technologies have been carefully crafted for each of the six scenarios and used as inputs to GCAM (Table 3.3). These combinations represent varying degrees of vulnerability

TABLE 3.3

GCAM Input Assumptions for the Six Scenarios

	POP6/MDG+	POP6/MDG–	POP9/MDG+	POP9/MDG–	POP14/MDG+	POP14/MDG–
	Sustainability and Equity	**Collapse**	**Consumerism**	**Muddling Through**	**Social Conservatism**	**Crowded Chaos**
Global Population (billion in 2100)	Low (5.5)	Low (5.5)	Medium (9.1)	Medium (9.1)	High (14.0)	High (14.0)
Per Capita GDP Growth (000 USD in 2100)	Fast (66)	Fast and Very Slow (13)	Medium (56)	Slow (36)	Medium/Slow (43)	Slow and Divided (30)
Global GDP (trillion USD in 2100)	Medium (367)	Low (74)	High (505)	Medium (323)	High (600)	High/Medium (416)
Per Capita Energy Service Demands	Lower	Higher	Higher	Higher	Lower	Higher
Nuclear/CCS Technology	Phase Out	Low Tech	Low Tech	Med Tech	High Tech	Low Tech
Renewable/End Use Technology	High Tech	Low Tech	Low Tech	Med Tech	High Tech	Low Tech
Fossil Fuel Extraction	Low Tech	High Tech	High Tech	Med Tech	Low Tech	High Tech
Crop Yield Improvement	High Tech	Low Tech	Med Tech	Med Tech	High Tech	Low Tech

and also maintain consistency with their storylines discussed in Section 3.1. For example, POP6/MDG+ (Sustainability and Equity) represents a low population, sustainability-oriented world characterized by social opposition to nuclear and CCS technologies, and greater emphasis on renewable resources than on fossil fuel. POP14/MDG− (Crowded Chaos) represents a high population, poor world with large income gap between regions, greater demand for energy services,* and greater reliance on fossil fuel. POP14/MDG+ (Social Conservatism) and POP9/MDG− (Muddling Through) are located about in the middle of POP6/MDG+ and POP14/MDG− in terms of population and per capita GDP. The other two scenarios, POP6/MDG− (Collapse) and POP9/MDG+ (Consumerism), have different population and economic growth assumptions with similar technology development assumptions.

3.2.6 RCPs and Climate Policy Assumptions

Each of our reference scenarios has been coupled with policy assumptions that deliver four different levels of radiative forcing that the world may want to limit by the century end. We limit radiative forcing to the four RCP target forcing levels (2.6, 4.5, 6.0, and 8.5 Wm^{-2}) and 3.7 Wm^{-2}, all in the year 2100.† Overshooting of radiative forcing above the targets before 2100 is allowed. We will refer to the level at which radiative forcing is limited and other characteristics associated with the limitation of emissions as Shared Policy Assumptions (SPAs) and the alternative levels as SPA8.5, SPA6.0, SPA4.5, SPA3.7, and SPA2.6 respectively. Note that each of the SPAs, except SPA3.7, corresponds directly to the RCPs run by the climate modeling community and could potentially be directly combined with any member of the CMIP5 climate model ensemble, which ran the corresponding RCP.

Our scenarios vary not only by the target forcing level, but also by the scope and characteristics of the global carbon market due to their potential effects on the cost and feasibility of the mitigation policy. We consider three alternative policy settings. The first is a simple, idealized policy environment in which the economic cost of limiting year-2100 radiative forcing to RCP levels is minimized. The idealized SPAs are run under the following conditions:

- All regions initiate emissions mitigation simultaneously in the year 2015 (Immediate Accession).

* The SSPs are separated by their pathways of service demands delivered by energy consumption. Because of more effective governance and greater emphasis on sustainability, people in SSP1 and SSP2a demand less of energy services, such as building, transportation, and industry services, than people in other SSPs. It is assumed that in the absence of the changes in service prices, per capita service demands in SSP1 and SSP2a at a given income level are less than those in other SSPs by 10% or less. In GCAM, this variation is implemented by employing different paths of income elasticity of demands for those energy services.

† SPA3.7 does not correspond to an RCP. It is included because it is a 550 ppm CO_2-e limit on radiative forcing, and sits halfway between SPA4.5 (650 ppm CO_2-e) and SPA2.6 (450 ppm CO_2-e).

- The price of carbon emissions—both from industrial activities and land use change—are equal in all regions and across all sectors in every period (universal carbon tax [UCT]).
- The price of carbon rises at the rate of interest plus the average rate of removal of carbon by oceans, the Hotelling–Peck–Wan path (Peck and Wan 1996).

The central principle of the UCT is that all greenhouse gas emissions are subject to tax, regardless of where or in which sector they are emitted. As such, the accumulation of carbon in the terrestrial system is rewarded at the same rate as fossil fuel and industrial emissions are taxed. This ideal setting characterizes the SPAs associated with POP6/MDG+ (Sustainability and Equity) and POP14/MDG+ (Social Conservatism).

We contrast the idealized policy with two alternative policy settings: (1) the less-ideal one with delayed accession to a global emission mitigation regime combined with carbon tax imposed universally (UCT) and (2) the least-ideal one with both delayed participation and carbon tax imposed only on fossil fuel and industry emissions (FFICT). The less-ideal policy setting is incorporated into the SPAs, based on POP9/MDG+ (Consumerism), and the least-ideal policy setting is incorporated into the SPAs, based on POP6/MDG– (Collapse), POP9/MDG– (Muddling Through), and POP14/MDG– (Crowded Chaos).

In the delayed accession case, Western Europe, Eastern Europe, and Japan join in 2015; the United States, China, Canada, Australia/New Zealand, and Korea join in 2030; India, Latin America, and Southeast Asia join in 2050; and Africa finally joins in 2070—the Middle East and the former Soviet Union never join the market. Because of the anticipated carbon price shock to regions that have so far not participated in the global carbon market, they are allowed to have 15 years until their regional carbon price gradually catches up with the globally prevailing carbon price. The delayed accession scheme is summarized in Table 3.4. Given all of these, since each of the six socioeconomic scenarios is experimented against five SPAs (combined with one of the three policy settings), totally 30 possible scenario combinations can be generated (Table 3.5).

TABLE 3.4

Delayed Accession to a Global Emissions Mitigation Regime as an Alternative Policy Setting

Year	Regions Joining Mitigation Regime	Years between Joining and Sharing the Common Carbon Price
2015	Western Europe, Eastern Europe, Japan	NA
2030	Australia/NZ, Canada, China, Korea, United States	15
2050	India, Latin America, South and East Asia	15
2070	Africa	15
Never	Former Soviet Union, Middle East	NA

TABLE 3.5

The GCAM Scenario Matrix

	POP6/MDG+	POP6/MDG−	POP9/MDG+	POP9/MDG−	POP14/MDG+	POP14/MDG−
	Sustainability and Equity	**Collapse**	**Consumerism**	**Muddling Through**	**Social Conservatism**	**Crowded Chaos**
Carbon Market Operation	Immediate Accession/ UCT	Delayed Accession/FFICT	Delayed Accession/UCT	Delayed Accession/ FFICT	Immediate Accession/ UCT	Delayed Accession/ FFICT
SPA8.5	N/A	N/A	N/A	N/A	N/A	Achievable
SPA6.0	N/A	N/A	Achievable	Achievable	Achievable	Achievable
SPA4.5	Achievable	Achievable	Achievable	Achievable	Achievable	Achievable
SPA3.7	Achievable	Achievable	Achievable	Achievable	Achievable	Not Feasible
SPA2.6	Achievable	Not Feasible	Not Feasible	Achievable	Achievable	Not Feasible

3.3 GCAM Integrated Assessment Model

GCAM* (Edmonds and Reilly 1985; Kim et al. 2006; Clarke et al. 2007) is an IAM that links a global energy–economy–agricultural land use model with a climate model of intermediate complexity.† GCAM has 14 global regions defined on geopolitical boundaries: the United States, Canada, Western Europe, Japan, Australia and New Zealand, the former Soviet Union, Eastern Europe, Latin America, Africa, the Middle East, China and the Asian Reforming Economies—India, South Korea, and the rest of South and East Asia. GCAM is a long-term model, typically operating in five-year time steps through the year 2095 (though the code is written to accommodate any arbitrary time-step or end year). As part of GCAM's modeling of human activities and physical systems, the GCAM tracks emissions and concentrations of the important greenhouse gases and short-lived species (including CO_2, CH_4, N_2O, NO_x, VOCs, CO, SO_2, BC, OC, HFCs, PFCs, and SF_6).

GCAM version 3.0 solves at 5-year time steps. It is disaggregated into 151 land use subregions around the globe. These land use subregions are based on a division of the extant agro-ecological zones (AEZs), which we derived from work performed for the GTAP project (Monfreda et al. 2009), within each of GCAM's 14 global geopolitical regions. These changes provide a substantial enhancement to the GCAM's ability to model crops and land-use decisions and implications in much more physical, technological, and spatial detail while maintaining tight integration with the rest of the GCAM.

GCAM models the energy, agriculture, and land use in an economically and physically consistent global framework. In each model period, the GCAM explicitly models markets and solves equilibrium prices in energy, agriculture and other land uses, and emissions; that is, the set of prices that ensures that supplies are equal to demands in all markets. The GCAM is a dynamic-recursive model, which means that it solves each period's market equilibrium sequentially.‡ The GCAM models energy and agriculture technologies explicitly.

3.3.1 Energy System in GCAM

GCAM models the complete energy system, tracking energy from its point of origin through alternative transformation processes to its final end use. The GCAM begins with potentially available resources, including both depletable (coal, gas, oil, uranium) and renewable (wind, solar, geothermal) resources, which are represented through graded resource curves. As more

* Note that GCAM was formerly known as MiniCAM.

† Documentation for GCAM can be found at http://www.globalchange.umd.edu/models/MiniCAM.pdf/

‡ In contrast, an inter-temporal optimization model would solve for all periods simultaneously.

energy is extracted and used, costs rise, though those cost increases can be ameliorated by technological improvement. The supply of bioenergy is determined by the agriculture and land use sub-model within the GCAM, which is discussed below.

GCAM models the transformation of primary energy resources into final energy forms (electricity, hydrogen, refined liquids, refined gas, coal, and solid bioenergy) through a set of conversion sectors, each of which may include a range of conversion technologies. For example, the GCAM includes multiple technologies for producing electricity from coal, natural gas, oil, bioenergy, wind energy, nuclear power, solar energy, hydropower, and geothermal energy. Final energy forms are consumed by three end use sectors (buildings, industry, and transportation). GCAM includes detailed representations of each of these demand sectors in the United States. It also includes a globally detailed representation of transportation demands. The consumption of energy in these sectors is determined by the demand for final energy services, as well as the characteristics of the technologies used to provide those services.

3.3.2 Resource Assumptions

Supply curves for fossil fuel resources, such as coal, natural gas, and oil (both conventional and unconventional), are specified for each GCAM region by a set of the cumulative amounts of individual resource grades available at specific price points. Resource supply curves for the resources were developed based on the hydrocarbon resource assessment by Rogner (1997) and available for each of the 14 GCAM regions at GCAM wiki (http://wiki.umd.edu/gcam/). Note that unconventional oil in the GCAM consists of an aggregation of shale oil, bitumen, and heavy oil. As the resource supply curves for unconventional oil do not include the cost of the energy used in extraction, the actual supply curves in each region have higher costs at all quantities. As GCAM assumes full global trade in all of the fossil fuel resources, the global resource supply curves are derived by adding up the individual regions' available resources at each price point. As a result, the price of primary resource delivered to any GCAM region in any time period equals the price specified by the global supply curve plus any region-specific costs associated with fuel transportation and delivery. To account for technological improvements that reduce the costs of resource extraction, we have assumed that extraction costs decline at 0.25, 0.50, and 0.75% per year in the Low Tech, Med Tech, and High Tech cases, respectively.

Regarding natural uranium resource, we use a supply curve based on a generalized simple crustal model (Schneider and Sailor 2007), with the relationship between uranium abundance and concentration fitted to the resource estimates and costs from the IAEA Redbook (IAEA 2003). This uranium supply curve provides the global availability of uranium in metric tons as a function of price ($/kgU). It is worth noting that the supply curve is assumed to be continuous and that significant amounts of natural uranium are available beyond those estimates in the Redbook data in lesser concentrations, but at higher costs. In GCAM, we assume that the uranium resources remain available if costs rise sufficiently high. Estimates of the practicality and extraction cost of speculative uranium resources vary widely, however.

Similar to depletable resources, renewable resources are represented as supply curves, although their quantities reflect annual production rather than cumulative totals. As renewable resource is not necessarily a commodity with a market price, its resource supply curve is used in combination with enabling energy technologies to represent the full cost of the technology that is expected to increase with deployment. Therefore, the total costs of renewable technologies include a resource cost, a technology cost, and if the technology is intermittent, any backup-related costs. Wind resource curves in each region account for transmission costs, siting costs, and higher costs due to decreasing capacity factors at less windy sites. The supply curve for the U.S. region is based on a detailed analysis documented in Kyle et al. (2007), and all other regions have the same shape, with the adjustment in magnitude according to the relative resource amounts in each region in IEA GHG (2000). No supply curve is used for the solar resource used in central electricity plants because the marginal costs are assumed not to increase with deployment, although full technology costs vary by GCAM region, based on the population-weighted average insolation received. However, for rooftop PV, GCAM employs regionally differentiated supply curves, based on the estimates from Denholm and Margolis (2008). Resource costs of geothermal electricity production in GCAM account for the phases of identification, exploration, and drilling, and the cost in the production phase is accounted separately at the technology level (Hannam et al. 2009). Resources in each region consist of the sum of the hydrothermal resource and the enhanced geothermal system (EGS) resource, if EGS is allowed. Supply curves for each are based on Petty and Porro (2007), with region-specific adjustments. Renewable resource supply curves for each region are presented in detail at the GCAM wiki (http://wiki.umd.edu/gcam/).

3.3.3 Agriculture, Forest, and Land Use Systems in GCAM

Land use interacts with mitigation, both as a supplier of bioenergy and as a source or sink of terrestrial emissions. For this reason, the GCAM includes a spatially-disaggregated land use model that models the land cover, land use, and the production of agricultural and forest products, as well as ecosystem types. Energy, agriculture, forestry, and land markets are integrated in the GCAM, along with unmanaged ecosystems and the terrestrial carbon cycle. GCAM determines the demands for, and production of, products originating on the land and the carbon stocks and flows associated with land use. GCAM 3.0 divides the globe into 151 land use subregions, based on a mapping of up to 18 AEZs (Monfreda et al. 2009) within each of the GCAM's 14 global geopolitical regions, as shown in Figure 3.10. These AEZs are defined

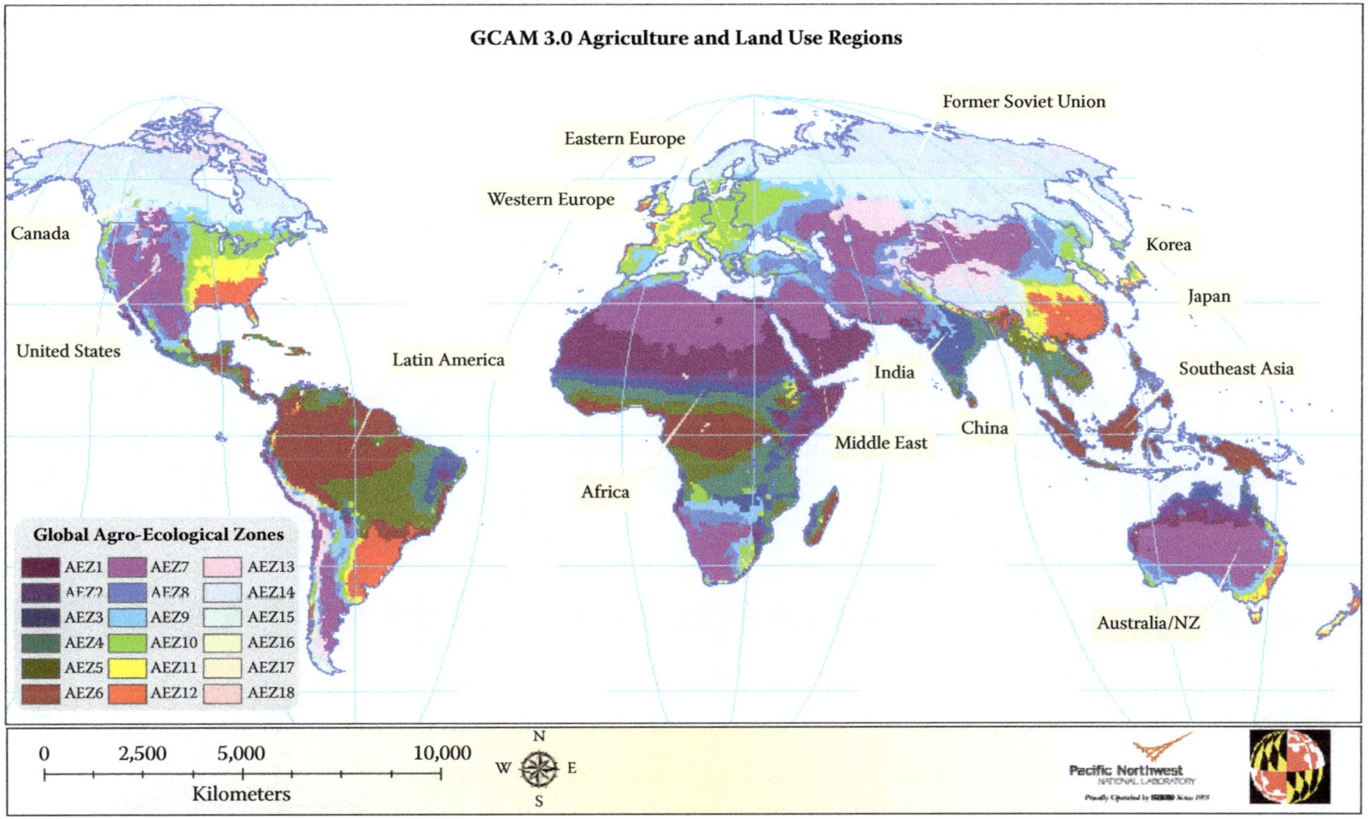

FIGURE 3.10
GCAM 3.0 geopolitical and land use region map. (Agro-Ecological Zones (AEZs) from Monfreda, C., Ramankutty, N., Hertel, T. W., In: Hertel, T. W., Rose, S., Tol, R. S. J. (eds). *Economic Analysis of Land Use in Global Climate Change Policy*. Routledge Explorations in Environmental Economics, Taylor & Francis Group, London and New York, 2009; map courtesy of JGCRI [Goode Homolosine projection].)

as zones with similar temperature and precipitation levels, and as such, are a useful division of land for modeling agriculture and other land use. A complete description of GCAM 3.0 modeling of agriculture and land use is provided in Wise and Calvin (2011) and Kyle et al. (2011).

Within each of these 151 subregions, the GCAM categorizes land into approximately a dozen types based on cover and use. Some of the land types in the subregions, such as tundra and desert, are not considered arable. Among arable land types, further divisions are made for lands historically in non-commercial uses, such as forests and grasslands, as well as commercial forestlands and croplands. Within each subregion, arable land is allocated across a variety of uses based on expected profitability, which depends on the productivity of the land, the non land costs of production (labor, capital, fertilizer, etc.), and the price of the product. The production of approximately 20 crops and commercial forest products is currently modeled, with yields of each specific to each of the 151 subregions. The model is designed to allow the specification of different options for future crop management for each crop in each subregion, and the model structure itself allows for other regional breakdowns besides the AEZs.

As with the GCAM energy system, the economic modeling approach for GCAM agriculture, forest, and land is that of an integrated economic equilibrium in the products, sectors, and factors that are modeled. Markets for products such as corn, wheat, wood, or bioenergy crops must be cleared so that supplies are equal to demands in each model period. Depending on the product or on user specifications, markets can be cleared globally, regionally, or across groups of regions.

The GCAM models the production of several types of bioenergy: traditional bioenergy (straw, dung, fuel wood, etc.), bioenergy from waste products (including crop residues, municipal solid waste, and black liquor from the pulp and paper industry), and purpose-grown bioenergy crops. Purpose-grown bioenergy crops, including perennial grasses like switch grass, and woody crops such as willow, are modeled as economically competing for land with all other agriculture, forestry, and other uses of land. Food crops, such as corn, soybeans, and sugar, can also be used as energy feedstocks to be supplied to GCAM's energy transformation and use sectors.

3.3.4 Climate System in GCAM

All IAMs must include some meaningful representation of global biogeophysical processes that govern the fate of greenhouse gas and other anthropogenic emissions. GCAM uses the MAGICC model (Wigley and Raper 2001) as its default biogeophysical representation. MAGICC provides a representation of important physical earth system elements: carbon cycle, atmospheric chemistry, ocean systems, and climate systems.

MAGICC operates by taking the anthropogenic emissions from the other GCAM components, converting these to global average concentrations (for gaseous emissions), then determining the anthropogenic radiative forcing relative to preindustrial conditions, and finally computing the global mean temperature changes. The MAGICC climate system model is an energy-balance climate model that simulates the energy inputs and outputs of key components of the climate system (sun, atmosphere, land surface, and ocean) with parameterizations of dynamic processes such as ocean circulations.

The carbon cycle in MAGICC is modeled with both terrestrial and ocean components. The terrestrial component includes CO_2 fertilization and temperature feedbacks; the ocean component is a modified version of the Maier-Reimer and Hasselmann (1987) model that also includes temperature effects on the terrestrial biosphere. Reactive gases and their interactions are modeled on a global-mean basis using equations derived from the results of global atmospheric chemistry models (Wigley et al. 2002).

Global mean radiative forcing for CO_2, CH_4, and N_2O are determined from GHG concentrations using analytic approximations. Radiative forcing for other GHGs is proportional to concentrations. Radiative forcing for aerosols (for sulfur dioxide and for black and organic carbon) is taken to be proportional to emissions. Indirect forcing effects, such as the effect of CH_4 on stratospheric water vapor, are also included. Given radiative forcing, global mean temperature changes are determined by a multiple box model with an upwelling-diffusion ocean component. Climate sensitivity is specified as an exogenous parameter.

3.4 GCAM Results: Climate System and Mitigation Effort

In the absence of mitigation policy, end-of-the-century radiative forcing in our six reference scenarios varies between 8.8 Wm^{-2} and 5.1 Wm^{-2} (Figure 3.11). Note that one of the six scenarios presents a radiative forcing level higher than

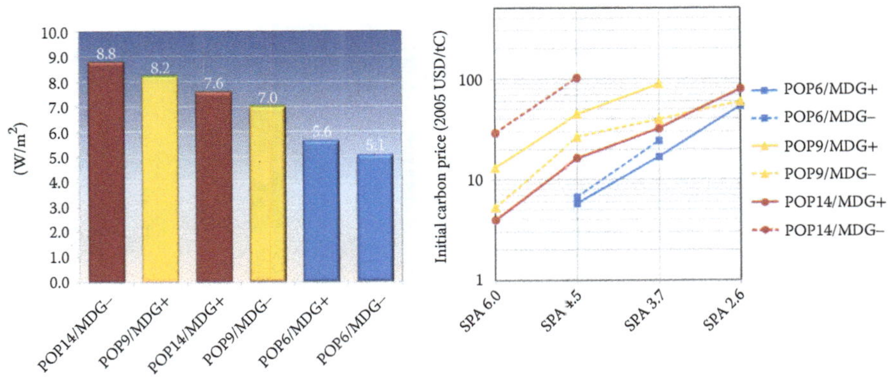

FIGURE 3.11
Year 2100 radiative forcing in the reference scenarios and initial carbon prices under SPAs.

the highest RCP, 8.5 Wm^{-2}. Although the higher population scenarios tend to have higher levels of the radiative forcing, POP9/MDG+ (Consumerism) still presents higher radiative forcing than POP14/MDG+ (Social Conservatism) because of the differences in technology portfolios that are deployed in the energy sector and social preferences for energy consumption. The relative level of the radiative forcing in any scenario to a particular radiative forcing target broadly represents the degree of mitigation effort required by the scenario to achieve the target.

When we constrain reference scenarios to radiative forcing levels taken from the RCPs, we use emissions taxes. The tax rates for different emissions is keyed to the carbon tax rate in a fixed proportion determined by the gas' global warming potential. Because all emissions taxes remain proportional throughout the analysis, rising and falling with the carbon tax, we use the initial carbon price to as an indicator of the cost that future society might face to achieve the target and of the feasibility of achieving the target.

The initial carbon price varies across the SPAs and across scenario families (Figure 3.11). Note that not all of the configurations have a carbon price. SPA8.5 presents a carbon price only under POP14/MDG– (Crowded Chaos), and, similarly, SPA6.0 does not give a carbon price under the two low population pathways, POP6/MDG+ (Sustainability and Equity) and POP6/MDG– (Collapse). The most stringent SPA2.6 is not achievable under any fossil-fuel-oriented scenario families combined with less-than-ideal carbon policy settings, such as POP14/MDG–, POP6/MDG–, and POP9/MDG+. POP14/MDG– world even fails to achieve SPA3.7 because of its high reference emissions and non-ideal climate policy setting.

The change in long-term, steady-state global mean surface temperature relative to preindustrial ranges from 2.7°C with a climate sensitivity (CS) of 2.0°C to 11.8°C with a CS of 5.0°C (Figure 3.12). The range in transient global mean surface temperature rise is far narrower, ranging between 2.1°C and 6.4°C, due to the presence of ocean thermal inertia, as shown in Figure 3.12. Climate

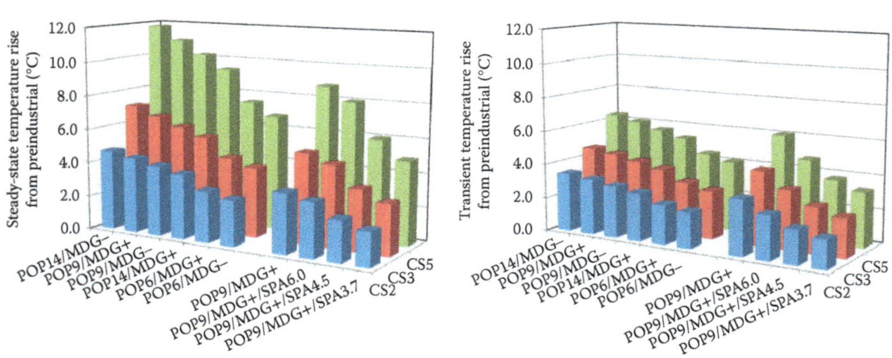

FIGURE 3.12
Year-2100 global mean surface temperature changes by scenario family and SPA (the left for steady-state temperature rise and the right for transient temperature rise from pre-industrial).

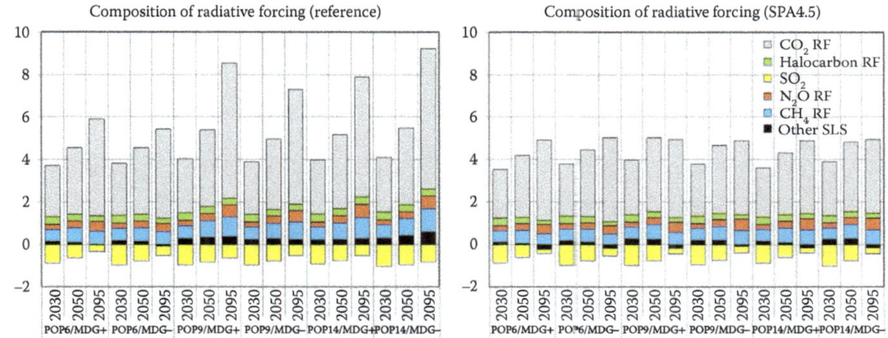

FIGURE 3.13
Contributions of the different greenhouse gases to total radiative forcing.

policy mitigates climate change by reducing greenhouse gases emitted over the course of the century. Still, the variety in future socioeconomic development presents a wide range in exposure and sensitivity to climate change.

Alternative socioeconomic pathways differ not only in radiative forcing and temperature rise over the century, but also in the contributions of climate forcers (Figure 3.13). Regardless of the scenarios, CO_2 constitutes the largest portion to total radiative forcing, followed by CH_4, N_2O, halocarbons, and other short-lived species (e.g., BC, OC, and tropospheric ozone). The relative contributions from non CO_2 gases remain nearly constant over time. SO_2 forcing is negative throughout the century, and its contribution decreases over time due to the increase in pollution control with income. Both POP9/MDG+ (Consumerism) and POP14/MDG− (Crowded Chaos) exhibit relatively greater forcing from CH_4 and other short-lived species and prolonged negative SO_2 forcing due to their reliance on fossil fuel, particularly coal, which has local environmental impacts.

3.5 GCAM Results: The Evolving Human Earth System

3.5.1 Energy System

The six reference scenarios cover a wide range in global primary energy demand throughout the century, and, except for the low population scenarios, primary energy demand continues to grow over time (Figure 3.14). By the end of the century, it will range from 570 exajoule (EJ) (POP6/MDG−) to 2010 EJ (POP14/MDG−), which is roughly the same range projected by IPCC Special Report on Emissions Scenario—514-2226 EJ in 2100 (SRES 2000). Note that a greater population or higher per capita income do not necessarily present higher global primary energy demand because of the variety in technology development and energy service preferences assumed in this study.

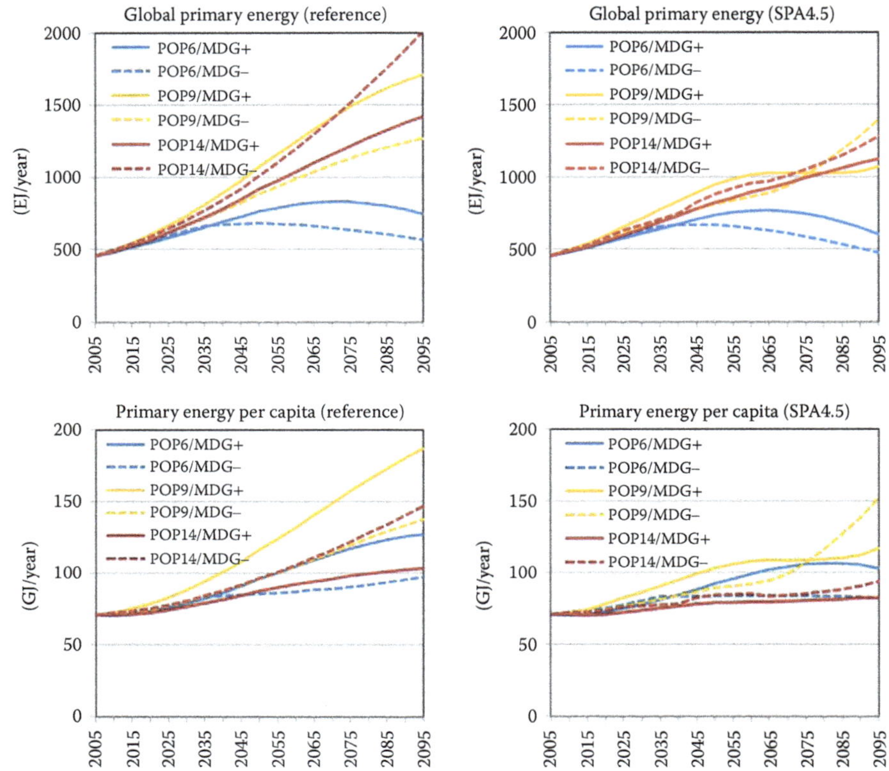

FIGURE 3.14
Total and per capita primary energy demands in the reference and SPA 4.5 scenarios.

For instance, despite its lower global GDP, POP9/MDG+ (Consumerism) demands more energy than POP14/MDG+ (Social Conservatism) because POP9/MDG+ requires more energy services that are delivered by more fossil fuel-intensive energy systems. Helped also by a relatively high per capita GDP, POP9/MDG+ presents the highest per capita energy demand among the six scenarios (Figure 3.14).

The mitigation policy raises the consumer price for energy service, leading to a decreased primary energy demand in all of the scenarios (Figure 3.14). However, the magnitude of the price effect varies because different scenarios achieving the same forcing target require different levels of carbon price (Figure 3.11). The price effect on per capita energy demand is the greatest in POP14/MDG– (Crowded Chaos) and POP9/MDG+ (Consumerism), as suggested by their high carbon prices (Figure 3.11).

Interestingly, in all of the six reference scenarios, the global energy system continues to rely heavily on fossil fuels, such as coal, oil, and natural gas (Figure 3.15). The continued dominance of fossil fuels is partly due to their limited long-term price increases (Figure 3.16). Because of relatively abundant unconventional oil resources, the global oil price does not continue to

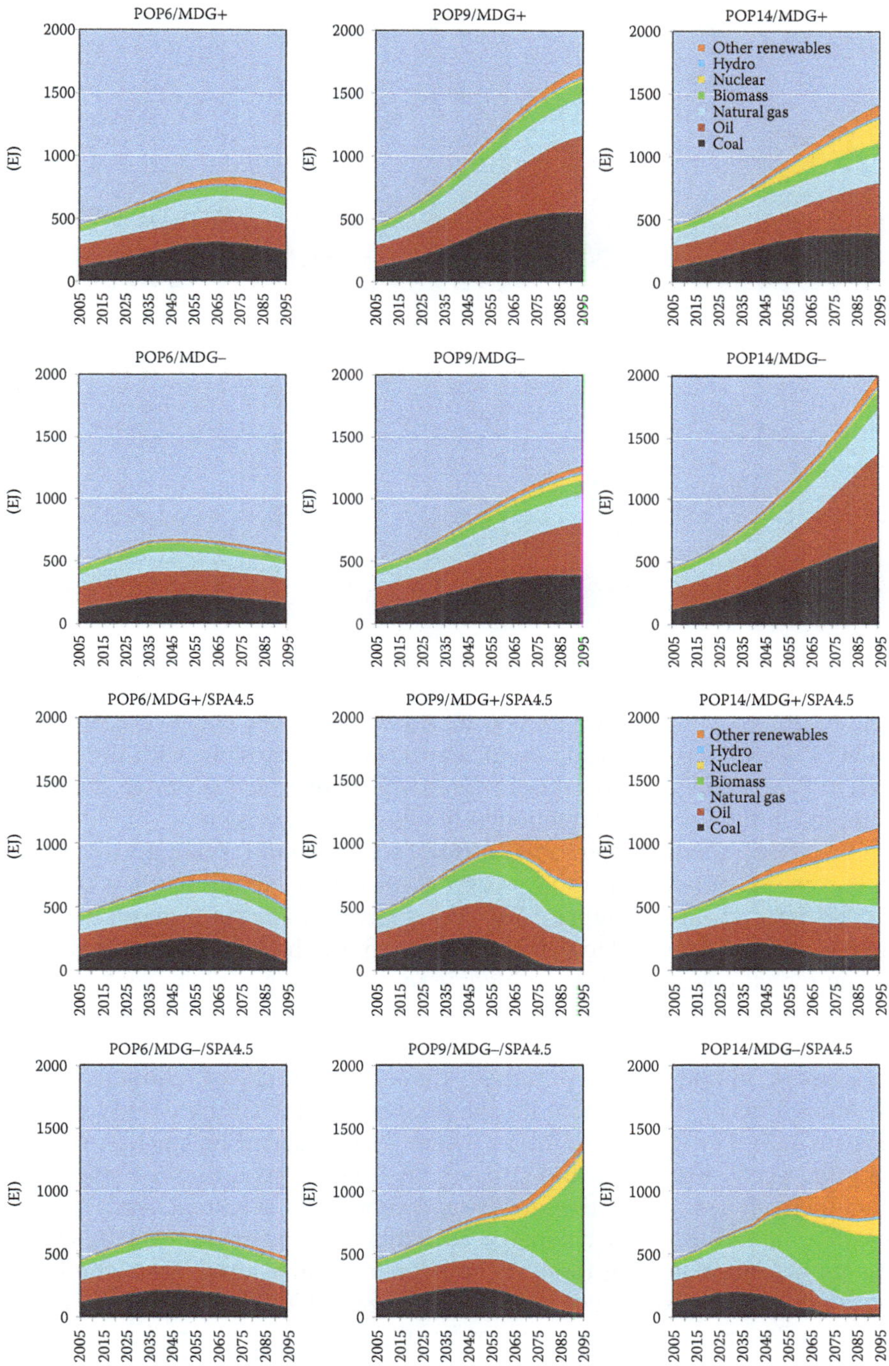

FIGURE 3.15
Global primary energy in the reference and SPA 4.5 scenarios.

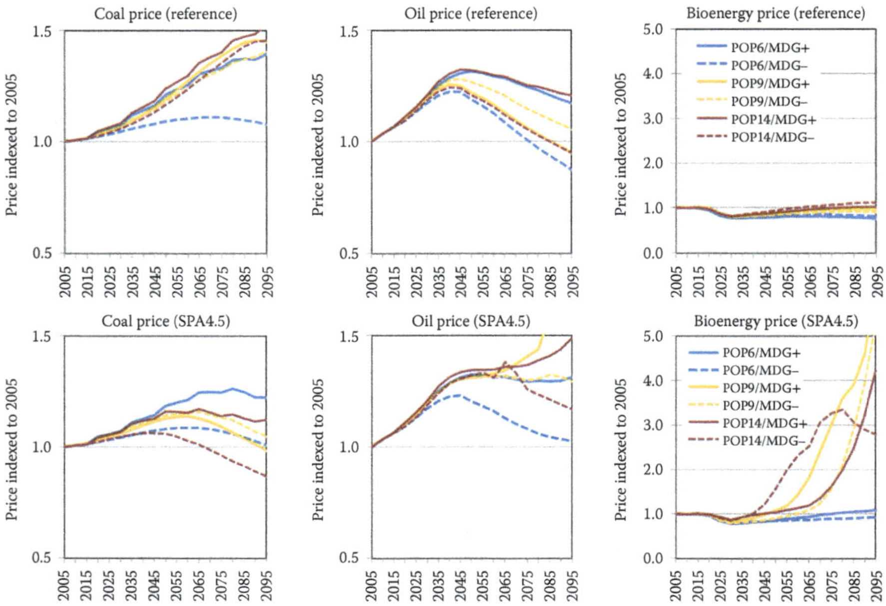

FIGURE 3.16
Supply prices of major energy commodities.

grow. Similarly, the abundance of coal implies relatively stable coal prices as well. Note that the expansion in nuclear power is very slow, with the exception of POP14/MDG+ (Social Conservatism), where nuclear power is widely accepted and fossil fuel extraction becomes most expensive.

The contribution of renewable sources, such as wind, solar, and geothermal, remains small even though they continue to expand as they become increasingly cheaper (Figure 3.15). The share of biomass remains slightly higher and less sensitive to the scenarios than the share of renewable energy in the global energy system, as people in developing regions switch away from the intensive use of traditional biomass toward more efficient commercial biomass and other modern fuels.

Given a climate forcing target, the specification of climate policy becomes a major determinant for the evolution of the future energy system (Figure 3.15). The scenario families where terrestrial carbon stock is not valued (i.e., scenarios with MDG−) have much faster expansion of bioenergy than the scenarios where the terrestrial carbon stock is valued. And, in such less-than-ideal scenarios, bioenergy accounts for a major share of the global energy system, replacing fossil fuel uses, particularly in the case where the demand for emissions mitigation is large (e.g., POP14/MDG− and POP9/MDG−). By contrast, in the scenarios families with the universal coverage of carbon emissions (i.e., scenarios with MDG+), renewable energy or nuclear power plays a more important role than bioenergy in lowering the carbon intensity of the energy system and fulfilling the mitigation requirement.

Similar trends are observed for the global power system (Figure 3.17). In the reference scenarios, the global power system remains fossil fuel-intensive, with coal and natural gas continuing to account for more than half of the global electricity production, except for POP14/MDG+ (Social Conservatism), where nuclear power technology improves most rapidly and fossil fuel extraction is most expensive. Variation of nuclear power in the fuel share comes from variation in our assumptions about nuclear power,* although renewable power—particularly wind and solar—plays a role in POP6/MDG+ (Sustainability and Equity).

The effect of mitigation policy on power production and fuel mix are shown in Figure 3.17. Land policy has a substantial effect on the deployment of bio-CCS technology. Bio-CCS technology deploys far more aggressively when terrestrial carbon is not valued (i.e., MDG– scenarios) than when terrestrial carbon is valued. Solar power, nuclear, and CCS can all provide power with dramatically reduced emissions. The relative role of each depends on our scenario assumptions. Nuclear and CCS technology assumptions in POP9/MDG+ (Consumerism) and POP14/MDG– (Crowded Chaos) lead to rapid market penetration for solar power, despite the relatively slow technical progress assumed for solar. In contrast, nuclear power plays a relatively larger role in the POP14/MDG+ (Social Conservatism) scenarios. POP9/MDG– (Muddling Through), in contrast, relies relatively heavily on bio-CCS to control net emissions.

The per capita final energy demand and the per capita electricity demand increase in all reference scenarios. See Figure 3.18. The share of electricity ranges between 25% and 32% of the global final energy at the century's end. This contributes to a decrease in final energy demand driven by the efficiency enhancement effect of electricity in end uses. Electrification occurs predominantly in the buildings and industry sectors, with less penetration in transport. Despite of its highest per capita income, POP6/MDG+ (Sustainability and Equity) does not present the highest per capita final energy demand because of its less energy-intensive lifestyle combined with a relatively high level of electrification. In contrast, the POP9/MDG+ (Consumerism) reference scenario exhibits the highest per capita final energy demand and the lowest electrification level because of its energy-intensive lifestyle, the availability of cheap fossil fuels, and relatively high per capita income.

When mitigation policies are applied, the final energy demand undergoes further electrification in all scenarios, reaching 25%–57% of electrification by the century's end (Figure 3.18). The carbon price raises the cost of fossil fuel commensurate with its carbon intensity, while electricity price increases are limited by the fact that power generation has non-emitting technology options, including the use of CCS technology with fossil fuel power generation. Increased electrification under the mitigation policy, as well as

* Note that GCAM places no a priori limits on nuclear power deployment. In the post-Fukushima period various governments have considered or implemented limits to nuclear deployment.

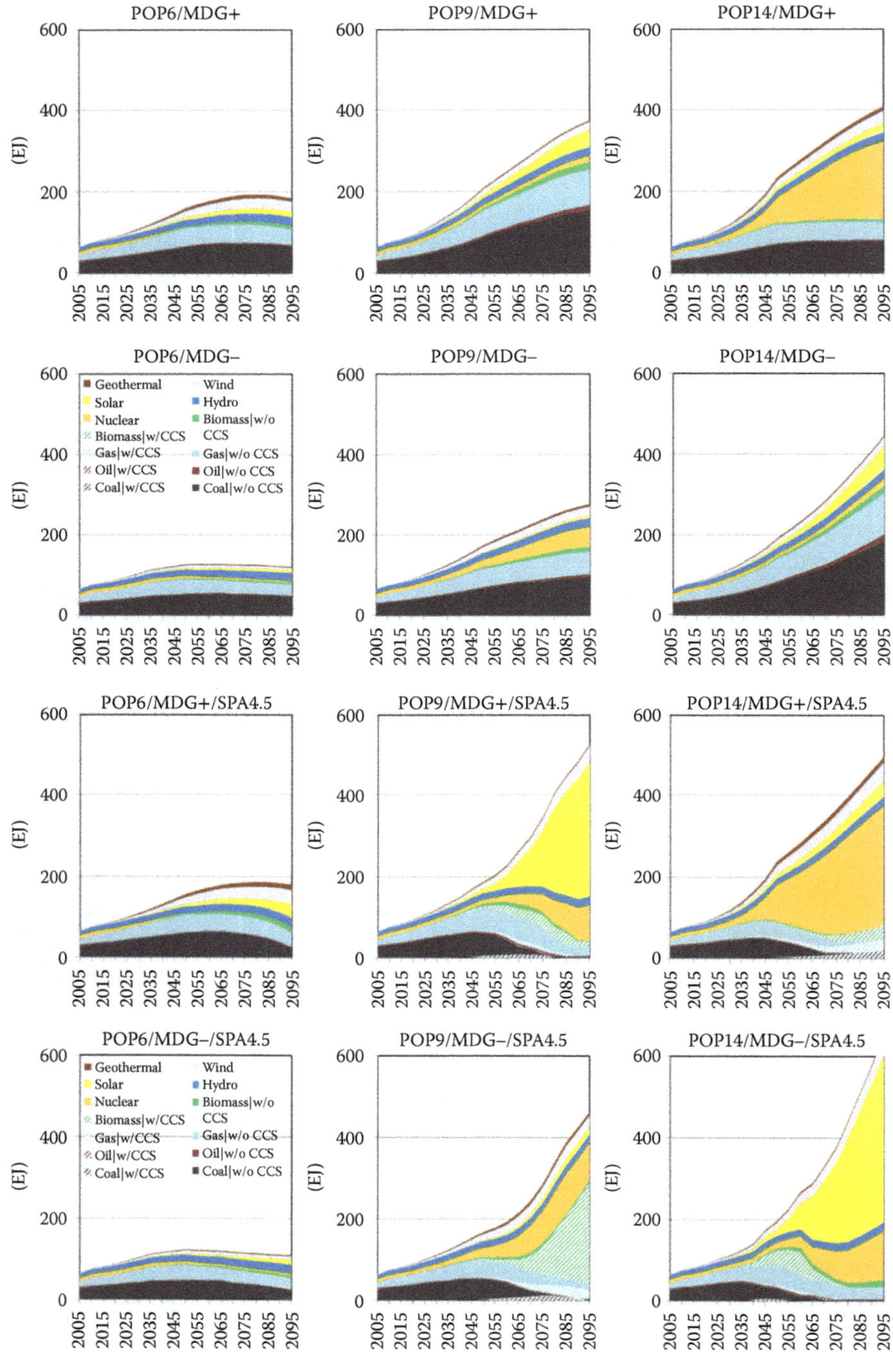

FIGURE 3.17
Global power generation by fuel in the reference and SPA scenarios.

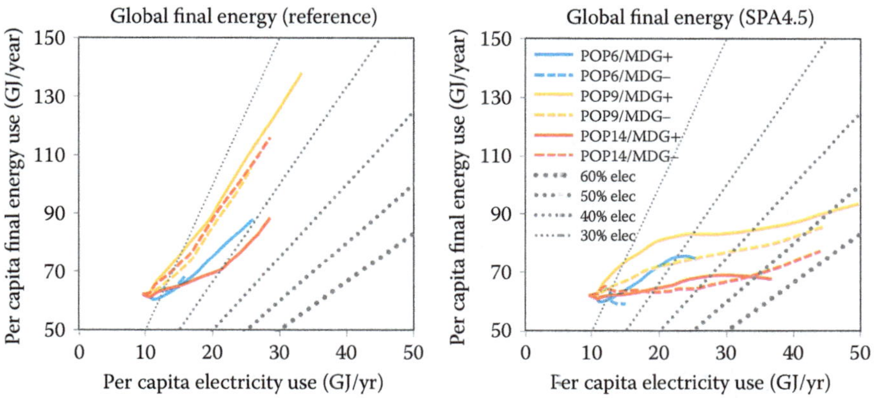

FIGURE 3.18
Per capita final energy use and per capita electricity use in the reference and SPA 4.5 scenarios.

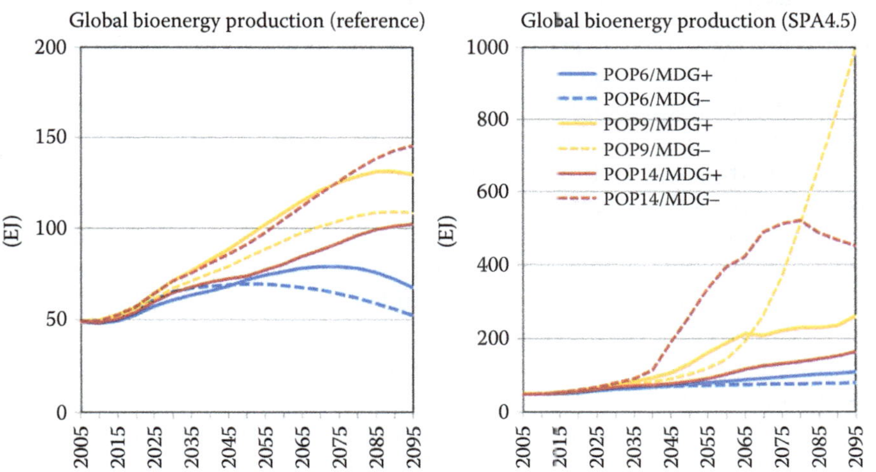

FIGURE 3.19
Global bioenergy production.

increased energy prices, lead to a decreased per capita final energy demand relative to the reference case, and the change is particularly pronounced in high population scenarios.

The production of bioenergy is sensitive to the alternative socioeconomic pathways (Figure 3.19). The sensitivity comes either from the demand side, such as the variations in global income and relative energy prices, or from the supply side, such as the variations in global population, attendant crop land requirement, and agricultural technologies. The overall trend and ordering across the scenarios generally follows those of primary energy demand (Figure 3.14), although POP14/MDG+ (Social Conservatism) demands less bioenergy than POP9/MDG– (Muddling Through) because of increased competition with crop land to feed the world's population.

In the carbon-constrained world, global bioenergy demand shows much more remarkable variety between the scenarios: the century-end bioenergy production in POP9/MDG– (Muddling Through) is nearly 20 times greater than the current level (Figure 3.19). This notable variation is mainly attributed to the difference in climate policy setting (i.e., FFICT vs. UCT) and its interactions with varying degrees of the mitigation effort and global crop demand. For example, although both POP14/MDG– (Crowded Chaos) and POP9/MDG– (Muddling Through) are better positioned to produce more bioenergy than other scenarios due to no-valuation on terrestrial carbon and high mitigation requirements, the growth of bioenergy production in POP14/MDG– levels out later in the century as crop demands remain strong and bio-CCS in the power sector becomes less attractive (Figure 3.17).

3.5.2 Passenger Transportation System

We disaggregate the delivery of passenger transport services into four major aggregates—light duty vehicles (LDV), bus, rail, and air. The composition of transport services depends on consumer preferences and the relative cost of delivering the service, dollars per passenger kilometer (PKM). Note that we include the value of time as part of the cost of travel, so that as incomes increase, modes that deliver the service more quickly become more attractive. The direct cost of delivering the service depends on fuel prices and vehicle costs (capital plus operation and maintenance) (Kyle and Kim 2011). Vehicle technologies that use refined liquids (petroleum oil and biofuel), gas, electricity, and hydrogen, are defined in GCAM. Note, of course, that each of the four major fuel types can be produced using more than one primary energy form.

The per capita demand for travel in the reference scenarios varies widely across the alternative socioeconomic pathways because they represent varieties in income growth, preferences for travel, and a set of fuel prices (Figure 3.20). For example, POP9/MDG+ (Consumerism) stands out owing to its relatively intensive demand for travel, high per capita income, and cheap access to transportation fuel (i.e., oil products). Regardless of the scenarios, LDVs continue to account for the major share of travel. The expansion of less energy-intensive modes, such as bus and rail, is restrained because they are time intensive, and hence grow more expensive as per capita income rise. Because it is less time-intensive, air travel use expands as per capita incomes rise over time. The expansion of alternative fuels, such as electricity, gas, biofuel, and hydrogen, is limited in the passenger transportation sector, regardless of socioeconomic assumptions. Energy supplied to LDVs, for example, is dominated by petroleum oil, although the role of the alternative fuels will continue to increase after 2020 (Figure 3.21).

The mitigation policy has modest price effects on the per capita demand for travel largely because energy accounts for a relatively small share of the

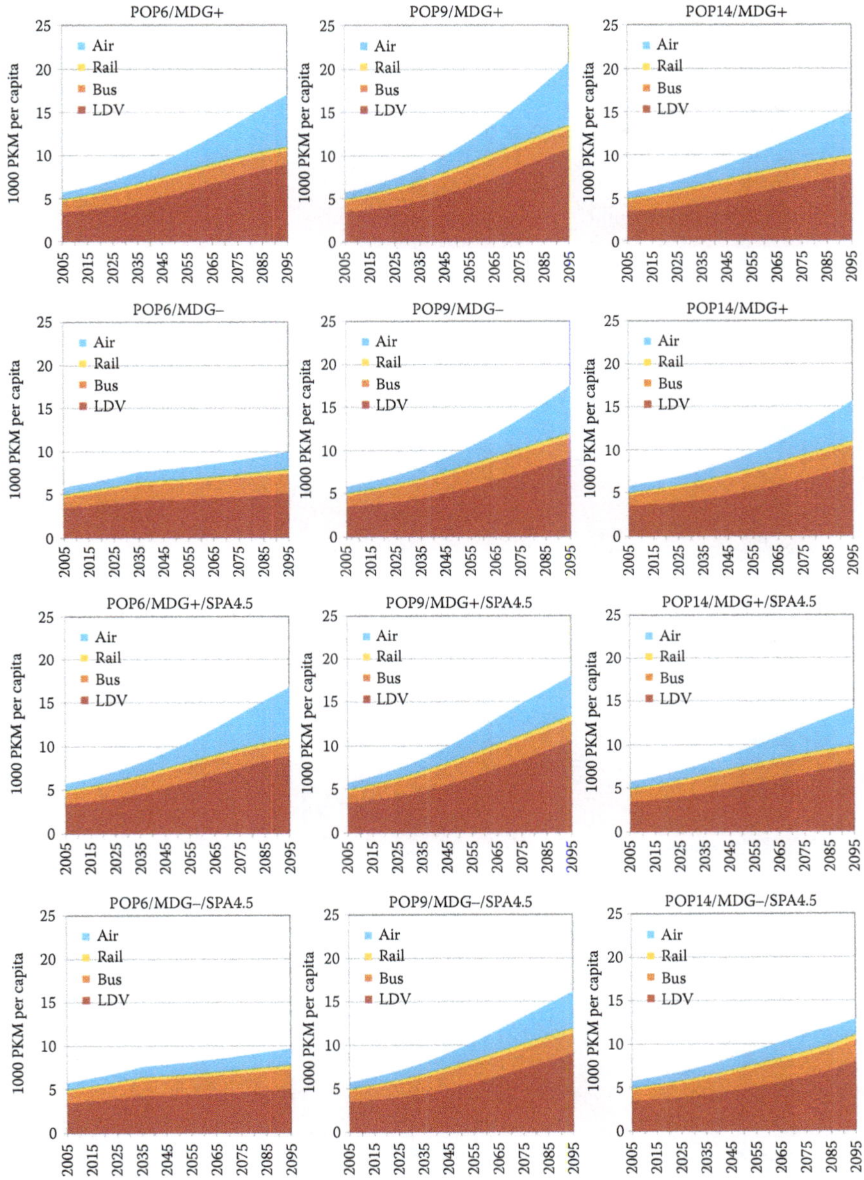

FIGURE 3.20
Per capita demand for travel by mode in the reference and SPA 4.5 scenarios.

cost of delivering transport services. The greatest demand reductions are found in POP14/MDG– (Crowded Chaos), POP9/MDG+ (Consumerism), and POP9/MDG– (Muddling Through) (Figure 3.20). This is because their relatively high carbon taxes imposed on oil products raise overall travel costs, discouraging the use of air travel in particular, which not only relies

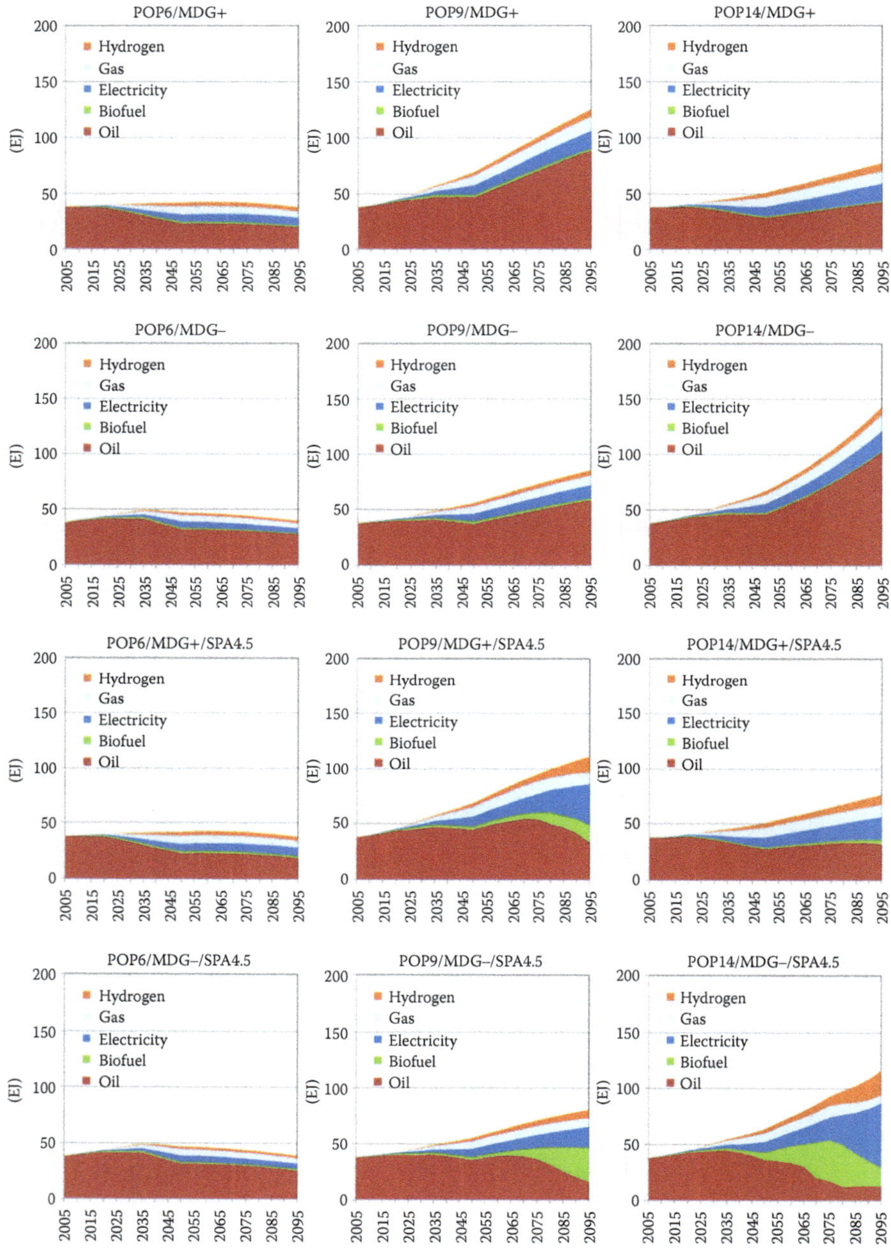

FIGURE 3.21
Final energy demands of light-duty vehicle transportation in the reference and SPA 4.5 scenarios.

heavily on petroleum oil, but also exhibit the highest travel energy intensity among the travel modes.

Mitigation does produce a shift in the fuel mix for transport. As carbon taxes rise, so do oil prices and the increased oil price promotes a shift toward alternative fuels, particularly toward electricity, biofuel, and hydrogen (Figure 3.21). Bioenergy becomes a prominent transportation fuel when the carbon price is high enough, such as in POP9/MDG– (Muddling Through) and POP14/MDG– (Crowded Chaos), and, to a lesser extent, POP9/MDG+ (Consumerism). Note that while in POP9/MDG–, bioenergy is predominantly demanded by the power sector, in POP14/MDG–, the transportation sector becomes the largest consumer as the fuel rapidly substitutes high-tax petroleum oil, helped by the gradually declining demand for bio-CCS in the power sector (Figure 3.17).

3.5.3 Agriculture and Land Use

Land allocation and the associated ecosystem stress vary across the six socioeconomic pathways (Figure 3.22). Low population worlds (POP6 scenarios) place the least stress on land use and ecosystems. The extent of crop land in these scenarios is relatively stable until mid-century, when the pace of crop yield improvements finally outpaces demands for new land driven by the population and income growth. In contrast, the high global populations (POP14 scenarios) generate intense pressure to expand crop lands, resulting in deforestation. In high population scenarios, demands for crop lands leave little room for the bioenergy production observed in the lower population scenarios, despite the larger scale of global energy system. As a result, higher population scenarios tend to have higher crop prices (Figure 3.23). Note that POP6/MDG+ (Sustainability and Equity) exhibits lower crop price than POP6/MDG– (Collapse) because the effect of faster yield improvement more than offsets the effect of higher income.

Mitigation policies create pronounced differences in land use patterns across the scenarios (Figure 3.22). In the scenarios where carbon sequestration in forests is rewarded—that is, in all MDG+ scenarios—afforestation occurs over the century, limiting the expansion of crop and bioenergy lands. The extent of unmanaged ecosystems actually increases relative to the reference scenarios in mitigation scenarios. For example, the POP9/MDG+ scenario exhibits rapid afforestation as Africa accesses to the global carbon market in 2070.

The expansion in unmanaged ecosystem means that less land is available for human activities. As the carbon price rises, that price is incorporated into the value and rental rate on land. The increased cost of land, in turn, is reflected in the price of all products originating from the land, such as crops. As demands for land to serve large populations' dietary needs go up, crop prices rise dramatically, stressing the most vulnerable part of society (Figure 3.23). However, if terrestrial carbon is not valued in the mitigation

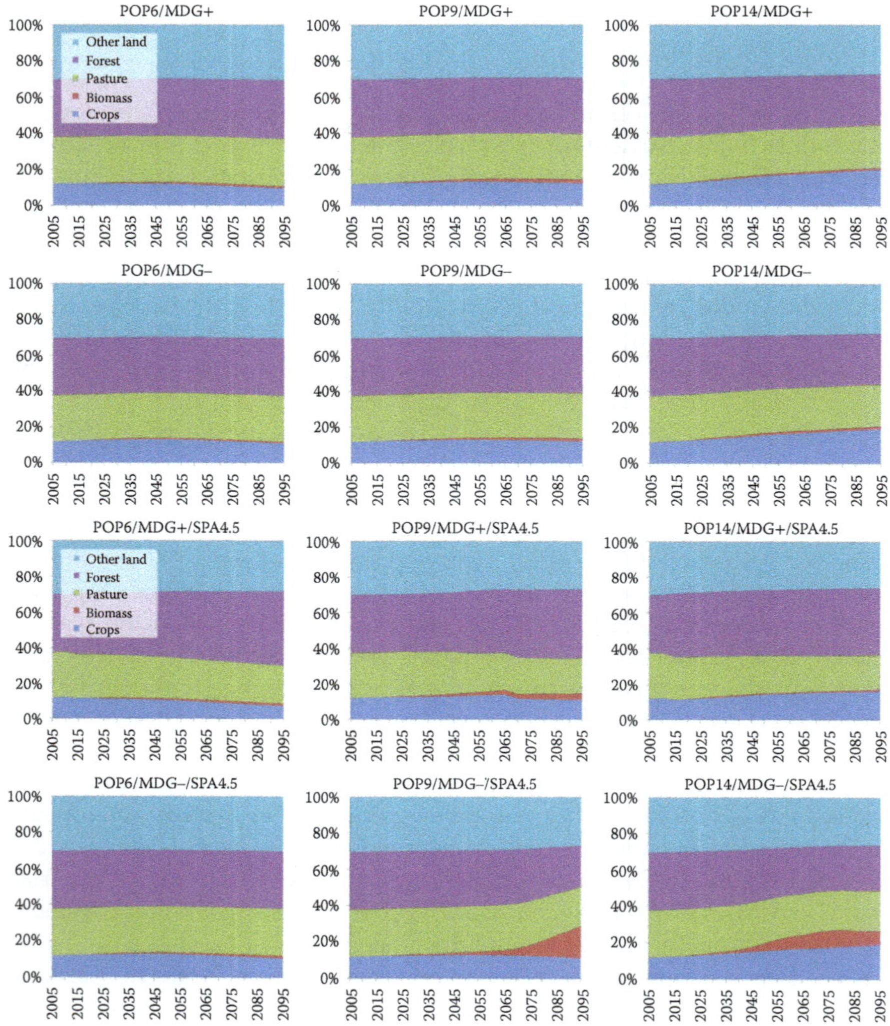

FIGURE 3.22
Global land allocation by use.

regime—that is, all MDG– scenarios—bioenergy and crop lands expand at the expense of rapid deforestation (Figure 3.22). Because the value of carbon is not incorporated into the value of land, and hence its rental rate, cropland supply becomes more elastic, and thus, crop prices remain lower than the case where carbon stock is rewarded (Figure 3.23).

Land use change emissions follow land use patterns (Figure 3.24). In the reference scenarios, land use change emissions track the size of the global population: a higher global population requires more crop lands, resulting in greater land use change emissions. Variation in the per capita income

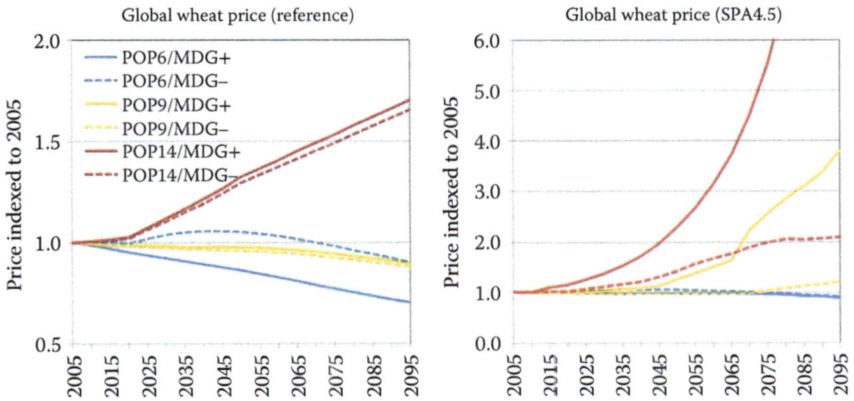

FIGURE 3.23
Global wheat price.

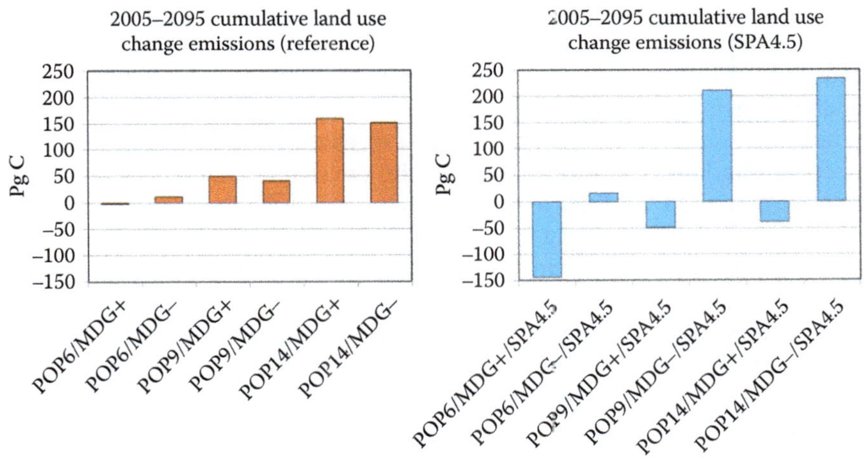

FIGURE 3.24
Land use change emissions.

and agricultural yields provide another source of variation in emissions. Under the mitigation policy, however, land use change emissions can either be accelerated or reversed—increased emissions are associated with deforestation and reduced emissions are associated with afforestation. If the carbon stock is valued, land use change emissions turn negative because of afforestation, whereas, if not valued, the emissions increase (become even greater than their reference scenario counterparts) as an increased demand for dedicated bioenergy lands accelerates deforestation. Note that, given any global mitigation target, greater land use change emissions place a greater mitigation burden on the global energy system, as represented by a higher price of carbon.

3.6 Challenges to Adaptation and Mitigation

So far, we have described scenarios that utilize assumptions about socioeconomic developments. We have crafted these scenarios with an eye toward exploring challenges to mitigation and to adaptation to climate change as well as uncertainty in potential future radiative forcing. We have already shown that the six reference scenarios include a scenario consistent with the highest radiative forcing scenario in the RCP set. Our climate policy scenarios contain scenarios that explore the low end of radiative forcing scenarios in the RCP set. To this point, however, we have not attempted to assess the degree to which we have been successful in exploring the uncertainties in challenges to mitigation and challenges to adaptation with our six scenarios. In this section, we test our scenarios in that regard.

In the literature, challenges to mitigation and to adaptation to climate change are not defined in measurable terms (Kriegler et al. 2011; CWT 2011). No metrics are either provided or recommended that would allow a scenario developer to place a scenario in the "challenges space." But, metrics are needed. In this section, we explore metrics and place our scenarios within the challenges space. We will use the initial carbon price in the first region undertaking emissions mitigation for a prescribed RCP replication as a proxy for challenges to mitigation and the per capita GDP and the percentage of unmanaged land use as proxies for challenges to adaptation.*

Kriegler et al. (2011) and CWT (2011) define challenges to adaptation in terms of a fixed climate change and challenges to mitigation in terms of a fixed degree of mitigation. In general, the challenges to adaptation become greater if exposure and sensitivity to climate change are higher and if adaptive capacity is lower. By contrast, the challenges to emissions mitigation are higher if the mitigation required to achieve any given radiative forcing target is greater and if the capacity of mitigation is lower (CWT 2011).

We employed the initial carbon price in the first region, initiating mitigation in an SPA4.5 as a proxy for the challenges to mitigation: the higher the level of the proxy, the greater the mitigative challenges. This is because the cost of mitigation, as expressed by the initial carbon price, captures the

* In future research, we will explore more sophisticated techniques such as the use a multivariative approach for ranking vulnerability that accounts for other factors such as environmental stress, access to safe water/sanitation, and income equity/disparity, as developed in the vulnerability-resilience indicator model (VRIM) (Moss et al. 2001; Brenkert and Malone 2005). VRIM integrates social and ecological aspects in a transparent way in order to provide a holistic picture of a region's ability to address climate issues. The VRIM, as it has been used at the global scale, identifies 17 factors that together assess the vulnerability of a society. Managed and unmanaged land, economic activities that are natural resource-intensive, and socioeconomic characteristics are represented. The VRIM methodology provides a hierarchical approach for indexing sensitivity and adaptive capacity of different sectors for a specific geography or jurisdiction and developing a single metric for comparing relative vulnerability under different socioeconomic and climate scenarios.

combined influence of various factors that make mitigation more challenging: the higher the population or economic scale, the slower the technology improvement, the less availability of mitigation options, and the less ideal the carbon policy setting (e.g., delays in global accession or no-valuation of terrestrial carbon stock). In particular, the proxy well captures the effect of alternative carbon policy settings, which makes the cost of mitigation variant even under the same technology assumptions.

For challenges to adaptation, we employed two different metrics, based on GDP/capita and the percentage of managed land in a region. Wealth generally provides access to markets, technology, and other resources that can be used to adapt to climate variability and change. By indexing each of the scenarios based on the inverse of GDP per capita, we obtain one measure of the challenges to adaptation, the absence of wealth to provide access to resources for adaptation.

Ecosystems and the functions they provide to society are sensitive to different levels of stress from human activities, as well as from variation and change in climate. The percentage of land under management is a proxy for the degree of intrusion of human activity into the natural landscape and the potential fragmentation of land, which would increase the sensitivity of ecosystems and the challenges to adaptation.

Note that these two metrics for adaptive challenges are not completely independent of the challenges to mitigation. A higher per capita GDP implies not only greater resources available to mitigate and adapt to climate change, but also a higher demand for energy services, which in turn place upward pressure on emissions and ecosystems. A greater portion of managed land may also increase mitigative challenges because as land use change increases, greater emissions reductions are required to achieve any radiative forcing target.

The six alternative scenario families present a wide variety in the challenges in mitigation and adaptation (Figure 3.25). Of interest, and in support

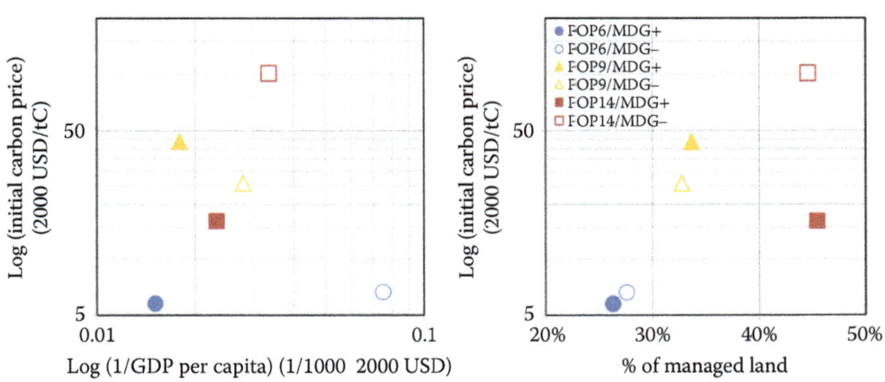

FIGURE 3.25
GCAM mapping of the six scenarios to the challenges space.

of the need for a systematic approach for assessing which scenarios and locations confront the greatest challenges to adaptation, the scenarios map out into different areas of this plot, including its off-diagonal portions, based on which metric is used for the evaluation. The per capita GDP seems to present a better split between the scenarios than the percentage of managed land: POP6/MDG+ (Sustainability and Equity) is characterized by low challenges to adaptation and mitigation, whereas POP14/MDG− (Crowded Chaos) exhibits high challenges to adaptation and mitigation; both POP9/MDG− (Muddling Through) and POP14/MDG+ (Social Conservatism) are in the middle; and two off-diagonal portions are populated by POP9/MDG+ (Consumerism; low adaptive, but high mitigative challenges) and POP6/MDG− (Collapse; high adaptive, but low mitigative challenges).

However, we note the limitations of using any single metric, including GDP per capita, as this captures only one dimension of the potential challenges to adaptation. Because any single metric focuses on only one dimension of a much more complex, multi-dimensional set of relationships that drive both the capacity for adaptation and mitigation, further research will be required to develop a more satisfying approach for the quantitative evaluation of socioeconomic scenarios in terms of the challenges they present to adaptation and mitigation.

In the case of challenges to adaptation, there is a relationship between the concept and measurement of vulnerability, the capacity to be harmed, and resilience, the ability to recover from a shock or extreme event (Turner 2003; Gallopin 2006). While vulnerability is typically decomposed into exposure, sensitivity, and adaptive capacity, the way this is operationalized in research can vary. For the purposes of analyzing and comparing the "challenge to adaptation" posed by different socioeconomic scenarios, we focus on sensitivity and adaptive capacity, leaving exposure to be determined by a force exogenous to the socioeconomic scenario itself—the level of climate change (which, in this case, will be defined by the radiative forcing level of the RCP). Sensitivity measures the degree of response of a system or "exposure unit" to a climate perturbation—for example, for a given level of climate change, the extent of changes in crop productivity, the frequency or extent of flooding, disturbance of an ecosystem, or location of disease vectors. Adaptive capacity measures the resources that are available to cope in the short term, with events such as natural disasters or to change infrastructure or practices in the long term—for example, the construction and maintenance of sea walls or water management infrastructure, of shifting crops. Including both components seems conceptually important, but complicates measurement and evaluation.

Much less explored is a similar approach to decomposing the elements of "challenges to mitigation." One could draw an analogy with vulnerability and seek to understand components similar to exposure, sensitivity, and (in this case) mitigative capacity. In the sense that exposure is a measure of the climate shock with respect to adaptation, it could be seen as a measure of the shock or perturbation to an economic and social system that will be posed by a particular emissions target relative to reference emissions. The bigger the gap between the two and the shorter the period of adjustment, the larger the exposure. Sensitivity measures responsiveness, so in the case of mitigation, we could approach this as the degree of responsiveness of the economic and social system to a given policy instrument, which would be influenced by the degree to which the current energy system and economy incorporate low emissions technologies and practices. If the baseline conditions already include the substantial implementation of mitigation measures, the sensitivity to future policies could be expected to be low, thus increasing the challenge to mitigation. Capacity measures the resources available for implementing emissions mitigation or sequestration measures—for example, the availability of capital, labor productivity and versatility, effectiveness of governance institutions, or ability to effect behavioral change. To the extent that these characteristics are lacking, one would expect the challenges to mitigation would be greater.

In both cases, one can identify measurable proxy variables for these different components. Another important area of improvement is to assess the differential vulnerability and challenges to adaptation and mitigation of different regions and even locations under different scenarios. These are potential approaches to be explored.

3.7 Summing Up

Scenarios in the domain of climate change research have moved from simplistic representations of fossil fuel and industrial CO_2 emissions to scenarios that seek to explore uncertainty in potential future radiative forcing as well as challenges to mitigation and challenges to adaptation. They have moved from being an input to carbon cycle models, to being a potentially unifying frame for the full climate assessment process (van Vuuren et al. 2011; Kriegler et al. 2011; CWT 2011). The six alternative socioeconomic scenarios and their associated RCP sets reflect the need for scenario sets that provide variety in the degree of mitigative and adaptive challenges. These scenarios, employing carefully crafted qualitative and quantitative assumptions and narrative storylines, highlight the following relevant characteristics: (1) variation in system scale (global population and global GDP); (2) variation in energy and agricultural technology (technical progress and availability); (3) variation in greenhouse gas emissions and century end radiative forcing (one reaching a level higher than RCP8.5); (4) variation in the cost of emissions mitigation; (5) variation in human vulnerability to climate change and mitigation (demand for and prices of energy and agricultural commodities); and (6) variation in environmental stress (land cover and land use change emissions).

3.8 Questions for Discussion

- What kind of socioeconomic conditions would make low population regions less likely to achieve any given radiative forcing target? Similarly, what would make high population regions easier to conduct major emissions mitigation?
- Suggest candidate proxies for global socioeconomic challenges to mitigation and adaptation. Would these also apply at the national or regional levels? If not, suggest alternative proxies.
- How would the variation in emissions across the six socioeconomic pathways change if they all had a single set of assumptions for technologies and climate policy implementation?
- How would climate change feedback, if fully taken into account, influence the above findings?
- What would the variety in water demand under supply constraints mean for the prices of energy and agricultural commodities and the challenges to mitigation and adaptation in the reference and climate policy scenarios?

References

Brenkert, A. and Malone, E. (2005). Modeling Vulnerability and Resilience to Climate Change: A Case Study of India and Indian States. *Climatic Change*, 72(1–2):57–102.

Bruinsma, J. (2009). The Resource Outlook to 2050: By How Much Do Land, Water, and Crop Yields Need to Increase by 2050? Expert Meeting on How to Feed the World in 2050, Food and Agriculture Organization of the United Nations.

Clarke, L., Edmonds, J., Jacoby, H., Pitcher, H., Reilly, J., Richels, R. (2007). Scenarios of Greenhouse Gas Emissions and Atmospheric Concentrations. Sub-report 2.1A of Synthesis and Assessment Product 2.1 by the U.S. Climate Change Science Program and the Subcommittee on Global Change Research. Department of Energy, Office of Biological & Environmental Research, Washington, DC, pp. 154.

CWT Core Writing Team on SSP. Arnell, N., Kram, T., Carter, T., Ebi, K., Edmonds, J., Hallegatte, S., Kriegler, E., Mathur, R., O'Neill, B., Riahi, K., Winkler, H., van Vuuren D. P., Zwickel, T. (2011). A Framework for a New Generation of Socioeconomic Scenarios for Climate Change Impact, Adaptation, Vulnerability, and Mitigation Research. Draft manuscript available at: http://www.isp.ucar.edu/sites/default/files/Scenario_FrameworkPaper_15aug11_0.pdf.

Dahowski, R. T., Davidson, C. L., Dooley, J. J. (2011). Comparing Large Scale CCS Deployment Potential in the USA and China: A Detailed Analysis Based on Country-Specific CO_2 Transport & Storage Cost Curves. *Energy Procedia*, 4:2732–2739, ISSN 1876-6102, DOI:10.1016/j.egypro.2011.02.175.

Decarre, S., Berthiaud, J., Butin, N., Guillaume-Combecave, J. L. (2010). CO_2 Maritime Transportation. *International Journal of Greenhouse Gas Control*, 4(5):857–864, ISSN 1750-5836, DOI:10.1016/j.ijggc.2010.05.005.

Denholm, P., Margolis, R. (2008). *Supply Curves for Rooftop Solar PV-Generated Electricity for the United States*. NREL/TP-6A0-44073, National Renewable Energy Laboratory, Golden, CO.

Dooley, J. J., Kim, S. H., Edmonds, J., Friedman, S. J., Wise, M. (2005). A First Order Global Geologic CO_2 Storage Potential Supply Curve and Its Application in a Global Integrated Assessment Model. *Greenhouse Gas Control Technologies*, 1:573–581.

Edmonds, J., Reilly, J. (1983). Global Energy and CO_2 to 2050. *The Energy Journal*, 4(3):21–47.

Edmonds, J., Reilly, J. (1985). *Global Energy: Assessing the Future*. Oxford University Press, New York.

Edmonds, J., Grubb, M., ten Brink, P., Morrison, M. (1994). *Carbon Dioxide Emissions, Fossil Fuels*. Encyclopedia of Energy Technology and the Environment, John Wiley & Sons, Inc, New York, pp. 504–530.

Eom, J., Calvin, K., Clarke, L., Edmonds, J., Kim. S. H., Kopp, R., Kyle, P., Luckow, P., Moss, R., Patel, P., Wise, M. (2012). Exploring the Future Role of Asia Utilizing a Scenario Matrix Architecture and Shared Socio-economic Pathways. *Energy Economics* (In Print).

Fisher, B., Nakicenovic, N., Alfsen, K., Corfee-Morlot, J., de la Chesnaye, F., Hourcade, J. -C., Jiang, K., Kainuma, M., La Rovere, E., Matysek, A. et al. (2007). Issues Related to Mitigation in the Long-Term Context. In: Metz, B., Davidson, O., Bosch, P., Dave, R., Meyer, L. (eds.) *Climate change 2007. Mitigation of climate change. Contribution of Working Group III to the Fourth Assessment Report of the Intergovernmental Panel on Climate Change*. Cambridge University Press, New York, pp. 169–250.

Gallopin, G. C. (2006). Linkages Between Vulnerability, Resilience, and Adaptive Capacity. *Global Environmental Change*, 16(3):293–303.

Hannam, P., Kyle, P., Smith, S. J. (2009). *Global Deployment of Geothermal Energy Using a New Characterization in GCAM 1.0 PNNL-19231*, Pacific Northwest National Laboratory, Richland, WA.

Häfele, W. (1981). *Energy in a Finite World*. Ballinger Publishing Company, Cambridge, MA.

IAEA (2003). *Uranium 2003: Resource, Production and Demand*, IAEA, Vienna, Austria and the OECD/NEA, Paris, France.

IEA GHG (2000). *The Potential of Wind Energy to Reduce CO_2 Emissions*. Report PH3/24, International Energy Agency Greenhouse Gas R&D Program, Cheltenham, UK.

IEA GHG (2011). *CO_2 Storage in Depleted Gas Fields*. 2009/01, IEA Greenhouse Gas R&D Programme, Cheltenham, UK.

IPCC (1996). Climate Change 1995: The Science of Climate Change. In: Houghton, J. T., Meira Filho, L. G., Callendar, B. A., Kattenberg, A., and Maskell, K. (eds.). *Climate Change 1995: The Science of Climate Change. The Contribution of Working Group I to the Second Assessment Report of the Intergovernmental Panel on Climate Change*. Cambridge University Press, Cambridge, UK.

IPCC (2001). *Climate Change 2001: Impacts, Adaptation, and Vulnerability. Contribution of Working Group II to the Third Assessment Report of the Intergovernmental Panel on Climate Change*. Cambridge University Press, Cambridge, England.

IPCC (2001). Climate Change 2001: The Scientific Basis. In: Houghton, J. T., Ding, Y., Griggs, D. J., Noguer, M., van der Linden, P. J., Xiaosu D (eds.) *Climate Change 2001: The Scientific Basis. The Contribution of Working Group I to the Third Assessment Report of the Intergovernmental Panel on Climate Change.* Cambridge University Press, Cambridge, UK, pp. 944.

IPCC (2007). Climate Change 2007: The Physical Science Basis. In: Solomon, S., Qin, D., Manning, M., Marquis, M., Averyt, K., Tignor, M. M. B., Miller, H. L., Chen, Z. (eds.). *Climate Change 2007: The Physical Science Basis. Working Group I Contribution to the Fourth Assessment Report of the Intergovernmental Panel on Climate Change.* Cambridge University Press, Cambridge, UK, pp. 996.

JASON (1979). *A Long-Term Impact of Atmospheric Carbon Dioxide on Climate,* (Technical Report JSR-78-07). SRI International, Arlington, Virginia.

Kim, S. H., Edmonds, J. A., Smith, S. J., Wise, M., Lurz, J. (2006). The Object-Oriented Energy Climate Technology Systems (ObjECTS) Framework and Hybrid Modeling of Transportation. *The Energy Journal,* (special issue no. 2), pp. 51–80.

Kriegler, E., O'Neill, B., Hallegatte, S., Kram, T., Lempert, R., Moss, R., Wilbanks, T. (2010). Socioeconomic Scenario Development for Climate Change Analysis. http://www.ipcc-wg3.de/meetings/expert-meetings-and-workshops/files/Kriegler-et-al-2010-Scenarios-for-Climate-Change-Analysis-Working-Paper-2010-10-18.pdf.

Kyle, P., Clarke, L., Pugh, G., Wise, M., Calvin, K., Edmonds, J., Kim, S. H. (2009). The Value of Advanced Technology in Meeting 2050 Greenhouse Gas Emissions Targets in the United States. *Energy Economics,* 31(Supplement 2):S254–S267, ISSN 0140-9883.

Kyle, P., Clarke, L., Rong, F., Smith, S. J. (2010). Climate Policy and the Long-Term Evolution of the U.S. Buildings Sector. *The Energy Journal,* 31(2):145–172.

Kyle, P., Patrick, L., Calvin, K., Emanuel, W., Nathan, M., Zhou, Y. (2011). *GCAM 3.0 Agriculture and Land Use: Data Sources and Methods.* Pacific Northwest National Laboratory. PNNL-21025. http://wiki.umd.edu/gcam/images/2/25/GCAM_AgLU_Data_Documentation.pdf.

Kyle, P., Smith, S. J., Wise, M. A., Lurz, J. P., Barrie, D. (2007). *Long-Term Modeling of Wind Energy in the United States.* PNNL-16316, Pacific Northwest National Laboratory, Richland, WA.

Kyle, P., Kim, S. H. (2011). Long-Term Implications of Alternative Light-Duty Vehicle Technologies for Global Greenhouse Gas Emissions and Primary Energy Demands. *Energy Policy,* 39:3012–3024.

Leggett, J., Pepper, W., Swart, R. J. (1992). Emissions Scenarios for the IPCC: An Update. In: Houghton, J. T., Callander, B. A., Varney, S. K. (eds) *Climate change 1992. The Supplementary Report to the IPCC Scientific Assessment.* Cambridge University Press, Cambridge, pp. 71–95.

Maier-Reimer, E., Hasselmann, K. (1987). Transport and Storage of CO_2 in the Ocean: An Inorganic Ocean-Circulation Carbon Cycle Model. *Climate Dynamics,* 2:63–90.

Monfreda, C., Ramankutty, N., Hertel, T. W. (2009). Global Agricultural Land Use Data for Climate Change Analysis. In: Hertel, T. W., Rose, S., Tol, R. S. J. (eds). *Economic Analysis of Land Use in Global Climate Change Policy.* Routledge Explorations in Environmental Economics, Taylor & Francis Group, London and New York.

Moss, R. H., Edmonds, J. A., Hibbard, K. A., Manning, M. R., Rose, S. K., van Vuuren, D. P., Carter, T. R., Emori, S., Kainuma, M., Kram, T., Meehl, G. A., Mitchell, J. F., Nakicenovic, N., Riahi, K., Smith, S. J., Stouffer, R. J., Thomson, A. M., Weyant,

J. P., Wilbanks, T. J. (2011). The Next Generation of Scenarios for Climate Change Research and Assessment. *Nature*, 463(7282):747–756.

Moss, R. H., Brenker, A. L., Malone, E. L. (2001). *Vulnerability to Climate Change: A Quantitative Approach*. Pacific Northwest National Laboratory, Richland, WA, PNNL-SA-33642.

Nakicenovic, N., Alcamo, J., Davis, G., de Vries, B., Fenhann, J., Gaffin, S., Gregory, K., Grübler, A., Jung, T. Y., Kram, T., La Rovere, E. L., Michaelis, L., Mori, S., Morita, T., Pepper, W., Pitcher, H., Price, L., Riahi, K., Roehrl, A., Rogner, H., Sankovski, A., Schlesinger, M., Shukla, P., Smith, S., Swart, R., van Rooijen, S., Victor, N., Dadi, Z. (2000). *IPCC Special Report on Emissions Scenarios*. Cambridge University Press, Cambridge, UK.

Niehaus, F., Williams, J. (1979). Studies of Different Energy Strategies in Terms of their Effects on the Atmospheric CO_2 Concentration. *Journal of Geophysical Research*, 84:3123–3129.

Nordhaus, W. D. (1979). *The Efficient Use of Energy Resources*. Yale University Press, New Haven, C. T.

Nordhaus, W. D., Yohe, G. W. (1983). Future Carbon Dioxide Emissions from Fossil Fuels. In: *Changing Climate*. Report of the Carbon Dioxide Assessment Committee. National Academy Press, Washington, DC, pp. 87–153.

NREL (2008). *20% Wind Energy by 2030*. NREL, DOE/GO-102008-2567. http://www.nrel.gov/docs/fy08osti/41869.pdf.

NREL (2010). *Western Wind and Solar Integration Study*. The National Renewable Energy Laboratory, Prepared by GE Energy, NREL/SR-550-47781. http://www.nrel.gov/wind/systemsintegration/wwsis.html.

Peck, S., Wan, S. (1996). *Analytic Solutions of Simple Optimal Greenhouse Gas Emissions Models. NATO ASI Series*. Vol. 14, Springer Verlag, Berlin, Heidelberg, pp. 113–121 (Chapter 6).

Petty, S., Porro, G. (2007). Updated U.S. Geothermal Supply Characterization. Conference Paper NREL/CP-640-41073, National Renewable Energy Laboratory, Golden, CO.

Rogner, H. (1997). An Assessment of World Hydrocarbon Resources. *Annual Review of Energy and Environment*, 22, 217–262.

Rotty, R. M. (1977). *Present and Future Production of CO_2 from Fossil Fuels*. ORAU/IEA(O)-77-15. Institute for Energy Analysis, Oak Ridge, TN.

Rotty, R. M. (1978). The Atmospheric CO_2 Consequences of Heavy Dependence on Coal. In: Williams J (ed.). *Carbon Dioxide, Climate and Society*. Pergamon, Oxford, UK, pp. 263–273.

Schneider, E., Sailor, W. (2008). Long Term Uranium Supply Estimates. *Nuclear Technology*, 162 (3): 379–387.

Turner, B. L., Kasperson, R. E., Matson, P. A., McCarthy, J. J., Corell, R. W., Christensen, L., Eckley, N., Kasperson, J. X., Luers, A., Martello, M. L., Polsky, C., Pulsipher, A., Schiller, A. (2003). A Framework for Vulnerability Analysis in Sustainability Science. *Proceedings of the National Academy of Sciences*, 100(14):8074–8079.

U.N. Millennium Project (2005). *Investing in Development: A Practical Plan to Achieve the Millennium Development Goals. Overview.* United Nations Development Programme, Washington D.C.

U.N. Population Division (2011). *World Population Prospects—the 2008 Revision: Long Range Projections Supplement*. United Nations Department of Economic and Social Affairs, Population Division, New York.

U.S. Census Projection (2008). 2008 National Projections, United States Census Bureau, http://www.census.gov/population/projections/

van Vuuren, D. P., Edmonds, J. A., Kainuma, M., Riahi, K., Weyant, J. P. (2011). A Special Issue on the RCPs. *Climatic Change*, 109(1–2):1–4.

van Vuuren, D. P., Meinshausen, M., Plattner, G. K., Joos, F., Strassmann, K. M., Smith, S. J., Wigley, T. M. L., Raper, S. C. B., Riahi, K., De La Chesnaye, F. et al. (2008). Temperature Increase of 21st Century Mitigation Scenarios. *Proceedings of the National Academy of Sciences USA*, 105:15258–15262.

van Vuuren, D. P., Riahi, K. (2011). The Relationship Between Short-Term Emissions and Long-Term Concentration Targets—A Letter. *Climatic Change*, 104(3–4):793–801.

van Vuuren, D. P., Edmonds, J., Kainuma, M., Riahi, K., Thomson, A., Hibbard, K., Hurtt, G. C., Kram, T., Krey, V., Lamarque, J., Masui, T., Meinshausen, M., Nakicenovic, N., Smith, S. J., Rose, S. K. (2011). The Representative Concentration Pathways: An Overview. *Climatic Change*, 109:5–31.

van Vuuren, D. P., Riahi, K., Moss, R., Edmonds, J., Thomson, A., Nakicenovic, N., Kram, T., Berkhout, F., Swart, R., Janetos, A., Rose, S. K., Arnell, N. (2012). A Proposal for a New Scenario Framework to Support Research and Assessment in Different Climate Research Communities. *Global Environmental Change*, 22:21–35.

Watson, R. T., Zinyowera, M. C., Moss, R. H. Editors (1997). *The Regional Impacts of Climate Change: An Assessment of Vulnerability*. Cambridge University Press, Cambridge (United Kingdom) and New York, NY, pp. 517.

Weyant, J. P., Davidson, O., Dowlatabadi, H., Edmonds, J., Grubb, M., Parson, E. A., Richels, R., Rotmans, J., Shukla, P. R., Tol, R. S. J. (1996). Integrated Assessment of Climate Change: An Overview and Comparison of Approaches and Results, In: Bruce, J. P., Lee, H., Haites, E. F. (eds.). *Climate Change 1995: Economic and Social Dimensions of Climate Change. The Contribution of Working Group III to the Second Assessment Report of the Intergovernmental Panel on Climate Change*. Cambridge University Press, Cambridge, UK, pp. 367–396.

Wigley, T. M. L., Raper, S. C. B. (2001). Interpretation of High Projections for Global-Mean Warming. *Science*, 293:451–454.

Wigley, T. M. L., Smith, S. J., Prather, M. J. (2002). Radiative Forcing Due to Reactive Gas Emissions. *Journal of Climate*, 15(18):2690–2696.

Wise, M. and Calvin, K. (2011). GCAM 3.0 Agriculture and Land Use: Technical Description of Modeling Approach. Pacific Northwest National Laboratory, Richland, WA, PNNL-20971. http://wiki.umd.edu/gcam/images/2.25/GCAM_AgLU_Documentation.pdf.

World Bank (2011). World Development Indicators. The World Bank.

Zurek, M. B., Henrichs, T. (2007). Linking Scenarios Across Geographical Scales in International Environmental Assessments. *Technological Forecasting and Social Change*, 74(8):1282–1295.

4

Understanding Sea Level Rise and Coastal Hazards

Ashish J. Mehta, Robert G. Dean, Clay L. Montague, Earl J. Hayter, and Yogesh P. Khare

CONTENTS

4.1 Introduction

Some *human* actions have a significant potential to adversely impact the future global environment. Besides environmental degradation and the quality of life, there is the possibility that global warming may increase storm frequency and intensity and alter storm paths. Due to inundation and resulting ecological changes, it is anticipated that globally, by the end of this century, about one-third of the coastal wetlands will be lost and infrastructure losses will increase substantially (IPCC 2007). This scenario, coupled with rapid infrastructure development along many shores, has already set the stage for significant needed changes in technical approaches and management policies on shore habitation and resources.

In the 1980s, a committee of the Marine Board of the U.S. National Research Council issued a report on future sea level rise (SLR), summarizing anticipated impacts and available engineering responses (Marine Board 1987). This was followed by a related study for the Department of Energy by the University of Florida in which the focus was on impacts of inundation and responses in the southeastern United States (Daniels et al. 1992). A somewhat more globally oriented summary based on these two studies was later published as part of the proceedings of a meeting on global warming sponsored by the National Institute for Global Environmental Change (Mehta et al. 1997). The present effort attempts to follow up on the 1997 work. During the 25 years since the Marine Board report, SLR and anticipated impacts have become more noticeable, and there has been an order of magnitude or more increase in the global literature on this topic and its ready availability. The number of impact-related research areas has risen hugely, making it impossible to summarize this research in the present context, except by updating those topics that were mentioned in the Marine Board (1987) report. The revised and expanded bibliographic section should permit an easier search of the broad range of published literature and reported data.

As mentioned in the Marine Board (1987) study, considerable uncertainty remains in predicting human actions in attempts to counter some of the impacts of SLR. In any event, consensual technical positions should be reasonable and justified by data, requiring a thorough scientific understanding of the expected responses to global warming. This chapter attempts to highlight specific topics in coastal science and engineering to improve the knowledge and prediction of sea level changes, their coastal impacts, and responses. Following a brief introduction to the SLR problem and comments on storm forcing and coastal engineering, we have used case studies of coastal events, their impacts, responses, and effectiveness to highlight key research needs consistent with the Marine Board study.

Sea level at a given location is the measured height of the local tide relative to a datum. Typically the mean sea level is the annual average of hourly values. A commonly used datum in the United States is the North American vertical datum, which is the mean sea level in 1988 (usually NAVD88, although other datums are also in use such as 2004.65 in Louisiana). Coastal response to sea level change in site-specific instances depends on the relative change, that is, the difference between the eustatic global change and any local change in land elevation. Depending on the direction and rates of global change and local change, the relative sea level may increase or decrease even as the eustatic level is rising. For example, in the present times, the relative sea level change along the Atlantic Coast near Miami, Florida, has been positive (water level is rising relative to land), but at Juneau, Alaska, it has been negative (water level is falling) (Boon 2004).

There are six noteworthy long-term causes of relative mean SLR, not all of which operate in every locality:

1. The globally averaged or eustatic component of sea level change is associated with the formation or melting of Antarctica and Greenland ice sheets and land-based glaciers, sedimentation in the ocean basins, and movements of continental masses or the mantle that change the volume of the basins. Melting of the floating ice shelves and the Arctic ice cap do not contribute to SLR, except in a very minor way owing to a slight decrease in the salinity of ocean water. The two most important causes of global rise are the melting of ice sheets and steric expansion of near-surface ocean water due to ocean warming. Steric refers to the specific volume of the medium, which generally expands when heated or shrinks when cooled (Douglas et al. 2000). Thermal expansion is likely to have contributed to about 2.5 cm of the eustatic rise of about 10 cm during the second half of the twentieth century. Over the twenty-first century, projection by the Intergovernmental Panel on Climate Change (IPCC 2007) is about 17–28 cm (±50%) because of thermal expansion.
2. Isostacy refers to the subsidence or uplift of the land surface because of tectonic warping of the earth's crust. There are five types of tectonic phenomena: subsidence of marginal belts (e.g., the United States southeastern seaboard); cooling crustal belts following tectonic rift (e.g., parts of the Gulf of California); subsidence in regions of long-term sediment loading (e.g., the Mississippi delta); uplift in regions of active crustal subduction (e.g., Puget Sound); and subsidence due to loading by volcanic eruptions (e.g., the Hawaiian Islands). The rising bottom at Juneau and the falling bottom at Miami are manifestations of the elastic rebound of the North American tectonic plate following the last ice age, which caused the original warping due to the expansion of the Arctic ice cap during the ice age (Gupta 2011).
3. Seismic uplift or subsidence of land occurs due to gradual tectonic movements or suddenly due to earthquakes. The latter phenomena can be dramatic; in Alaska, a devastating earthquake in 1964 resulted in vertical displacements of land up to 11.5 m over an area of 250,000 km^2 (National Research Council 1968). The 2010 earthquake in south-central Chile lifted part of the coastal land by more than 2.4 m (Lorito et al. 2011).
4. Subsidence owing to consolidation of initially soft sediment deposits. This process operates along barrier islands as they roll over onto back-barrier lagoon deposits (Davis and Fitzgerald 2004). Examples are shoreline recession along the coast of Florida, exposing the

remnants of formerly healthy mangroves that were once on the lagoonal side of the barrier island.

5. Man-made subsidence due to structural loading, diversion of sediment from the river by levee construction, and also groundwater, oil, and gas extraction. Of the different types of subsidence processes, only this cause can be reversed or arrested by, for example, fluid recharge (USGS 2000).
6. Sea-level changes due to climatic fluctuations. These arise from oceanic factors, including El Niño–Southern Oscillation (ENSO) effects in the Pacific, and are related to changes in the size and locations of subtropical high pressure cells (Philip and Van Oldenborgh 2006). However, the effect of these cells is felt globally. In a study of the Russian sector of the Arctic Ocean for the 1954–1989 period, it was found that 30% of the rise was due to an ongoing decrease of sea level atmospheric pressure over the study region and an additional 10% due to increased cyclonic wind activity (Proshutinsky et al. 2004).

Among these causes of relative SLR, only the eustatic rise is a global effect. For any one area, the other processes come into play in various proportions. Although it appears that no national survey of the local extent of the processes has been made, it is clear that the variations are highly regional. To forecast sea level changes over the coming decades requires a predictive capability for the eustatic change and all important regional or local variations.

4.2 Global Sea Level

Sea level fluctuates over a wide range of frequencies from short-term but devastating storm tides up to 6 m in height to variations on a geological timescale exceeding 100 m. These variations have historically affected human patterns of coastal habitation (Khare and Mehta 2011) and will undoubtedly do so in times to come. Figure 4.1 shows the sea level over the past ~22,000 years since the end of the last ice age (glacial episode). The data were compiled by Rohde (2009a) using values from Fleming et al. (1998), Fleming (2000), and Milne et al. (2005). The trend line, based on a least-squares analysis, shows numerous fluctuations. Meltwater pulse 1A was a unique period of significantly rapid deglaciation. The lowest point of sea level during the last glaciation was believed to have been about 130 ± 10 m below present, approximately 22,000 ± 3,000 years ago. This time is more or less equivalent to the last glacial maximum. Before that the ice sheets were still increasing, and sea level was decreasing almost continuously over a period of approximately 100,000 years.

The anthropocene period of 6000–7000 years before present (BP) has been one of relative stability, yet coastal habitation has been sensitive to short-term

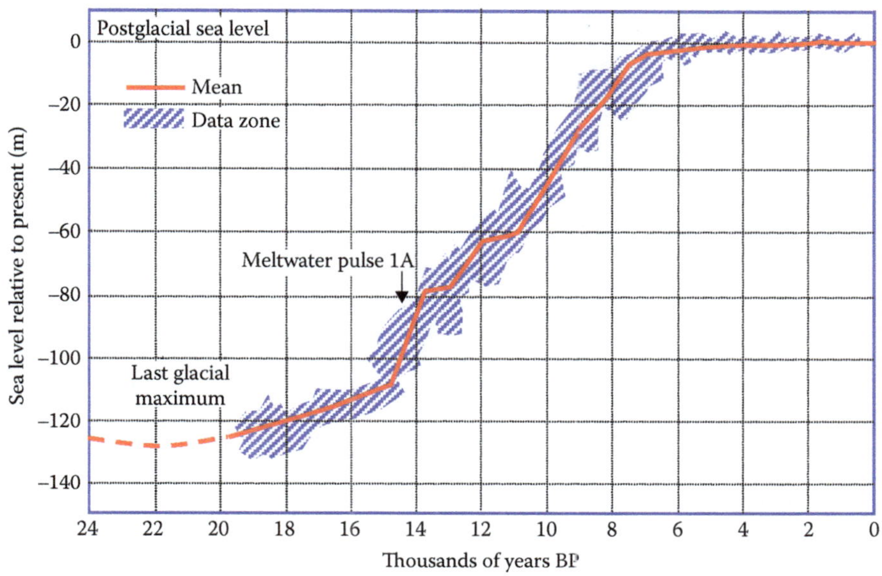

FIGURE 4.1
Sea level rise since the last glacial maximum. (Adapted from figure by Robert A. Rohde, 2009, http://en.wikipedia.org/wiki/File:Post-Glacial_Sea_Level.png.)

variations. A case in point is the Gujarat region of western India (Figure 4.2), where the locations of ancient shorelines landward of their present position have been used to infer that 3000 years BP, the relative sea level was higher than now owing to tectonic causes (Gaur and Vora 1999).

In general, paleo-trends in SLR have been derived mainly from carbon dating and indirectly from other evidence including tree rings and so on, which help establish the historic air temperature time series (D'Arrigo et al. 2003), because tide gage data became available only in recent centuries.

In the past couple of decades, satellite altimetry data have become available (Church et al. 2004). In spite of some issues of data interpretation from different satellites and different periods (e.g., the TOPEX/Poseidon satellite data investigated by Mitchum [1998]), the measurements are more accurate than tide gages and spatially extensive, but the time series are too short in relation to the 200–300 years (and longer) average response time between global temperature rise and global water level rise. This delayed response is significantly due to the large volume and depth of water on the earth (Grinsted et al. 2009) and complicates the interpretation of the causes of present-day sea level behavior. Tide gage data are longer than satellite altimetry data, but only a small number is long-term (>80 years) and practically continuous (missing data <5%). In the United States, only 18 gages qualify (Houston and Dean 2011) (see also Rahmstorf and Vermeer [2011] for a discussion of the results reported in the study by Houston and Dean).

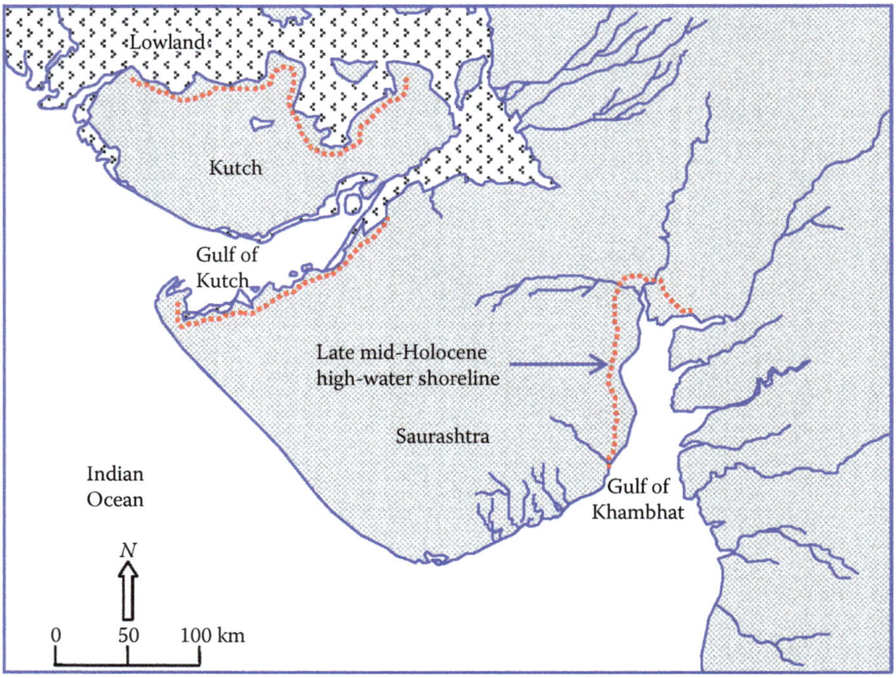

FIGURE 4.2
Late mid-Holocene (high-water) shorelines (dotted lines) in Gujarat, western India. (Adapted from Gaur, A. S., Vora, K. H., *Current Science*, 77(1), 180–185, 1999.)

In addition to the unknown trends in the future climate behavior, gaps in the data and the fact that identification of the mean trend is not independent of the method of estimation increase the ambiguity in sea level prediction. SLR has been commonly treated as a linear trend over the historical data set. In his analysis of data from four tide gages in Florida, Walton (2007) allowed for acceleration (or deceleration) in SLR and showed that this resulted in a greater predicted (2006–2080) rise than forecasted by the linear trend assumption. Figure 4.3 shows the mean monthly relative sea level from October 1948 through May 2004 for Pago Pago in American Samoa. Linear regression indicates an average rise of 1.97 mm·year^{-1} over the 54.67-year period. In contrast, the polynomial fit suggests that there may be accelerated relative SLR since about mid-1997 (Gilman 2005). Similar uncertainties with the interpretation of temperature data (short records, locations in urban areas, the northern hemisphere, etc.) further limit predictive accuracy for the rate of SLR. This is illustrated later in regard to some of the contemporary modeling efforts for SLR.

Furthermore, the different factors contributing to SLR vary in their individual response times. For instance, a volume of glacier ice melts more rapidly than the polar ice sheet. Inasmuch as the effects of the melting of ice sheets and the steric effect are dominant but with different response times,

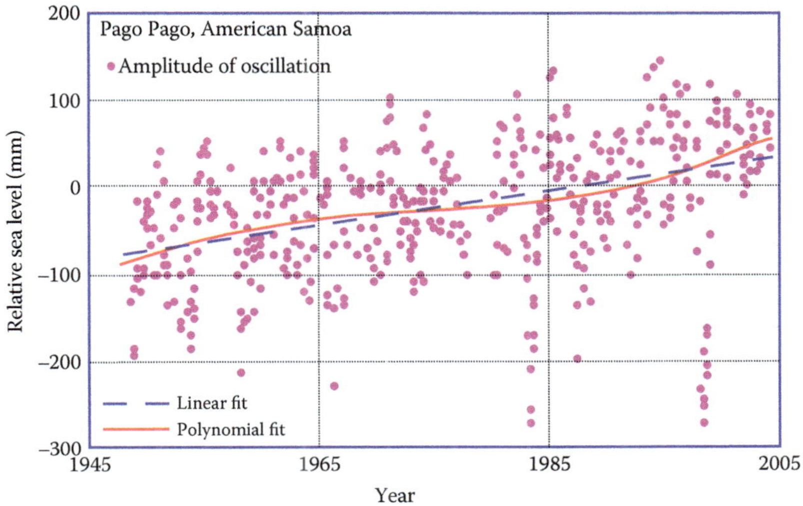

FIGURE 4.3
Mean monthly relative mean sea level from a tide gage at Pago Pago, American Samoa, October 1948–May 2004. Data sources: Permanent Service for Mean Sea Level and the University of Hawaii Sea Level Center Joint Archive for Sea-Level and GLOSS/CLIVAR databases. (Based on Gilman, E., Contribution to the Barbados+10 United Nations International Meeting on Sustainable Development of Small Island Developing States, Port Louis, Mauritius, January 10–14, 2005.)

a commonly used approach for modeling purposes is to assume that all the causes can be lumped together so that there is only one relationship between global temperature and global sea level. For prediction over, for example, a century, such a single-relationship approximation is necessary because the individual response times are not well understood or quantified.

Figure 4.4 shows time series of the reconstructed global mean temperature anomaly (difference relative to the year 2004). The past two millennia are represented by using data from ten separate studies. Observe that there is considerable uncertainty among the records, even though the data for the twentieth century and later show better agreement. It has been pointed out (Grinsted et al. 2009) that the medieval warming period and especially the little ice age may continue to exert influences on sea level change and affect its prediction.

4.3 Future SLR

An approach taken by Grinsted et al. (2009) illustrates analytic difficulties in the prediction of global mean SLR S in the present century based on the predicted rise in the global mean temperature T (Figure 4.4).

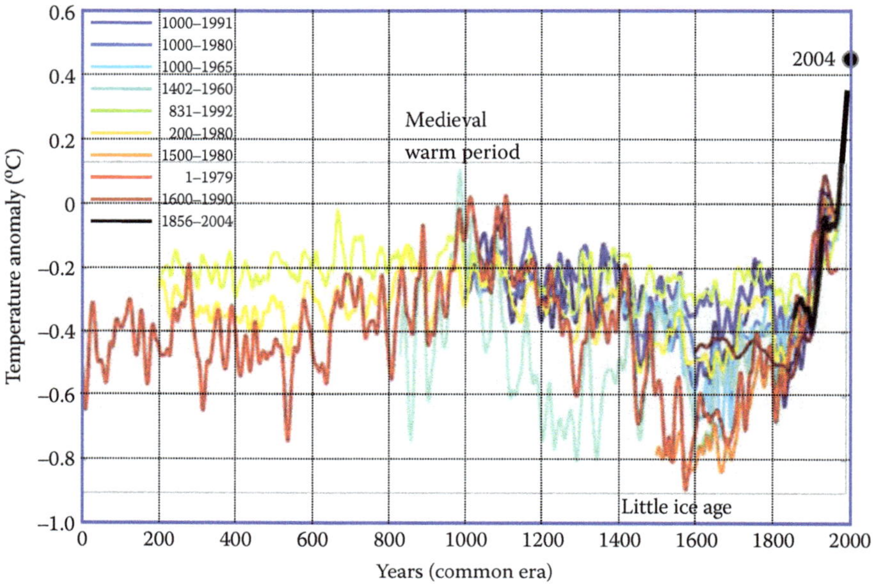

FIGURE 4.4
Global temperature anomalies during the past two millennia derived from ten different investigations. Note the medieval warm period and the little ice age. (Adapted from figure by Robert A. Rohde, 2009b, http://commons.wikimedia.org/wiki/File:2000_Year_Temperature_Comparison.png.)

In the analysis, by considering changes with time *t* in the sea level *S* and temperature *T* ($\partial S/\partial t$ and $\partial T/\partial t$, respectively), difficulties in calculating the relative mean sea level, which is local, are avoided. It is assumed that for every value of *T*, there is an equilibrium sea level S_{eq} obtained from the linear relationship

$$S_{eq} = aT + b \tag{4.1}$$

where *a* and *b* are empirical coefficients. In response to a change in *T*, the rate at which *S* approaches S_{eq} is assumed to be proportional to the difference $S_{eq} - S$, that is,

$$\frac{\partial S}{\partial t} = \frac{S_{eq} - S}{\tau} \tag{4.2}$$

where τ is the (empirical) rate constant.

Based on a global temperature record reconstructed by Jevrejeva et al. (2006), Grinsted et al. (2009) have highlighted the significance of differences among the data sets and analytic assumptions used. This is evident from the wide ranging estimates (mean values and 5%–95% confidence interval) of *a*, *b*, and τ (Table 4.1).

TABLE 4.1

Values and Ranges of Parameters in Equations 4.1 and 4.2

Method	a (m·C^{-1})	b (m)	τ (year)
Moberg et al. (2005)[a]	1.29 ± 0.36	0.77 ± 0.25	208 ± 67
Jones and Mann (2004)[b]	2.56 ± 0.90	1.36 ± 0.52	317 ± 117
Historical[c]	3.10 ± 1.64	3.68 ± 1.15	1193 ± 501

Source: With kind permission from Springer Science+Business Media: *Climate Dynamics*, Reconstructing sea level from paleo and projected temperatures 200 to 2100AD, doi.org/10.1007/s00382-008-0507-2, 2009, Grinsted, A., Moore, J. C., Jevrejeva, S., Copyright © 2009.

[a] Sea level reconstructed since Common Era (CE) 0.

[b] Sea level reconstructed since CE 200.

[c] Using the HadCRUT3v temperatures and post-1850 global sea level only.

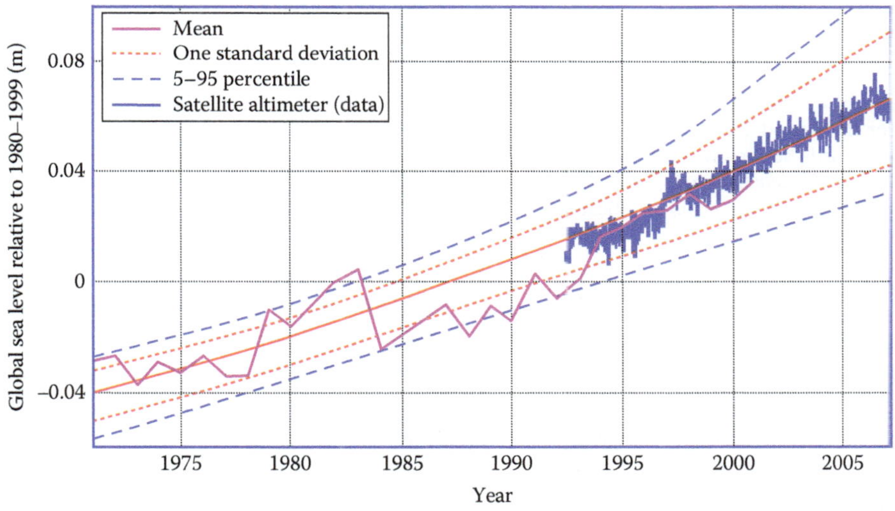

FIGURE 4.5
Measured and simulated sea levels. (With kind permission from Springer Science+Business Media: *Climate Dynamics*, Reconstructing sea level from paleo and projected temperatures 200 to 2100AD, doi.org/10.1007/s00382-008-0507-2, 2009, Grinsted, A., Moore, J. C., Jevrejeva, S., Copyright © 2009.)

Figure 4.5 shows the simulated sea level (thin smooth curve), one standard deviation (dotted line band), 5–95 percentiles (dashed line band), reconstructed global sea level (thick curve) (Jevrejeva et al. 2006), and satellite altimetry-based sea level (jagged) (Leuliette et al. 2004). The simulation was calibrated for the post-1990 period of measured data during which the reconstructed curve and the satellite-derived values were close to each other. Based on this, the predicted curve for the present century is shown in Figure 4.6. IPCC monitors efforts and develops/revises high-, moderate-, and

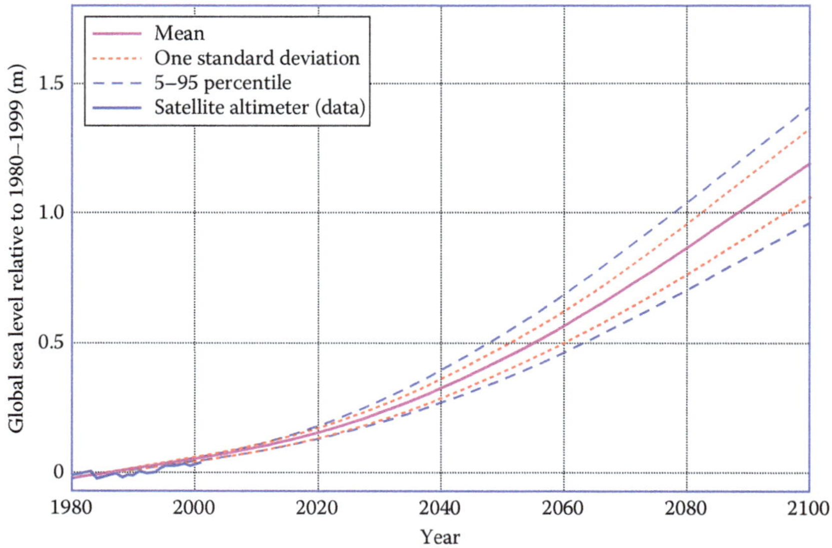

FIGURE 4.6
Predicted sea level rise (continuous curve) relative to satellite altimetry data (thick blue curve) for the period 1980–1999. Note the one standard deviation (dotted line band) band and 5–95 percentile bounds (dashed line band). (With kind permission from Springer Science+Business Media: *Climate Dynamics*, Reconstructing sea level from paleo and projected temperatures 200 to 2100AD, doi.org/10.1007/s00382-008-0507-2, 2009, Grinsted, A., Moore, J. C., Jevrejeva, S., Copyright © 2009.)

low-intensity scenarios of future rise in global temperature and periodically reports its findings (IPCC 2007). For the calculations in Figure 4.6, the IPCC's A1B scenario was used as the temperature curve. Among the six scenarios, this is an intermediate case in which the most likely estimate of mean temperature rise in 2090–2099 relative to 1990–1999 is 2.8°C.

Based on their interpretation of measurements, Church and White (2006) determined that the most likely rise of the global sea level (S) from January 1870 to December 2004 was 195 mm, a twentieth century rate of SLR of 1.7 ± 0.3 mm·year^{-1} and acceleration of 0.013 ± 0.006 mm·year^{-2}. The normalized coefficient of variation of $0.006/0.013 = 0.46$ is a large fraction of the mean and is commensurate with the great uncertainties in every compilation of sea level change. A common feature of predictions also is that any acceleration in SLR will occur less rapidly at first but more rapidly in future. In other words, the curves are concave upward as in Figure 4.6. The uncertainty bands at the end of this century indicate a range of sea level estimate between about 0.9 and 1.3 m. The fact that the uncertainty bands are wide and become wider with every decade suggests that several decades may pass before it will be possible to obtain a more accurate estimate of the expected rise. It is interesting to note that due to the lag between temperature and sea level, a rise is expected to occur even if the temperature does not increase further. To counter this rise, the required decrease in temperature is estimated to be about −0.6°C (Grinsted et al. 2009), although it is uncertain if this would fully account for any lagged effect of the combustion of fossil fuels should it become clearly detectable and prominent.

We briefly look at the role of climate subject to global warming, followed by scientific and engineering research issues related to responses and prediction of impacts.

4.4 Climate Impacts

4.4.1 Coastal Storms

Vulnerability-and-risk analysis is the first step in any coastal impact assessment for considering appropriate site-specific responses associated with a particular rise in sea level. In the United States, several agencies are now directly involved in providing information necessary for impact assessment (NOAA-NOS 2010). There are likely two reasons for increased storm impacts because of SLR. One is that the elevations of coastal infrastructure such as roads and buildings on barrier islands become lower relative to the rising sea level and so more susceptible to coastal flooding. The second reason is increased storminess. Physical evidence that global warming may bring about a significant increase of coastal storminess is not yet unequivocally established. However, if this effect is occurring, it should be of critical importance to all matters related to coastal defense. Under the IPCC's A1B scenario, hurricane intensity has been predicted to increase between 2% and 11% in the latter part of this century, which is not substantial at the low end; however, some scenarios predict large increases.

The potential for damage to coastal infrastructure may eventually increase by as much as 30% (Knutson 2011; Knutson and Tuleya 2004) in monetary terms. Considering the storm climate to remain the same at least in the coming years, it is relatively straightforward to recalculate, for example, the return period of a given storm tide level for a given SLR. Figure 4.7 shows the calculated cumulative return period distribution of storm tide level in Lee County, Florida. An approximately 0.5 m increase in mean sea level, with storm-tide statistics (i.e., normalized frequency distribution of storm tide level) remaining unchanged, would decrease the return period of a presently 100-year storm tide level to about 55 years. If the storm statistics change substantially, plots such as Figure 4.7 will have to be redrawn based on the output of numerical models for storm tide level that permit nonlinear coupling between waves, astronomical tide, and storm surge in the shallow water environment complicated by the shoreline and bottom geometry.

With increased coastal development, major storms can cause potentially significant damages and financial losses. The Saffir–Simpson scale (Table 4.2) attempts to classify damage to property due to hurricanes (Saffir 1977; Simpson 1979). The damage categories (from minimal to catastrophic)

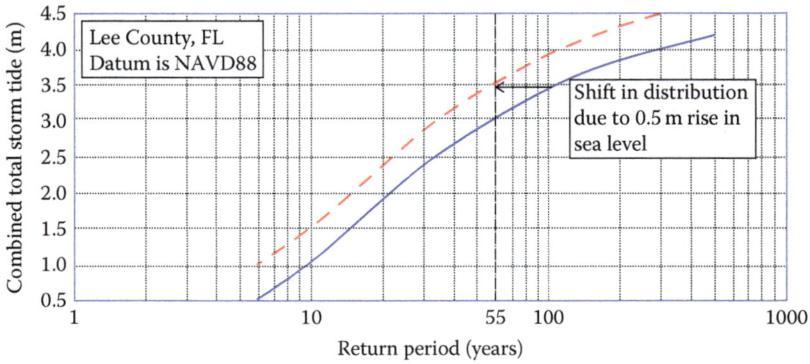

FIGURE 4.7
Calculated storm tide (storm surge + astronomical tide) relative to the standard NAVD88 datum. (Adapted from Dean, R. G., Chiu T. Y., Wang, S. Y., 1990. *Combined Total Storm Tide Frequency Analysis For Lee County, Florida,* Report submitted to the Florida Department of Natural Resources, Beaches and Shores Center, Tallahassee.)

TABLE 4.2
Saffir–Simpson Scale for Hurricane Damage

Scale	Pressure (millibar)	Wind (km·h^{-1})	Surge (m)	Damage
1	≥980	118–153	1.2–1.5	Minimal
2	965–980	154–177	1.8–2.4	Moderate
3	945–965	178–209	2.7–3.7	Extensive
4	920–945	210–249	4.0–5.5	Extreme
5	<920	≥250	>5.5	Catastrophic

Sources: Data from Saffir, H. S., ASCE Preprint no. 2830, 1977; Simpson, R. H., *Minutes of the Eighth NOAA-NWS Hurricane Conference,* Miami, 1979; Dean, R. G., Dalrymple, R. A., *Coastal Processes with Engineering Applications.* Cambridge, U.K.: Cambridge University Press, 2002.

are relative as the density and value of infrastructure vary along the coast. Therefore, site-specific calibration is required, such as by identifying, say, "extreme damage" in terms of property loss in dollars.

The Dolan–Davis scale for northeasters (Table 4.3) is based on wave heights (H) and storm duration (t_d) and uses the product H^2t_d for storm intensity risk index I. The damage potential (minor to extreme) can be high if H^2, which represents wave energy, is high or if the storm duration t_d is long. This scale as developed is based on limited data from Cape Hatteras, NC, and is not shown to be globally accurate. Further examinations of the relationship between such scales and damage have revealed that storm intensity as defined does not always correlate well with damage. In a preliminary effort, Kriebel et al. (1997) showed that the definition of I should include, in addition to H and t_d, the storm surge height as an independent variable. This revised index has not been calibrated extensively.

TABLE 4.3

Dolan–Davis Scale for Northeasters

Risk Index *I*	H^2t_d (m² hr)	Damage
1	70	Minor
2	160	Moderate
3	930	Significant
4	2300	Severe
5	>2300	Extreme

Sources: Data from Dolan, R., Davis, R. E., *Journal of Coastal Research*, 8(4), 840–853, 1992; Dean, R. G., Dalrymple, R. A., *Coastal Processes with Engineering Applications*. Cambridge, U.K.: Cambridge University Press, 2002.

FIGURE 4.8
A coastal home on piles for protection against storm surge.

In the U.S. coastal states such as Florida, several measures are now in place to make the structures storm-resistant. There are three primary structural characteristics that are required to make a building storm-resistant: (1) The structure must be built on piles so that the main horizontal elements are elevated above the reach of the wave crest during the design storm (Figure 4.8); (2) The piles should be sufficiently deep so that vertical erosion of land during storm will leave intact the foundation's integrity; and (3) Structural connections must be designed for high wind loads. As an illustration of the effectiveness of these measures, soon after Hurricanes Hugo (maximum sustained winds at peak 258 km·h^{-1}; i.e., category 5 on the Saffir–Simpson scale) in 1989, inspections were carried out in South Carolina to document erosion and structural damage. It was found that no structures built to the prescribed standards suffered major damage. Some states such as Florida require that single-family homes be set back from the shoreline to a distance equal to 30 times the annual erosion; for multi-family or commercial buildings, the factor is increased to 60 (Dean 1991).

Another way in which hazards due to increased development along the shore has been addressed is by requiring those incurring the risk to bear the financial burden through participation in the Federal Emergency Management Agency (FEMA) insurance program. To do so, FEMA presently uses the HAZUS-MH hurricane wind model to estimate hurricane-induced damage to buildings, essential facilities, transportation, and utility lifelines (HAZUS-MH MR4 Hurricane Manuals 2012).

These observations suggest that (1) calibration of the construction code over a wide set of coastal physical conditions, (2) code implementation, and (3) code enforcement are the main factors responsible for preventing continued loss of coastal property from storms. On the other hand, the potential for damage from more intense and frequent storms will require consideration in the coming years.

In 1996, a study in Japan used land mass, population, and property as risk indicators to assess the impact of SLR because of a severe storm or tsunami surge. It was revealed that even under non-storm conditions, more than 860 km^2 of land was already below mean high water level. Over 2 million people live in this zone. If the sea level was to rise by 1 m, this zone would expand to 2300 km^2. The affected population would be over 4 million and property loss exceeding a trillion dollars. The present flood-prone area of about 6300 km^2 would expand to 8900 km^2 and the lives of more than 15 million people would be affected (CGER 1996). In fact, the 2011 tsunami on the Pacific coast of Tōhoku hugely exceeded any predicted scenario; the surge rose to heights of up to 40.5 m in Miyako (Iwate Prefecture) and traveled up to 10 km inland in the Sendai area. Meltdown occurred at three earthquake-damaged reactors in the Fukushima-I nuclear power plant and required the evacuation of hundreds of thousands of residents. Human casualty was well in excess of 10,000 across 18 prefectures and well over 100,000 buildings were damaged or destroyed. Table 4.4 projections suggest that tsunami damage to land area, loss of life, and property destruction will rise rapidly due to global SLR.

Apart from low-lying small countries such as the Pacific Islands, large countries will experience high losses of low-lying coastal lands. Along the coast of the Indian peninsula, a 1 m rise of sea level will inundate 1810 km^2 of

TABLE 4.4

Predicted Impacts of Storm or Tsunami Surge in Japan

Sea Level (m)	Land Area Affected (m^2)	Population Affected (Millions)	Property Affected (Trillion Dollars)
0 (1996)	6300	11.7	3.7
0.3	6700	12.3	3.9
0.5	7600	13.6	4.2
1.0	8900	15.4	4.8

Source: Data from CGER, *Data Book on Sea-Level Rise*, Center for Global Environmental Research, Environment Agency of Japan, Tokyo, 1996.

land in the state of Gujarat, 410 km^2 in Maharashtra, 160 km^2 in Goa, 290 km^2 in Karnataka, 120 km^2 in Kerala, 670 km^2 in Tamil Nadu, 550 km^2 in Andhra Pradesh, 480 km^2 in Orissa, and 1220 km^2 in West Bengal. This amounts to a total of 5710 km^2, a large land area (Das 2012).

4.5 Response to SLR

4.5.1 Understanding Processes

The variety of climate impacts implies that the response to a rise in the sea level will vary widely with the rate of rise and location. Many types of responses have often been successful over the past centuries, although the extent of protection required for large land areas has often proved to be impractical, expensive, or environmentally damaging. Feasible responses are expected to become more critical as SLR accelerates further in the latter half of this century. The cost of fuel will also play into this if it continues to rise. Let us consider some coastal issues in terms of available measures and unknowns for which critical technical research is required now and in the coming decades.

4.5.2 Sea Level, Storm Surge, and Waves

Resolution of the different components of the relative SLR requires use of a high-precision geodetic reference system with the aim to remove land movement signals from sea level records. Teferle et al. (2006) describe the development and implementation of techniques based on the Global Positioning System (GPS) for measuring vertical land movements (VLM) at tide gages. Also discussed is the use of absolute gravity as an independent technique for measuring VLM at tide gages. Although tide gages are now found in numerous locations, the global distribution of long-term tide gages that are reliable continues to be nonuniform (Figure 4.9) and prone to several biases due to historic uncertainties or errors in benchmarking, local subsidence, and so on.

On the other hand, considerable advances have taken place over the past 10–15 years with GPS and also DORIS (Doppler Orbitography and Radiopositioning Integrated by Satellite) for monitoring VLM at tide gages. In fact, the quality of tide gage data has already improved in cases where a recording GPS unit is located near the gage. These so-called collocated GPS units are currently present at approximately one-seventh of the long-term tide gages of the world and allow removal of the vertical land motion from the measurements of relative sea level. The GPS data including proximity to the tide gage are available at http://www.sonel.org/-GPS-.html. Proximity can range from 1 m to up to 15 km or so, and durations of the GPS gages vary from a few years to up to approximately a decade.

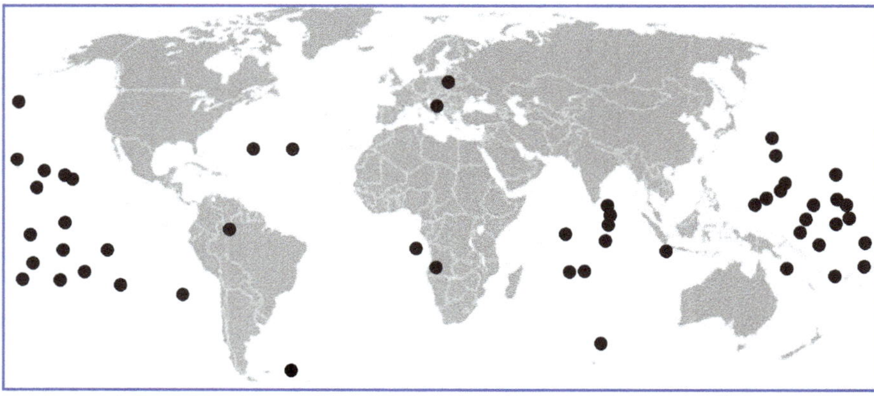

FIGURE 4.9
Locations of tide gage data used by Mitchum (1998) in analysis of time-dependent drift in satellite altimeter data on sea surface height. (Modified from Mitchum, G. T., *Journal of Atmospheric and Oceanic Technology*, 15(3), 721–730, 1998.)

Calculation of the storm tide level requires information on waves generated by wind. Modeling wind–waves in deep and shallow water is now fairly advanced. However, there is a lack of detailed, high-quality wind and wave data with which to verify the models. Also, many advanced models are workbench versions that are not easily accessible. A commonly used public domain model is SWAN (Booij et al. 1999), which has good directional capabilities and well-tested parameterizations of wave physics: shoaling, refraction, diffraction (in the most recent release), wind input, white-capping, depth-limited breaking, bottom friction, and four-wave nonlinear interactions. However, SWAN's description of the essential shallow water mechanism of wave–wave interaction has significant deficiencies (among other problems, it does not describe long-wave generation).

In the United States, one of the greatest recorded storm surges was generated by Hurricane Katrina in 2005, producing a maximum storm surge of 8.5 m at Pass Christian, MS (FEMA 2005). In general, improvements in modeling storm surge and wind–waves require research to improve our understanding of the basic geophysical phenomena. There have been intensive theoretical and numerical studies of storm surge over the past three decades, and several sophisticated numerical models for storm surge prediction exist. However, along large sections of the world's shorelines, adequate field measurements of hurricane and extratropical storm surges with which to calibrate and validate these models are sparse or absent and predictions are based on numerical predictions alone (Unnikrishnan et al. 2006). Field work is needed to shore up these gaps in data and further improve surge prediction techniques. It is also essential to examine storm level statistics and spatial distribution including landfall locations from the historical records (Figure 4.10), so that a reliable baseline can be established with which future

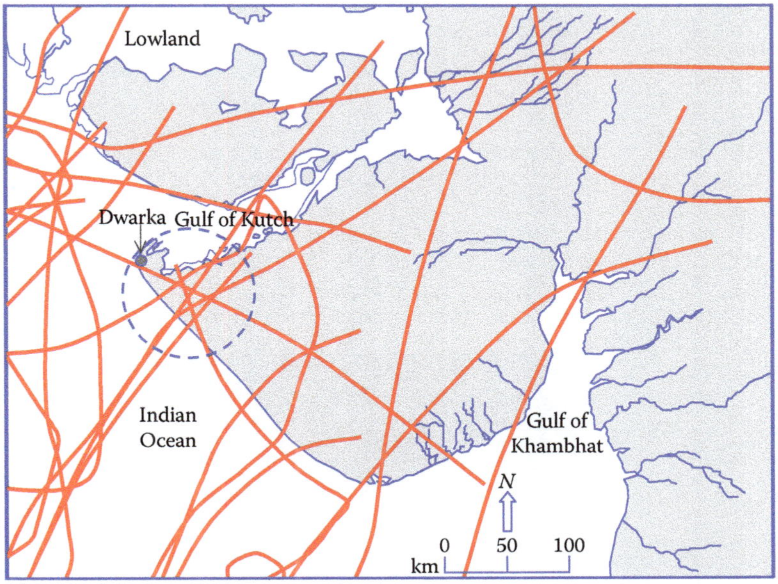

FIGURE 4.10
Cyclonic tracks in Gujarat vicinity, 1970–2005. Dotted circle encompasses a zone of high activity and track convergence. (Adapted from Khare, Y. P., Mehta, A. J., *Gujarat and the Sea*, L. Varadarajan ed., Darshak Itihas Nidhi, Greater Noida, 33–48, 2011.)

records can be compared. Some work of this nature has already been done in conjunction with FEMA sponsored studies for coastal counties, especially after Hurricane Katrina (Bunya et al. 2010; Dietrich et al. 2010).

4.5.3 Shoreline Erosion

Although shoreline erosion has been researched extensively since the mid-twentieth century, there remain significant limits to modeling the rate of erosion accurately; limitations will continue until much more is known about sand grain movement in turbulent flows. Equally important is the need to use appropriate models with respect to their domains of applicability, strengths, and weaknesses. Not doing so is a reason why predictions sometimes turn out to be incorrect. Consider the following.

For erosion of a sandy beach due to waves, a widely used method relating shoreline recession R to SLR S is the Bruun Rule

$$R = S\frac{W_*}{h_* + B} \tag{4.3}$$

in which W_* and $h_* + B$ are the horizontal and vertical dimensions, respectively, of the "active" beach profile over along which sediment movement

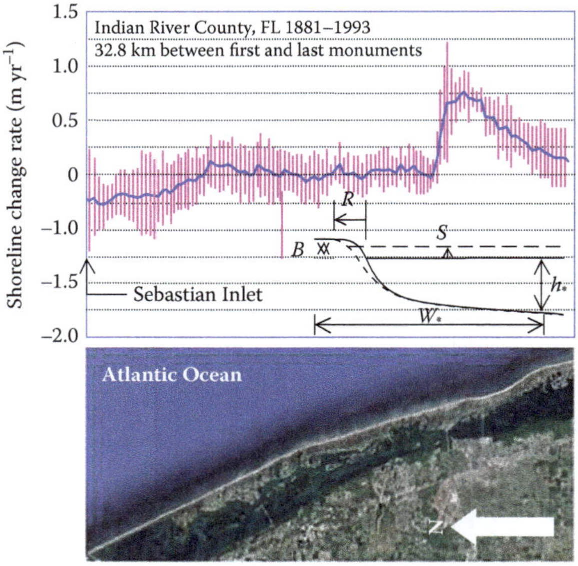

FIGURE 4.11
Shoreline change rate (mean and standard deviation) at survey monuments along the Indian River County shoreline, Florida, for the period 1881–1993. (Adapted from Dean, R. G., Dalrymple, R. A., *Coastal Processes with Engineering Applications,* Cambridge, U.K.: Cambridge University Press, 2002.)

molds the profile (Figure 4.11, inset) (Bruun 1962; Dean and Dalrymple 2002). The basis of this relationship does not include any net gain to, or loss from the profile, but only sediment shifted seaward or landward over the active profile to maintain an equilibrium shape. The factor $W_*/(h_* + B)$ is usually in the range of 50–100. Based on Equation 4.3, one would anticipate that doubling the SLR rate would double the shoreline recession rate.

A significant limitation of the Bruun rule is the presupposition that all sandy beaches experiencing SLR must recede. However, from shoreline data over the past century, it is known that not all beaches have eroded, and at a single beach, there can be persistent zones of sediment erosion as well as accretion. Figure 4.11 is based on long-term (1881–1993) shoreline position data for Indian River County on the east coast of Florida (Dean and Dalrymple 2002). At every survey monument, the shoreline has oscillated about its mean position, with vertical bars representing standard deviation relative to the mean position. On average, the northern reach of the shoreline moved landward due to the presence of Sebastian Inlet: the middle reach was more or less stable and the southern reach advanced seaward.

Depending on the littoral sand drift, the presence of a tidal inlet, especially one trained by jetties and regularly dredged, usually leads to accumulation of sand updrift and erosion downdrift of the inlet (e.g., Figure 4.12

FIGURE 4.12
Storm erosion south of Jupiter Inlet at the Atlantic coast of southern Florida. (Courtesy of Mike Grella, Jupiter Inlet District.)

showing erosion downdrift of Jupiter Inlet). At most inlets in Florida (and numerous other locations), sand drawn from the littoral drift is deposited in the interior waters. Based on shoreline changes at 1600 locations along the east coast of Florida, long-term changes have been found to vary from inlet-induced recession rate of 9 m year^{-1} to an advancement rate of 6 m year^{-1}. It is also interesting that on average, there has been shoreline accretion of 0.18 m year^{-1} over a period exceeding 100 years (Dean and Dalrymple 2002).

To model beach retreat and advancement due to SLR, it is essential to include the effects of net gain and loss of sand in the Bruun rule. It is assumed that the present recession rate R_1 is related to the present SLR rate S_1 according to Bruun rule plus a site-specific deviation, G_i

$$R_1 = S_1\left(\frac{W_*}{h_* + B}\right) + G_i \quad (4.4)$$

where both G_i and R_1 can be either positive (erosion) or negative (accretion). A case of a negative G_i value would be the arrival of sediment from offshore sources by natural shoreward transport over the timescale of SLR. For a different SLR rate S_2, we assume that G_i will be unchanged. Thus, the recession rate becomes

$$R_1 = (S_2 - S_1)\left(\frac{W_*}{h_* + B}\right) \tag{4.5}$$

Essentially, G_i amounts to calibration of the Bruun rule with respect to the local condition, correcting the "erosion only" limitation. Equation 4.5 has been applied to project coastal land loss over this century at sites based on factors that would make these sites vulnerable to inundation at higher stands of sea and storm surges. However, given the great concern for the expected effects of global warming, the need for more accurate estimates of erosion through process-based modeling has been pointed out (Ranasinghe et al. 2011). Process-based modeling can account for shoreline changes because of alongshore transport, which is not included in Equation 4.5.

Eroding beaches that are not maintained adequately, such as the many eastern shoreline segments of peninsular India, face a rise in sea level coupled with increased severity of the wave impacts (Indu et al. 2010). Along a 60 km stretch of the beach near Cochi (Cochin) in southwest India, shore erosion is expected to increase by about 15% over the present rate (118.75 m^3 per meter length of the beach), which may be an underestimate. In addition, a substantial increase in inland flooding because of storm surge is expected because beach dune elevations are low (Kumar 2006). Large human populations along the Indian southwest coast derive their livelihood from the sea and the lagoons between the barrier beaches and the mainland. A special consideration is required to prevent erosion and destruction of the fragile ecosystem of the lagoon behind the barrier beach. A very rough estimate is that capital and maintenance costs of protecting these beaches could approximately double over the present in the event of a meter rise in sea level (Das 2011). However, if built infrastructure along the coast (which is presently only low to moderate in density outside the main urban areas) increases, this cost estimate will be substantially exceeded.

A broad-ranging capability to reliably predict shoreline response to future SLR requires accurate simulations of cross-shore and alongshore sediment transport and applications including mathematical modeling (Dean and Dalrymple 2002). Strides in modeling during the past decades have been impressive but, predicting morphodynamic changes due to SLR, for example, change in the location, areal shape, and topography of a barrier island or the bathymetry of a lagoon, remains largely out of reach. This is only in part due to the lack of high-quality and long-term hydrodynamic as well as meteorological data. Rules for morphodynamic modeling are not well understood, despite scientific advancements over the recent decades. In turn, this problem has a bearing on the reliability of coastal ecological modeling.

Consider the hierarchy of coastal beach features starting with the sand grain at the smallest scale and the barrier beach at the other end, which is several orders of magnitude larger (Figure 4.13). The grain's linear dimension is $O(10^{-4}$–10^{-3} m) and its rolling motion in turbulent flow has a period

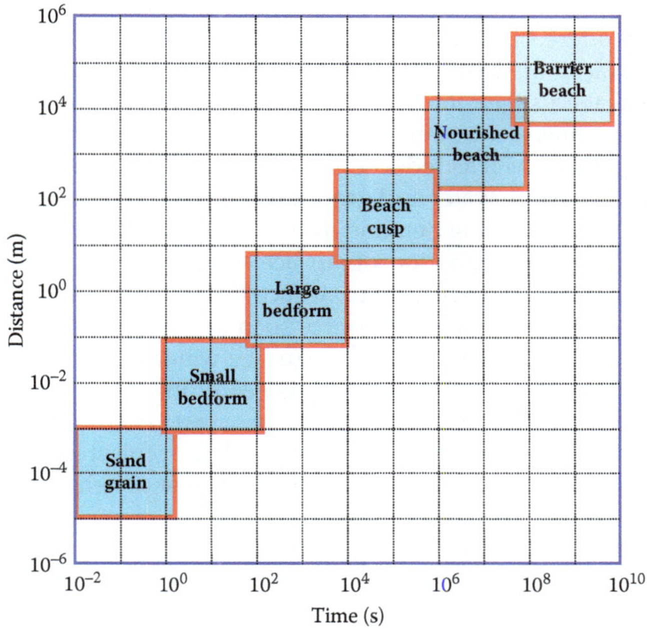

FIGURE 4.13
Continuous space–time hierarchy of beach features.

of O(10^{-2}–100 s). In contrast, the length and the evolutionary timescales for the barrier beach are O(10^4–10^6 m) and O(10^7–10^{10} s), respectively. Such a space–time hierarchical classification helps us recognize that we must be able to model the role of sand grains in the development of large landforms from a basic knowledge of grain motion together with iterative interactions between landforms at all intermediate scales and the respective forces. Morphodynamic modeling attempts to achieve this aim by means of co-adjustments between forces, processes, and forms as they evolve over different space and timescales. Without these co-adjustments, the space–time pathway defined by the overlapping boxes may not be sustained in the course of model iterations. For these reasons, processes occurring over short distances (meters to tens of meters) and short times (seconds to perhaps days) are modeled more accurately than over greater distances and times (Wright and Thom 1977).

4.5.4 Saltwater Intrusion

The effect of an increase in saltwater intrusion in estuaries due to SLR is similar to increased penetration of saltwater when a ship channel is deepened. An historic example and an ongoing environmental issue has been the deepening of the 60 km navigation channel connecting Lake Maracaibo in Venezuela to the Gulf of Venezuela. Capital dredging to deepen the channel for oil tankers starting in the 1930s has made the lake water brackish and substantially altered the ecosystem (Laval et al. 2005). Such an impact can be further exacerbated by reductions in river runoff, which is generally expected as water usage continues to rise, increasing salt penetration by tens of kilometers. Thus, a substantial rise in the eustatic sea level coupled with bottom dredging as well as reduced freshwater outflow can be severe.

Cities rely on surface water and groundwater to meet domestic and industrial needs. With increasing demand for water together with SLR, the potential for saltwater contamination of surface water as well as the aquifer is expected to increase. With respect to surface water, where saltwater penetration is confined to a well defined source such as a narrow channel, salt barriers have been installed to minimize surface water saltiness, even though groundwater intrusion into the aquifer may continue. An example of surface water barrier is the underwater sill structure constructed in the Mississippi River south of New Orleans (Figure 4.14). This project of the Corps of Engineers has had a good degree of success, albeit at a high cost of construction (Fagerburg and Alexander 1994).

In the Netherlands, increasing salt intrusion into the rivers and seepage of saltwater into inland polders have been matters of great concern for decades if not centuries (de Ronde 1991). Cost of installation and maintenance, rising cost of energy, the effectiveness of the arrangement, and ecological consequences are issues that have to be considered in arriving at the outcome, all against the backdrop of a continued rise in the sea level. The cost factor is such that where large areas are being inundated and there is insufficient economic justification for defending the coast, abandonment, or "managed retreat"

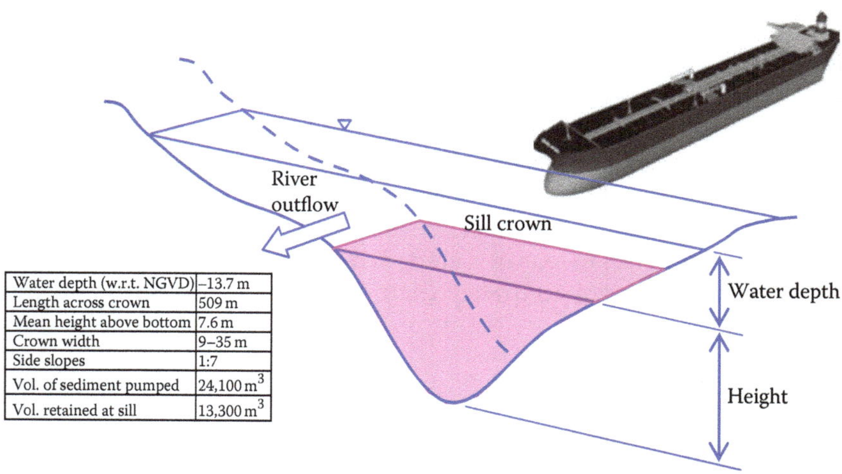

Water depth (w.r.t. NGVD)	–13.7 m
Length across crown	509 m
Mean height above bottom	7.6 m
Crown width	9–35 m
Side slopes	1:7
Vol. of sediment pumped	24,100 m^3
Vol. retained at sill	13,300 m^3

FIGURE 4.14
Schematic drawing of salt barrier sill in the Mississippi River. (Adapted from Fagerburg, T. L., Alexander, M. P., *Miscellaneous Paper HL-94-1,* U.S. Army Engineer Waterways Experiment Station, Vicksburg, MS, 1994.)

when planned ahead of time, appears to be the only realistic option. Tam (2009) identifies managed retreat as a way to safely remove settlement from encroaching shorelines, allowing the water to advance unimpeded. It involves abandoning, demolishing, or moving existing buildings and infrastructure to higher ground. It also includes banning new development in areas likely to be inundated. It is to be implemented when coastal armoring and other shoreline protection efforts become expensive or are judged to be a losing battle against protection from the rising sea. The "managed" part of retreating involves establishing thresholds to trigger activities such as demolishing buildings or abandoning efforts to control coastal erosion. There are numerous locations in the world where this approach may be essential, with some countries such as the United Kingdom, the Netherlands, and Spain having developed the outlines of step-by-step approaches. In the United States, planning efforts have been made such as for the city of San Francisco, which has low-lying areas along the San Francisco Bay (Tam 2009). When very large low-lying areas are involved such as in coastal Louisiana, managed retreat from all but a few of urban zones is expected to be the only option (Dean 2005).

Given the ambient flow conditions in a coastal aquifer, its transmissive properties, porosity, and various scenarios of extraction demands, it appears that salinity intrusion, including any time dependencies, can be predicted with reasonable accuracy using available mathematical models. Unfortunately, effects of SLR on saltwater intrusion, particularly in groundwater, have been examined largely for special cases. Fortunately, for further improvements in predictive modeling, the basic research issues are less related to sea level change than to marine physics, chemistry, and biology as they interact and influence coastal and estuarine processes. When funding for basic research is available, this has provided the opportunity to improve applied modeling.

4.5.5 Impacts on Estuaries and Wetlands

The coastal tidal range increases with sea level when the dominant change is a reduction in the bottom resistance to flow due to deepening. A consistent trend of rising tidal range over the past century has been recorded at ten stations along the coast in the German Bight (Gornitz 1995). This effect is not merely due to long-term changes in meteorological conditions, but in part arises from the morphology of the North Sea, a shallow basin in which the influence of global rise of mean water level is amplified by the so-called standing wave effect. This further suggests a case in which the natural frequency of the sea approaches the frequency of astronomical tide as the water depth continues to increase and changes the volume and distribution of water in the basin. In the estuary proper, these amplifications are further enhanced by reduced bottom friction in deeper water coupled with the so-called funneling effect due to decreasing estuarine cross section with increasing upstream distance (Valle-Levinson 2010; Hardisty 2007).

By altering the biological processes as the sea level and perhaps the tidal height itself rises, estuarine intertidal and supratidal zones may extend further inland, resulting in changes in the existing ecological setup. The limit of the landward margin will cause a vertical rise, resulting in waterlogging and ultimately killing the mangroves and dependent biota. In a review of information on mangroves along the coast of the Indian subcontinent, Jagtap and Nagle (2007) note that while mangroves may be able to cope with a SLR rate of 8–9 cm 100 $year^{-1}$, they would be under environmental stress at rates between 9 and 12 cm 100 $year^{-1}$. At a rise rate of 12 cm 100 $year^{-1}$ and more, the mangroves will neither grow nor may form peat. As sediment discharge increases in the wider and longer estuaries and their river reaches, the mangroves are expected to be choked by pneumatophores, leading to very low (or no) rate of survival. Heavy sediment loading will result in the shoaling of waterways, inundation of the floodplains, and waterlogging harmful to mangroves. At a SLR in excess of about 50 cm 100 $year^{-1}$, all low-lying mangrove areas along the Indian coast are expected to vanish. However, it must be pointed out that when sediment supply is not properly accounted for because projections are difficult to forecast, predictions of total loss of coastal vegetation must be tempered by this limitation in modeling.

In some situations, shallow bays surrounded by wetlands expand rapidly in response to SLR both because of the gentle slope and the deterioration of the marshes because of salinity increase coupled with other major factors such as decreasing sediment supply if it occurs. In that event, the affected salt marshes will be unable to maintain their elevations and are expected to be inundated because land subsidence exacerbates the rise in water depth. Inundation of vast areas previously dominated by salt marshes has already been occurring along the coast of Louisiana, where owing to soil compaction, the relative sea level has been rising at a rate that is an order of magnitude greater than the eustatic rise. Barataria Bay has increased its surface area by about 10% to 15% over the last century in response to about 1-m local relative SLR. Land loss in this area is predicted to be significant even for moderate SLR scenarios (Blum and Roberts 2009; Fitzgerald et al. 2004). Such projections would not apply to locations where sediment supply is not significantly reduced as water level rises.

4.5.6 Role of Coastal Structures

Many breakwaters and sea walls can be retrofitted or modified to perform their functions adequately in response to relative SLR and changes in storm frequency and intensity. An example is the King Harbor breakwater at Redondo Beach in California. This coastal area had subsided in excess of a meter over decades. On January 17–18, 1988, it experienced storm waves rarer than a once-per-century event. Repair and remedial work was successfully done on the breakwater and the buildings at the water's edge within the harbor. The sea wall, a revetment between a hotel and the water, was strengthened with additional rock, and a concrete "wave splash wall" was constructed on top of the revetment and some areas graded and paved (Dill et al. 2006).

Several low-lying coastal areas around the world are protected by dikes. Some of these works have been in place for several hundred years. Essential parts of present protection structures include storm surge barriers, sea walls, and so on, as in Germany and the Netherlands. Appropriate planning and design of these features requires forecasting changes in the area seaward of the structures such as a tidal flat, a salt marsh, a sandy beach, and so on. There is greater confidence in being able to arrive at rough estimates of shoreline and slope changes at sandy beaches and tidal flats than where wetland vegetation intervenes between the shoreline and open water. Success in the latter instances has been generally poor except where model output at the wetland boundary of the sea is tweaked by significant site-specific calibration. Improved forecasting methods are also required for decisions on the development of the coastal zone for preservation of nature including wildlife, recreation, and so on (National Research Council 2010).

The natural rise of the mean high water predicted by using historic tide data has been taken into account in coastal protection works along the North Sea coast, for example, the Wadden Sea shoreline in Germany. There the Trilateral Expert Group (2010) has considered a 25 cm rise by 2050, the most realistic scenario. Coastal protection design must be adjusted when an accelerating rise of the relative mean sea level can be established, and for this, research on natural shore processes is essential. For example, an accurate model of the littoral sand drift in the surf zone capable of reproducing and predicting beach response to different storms and structures would be the tool used to redesign or retrofit these structures for future SLR. Until we have a better understanding of the linked processes of particle movement and morphodynamics (Figure 4.13), there is little reason to expect accurate prediction of the shoreline response to SLR.

4.5.7 Impact on Ports

Although the elevations of ports and harbors are established to provide efficient operation within a certain range of sea level fluctuations, the economics of these systems is such that increasing the working surface elevations of land- or pile-supported structures every 50 or so years should not present insurmountable problems as the century progresses. In fact, increased water depths resulting from SLR may provide some benefits in terms of reduced dredging. An historic example is the San Francisco Bay, where SLR has countered the effect of shoaling due to sediment from the San Joaquin-Sacramento River system during the second-half of the nineteenth century. This was the period during hydraulic mining for gold, which transported the tailings from the upstream mining operation into the bay (Krone 1979). SLR has created shipping benefits as navigation channels have not been completely blocked by the deposit.

Technical assessments of the vulnerability of soon to be installed infrastructure to expected rise in sea level is now essential. Due to the growth of the oil and gas industry and expected rise in sea level, an oil refinery in Trinidad and Tobago in the Caribbean Sea was required to move its port facility from its location at Point Fortin to a site on Irois Bay in the Gulf of Paria. Singh et al. (2011) reported an investigation in which coastal erosion was evaluated for different SLR scenarios for stabilization of the shore slope at the facility. It was recommended that the port facility be constructed to account for a 1.5 m rise in the sea level.

4.5.8 Impacts on Small Islands

As a sizeable fraction of the human population resides in low-lying coastal areas, the effects of a rise in the sea level in regions where resources are stringent is being given special consideration by the United Nations and other agencies concerned with the global environment. One category includes small islands with low tidal ranges, a virtually nonexistent continental shelf and a low population. A few islands have been experiencing a decrease in the relative sea level due to tectonic uplift. However, concern is for numerous islands where the relative mean SLR rate is positive (Kench and Cowell 2010). Examples are coral islands such as the Maldives, the Marshalls, Tokelau, Funafati, and so on, which are merely a meter or two above the present sea level. Several islands such as the Maldives (in the Indian Ocean) are entirely flat with elevations nowhere exceeding 3 m. A study at Funafati Island based on SLR for the 60-year period (1950–2009) found that the relative mean SLR has been three times the eustatic value due to tectonic subsidence (Dickinson 1999; Becker et al. 2012). Technical responses to such eventualities are available today, but techno-economically speaking, these responses are not viable everywhere (Khan et al. 2002; Church et al. 2006).

4.5.9 Impacts on River Deltas

Significant impacts from the population viewpoint have been occurring in regions encompassing the deltas of major rivers such as the Huang He, Changjiang, Mekong, Ganga–Brahmaputra, Sindhu, Congo, Mississippi, and Amazon. These deltas have high river and sediment discharges, although in some, the sediment load is now reduced due to human intervention. The Nile is somewhat different as there is practically no river discharge now in the delta region, which also means that the delta is starved of alluvial sediment. Subsidence of the delta has been exacerbated by construction of canals, which have redistributed sediment previously available for countering subsidence to agricultural areas (Stanley 1996).

Similar to events at the Nile Delta, damming the upland tributaries as well as pumping water for agriculture has reduced river discharge at the mouth of the Mississippi. This, in turn, has resulted in a shortfall of sediment needed to counter subsidence as consolidation of the deltaic deposit continues (Boesch et al. 1994). The mean rate of land loss in the delta region has been 42 km^2 $year^{-1}$ (Couvillion et al. 2011). This type of land loss is globally widespread, affecting both large and small rivers. Smaller deltas will sink entirely due to ongoing consolidation. The delta of the Frazer River in Canada is expected to sink by as much as 2 m by the end of the century (Mazzotti et al. 2009).

Extensive dike networks protect many low-lying deltaic areas against water levels that may occur with return periods of 25 to 50 years. A rise in sea level involves raising and strengthening these networks. Raising the dikes is necessarily accompanied by widening the cross section; however, the cost tends to go up as the square of the height of the dike, and more, if one gives consideration to the likely increase in the frequency and intensity of hurricanes and tropical cyclones resulting from global warming.

4.5.10 Coastal Ecosystems

In recent decades, pioneering scientific work has led to important methods to assess the value of ecosystems based on energy analysis and techniques in natural resource economics if specific fish, wildlife, ecological diversity, and aesthetic values can be included (Odum 1983). Yet, overall the level of scientific understanding of the coastal ecosystem function remains generally poor with vast consequences for the protection of coastal ecosystems such as the marshes. In some locations where the marshes have been generally stable in the past, future SLR scenarios predict major losses. Along the Georgia coast, presently available modeling approaches together with field and laboratory measurements and geographic information systems suggest that salt marshes may decline in area by 20% and 45%, respectively, under the mean and maximum IPCC estimates of SLR by the end of this century. The area of tidal freshwater marshes may decline by as much as 40% (Craft et al. 2009). The loss may be lower if it is found that one can expect excess sediment to be available at the desired rate to maintain marsh elevations.

Loss of intertidal habitat has great significance for a wide range of aquatic species, birds, and so on. An investigation of beaches of the Caribbean Island of Bonaire suggested that a SLR of about 0.5 m would result in the loss of about one-third of the present beach area and impact the population of nesting Hawksbill and Loggerhead turtles (Fish et al. 2005).

Uncertainty in model predictions over a period of nearly 100 years remains high. Underpinning this is the substantial ambiguity in ecological generalizations that are difficult to formulate because biological systems tend to have exceptionally specialized responses. Species coexist because they respond to the same macroenvironment of a region differently than other species do. Accordingly, the diversity of species in coastal areas represents an equivalent diversity of responses to the same basic environmental conditions. Detailed biological and ecological study of animals and plants may be required to adequately represent in a model the response of specific organisms to SLR. Because of the long response times and lack of adequate research focus, predictions from ecological models usually remain untested. The models are best thought of as hypotheses based on state-of-the-art understanding—a state that is highly incomplete.

Fortunately, as modeling improves, it has become increasingly possible to explore uncertainty in ecological responses by sensitivity analysis of mathematical models. In many cases, in addition to the coefficients, the selection of the governing relationships must be checked by sensitivity analysis. Common sense dictates that simple concepts should never be overlooked. Loss of nesting beach, for example, will surely reduce sea turtle and tern production at least in proportion.

Mention must be made of the development of the ecological/ecosystem modeling software Ecopath with Ecosim (EwE). This code, which is used widely, has three main components: (1) Ecopath: this is a static, mass-balanced snapshot of the system; (2) Ecosim: this is a time-dependent (dynamic) simulation module for policy exploration; (3) Ecospace: this is a spatial and temporal dynamic module for exploring impact and placement of marine-protected areas while accounting for the effects of advection and dispersion (Heymans et al. 2012; Christensen and Walters 2004). Pitcher and Cochrane (2002) contain numerous papers that describe applications of EwE, and Polovina (2002) provides the application of Ecosim to evaluate the effect of lobster fishery on endangered monk seals. Because of unknown ranges of the numerous coefficients of the equations of processes modeled and inadequate state-of-the-art understanding of physical–chemical–biological interactions, these models are at best run in the diagnostic mode with limited capabilities for long-term prediction.

4.6 Delta Committee Recommendations

The predicted rise in sea level and fluctuations in river discharge led the Dutch government to appoint an ad hoc Sustainable Coastal Development Committee, commonly called the Delta Committee, after an earlier body of the same name formed following the disastrous floods in 1953. The new committee was mandated to formulate a vision of long-term protection of the Dutch coast and hinterland. It was asked to develop recommendations not because a disaster had occurred, but to avoid one or more in the future based on an integrated assessment for the Netherlands. The central question posed was how the nation must ensure that future generations continue to find the country an attractive place in which to live and work, to invest and find personal fulfillment.

The committee's report (Delta Committee 2008) had two sets of recommendations: measures to be implemented by 2050 and post-2050 (going beyond 2100 where feasible) when SLR is expected to accelerate. Several recommendations are specific in terms of the tasks to be performed, location, and timing. They can be summarized here in somewhat general terms to illustrate the determination of this nation to use national planning that is critical to its survival.

1. For planning and design purposes, the committee selected a predicted relative SLR of 40 cm by 2050 and 0.65 to 1.3 m by 2100. It further indicated an expected rise of 4 m by 2200. These estimates for the low-lying areas take into account a land subsidence rate of 10 cm 100 year^{-1}.
2. For the flood defenses, it was recommended that the present safety level be increased by a factor of 10 before 2013. In some areas where even better protection is needed, the so-called Delta Dike concept is considered promising. These dikes are either so high or so wide and massive that there is virtually no likelihood that the dike will suddenly and uncontrollably fail. All measures to increase the flood protection levels are to be implemented before 2050.
3. Off the coasts of Zeeland, Holland, and the Wadden Sea Islands (Figure 4.15), flood protection must be maintained by beach nourishment (Dean 2003), possibly with relocation of the tidal channels. Beach nourishment must be done in such a way that the coast can expand seaward in the next century to provide an added value to the society.

FIGURE 4.15
The Netherlands and vicinity.

4. The massive storm-surge barrier at the mouth of the Eastern Scheldt estuary build after the 1953 floods should be maintained until 2050. A disadvantage of the barrier is the restriction it has placed on tidal movement and, as a result, the loss of the intertidal zone. This must be countered by nourishment using sand from the outer delta of the estuary. After 2050, the lifespan of the storm-surge barrier must be extended by engineering. This should be done up to a SLR of about 1 m (2075 at the earliest). If the barrier is no longer adequate, a "managed retreat" solution may be sought that largely restores the tidal dynamics with its natural estuarine regime while maintaining safety against flooding.
5. The Western Scheldt estuary should remain an open tidal system to maintain its ecological value and the navigation route to the major port of Antwerp (Belgium). Safety against flooding must be maintained by reinforcement of the existing dikes.
6. Subject to cost-effectiveness, measures must be initiated now to accommodate an increased discharge of 18,000 $m^3 \cdot s^{-1}$ from the Rhine River and 4,600 $m^3 \cdot s^{-1}$ from the Meuse River. These measures must be completed during the 2050–2100 period.
7. It was recommended that the level of the Lake Ijssel be raised by a maximum of 1.5 m after 2050. This will allow gravity-drainage from this lake into the Wadden Sea beyond 2100. This lake retains its strategic function as fresh water reservoir for the northern Netherlands, northern Holland and, in view of the progressive penetration of salt wedge in the Nieuwe Waterway (in Holland), for the western Netherlands. However, in the adjoining Lake Marker, no level change is recommended.
8. Implementation of the overall plan to sustain the quality of life was estimated to be $1.6 to $2.1 billion per year until 2050 and $1.2 to $2.0 billion per year between 2050 and 2100.

4.7 Points to Note

As a summary of the broad subject of SLR and coastal hazards (impacts), the following main points should be noted:

1. A eustatic rise in sea level (water volume increase, e.g., due to melting of polar ice and water density decrease due to temperature rise) is presently occurring and it is expected that it will continue to do so through the end of this century and beyond. This trend is predicted even in the absence of any further global warming either because

of natural causes or because of anthropogenic contributions. Since anthropogenic effects have much to do with the burning of fossil fuels, greenhouse gases, and so on, if significant, they will be additive to the eustatic rise (Held and Soden 2000). A reduction in fossil fuel use may occur (Odum and Odum 2001), although its impact on sea level may not become apparent during this century.

2. The main interest to coastal protection is related to the relative SLR, which has been increasing in many places and decreasing elsewhere. The main cases of relative sea level change are eustatic change, isostacy (tectonic warping of the earth's crust), and subsidence due to natural and anthropogenic factors.
3. If storm frequency and intensity increase, they will exacerbate the effects of relative SLR on the coast.
4. The relationship between the rate of relative sea level change in the anthropocene era, the lifespan of humans, and the durability of coastal infrastructure has been such that engineering response to impacts has been accomplished with a reasonable degree of success wherever resources have been available and at least some advanced planning has been carried out. At present planning, for example, increased protection by structures or managed retreat, has been envisaged in numerous countries over the 100-year timescale.
5. As the sea level rises, it may not be feasible to defend long coastlines such as large segments of the bird-foot delta of coastal Louisiana, against encroachment by sea. At the global scale even for much smaller areas, the cost will be high for defending low-lying, densely populated, and heavily urbanized locations where coastal impacts have not been assessed in advance.
6. The Marine Board (1987) report was cautionary in regard to needed action because the rise estimates over the twenty-first (this) century were generally lower than more recent predictions of Grinsted et al. (2009). However, over the past couple of decades, there is evidence that SLR is accelerating. Considering the limitation we have noted in available data, understanding of the processes and modeling, high-quality basic and applied research is needed starting immediately to implement the most appropriate technical responses as the century progresses. A significant role of ecologists in guiding ecosystems analysis for sea level impact assessment is reviewing available predictions of the major site-specific physical and chemical environmental changes such as loss of shoreline, watershed changes in precipitation, future availability of sediment influx, and salinity intrusion. Reviews and ensuing evaluations will lead to the development of much needed refinements in the scientific framework within which better educated choices for further analysis including modeling can be made.

Exercises

1. Local relative sea level in a region of the Northern Gulf of Mexico has risen 2.5 m over the past 1000 years.
    a. If the eustatic rise is 0.5 m over the same time period, calculate the local long-term subsidence rate.
    b. The calculated subsidence rate is much less than the rate calculated from local tide gages. Provide two alternative interpretations for this.

   The Sacramento-San Joaquin River Delta in California covers more than 2980 km^2 and is the largest estuary on the west coast of the United States. Water resource managers of the Delta need to evaluate the potential effects of sea level rise (SLR) on natural and managed ecosystems. To accomplish this, they need to develop two SLR scenarios: a low SLR estimate and a high SLR estimate. In the following two problems, you will have to calculate these two estimates.

2. Predict the local mean sea levels in the years 2050, 2075, and 2100 by fitting linear and quadratic equations to the annual mean sea level data at the following locations: (a) Townsville, Australia, (b) Juneau, Alaska, and (c) Vlissingen, Netherlands (1890–2010). Annual mean sea level data can be downloaded from http://www.psmsl.org/data/obtaining.
3. The low SLR scenario is to be developed using historically measured data at the San Francisco tide gauge (Figure 4.16). The solid vertical line seen between 1900 and 1910 indicates the 1906 earthquake.

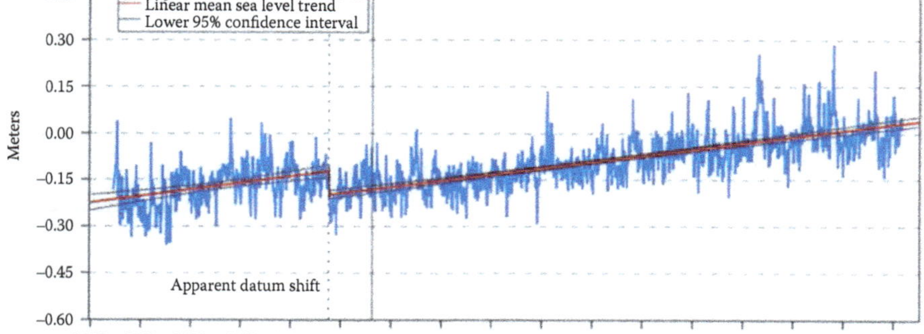

FIGURE 4.16
Sea level trend at the San Francisco tide gauge. (From NOAA, *Technical Report NOS CO-OPS 523*, U.S. Department of Commerce, National Ocean Service, Center for Operational Oceanographic Products and Services, Silver Spring, MD, 2009.)

Using this tide record, calculate the rate of SLR for the time of record after the apparent datum shift in 1897 and fill in the appropriate cells in Table 4.5. The red line in this figure shows the linear mean sea level trend. In 1897, the y-coordinate of the red line was −0.21 m, and in 2010, it was 0.04 m.

4. The high SLR scenario is to be developed using the modified NRC Curve III (USACE 2011) given below.
   a. Calculate the 50-year and 100-year SLR values for this scenario and fill in the appropriate cells in Table 4.5.

$$E(t_2) - E(t_1) = 0.0017(t_2 - t_1) + b(t_2^2 - t_1^2) \tag{4.6}$$

   where $E(t)$ = eustatic sea level change (in meters); t_2 = time between the projected time and 1986; t_1 = time between current time and 1986; and $b = 0.0001005$ m year^{-2} for the high SLR scenario.

   b. Compare the calculated low and high SLR values for San Francisco with the values given in Table 4.6.
5. Using data from the United States National Water Level Network (NWLOG), Zervas (2009) found the following regression relation between the ±95% confidence interval of linear MSL trends versus range of data in years:

$$y = 395.5x^{-1.643} \tag{4.7}$$

where y = ±95% confidence interval (mm year^{-1}), and x = year range of data. Calculate the ±95% confidence interval for the post-1897 tide

TABLE 4.5

Calculated Sea Level Rise Values for the Low and High SLR Scenarios

Sea Level Rise Scenario	50-Year Rise (m)	100-Year Rise (m)
Low		
High		

TABLE 4.6

Projected Sea Level Rise Values

Source	100-Year SLR Range (m)
California Climate Change Center–Projecting Future Sea Level Rise (CCCC 2006)	0.13–0.89
International Panel on Climate Change–Synthesis Report (IPCC 2007)	0.18–0.59
Delta Risk Management Strategy (DRMS)–Climate Change (DRMS 2008)	0.20–1.40

record at San Francisco shown in Figure 16. How should this estimated confidence interval be used by the water resource managers for the Sacramento–San Joaquin River Delta?

6. Equilibrium beach profile at some location follows $h = Ax^{2/3}$, where h = depth of active profile (m) at a seaward distance, x (m) and A = particle size dependent constant ($m^{1/3}$). How much will the shoreline recede in response to mean, 5 and 95 percentile sea level rise values in Figure 4.6, if berm height is 1.5 m and depth of active profile is 5 m? Assume $A = 0.153\ m^{1/3}$.

References

Becker, M., Meyssignac, B., Letetrel, C., Llovel, W., Cazenave, A., Delcroix, T., 2012. Sea level variations at tropical Pacific islands since 1950. *Global and Planetary Change*, 80/81, 85–98.

Blum, M. D., Roberts, H. H., 2009. Drowning of the Mississippi Delta due to insufficient sediment supply and global sea-level rise. *Nature Geoscience*, June 28, 2009, doi:10.1038/ngeo553.

Boesch, D. E., Josselyn, M. N., Mehta, A. J., Morris, J. T., Nuttle, W. K., Simenstad, C. A., Swift, D. J. P., 1994. Scientific assessment of coastal wetland loss, restoration and management in Louisiana. *Journal of Coastal Research*, SI 20.

Booij, N., Ris, R. C., Holthuijsen, L. H., 1999. A third generation wave model for coastal regions: I. Model description and validation. *Journal of Geophysical Research*, 104, 7649–7666.

Boon, J. D., 2004. *Secrets of the Tide: Tide and Tidal Current Analysis and Applications, Storm Surges and Sea Level Trends*. Horwood, West Sussex, UK.

Bruun, P., 1962. Sea level rise as a cause of shore erosion. *Journal of the Waterways Harbors and Coastal Engineering Division*, ASCE, 88(1), 117–130.

Bunya, S., Dietrich, J. C., Westerink, J. J., Ebersole, B. A., Smith, J. M., Atkinson, J. H., Jensen, R., Resio, D. T., Luettich, R. A., Dawson, C., Cardone, V. J., Cox, A. T., Powell, M. D., Westerink, H. J., Roberts, H. J., 2010. A high resolution coupled riverine flow, tide, wind, wind wave and storm surge model for southern Louisiana and Mississippi: partI-model development and validation. *Monthly Weather Review*, 138, 345–377.

California Climate Change Center, 2006. Projecting future sea level, *White Paper*.

CGER, 1996. *Data Book on Sea-Level Rise*, Center for Global Environmental Research, Environment Agency of Japan, Tokyo.

Christensen, V., Walters, C. J., 2004. Ecopath with Ecosim: methods, capabilities and limitations. *Ecological Modeling*, 172(2–4), 109–139.

Church, J. A., White, N. J., Coleman, R., Lambeck, K., Mitrovica, J. X., 2004. Estimates of regional distribution of sea level rise over the 1950–2000 period. *Journal of Climate*, 17, 2609–2625.

Church, J. A., White, N. J., 2006. A 20th century acceleration in global sea-level rise. *Geophysical Research Letters*, 33, L01602, doi:10.1029/2005GL024826.

Church, J. A., White, N., Hunter, J., 2006. Sea level rise at tropical Pacific and Indian Ocean islands. *Global Planet Change*, 53, 155–168.

Couvillion, B. R., Barras, J. A., Steyer, G. D., Sleavin, W., Fisher, M., Beck, H., Trahan, N., Griffin, B., Heckman, D., 2011. Land area change in coastal Louisiana from 1932 to 2010. *Pamphlet*, U.S. Geological Survey Scientific Investigations Map 3164 (scale 1:265,000).

Craft, C., Clough, J., Ehman, J., Joye, S., Park, R., Pennings, S., Guo, H., Machmuller, M., 2009. Forecasting the effects of accelerated sea-level rise on tidal marsh ecosystem services. *Frontiers in Ecology and the Environment*, 7, 73–78.

Daniels, R. C., Gornitz, V. M., Mehta, A. J., Lee, S. C., Cushman, R. M., 1992. Adapting to sea-level rise in the U.S. southeast: the influence of built infrastructure and geophysical factors on the inundation of coastal seas, *Report ORNL/CDIAC-54*, Environmental Science Division Publication No. 3915, Oak Ridge National Laboratory, Oak Ridge, TN.

D'Arrigo, R. D., Cook, E. R., Mann, M. E., Jacoby, G. C., 2003. Tree-ring reconstructions of temperature and sea-level pressure variability associated with the warm-season Arctic Oscillation since AD 1650. *Geophysical Research Letters*, 30(11), 1549, doi:10.1029/2003GL017250.

Das, N., 2012. Sea level rise and inundation of coastal India, blog.ivarta.com/Global-warming-Sea-level-rise-inundation-coastal-India/blog-206.htm.

Das, S., 2011. Examining the storm protection services of mangroves of Orissa during the 1999 cyclone. *Economic & Political Weekly*, xlvi(24), 60–68.

Davis, R. A., Fitzgerald, D. M., 2004. *Beaches and Coasts*. Blackwell Publishing Co., Victoria, Australia.

Dean, C., 2005. Hard choices seen in efforts to help Louisiana wetlands, *The New York Times*, http://www.nytimes.com/2005/11/10/national/nationalspecial/10marsh .html?_r=1. Accessed November 2, 2012.

Dean, R. G., Chiu T. Y., Wang, S. Y., 1990. *Combined Total Storm Tide Frequency Analysis For Lee County*, Florida, Report submitted to the Florida Department of Natural Resources, Beaches and Shores Center, Tallahassee.

Dean, R. G., 1991. Impacts of global change: engineering solutions. Our changing planet: joining forces for a better environment. *Proceedings of Symposium in Commemoration of the 20th Anniversary of the Graduate College of Marine Sciences*, University of Delaware, Newark, 13–17.

Dean, R. G., Dalrymple, R. A., 2002. *Coastal Processes with Engineering Applications*. Cambridge, UK: Cambridge University Press.

Dean, R. G., 2003. *Beach Nourishment: Theory and Practice*, Advanced Series on Ocean Engineering, World Scientific.

Delta Committee, 2008. Working together with water: A living land builds for its future: Findings of the Deltacommissie 2008, The Netherlands, www .deltacommissie.com.

Delta Risk Management Strategy (DRMS), 2008. Climate Change, *Technical Memorandum*, URS Corporation.

de Ronde, J. G., 1991. Rising waters: impact of the greenhouse effect for the Netherlands, *Report gwao 90.026*, Tidal waters Division, Rijkswaterstaat, The Hague.

Dickinson, W. R., 1999. Holocene sea-level record on Funafuti and potential impact of global warming on Central Pacific atolls. *Quaternary Research*, 51(2), 124–132.

Dietrich, J. C., Bunya, S., Westerink, J. J., Ebersole, B. A., Smith, J. M., Atkinson, J. H., Jensen, R., Resio, D. T., Luettich, R. A., Dawson, C., Cardone, V. J., Cox, A. T., Powell, M. D., Westerink, H. J., Roberts, H. J., 2010. A high resolution coupled riverine flow, tide, wind, wind wave and storm surge model for southern Louisiana and Mississippi: partII-synoptic description and analyses of hurricanes Katrina and Rita. *Monthly Weather Review*, 138, 378–404.

Dill, R., Elwany, M. H., Johnson, J., Marshall, N., 2006. Subsidence of King Harbor breakwater at Redondo Beach, CA. *Proceedings of the 30th International Conference on Coastal Engineering*, San Diego, CA, World Scientific, 5221–5228.

Dolan, R., Davis, R. E., 1992. An intensity scale for Atlantic Coast northeast storms. *Journal of Coastal Research*, 8(4), 840–853.

Douglas, B. C., Kearney, M. S., Leatherman, S. P., 2000. *Sea Level Rise: History and Consequences, Volume 1*. Academic Press, San Diego, CA.

Fagerburg, T. L., Alexander, M. P., 1994. Underwater sill construction for mitigating salt wedge migration on the Lower Mississippi River, *Miscellaneous Paper HL-94-1*, U.S. Army Engineer Waterways Experiment Station, Vicksburg, MS.

FEMA, 2005. Mississippi Hurricane Katrina surge inundation and advisory base flood elevation map panel overview, Federal Emergency Management Agency, http://www.fema.gov/pdf/hazard/flood/recoverydata/katrina/ms_overview.pdf. Accessed November 2, 2012.

Fish, M. R., Côté, I. M., Gill, J. A., Jones, A. P., Renshoff, S., Watkinson A. R., 2005. Predicting the impact of sea-level rise Caribbean sea turtle nesting habitat. *Conservation Biology*, 19(2), 482–491.

Fitzgerald, D. M., Kulp, M., Penland, S., Flocks, J., Kindinger, J., 2004. Morphologic and stratigraphic evolution of muddy ebb-tidal deltas along a subsiding coast: Barataria Bay, Mississippi River delta. *Sedimentology*, 51(6), 1157–1178, doi: 10.1111/j.1365-3091.2004. 00663.x.

Fleming, K., Johnston, P., Zwartz, D., Yokoyama, Y., Lambeck, K., Chappell, J., 1998. Refining the eustatic sea-level curve since the Last Glacial Maximum using far- and intermediate-field sites. *Earth and Planetary Science Letters*, 163(1–4), 327–342, doi:10.1016/S0012-821X(98)00198-8.

Fleming, K. M., 2000. Glacial rebound and sea-level change constraints on the Greenland Ice Sheet, PhD thesis. Australian National University, Canberra.

Gaur, A. S., Vora, K. H., 1999. Ancient shorelines of Gujarat, India, during the Indus civilization (Late Mid-Holocene): A study based on archaeological evidences. *Current Science*, 77(1), 180–185.

Gilman, E., 2005. Assessing and managing coastal ecosystem response to projected relative sea-level rise and climate change, Contribution to the Barbados+10 United Nations International Meeting on Sustainable Development of Small Island Developing States, Port Louis, Mauritius, January 10–14, 2005.

Gornitz, V., 1995. Sea-level rise: a review of recent past and near-future trends. *Earth Surface Processes and Landforms*, 20, 7–20.

Grinsted, A., Moore, J. C., Jevrejeva, S., 2009. Reconstructing sea level from paleo and projected temperatures 200 to 2100AD. *Climate Dynamics*, doi.org/10.1007/s00382-008-0507-2.

Gupta, H., ed., 2011. *Encyclopedia of Solid Earth Geophysics*. Springer, Dordrecht, the Netherlands.

Hardisty, J., ed., 2007. *Estuaries: Monitoring and Modelling the Physical System*. Blackwell Publishing Co., Victoria, Australia.

HAZUS-MH MR4 Hurricane Manuals, 2012. Federal Emergency Management Agency, U.S. Department of Homeland Security, Washington, DC.

Held, I. M., Soden, B. J., 2000. Water vapor feedback and global warming. *Annual Review of Energy and the Environment, Annual Reviews*, 25, 441–475.

Heymans, J. J., Coll, M., Christensen, V., Libralato, S., 2012. Ecopath theory, modelling and application to coastal ecosystems, Volume 9 of *Treatise on Estuarine and Coastal Science*, E. Wolanski, D. McLusky, D. Baird, A. Mehta, eds., Elsevier, 93–113.

Houston, J. R., Dean, R. G., 2011. Sea-level acceleration based on U.S. tide gauges and extensions of previous global-gauge analyses. *Journal of Coastal Research*, 27(3), 409–417.

Indu, J., Rao, A. D., V Jitendra, V., Dube, S. K., 2010. Computation of expected total water levels along the east coast of India. *Journal of Coastal Research*, 26(4), 681–687.

IPCC, 2007. *Climate Change 2007 Synthesis Report*, Contribution of Working Groups I, II and III to the Fourth Assessment Report of the Intergovernmental Panel on Climate Change, R. K. Pachauri, A. Reisinger, eds., Geneva. Cambridge University Press, Cambridge, UK.

Jagtap, T. G., Nagle, V. L., 2007. Response and adaptability of mangrove habitats from the Indian Subcontinent to changing climate. *Ambio*, 36(4), 328–334.

Jevrejeva, S., Grinsted, A., Moore, J. C., Holgate, S. 2006. Nonlinear trends and multiyear cycles in sea level records. *Journal of Geophysical Research*, 111, doi:10.1029/2005JC003229.

Jones, P. D., Mann, M. E., 2004. Climate over Past Millennia. *Reviews of Geophysics*, 42, RG2002, doi: 10.1029/2003RG000143.

Kench, P. S., Cowell, P. J., 2010. The morphological response of atoll islands to sea level rise. Part 2, Application of the modified shore face translation model. *Journal of Coastal Research*, 34, 645–656.

Khan, T. M. A., Quadir, D. A., Murty, T. S., Kabir, A., Aktar, F., Sarker, M. A., 2002. Relative sea level changes in Maldives and vulnerability of land due to abnormal coastal inundation. *Marine Geodesy*, 25(1–2), 133–143.

Khare, Y. P., Mehta, A. J., 2011. Sea level change and seashore settlement, *Gujarat and the Sea*, L. Varadarajan ed., Darshak Itihas Nidhi, Greater Noida, 33–48.

Knutson, T. R., 2011. Global warming and hurricanes: an overview of current research results, Geophysical Fluid Dynamics Laboratory, NOAA, September 3, 2008; Last Revised August 26, 2011, http://www.gfdl.noaa.gov/global-warming-and-hurricanes. Accessed November 2, 2012.

Knutson, T. R., Tuleya, R. E., 2004. Impact of CO2-induced warming on simulated hurricane intensity and precipitation: sensitivity to the choice of climate model and convective parameterization. *Journal of Climate*, 17(18), 3477–3495.

Kriebel, D., Dalrymple, R., Pratt, A., Sakovich, V., 1997. A shoreline risk index for northeasters. *Proceedings of the Conference on National Disaster Reduction*, ASCE, New York, 251–252.

Krone, R. B., 1979. San Francisco Bay: an urbanized estuary, *Sedimentation in the San Francisco Bay System*, T. J. Conomos ed., American Association for the Advancement of Science, San Francisco, 85–96.

Kumar, P. K. D., 2006. Potential vulnerability implications of sea level rise for the coastal zones of Cochin, southwest India. *Environmental Monitoring and Assessment*, 123, 333–344.

Laval, B. E., Imberger, J., Findikakis, A. N., 2005. Dynamics of a large tropical lake: Lake Maracaibo. *Aquatic Sciences*, 67(3), 337–349.

Leuliette, E. W., Nerem, R. S., Mitchum, G. T., 2004. Calibration of TOPEX/Poseidon and Jason altimeter data to construct a continuous record of mean sea level change. *Marine Geodesy*, 27, 79–94.

Lorito, S., Romano, F., Atzori, S., Tong, X., Avallone, A., McCloskey, J., Cocco, M., Boschi, E., Piatanesi, A., 2011. Limited overlap between the seismic gap and coseismic slip of the great 2010 Chile earthquake. *Nature Geoscience*, 4, 173–177.

Marine Board, 1987. *Responses to Changes in Sea Level: Engineering Implications*. National Academy, Washington, DC.

Mazzotti, S., Lambert, A., Van der Kooij, M., Mainville, A., 2009. Impact of anthropogenic subsidence on relative sea-level rise in the Fraser River delta. *Geology*, 37(9), 771–774, doi:10.1130/G25640A.1.

Mehta, A., Dean, R., Kunz, H., Law, V., Lee, S. -C., Tarapore, Z., 1997. Sea level rise and coastal hazards: an assessment of impacts and coastal engineering research needs, *Engineering Response to Global Climate Change: Planning a Research and Development Agenda*, R. G. Watts ed., Ch. 6, Lewis Publishers, Boca Raton, FL, 291–338.

Milne, G. A., Long, A. J., Bassett, S. E., 2005. Modelling Holocene relative sea-level observations from the Caribbean and South America. *Quaternary Science Reviews*, 24(10–11), 1183–1202, doi:10.1016/j.quascirev.2004.10.005.

Mitchum, G. T., 1998. Monitoring the stability of satellite altimeters with tide gauges. *Journal of Atmospheric and Oceanic Technology*, 15(3), 721–730.

Moberg, A., Sonechkin, D. M., Holmgren, K., Datsenko, N. M., Karlén, W., 2005. Corrigendum: Highly variable Northern Hemisphere temperatures reconstructed from low- and high-resolution proxy data. *Nature*, 443, 613–617, doi:10.1038/nature03265.

National Research Council, 1968. *The Great Alaska Earthquake of 1964, Volume1, Part1*. Committee on Alaska Earthquakes, National Academies Press, Washington DC.

National Research Council, 2010. *Adapting to the Impacts of Climate Change*. The National Academies Press, Washington DC.

NOAA, 2009. Sea level variations of the United States 1854–2006. *Technical Report NOS CO-OPS 523*, US Department of Commerce, National Ocean Service, Center for Operational Oceanographic Products and Services, Silver Spring, MD.

NOAA-NOS, 2010. Technical considerations for use of geospatial data in sea level change mapping and assessment, *Technical Report*, National Oceanic and Atmospheric Administration, National Ocean Service, U.S. Department of Commerce.

Odum, H. T., 1983. *Systems Ecology: An Introduction*. Wiley, New York.

Odum, H. T., Odum E.C., 2001. *A Prosperous Way Down*. University Press of Colorado, Boulder.

Philip, S., Van Oldenborgh, G. J., 2006. Shifts in ENSO coupling processes under global warming. *Geophysical Research Letters*, 33(11), doi:10.1029/2006GL026196.

Pitcher, T. J., Cochrane, K., eds., 2002. The use of ecosystems models to investigate multispecies management strategies for capture fisheries. *Research Report 10(2)*, University of British Columbia Fisheries Centre, Vancouver.

Polovina, J. J., 2002. Application of Ecosim to investigate the impact of the lobster fishery on endangered monk seals in Hawaii. *Proceedings of the INCO-DC Conference Placing Fisheries in their Ecosystem Context*, V. Christensen, G. Reck, J. L. Maclean, eds., *ACP-EU Fisheries Research Report 12*, Galápagos Islands, Ecuador, Brussels, 4–8 December 2000, 24–25.

Proshutinsky, A., Ashik, I. M. Dvorkin, E. N., Häkkinen, S., Krishfield R. A., Peltier, W. R., 2004. Secular sea level change in the Russian sector of the Arctic Ocean. *Journal of Geophysical Research*, 109, C03042, doi:10.1029/2003JC002007.

Rahmstorf, S., Vermeer, M., 2011: Discussion of: Houston, J. R., Dean, R. G., 2011. Sea-level acceleration based on U.S. tide gauges and extensions of previous global-gauge analyses. *Journal of Coastal Research*, 27(4), 784–787.

Ranasinghe, R., Callaghan, D., Stive, M. J. F., 2011. Estimating coastal recession due to sea level rise: beyond the Bruun rule. *Climate Change*, doi:10.1007/s10584-011-0107-8.

Rohde, R. A., 2009a. Post glacial sea level, http://en.wikipedia.org/wiki/File:Post-Glacial_Sea_Level.png. Accessed November 2, 2012.

Rohde, R. A., 2009b. File: 2000 year temperature comparison.png, http://commons.wikimedia.org/wiki/File: 2000_Year_Temperature_Comparison.png.

Saffir, H. S., 1977. Design and Construction Requirements for Hurricane Resistant Construction, Preprint no. 2830, ASCE.

Simpson, R. H., 1979. A proposed scale for ranking hurricanes by intensity, *Minutes of the Eighth NOAA-NWS Hurricane Conference*, Miami.

Singh, B., El Fouladi, A., Ramnath, K., 2011. Vulnerability assessment of the port and coastal infrastructure facilities of TRINMAR, on the Gulf of Paria, Trinidad, to sea level rise. *Ecology and the Environment*, (88), doi:10.2495/CENV060441.

Stanley, D. J., 1996. Nile delta: extreme case of sediment entrapment on a delta plain and consequent coastal land loss. *Marine Geology*, 129 (3–4), 189–195.

Tam, L., 2009. Strategies for managing sea level rise. *The Urbanist*, Nov/Dec issue, http://www.spur.org/publications/library/report/strategiesformanagingsealevelrise_110109. Accessed November 2, 2012.

Teferle, F. N., Bingley, R. M., Williams, S. D. P., Baker, T. F., Dodson, A. H., 2006. Using continuous GPS and absolute gravity to separate vertical land movements and changes in sea-level at tide-gauges in the UK. *Philosophical Transactions of the Royal Society A*, 364(1841), 917–930.

Trilateral Expert Group on Sustainable Coastal Defense of the Wadden Sea, (2010). Coastal protection and sea level rise, *CPSL Report III*, Common Wadden Sea Secretariat, Wilhelmshaven, Germany.

Unnikrishnan, A. S., Kumar K. R., Fernandes, S. E., Michael, G. S., Patwardhan, S. K., 2006. Sea level changes along the Indian coast: observations and projections. *Current Science*, 90(3), 362–368.

U.S. Army Corps of Engineers, 2011. Water resources policies and authorities incorporating sea-level change considerations in civil works programs. *EC 1165-2-211*, U.S. Army Corps of Engineers, Department of the Army, Washington, DC.

USGS, 2000. Land subsidence in the United States, *USGS Fact Sheet 165-00 (December)*, Ground Water Resources of the Future, U.S. Geological Survey, http://water.usgs.gov/ogw/pubs/fs00165. Accessed November 2, 2012.

Valle-Levinson, A., ed., 2010. *Contemporary Issues in Estuarine Physics*, Cambridge, UK: Cambridge University Press, 326.

Walton, T. L., 2007. Projected sea level rise in Florida. *Ocean Engineering*, 34, 1832–1840.

Wright, L. D., Thom, B. G., 1977. Coastal depositional landforms: a morphodynamic approach. *Progress in Physical Geography*, 1(3), doi:10.1177/030913337700100302.

Zervas, C., 2009. Sea level variations of the United States 1854-2006. *NOAA Technical Report NOS CO-OPS 053 and updates*. http://tidesandcurrents.noaa.gov/publications/Tech_rpt_53.pdf. Accessed November 2, 2012.

5

Water Resources

William H. McAnally, Phillip H. Burgi, Richard H. French, Jeffery P. Holland, James R. Houston, Igor Linkov, William D. Martin, Bernard Hsieh, Barbara Miller, Jim Thomas, James R. Tuttle, Darryl Calkins, Jose E. Sanchez, Stacy E. Howington, William R. Curtis, and Matteo Convertino

CONTENTS

5.1 Introduction

Current estimates of climatic change have significant implications for the hydrologic cycle and water resource system. This chapter provides an overview of the problems and issues likely to result from potential changes in the magnitude and distribution of water as a result of global climate change. Potential response strategies to cope with current hydrologic variability, as well as projected changes, are also examined. Finally, research and development needs are outlined to improve the flexibility, resiliency, and robustness of water resource systems to deal with the projected, but uncertain, climatic changes and present variability that stresses existing systems.

5.2 Overview

Global climate change is projected to alter the temporal and spatial patterns and ranges of temperature, precipitation, and evaporation. Changes in other important climate variables, such as solar radiation, cloud cover, wind, and humidity, are also predicted. These changes and their interactions could significantly alter key components of the hydrologic cycle, including runoff, soil moisture, groundwater recharge, and snowmelt patterns. The magnitude, distribution, and frequency of extreme climatic events, such as droughts and floods, may also be modified.

Currently, water resources projects are planned, designed, and operated based on historical patterns of water availability, quality, and demand, assuming the climate is statistically stationary. Changes outside the expected normal range of variation in these parameters can stress the ability of hydraulic structures to perform safely as designed and the ability of water resource systems to meet designated, and often competing, uses. Some examples of the potential effects of climatic change on water resources are described in Sections 5.2.1 through 5.2.2.

5.2.1 Consumptive versus Nonconsumptive Uses

Any discussion of water resources must recognize the substantial differences between the consumptive use of water, in which little or no flow returns directly to the environment, and the nonconsumptive uses, in which most of the water is returned to the local environment. Agriculture is primarily consumptive, since crops transpire water to the atmosphere. The water is still in the system, but is not locally available for reuse. Municipal use is primarily nonconsumptive, as most drinking, bathing, and cleaning water is returned and available as treated effluent.

5.2.2 Water Resources by Sector

Water resources are typically defined by use sectors—water supply for municipal, industrial, and agricultural uses; transportation by waterborne commerce; power generation; flood and storm damage reduction; recreation; water quality; and habitat preservation. Figure 5.1 illustrates these sometimes competing uses. Two other cross-cutting sectors influence our use of water—the infrastructure that contains, conveys, and otherwise makes water useful, and the institutions that regulate, preserve, and provide water quantity and quality. Each is examined briefly in this section.

Climate change will affect water resources through its impact on the quantity, variability, timing, form, and intensity of precipitation. There is already evidence that this is happening in some regions through the increased frequency of floods, droughts, and changes in long-term precipitation trend

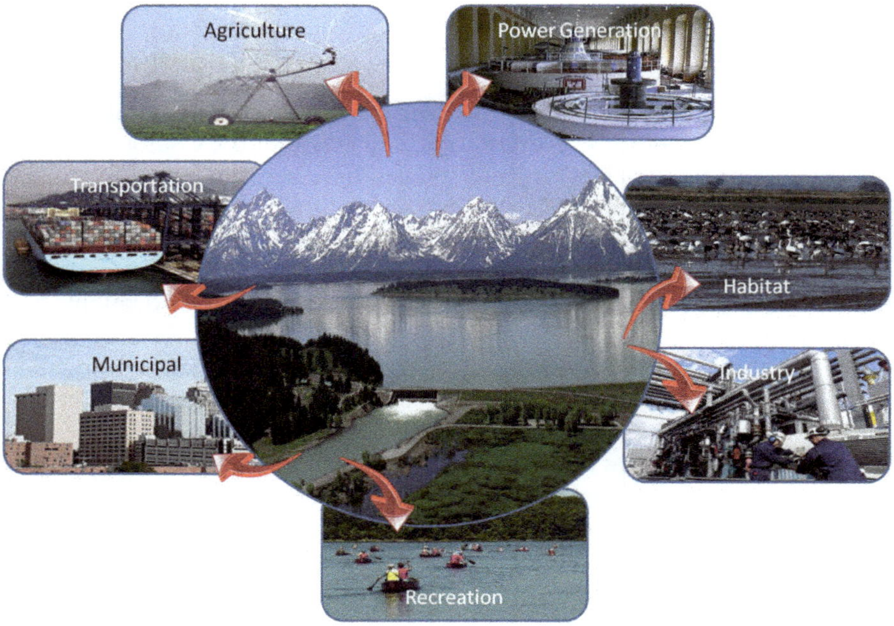

FIGURE 5.1
The many competing uses of water. (Photos courtesy of U.S. Bureau of Reclamation, Geological Survey, Corps of Engineers, Department of Agriculture, National Park Service, and Transportation Security Administration.)

(U. N. 2011). The ability to anticipate and efficiently prepare for future water resource management challenges is currently limited by imprecise regional climate change models and long-term weather forecasts. Uncertainty about future climate conditions and our will to adapt also make it difficult to prepare for and adapt to changes in water availability and quality.

5.2.2.1 Water Supply for Municipal, Industrial, and Agricultural Uses

Although increasing human demand for freshwater is the largest challenge facing water resource managers, substantially altered hydrologic cycles as a result of future climate change can make their task even more difficult and uncertain. Under most climate change scenarios, both the supply and demand of water will change, creating potential imbalances between the two or exacerbating existing imbalances. Reservoirs that store surface runoff from snowmelt or precipitation input may be either too small to accommodate the increase in demand or oversized for the change in requirements. The recharge and withdrawals from groundwater basins may be altered. Transmission and distribution systems (canals and conduits) that convey the water to the end user may be similarly impacted. Spillways that protect reservoirs may also need to be resized to accommodate the additional inflow or changes in the timing of the inflow hydrograph. Water users will need to adapt to more frequent and severe droughts, such as by shifting limited water supplies towards higher-value uses. Such shifts could be from low- to high-value crops, or from agricultural and industrial to environmental and municipal uses. Gradual changes in the frequency and severity of drought will be difficult to distinguish from normal variations in precipitation. Therefore, climate change adaption will likely be delayed.

Global food production depends on water not only in the form of precipitation, but also, and critically so, in the form of available water resources for irrigation. Indeed, irrigated land, representing a mere 18% of global agricultural land, produces 1 billion tons of grain annually, or about half the world's total supply; this is because irrigated crops yield, on average, two to three times more than their rain-fed counterparts (IPCC 2011).

The projected increase in world population, combined with a change in the climate, will challenge existing food production regions to increase their output. This may result in an increased demand for irrigation in some regions. If the climate shifts or changes, existing irrigation systems may not be adequate to accommodate the increased demand. Resource allocations to these systems will have to be adjusted, which will be difficult without accurate hydrologic predictions. In addition, agricultural areas that have traditionally not required irrigation may, under changed climatic conditions, require irrigation. Water banking—using groundwater storage or water transfer banks—will become more popular and necessary in mitigating economic impacts by increasing the reliability of water supply or facilitating the short term reallocation of water.

Existing conjunctive use of ground and surface water may be altered. Groundwater withdrawal in many areas has been shown to cause dramatic ground subsidence. High water withdrawal rates near the coastal regions often induce salinity intrusion, and any sea-level rise, combined with increased groundwater pumping, will magnify these problems. Groundwater recharge areas will also be affected by changes in the amount of precipitation, its form, and distribution.

5.2.2.2 Transportation

Land transportation is vulnerable to water impacts at crossings and near the margins of seas and lakes. Flooding due to sea-level rise and increases in the intensity of extreme weather events will pose threats to land transportation networks in some areas. These include localized street flooding, flooding of subway systems, and flood and landslide-related damages to bridges, roads, and railways (IPCC 2011).

Waterborne transportation relies on an inland and coastal network of waterways and structures that facilitate the passage of people and goods. Navigation channels and ports have been designed with channel depths and widths based on historical water levels, flows, and sedimentation rates.

Climate-induced changes could potentially alter these physical conditions, and thus affect channel and port navigational capacity. The result would be an alteration of operation and maintenance costs and a reduction of the confident use of the port, in addition to making the identification and justification of new facilities difficult.

See Chapter 9 for a further discussion of transportation impacts.

5.2.2.3 Flood Damage Reduction

Flood control structures (channels, dams and spillways, levees, retention and detention basins, and floodwalls) are designed to provide protection against events with an occurrence frequency based on statistics derived from historical data. For example, many levee systems are designed to contain a flood that happens, on average, every 100 years. However, potential climate changes may alter both the frequency distribution of storms that produce flood events and the magnitude of probable maximum floods; and thus, existing projects may have future protection requirements that are quite different from the initial design requirements.

Ongoing research suggests that flood event magnitude and frequency in a given location may be so altered by climate change as to render the period of record non-homogeneous and, therefore, of little use in predicting the future magnitude and frequency of extreme and non extreme events. It is predicted that climate change will increase the frequency and magnitude of droughts, floods, and destructive storms in specific regions. Floods are likely to become more problematic in many temperate and humid regions, necessitating advanced planning, flood forecasting, and even greater attention to well-developed emergency response networks to avoid significant loss of life and property.

Risk reduction not only saves lives; it is also less expensive than responding to a disaster. A number of countries have reduced the impact of disasters by investing in measures such as flood control, hurricane-proof building design, and protection of coastal ecosystems, including mangroves and coral reefs.

Society has become more vulnerable to natural hazards. Although floods are natural phenomena, human activities and human interventions into the processes of nature, such as alterations in the drainage patterns from urbanization, agricultural practices, and deforestation, have considerably changed the situation in whole river basins. At the same time, exposition to risk and vulnerability in flood-prone areas have been growing (U. N. 2000).

5.2.2.4 Power Production

Hydropower generation is likely to be directly affected by climate change because it is sensitive to the amount, timing, and geographical pattern of precipitation and temperature. Further, hydropower needs may increasingly conflict with other priorities, such as salmon restoration goals in the Pacific Northwest. However, changes in precipitation are difficult to project at the regional scale, which means that climate change may affect hydropower either positively or negatively, depending on the region. Hydropower is a nonconsumptive use in that the water used immediately returns to the stream. Fossil fuel and nuclear power are consumptive uses in which much of the cooling water is evaporated into the atmosphere.

Infrastructure for energy production, transmission, and distribution could be affected by climate change. For example, if a warmer climate is characterized by more extreme weather events, such as windstorms, ice storms, floods, tornadoes and hail, the transmission systems of electric utilities may experience a higher rate of failure, with attendant costs (IPCC 2007).

Power plant operations can be affected by extreme heat waves. For example, intake water that is normally used to cool power plants can become warm enough during extreme heat events that it compromises power plant operations.

Renewable energy sources will likely be affected by climate change, although these changes are very difficult to predict. If climate change leads to increased cloudiness, solar energy production could be reduced. Wind energy production would be reduced if wind speeds increase above or fall below the acceptable operating range of the technology. Changes in growing conditions could affect biomass production—a transportation and power plant fuel source that is starting to receive more attention (CCSP 2008).

5.2.2.5 Water Quality

Freshwater bodies have a limited capacity to process the pollution stemming from expanding urban, industrial, and agricultural uses. Climate-induced changes in the magnitude, timing, and quality of runoff will impact waste assimilative capacity, nutrient levels, water temperature, salinity, turbidity, and dissolved oxygen levels, which in turn affect riverine, reservoir, estuarine, wetland, and lake environments. The balance in the environmental conditions for some sensitive biota is very delicate. Waterfowl, raptors, and other higher trophic species are affected by water quality conditions.

One of the most significant sources of water quality degradation results from an increase in water temperature. The increase in water temperatures can lead to a bloom in microbial populations, which can have a negative impact on human health. Additionally, the rise in water temperature can adversely affect the different inhabitants of the ecosystem due to a species' sensitivity to temperature. The health of a body of water—such as a river—is dependent upon its ability to effectively self-purify through biodegradation, which is hindered when there is reduced dissolved oxygen. This occurs when water warms and its ability to hold oxygen decreases.

5.2.2.6 Habitat

Habitat, a species-dependent variable, is defined as the portion of landscape that is potentially suitable for the species considered (Storch et al. 2007; Convertino et al. 2009). A large number of studies have been published on the influence of climatological variables on ecosystems and singles species, but in this chapter, we focus on the effect of hydrological variables on biodiversity patterns for three water-controlled ecosystems.

Climate change will have a substantial effect on the habitats of species for a variety of taxa (UNESCO 2010; Westervelt and Hargrove 2010; Lozar et al. 2010a, b). Every species on earth is affected by the water cycle, either directly or indirectly, in that the entire food chain is directly affected by changes in water resources. For example, heavy rain associated with strong tropical cyclones will affect bird metapopulations (Convertino et al. 2010, 2011) and alter vegetation and pool-dependent invertebrate communities. Moreover, the portion of the rain that infiltrates underground will potentially affect bacteria communities. The connections and feedback within the food chain are enormous.

While changes in temperature are causing the expansion and contraction of many species' habitats (UNESCO 2010), together with phonological changes like breeding and flowering time, long-term changes in water resources are much more dangerous for species and biodiversity (Rodríguez-Iturbe et al. 2009). Changes in water resources have proven to modify species' composition and abundance at the landscape scale (Convertino et al. 2009). On the contrary, temperature variations do not produce perennial changes for species, and these changes determine spatiotemporal shifts that do not influence the whole ecosystem.

Water-controlled ecosystems (Figure 5.2) are complex, evolving structures whose characteristics and dynamic properties depend on many interrelated links between climate, soil, and vegetation. Different water variables are important for different ecosystems and species habitats (Rodríguez-Iturbe et al. 2009). To illustrate the relationship between water resources and habitat characteristics, we provide the examples of river networks, forests, and arid ecosystems, in which river runoff, rainfall, and soil moisture, are respectively the key water variables shaping habitats and biodiversity patterns.

In river networks (Figure 5.2a), the local habitat capacity (i.e., the number of individuals at each link) is directly proportional to the river runoff (Muneepeerakul et al., 2008). Though the runoff is certainly not solely responsible for all the changes in the species habitat, the topology of the species dispersal and the network connectivity are also key variables; variations in the river runoff consistently change species' habitat extent and abundance. For example, this has been verified for the entire Mississippi-Missouri River System (MMRS), where biodiversity patterns have been faithfully reproduced. For a more uniform river runoff, the local species richness (of freshwater fish and riparian vegetation) is, on average, very homogeneous, while

FIGURE 5.2
Water-controlled ecosystems: (a) River basin ecosystem in which the controlling driver for species (e.g., freshwater fishes and riparian vegetation) is the river runoff (Muneepeerakul et al. 2008) (Ichetucknee River). (b) Forest ecosystem in which the controlling variable is the average annual rainfall (Konar et al. 2010) (Pennsylvania forest new Delaware Water Gap). (c) Arid ecosystem in which the main controlling variable is the soil moisture (Rodríguez-Iturbe et al. 2001, 2009) (PaloDuroCanyonStatePark). (Photos courtesy of M. Convertino).

for a very heterogeneous runoff, the local species richness is smaller and the species' habitats are highly fragmented. Fragmentation causes an increase in the risk of extirpation (Akcakaya et al. 2004), increasing the probability that an epidemic, or even an extreme discharge will completely destroy the habitat of a species.

For forest ecosystems (Figure 5.2b), the average annual precipitation is the hydrological variable controlling biodiversity patterns (Konar et al. 2010; Convertino 2011). In forests, species dispersal is an intrinsic biological property that is unlikely to change as a function of climate change because evolutionary changes like these usually occur at much slower speeds than that of climate change. The change of habitat for species in a forest ecosystem has been simulated in the Mississippi-Missouri River System. Spatially explicit estimates of the impact of climate change under "species-poor scenarios" on region-averaged local species richness were calculated and resulted in a decreasing trend in the percentage of species lost from west to east of the MMRS. However, regions west of the 100° W meridian are composed of species-poor sub-basins, while those to the east encompass species-rich sub-basins, resulting in the increasing trend from west to east in the absolute loss in mean local species richness. The largest decrease in region-averaged local species richness occurs in the south, with 6.3 species, on average, lost per sub-basin across the region.

For arid ecosystems (Figure 5.2c), soil moisture is the key variable that synthesizes the action of climate, soil, and vegetation on the water balance and the dynamic impact of the water balance on plants (Rodríguez-Iturbe et al. 2001, 2009). Moreover, many ecosystems of tropical and subtropical latitudes suffer water stress, which is, in turn, controlled by the temporal fluctuations of soil moisture. Although other sources of stress (fire, grazing, nutrient availability, etc.) are certainly present, in many of the world's arid ecosystems, soil moisture is the most important resource affecting vegetation structure and organization. Changes in soil moisture patterns have proven to dramatically change patterns of plant biodiversity (e.g., from uniform to clustered patterns), thus having a profound impact on the spatial organization of single species habitats and their abundance. Like river and forest ecosystems, the fragmentation of arid ecosystems due to more frequent droughts increases the risk of species extirpation.

These three examples underline the importance of water resources on single species' habitats and ecosystem composition. Thus, in a changing climate, it is important to predict the potential effects of changes in water-controlling variables on species. While it is impossible to substantially modify the climate, it is possible to adopt sustainable plans that mitigate the effects of changing water resources on the earth's ecosystems. We reviewed the cases of river, forest, and arid ecosystems; however, coastal (affected by sea-level rise), sub-surface (affected by base flow), and many other types of ecosystems are also affected by changes in water-controlling variables that need to be managed.

5.2.2.7 Recreation

Water is critical for many recreation activities—from boating and fishing to mountain biking and backpacking. Impacts on water resources from climate change are pervasive and will vary by region and by season. They will likely alter winter snowpack, impact fishing and boating due to higher or lower stream flows and reservoir levels (depending on region), and change hunting and wildlife viewing opportunities.

In recent years, public demands have required those agencies with water control authority to allocate water resources for boating, fishing, rafting, swimming, and other recreational uses in addition to the original water use requirement of the designed facility. With climate changes, additional pressures for continued or increased recreational uses will create more demand for sustainable water supplies.

5.2.2.8 Infrastructure

An increase or decrease in the demand for water from surface supplies will impact both the water supply facilities and wastewater treatment facilities. Changes in water quantity and quality could require modified or new strategies for treatment.

Cities and towns are very vulnerable to climate change. Hundreds of millions of people in urban areas across the world will be affected by rising sea levels, increased precipitation (in some areas), inland floods, more frequent and stronger cyclones and storms, periods of more extreme heat and cold, and the spread of diseases. Climate change will likely have a negative impact on the sustainability of infrastructure and worsen access to basic urban services and the quality of life in cities.

The last century has seen the rapid urbanization of the world's population, as the global proportion of urban population rose from 13% (220 million) in 1900, to 29% (732 million) in 1950, to 49% (3.2 billion) in 2005. By 2050, over 6 billion people—two-thirds of humanity—will be living in towns and cities.

The changes in the average precipitation and seasonal distribution of rainfall and runoff due to climate change make the job of urban planners very difficult. In some places, the dry months get even drier and the wet months get even wetter. The engineering profession has serious challenges as it faces an uncertain future in how to adapt the infrastructure system to the long-term effects of climate change and the short-term shocks of extreme weather.

Today, the cities with the highest value of property and infrastructure assets exposed to coastal flooding caused by storm surge and damage from high winds are primarily in developed countries. However, the rapid economic development expected in the developing nations means that, in the future, the highest exposure becomes more concentrated in Asian cities, with eight of the top ten situated in this region. Over the coming decades, the unprecedented growth and development of the Asian megacities will be a key factor in driving the increase in coastal flood risk globally. Climate change could triple the population at risk from coastal flooding by 2070 (Nicholls and Hanson 2007).

The increasingly extreme weather—in particular, heavy storms and flooding—is severely affecting the living standards of millions globally. In 2011 alone, megafloods inundated one-fifth of the total land area in Pakistan and vast stretches in Queensland and Victoria in Australia, and the overflowing Mississippi River in the United States. It is estimated that since 1970, storms and floods were responsible for more than 90% of the economic costs of extreme weather-related events worldwide.

It is clear that if such "unusual" climatic events are visited upon us ever more regularly, then there will be practical limits to the adaptation, or at least exponentially rising costs involved in coping. There are a few general statements that have become true in today's climate uncertainty (Khor 2010):

- When natural calamity strikes, it can be—and nowadays, more often than not, it is—devastating.
- Climate crisis is for real, and its severe manifestations are more evident. It is often not easy to ascertain if an extreme weather event is due to climate change.
- Recent floods show again why climate change is an economic and social issue, though the cause may be environmental. As an example, recent floods in Pakistan have set back its development prospects by many years. Its leaders have estimated that the floods caused $43 billion in damage.
- The Pakistan case illustrates an acute deficit in the international approach to climate change. Despite the legal commitment of developed countries, and years of talks in U.N. Climate Change Conventions since 1992, there is still no international system for adequately or predictably financially assisting developing countries that have been affected by climate-change-related catastrophes.

5.2.2.9 Institutions

The competitive use of water is increasing, and with a change in climatic conditions, institutions and the legal and social framework for allocating water and resolving conflict may be forced to change. The pressure to meet competing interests and supply water for multiple uses can increase as a result of any climate change, producing contention and water wars, such as the challenges faced by the California Department of Water Resources in diverting water from the Sacramento-San Joaquin Delta to southern California, the Federal Bureau of Reclamation in apportioning Colorado River among the several western states that depend on it, and Corps of Engineers' management of Lake Lanier, which provides water supply to Atlanta and environmental flows to Georgia, Alabama, and Florida. Thousands of local water management districts and water utilities must deal with such challenges within a changing environment.

In a few places, basin-wide institutions are in place to manage water across U.S. political boundaries—the Tennessee Valley Authority system crosses seven states and the Delaware Basin Commission four states. State compacts cover a number of interstate waters, such as the Great Lakes Compact, without full management authority, but a great improvement over management by litigation.

5.2.2.10 Holistic Management of Water Resources

Multiple organizations are trying to move beyond the single project, single sector perspectives that have too long dominated water resources management. Projects have consequences at great distances, as demonstrated by disputes over Lake Lanier water among the states of Georgia, Alabama, and Florida, and federal agencies ranging from the Corps of Engineers to Fish and Wildlife Service. Disputes between advocates for economic development and the endangered pallid sturgeon of the southeast provide grist for multiple court disputes and public debates as advocates for single sectors make their respective cases.

As encouraging as these efforts to use a broader, more inclusive perspective are, they also raise the question of where to draw the boundaries of our broader examination. Aside from questions of scalability, we must balance our need to see the bigger picture with our ability to properly grasp what we are seeing. If we begin by paraphrasing Jacob Marley's cry* to say that the "whole earth is our business," we have properly recognized that the earth is a interconnected ecosystem, but have also overstepped our abilities to manage or even fully understand how it works.

The proper perspective for water-related management is the hydrologic footprint, or aquascape—the watershed plus water spread in the ocean, the landscape over which water flows to the ocean, and the coastal and ocean zone over which that water spreads and carries the material acquired during that journey. An aquascape perspective supports and reinforces integrated watershed management in its many forms, plus marine spatial planning, and the ecosystem approach to management, as discussed below. We use the phrase "holistic aquascape management" to denote the practice and process of achieving sustainable water resources use for the benefit of humans and the natural environment throughout the hydrologic footprint.

The word "holistic" has often been misused, but is so uniquely descriptive of the need that we are compelled to use it here. It is derived from the Greek *holos*, meaning "altogether" or "entire," which was defined

* "Mankind was my business!" (Dickens 1843)

by Aristotle (350 BCE) as, "the whole is greater than the sum of the parts." Jan Smuts* (1926) is credited with coining the English term "holism," which he described as "the tendency in nature to form wholes that are greater than the sum of the parts through creative evolution." The definition has been refined and applied in diverse fields, most vividly by Douglas Adams (1987) as the "fundamental interconnectedness of all things." Adams' definition helps to remind us, first, that economic development and a healthy ecosystem are fundamentally connected as interacting contributions to the quality of life, and second, that what happens in one part of an aquascape affects other, often unseen aspects and areas of the aquascape.

- Smuts' concept of holism was much more than interconnectedness. He saw it as an active force in the universe, responsible for organizing "wholes." He defined wholes as "… composites which have an internal structure, function, or character which clearly differentiates them from mere mechanical additions or constructions …." He described as wholes a water molecule (more than a simple mixture of hydrogen and oxygen atoms), cells (more than a collection of water, minerals, and organic molecules), an organism (more than a collection of cells), and the universe. We might add ecosystems, societies, and aquascapes to his list of wholes.
- Smuts presented holism as the "… ultimate synthesizing, ordering, organizing, regulating activity in the universe …." Examples of the interconnectedness of Smuts' wholes abound.
- Paine (1966) reported on a set of coastal ecosystems in which 15 large species existed in relative equilibrium. Removing the starfish from some of the systems resulted in a crash so severe that one year later, only eight species dominated, while the control systems remained in balance.
- Savory (1999) describes a lush, wildlife-rich Luangwa Valley in Zambia that was converted to a national park and game preserve by relocating local hunting and farming villages. Within a few decades, the landscape became denuded of vegetation, serious riverbank erosion occurred, and game species all but disappeared, because villagers were replaced by park employees and tourists.
- Weins and Roberts (2003) attribute the decline of bottomland hardwood wetlands along the Wolf River in Tennessee to headcutting, a stream erosion process that moves from a downstream disturbance (such as channelization) to upstream areas far from the original disturbance.

* Smuts was a military leader, statesman (the only person to sign the charters of both the League of Nations and United Nations), and scholar (Albert Einstein said that Smuts was one of only 11 people in the world who understood the Theory of Relativity).

Holistic management of water resources is related to concepts and terms such as "integrated water resources management," "total water management," "watershed management," and "regional management." Total water management is defined by the American Water Works Association Research Foundation (AWRA 1996) as

> ... the exercise of stewardship of water resources for the greatest good of society and the environment. A basic principle of total water management is that the supply is renewable, but limited, and should be managed on a sustainable-use basis.

The AWRA definition includes the concept of sustainability, which the American Society of Civil Engineers defines for water resources as, "Sustainable water resource systems are those designed and managed to meet the needs of people living in the future as well as those of us living today."

A frequent criticism of the sustainability concept is that it is idealistic and impossible—any use of resources is bound to decrease the amount available to future generations. However, that criticism is no more valid than saying that we need not strive for safety, since perfect safety is never achieved. Absolute environmental sustainability can be an ideal goal that is balanced with economic development and the cultural fabric of a region, which are implicitly included in the above sustainability definition.

The Corps of Engineers (USACE 2000) defines watershed perspective planning as

> ... accomplished within the context of an understanding and appreciation of the impacts of considered actions on other natural and human resources in the watershed. In carrying out planning activities, we should encourage the active participation of all interested groups and use of the full spectrum of technical disciplines in activities and decision-making. We also should take into account: the interconnectedness of water and land resources (a systems approach); the dynamic nature of the economy and the environment; and the variability of social interests over time. Specifically, civil works planning should consider the sustainability of future watershed resources, specifically taking into account environmental quality, economic development and social well-being.

The Tennessee Valley Authority (TVA) is often cited as the model for managing a watershed for multiple purposes. Chartered by the federal government in 1933, its intended purpose was "... in the interest of the national defense and for agricultural and industrial development, and to improve navigation in the Tennessee River and to control the destructive flood waters in the Tennessee River and Mississippi River Basins, ..." (U.S. 1933). TVA became an engine for not just economic growth, but also education, cultural preservation, and environmental stewardship, all centered around water management.

Another term that leads to many of the same conclusions as the watershed perspective is "systems," sometimes expressed as systems thinking, systems engineering, and so on, and appears in the Corps of Engineers' definition above.

5.3 Hydrometeorologic Cycle

As the atmosphere warms, its ability to hold water vapor will increase at an approximate rate of 5%–6% per °C (Rosenberg et al. 1989). Higher air temperatures will also increase surface evaporation. To maintain equilibrium in the moisture budget, the precipitation rate must also increase (MacCracken and Luther 1985).

The global average increases in precipitation and evaporation are presently estimated at 7%–15% for a doubling in CO_2 concentrations. These increases, however, will vary both spatially and temporally. Individual regions may actually experience reduced precipitation. The changes in the seasonal distribution of precipitation will also vary from region to region (Rind and Lebedeff 1984; Waggoner 1990). Moreover, recent studies indicate that when complex plant–climate interactions are considered, changes in evapotranspiration can vary from –20 to +40% in specific river basins (Martin et al. 1989).

The predicted changes in precipitation and evaporation are likely to be coupled with changes in other climatic variables, such as solar radiation, cloud cover, wind, and humidity. The range of these changes on a global basis are shown in Figure 5.3. Projected changes in these meteorological variables could significantly alter key components of the hydrologic cycle, including runoff, soil moisture, groundwater recharge and discharge, and snowmelt patterns. Due to the nonlinear relationship between precipitation and runoff, small changes in rainfall and evaporation can produce significant changes in runoff and regional water availability (Gleick 1986, 1989). Nemec and Schaake (1982) used hydrologic watershed models to show that a 10°C temperature increase, combined with a 10% decrease in precipitation, can cause a 25% reduction in average annual runoff in a humid basin and a 50% reduction in an arid basin. Using statistical correlations in the Colorado River Basin, Revelle and Waggoner (1983) estimated that a 20°C rise in temperature, even when coupled with a 10% increase in average annual precipitation, could still produce an 18% reduction in annual runoff due to increased evapotranspiration. Although more complex studies and further refinements in hydrologic analysis procedures may alter the magnitude of these results, the sensitivity of runoff to changes in meteorology remains an important consideration.

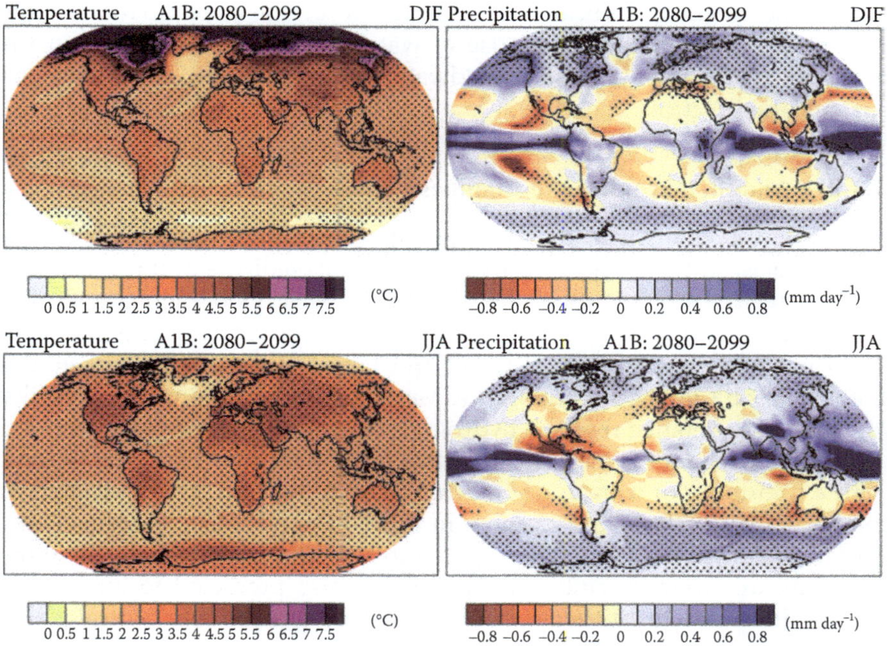

FIGURE 5.3
Changes in temperature and precipitation for 2080–2099 from two GCM. (Adapted from Solomon, S., et al., *Contribution of Working Group I to the Fourth Assessment Report of the Intergovernmental Panel on Climate Change*, Cambridge University Press, Cambridge, United Kingdom and New York, NY, 2007.)

5.3.1 Surface and Ground Water

Groundwater, in general, exists as either a perched water table or in an aquifer connected by some geologic means to the surface. Perched water typically has infiltrated deep into the ground and encountered an impermeable rock formation, properly formed to act as a large receptacle. There, the water collects and can remain for hundreds of years. The recharge of such formations is so slow as to be almost negligible. Therefore, once they are depleted, they are gone. Connected aquifers are recharged at a faster rate. Those connected directly to large rivers may recharge very rapidly. Those connected to the surface in remote areas require the recharge water to travel sometimes long distances before being available for use in the aquifer. The reliable yield rate for such aquifers can be calculated and water withdrawals matched to recharge rates. Climate change that alters the rainfall patterns over recharge areas will affect the safe rate of water withdrawal. Unfortunately, populations and land use have been established based on previously available ground water. The shifts to surface water use could start a domino effect of consequences. Alternately, surface water may not be a viable alternative.

This could result in radical land use change or the depopulation of the region. The most consumptive use of water is irrigation. Changes in the availability of surface and/or groundwater could have a profound effect on agricultural practices.

5.3.2 Variability, Trends, and Changes

In addition to changes in the average annual values of hydrometeorological variables, shifts in regional distribution, seasonality, extremes, variability, and recurrence frequency are also likely. Most global climate models (GCMs) predict distinct changes in the regional distribution and seasonality of key hydrologic variables (Rind and Lebedeff 1984; MacCracken and Luther 1985; IPCC 2007). Although GCMs currently have a limited capability for predicting detailed regional hydrologic effects, some general patterns are apparent. Surface air is expected to warm faster over land than over oceans, with a minimum of warming predicted to occur around Antarctica and in the northern North Atlantic region. In other high northern latitudes, however, warming in the winter is projected to be 50% to 100% greater than the global mean. Average winter precipitation is also predicted to increase in the middle and high latitude continents, including areas over central North American and southern Europe. The predicted changes in summer precipitation are variable, with recent models showing decreases in central North America and southern Europe and increases over areas such as Australia and southeastern Asia (IPCC 2011). Other investigators suggest that elevated temperatures and evaporation may be accompanied by decreased summer precipitation in the lower latitudes (Shiklamanov 1987; World Meteorological Organization 1988).

In the northern and western United States, where runoff is largely derived from snowmelt, distinct shifts in the relative amount of rain and snow, as well as earlier snowmelt resulting from warmer winters, are likely. The resulting changes in runoff patterns will alter the magnitude, timing, and probability of flooding patterns. The availability of water during peak demand periods, such as the irrigation season, would also be impacted. In the southeastern United States, where runoff is largely precipitation driven, some GCMs also project shifts in current seasonal patterns (Frederick and Gleick 1988; Smith and Tirpak 1988; Hains and Henry 1989), with attendant changes in vegetative land cover.

Changes in variability, or inter- and intra-annual deviations from mean conditions, can impact the adequacy and reliability of water resource projects. IPCC (2011) suggests that precipitation and seasonal runoff variability will increase in the future. For some parameters, the predictions are more uncertain and variability may stay relatively constant about a rising mean. For example, in the case of temperature, this latter case implies that changes in mean conditions would shift the entire temperature distribution upward, resulting in more days above some critical temperature on the high end and fewer days with temperatures on the low end of the distribution; but the deviation from the mean would remain constant. Other investigators have also found that, despite considerable noise in GCM results, the standard deviations of surface temperatures appear to be more likely to decrease than increase under global warming scenarios (Rind et al. 1989). These impacts on variability, however, are surmised from equilibrium-double CO_2 conditions after new mean conditions have been established. During the transition from current to new equilibrium conditions, variability (more higher highs and lower lows) may well increase.

Based on statistical reasoning, several investigators argue that small shifts in mean values can imply large changes in the frequencies of extreme events such as droughts and floods (Mearns et al. 1984; Waggoner 1990). The analysis of tropical cyclones and their relationship to warm low latitude oceans suggests that severe flood frequency could increase in a greenhouse-enriched environment (Michaels 1989). Some modeling evidence suggests that hurricane intensities would also increase with climatic warming (Emanual 1987; IPCC 2007), but the climate models do not give a consistent indication of whether tropical storms will increase or decrease in either frequency or intensity as the climate changes. At present, tropical storms, such as typhoons and hurricanes, develop over seas that are warmer than approximately 26°C. Although the areal extent of seas exceeding this critical temperature will increase in a warming scenario, the critical temperature itself might change under warmer conditions (IPCC 1990).

5.3.3 Impacts on Water Use

In the long term, alterations in the hydrologic cycle could be complicated by climate change-induced effects on plant growth and land use. Elevated CO_2 concentrations have been shown to increase photosynthesis and plant growth potential in laboratory studies. In these CO_2-enriched environments, plant transpiration is also generally reduced due to increased stomatal resistance (Waggoner 1990). Under natural conditions, the net effect of these plant responses to increased CO_2 levels will vary, depending on plant type, relative changes in a range of climatic variables, and individual ecosystems. Complex plant–climate interactions may moderate or augment the predicted increases in evapotranspiration, with consequences for plant growth and development (Rosenberg et al. 1989; Martin et al. 1989). These effects, coupled with changes in the hydrologic cycle, could ultimately alter vegetative patterns. Changes in the extent and type of vegetative cover could modify the water content of soil layers, thereby further influencing runoff and water availability (Abramopoulos et al. 1988).

Land use could be further impacted by changes in agriculture, forestry, population distributions, and industrial activity, with associated feedback implications for the hydrologic cycle. Higher temperatures, modified rainfall patterns, and increased CO_2 concentrations would affect agricultural practices, the type and distribution of crops, and irrigation needs. Similarly, the distribution and composition of forests, wildlife, and other natural ecosystems could be altered. Smith and Tirpak (1988) project potential demographic shifts, as well as possible changes in the location of industrial and agricultural centers. Such conversions in land use would not only affect the relative balance between water supply and demand, but would also impact the soil and land surface characteristics, thereby influencing important components of the hydrologic cycle, such as infiltration, interception, evapotranspiration, and groundwater recharge.

It is important to recognize that many of these impacts represent the potential consequences of an equilibrium-double CO_2 scenario. Although abrupt changes and large transients are possible (Schneider 1989), actual changes could occur more slowly over time or be moderated by other climatic factors. Changes in the hydrologic cycle may also work to the benefit or detriment of individual regions. For example, the IPCC (2011) predicts with high confidence that semi-arid areas of the western United States will experience a decrease in water availability.

5.3.4 Water Resources Demand

Water resource projects are generally planned, designed, operated, and maintained to accommodate historical ranges and patterns of climatic variability, based on the assumption of a statistically stationary climate. Current, as well as reasonable projected (based on past experience), patterns of relative water supply and demand are also incorporated in system designs. The projected changes in the magnitude, timing, and distribution of hydrometeorologic parameters—particularly if coupled with demographic shifts and changes in industrial and agricultural activity—could impact the safety of hydraulic structures, as well as the ability of water resource systems to effectively balance available supplies against competing water uses. This section provides representative examples of the potential beneficial and adverse impacts of climatic changes on water resources.

Climate change is likely to impact both water supply and water demand. The predicted changes in the hydrologic cycle—changes in precipitation patterns, evapotranspiration rates, temporal and spatial distributions in magnitude of runoff, and frequency and intensity of severe storms—will affect the quantity and quality of water supplies (Frederick and Gleick 1988). Long-term climatic impacts on demographics, industrial development, agricultural production, irrigation needs, natural systems, and energy use would ultimately impact water demand patterns (Smith and Tirpak 1988).

To address basin vulnerability to climate change with regard to water supply, one must consider the relative demand, or the ratio of demand (consumptive depletions, including consumptive use, water transfers, evaporation, and groundwater overdraft) to annual mean renewable supply. Water is considered a critical factor in economic development when the relative demand exceeds a value of 0.20 (Szesztay 1970). High relative demand ratios indicate that existing supply is susceptible to stress from growing populations, increased industrial and commercial demand, and climatic fluctuations. The Alaskan basin, for example, has essentially no storage, but as relative demand is also extremely low, the basin is not susceptible to drought for domestic or industrial consumption. Conversely, the Lower Colorado River Basin has a high relative storage capacity, but as consumptive use is 96% of renewable supplies, the users of runoff are vulnerable to drought. Basins such as the Upper and Lower Colorado, Rio Grande, Great Basin, and Missouri are vulnerable to climate change-induced reductions in supply (Frederick and Gleick 1988).

While relative storage and demand give some indication of the vulnerability of large river basins to climatic change, the range of potential impacts is dependent on other local factors, including legal regulations, compact, and treaties. The relative importance of in-stream water uses, such as hydropower production, recreation, navigation, wildlife, and aquatic habitat, can influence the flexibility a water resource system has to meet competitive water use demands and to reallocate water uses during times of scarcity. The climatic impacts on snowmelt and the timing of runoff would also influence the magnitude of impacts. For example, in the Sacramento-San Joaquin River Basin, Lettenmaier et al. (1989) found that the increased temperatures in four GCM scenarios produced major reductions in snow accumulation, resulting in increased winter runoff, but reduced spring and summer runoff. Although the total volume of water increased in these scenarios, Sheer and Randall (1989) found that the shift in seasonality resulted in an increased probability of spring flooding and substantially reduced water deliveries to consumers in the California Central Valley in the summer. Consequently, under current operating constraints and water allocation policies, water supplies during peak summer demand periods could become a problem.

5.4 Climate Change Issues

5.4.1 Water Supply for Municipal, Industrial, and Agricultural Uses

Changes in temperature, runoff, snowmelt, evapotranspiration, and other hydrological and meteorological factors could have major effects on irrigated agriculture. Peterson and Keller (1990) examined the potential effects of global climate change on irrigation in both the Western and Eastern parts of the United States. They computed a potential Net Irrigation Requirement (NIR) for four scenarios: (1) present conditions; (2) climate change of +30°C; (3) climate change of +30°C and +10% precipitation; and (4) climate change of +30°C and −10% precipitation. The NIR increased under the latter three scenarios compared to present conditions. The implication is that the percentage of cultivated land requiring irrigation will increase. Western states will find it increasingly difficult to maintain the present level of irrigation without developing new water sources through conservation and improved delivery system efficiencies. However, irrigation may well continue to increase in the east. Similar scenarios are likely in other parts of the world.

5.4.2 Flood Damage Reduction and Dam Safety

Flooding—water flowing beyond its normal confines, especially onto usually dry land—can result from a variety of hydrometeorologic events. Floods result from a combination of heavy precipitation from severe storms, snowmelt, combinations of rain and snowmelt, storm surge and/or wave effects; the physical characteristics of drainage basins and coastal reaches; and modifications to drainage basin characteristics. Both high intensity, short duration and low intensity, long duration precipitation events can exceed the storage capacity of a watershed, resulting in downstream flooding. The melting of either snow or ice (e.g., a glacier) in a watershed can also result in elevated downstream discharges, and warm rain on snow has often resulted in epic downstream flooding. In coastal areas, storm surges can result in flooding as the abnormal rise of water generated by a storm is over and above the predicted astronomical tide. In semi- and arid environments, the flood hazard is generally more related to the quickness and ferocity of the event than the magnitude. Flooding depends not only on the nature of the causative hydrometeorologic event, but also on the physical characteristics of the watershed, and the anthropogenic modifications that have been made in the watershed.

Engineered flood mitigation is typically based on estimating the magnitude of a design event with a specified frequency; for example, the 100-year event. The quantification of design events, whether from peak flow data or rainfall-runoff modeling, is based on statistical analysis that assumes a stationary time series. Therefore, climate change, by definition, will invalidate the theoretical basis of the analysis on which the design of all engineered flood mitigation structures is based. However, flood mitigation projects are designed with safety factors; therefore, climate change trends toward higher peak flows will, to a point, only reduce the margin of safety, not eliminate it. In addition, climate change may modify the time base of the design flood event hydrograph; for example, shorter or longer durations at or near peak flow. New, more flexible design and operation procedures will be needed. The innovative use of design concepts and policies that include overtopping embankments, emergency off-channel storage, fuse plugs, labyrinth spillway designs, and aggressive water harvest in urban and suburban areas are examples of flexible and robust designs and policies for adapting to the challenges of climate change.

Downstream flood protection can be maintained by incorporating expected climate change scenarios into reservoir operating rules, such as keeping reservoir water levels lower or higher. Changing reservoir operating procedures can have detrimental effects on multipurpose projects such as those incorporating hydropower production, recreation, water supply, and navigation. Conversely, reduced runoff and flood risk could facilitate the reallocation of flood storage space for other uses, such as raising reservoir levels to increase hydropower generation and recreational opportunities. These higher levels could have negative impacts on recreation as docking facilities and boat ramps are inundated.

This, as well as other discussions of the effects of climate change on flood hazard identification and mitigation, does not address three potential issues that could be important, but cannot be quantified at the present time. First, it is tacitly assumed that the fluid primarily responsible for flood hazard will remain Newtonian—that is, a water-sediment mixture—rather than non-Newtonian fluids associated with mud and debris floods. Second, it is assumed that climate change will result in a change of the average depth of precipitation; however, it will also likely result in a change in precipitation variance and temporal distribution. Any of these changes could have an effect on runoff, and hence, flood hazard and mitigation design. For example, under current climate conditions, Leopold (1951) and Bull (1964) demonstrated that a small change in the annual temporal distribution of precipitation caused the landscape to become unstable, with an increase in runoff and erosion. Therefore, the implications of changes in variance and temporal distribution must be considered. Third, flooding, which under current climate conditions is more an annoyance than a hazard, may become a serious concern. For example, there are major urban areas (Salt Lake City) and critical infrastructure (Edwards Air Force Base) located near terminal lakes that are currently subject to infrequent flooding, but could be a risk in the future with a changed climate.

5.4.3 Transportation

The frequency and duration of extremes, if increased due to climate change, will adversely impact the navigation industry. Extreme high flows creates problems with bridge clearances, produce wave damage to waterside structures, increase project maintenance (dredging, bank stabilization, structures, etc.), potentially terminate navigation for short periods of time, create shoaling problems in channels due to increased sediment inflow, and may require a modification of dam capacity for lock and dam structures to handle higher-than-anticipated flows effectively, or will lead to more periods of suspended navigation. High flows also increase fuel consumption of vessels as extra power is required for travel both with and against an increased current speed.

Extreme low flows will result in possible long-term channel closures, with enormous negative economic impacts. More hazardous navigation conditions resulting from low flows cause delays, and subsequent increased travel time from port to port. Increased maintenance of navigation aids will be required to insure safe operation within navigable waters. Shoaling patterns and shoaling rates can change, requiring an increase in maintenance dredging. Low flows in the winter increases the likelihood of an earlier ice cover and, depending on the temperature regime, a thicker ice sheet could be expected in cold regions. However, costs to the shipping industry for the Great Lakes due to a climate warming scenario has been investigated by Crissman (1988), who found that the duration of the lake and connecting river ice covers would decrease, permitting a longer shipping season without the hindrance of the ice. This results in an economic gain for this particular transportation mode in this region because of lower operating costs.

Increased rainfall amounts or changes in storm intensities will affect design parameters at highway, pipeline, and rail crossings of water gaps. The designs typically provide minimum clearance of the bridge low chord above a 25-, 50-, or 100-year flood profile. Changes in rainfall patterns and intensities may result in some structures being over-designed. This is of no particular safety concern, but replacement structures should reflect changed conditions. Conversely, other structures may be under-designed and experience flows that encroach on the bridge. This will lead to higher flowlines upstream, increased scour at the bridge piers and abutments, and possible failure/overtopping, threatening traffic and endangering lives. Additionally, utility crossings such as water, gas, oil and telecommunications lines may be similarly adversely affected. Failure of these transportation and communication systems due to scour undermining them or removing protective overburden and allowing secondary breaks will lead to the disruption of services and potential environmental pollution. River and estuarine ports may experience increased periods of time where facility loading/unloading operations are shut down due to higher or lower water levels than were originally anticipated when they were designed. There may be a need to design bridges (highway and rail) to withstand the higher runoff and sediment loads to which they will be exposed. Better data sets of present scour and channel capacities will be needed to estimate future conditions properly.

5.4.4 Power Production

Shifts in the magnitude and seasonality of streamflows would directly impact hydropower generation and capacity, as well as system flexibility and reliability. In a study of the TVA reservoir system (Miller and Brock 1988), a warm and wet climate change scenario (4°C increase in average annual temperature and 31% increase in average annual runoff, with monthly variations in runoff ranging from +73% in March to −28% in November) increased average annual hydropower generation by 16%.

Miller and Brock (1988) also evaluated the impacts of a dry climate change scenario (31% reduction in average annual stream flow) on reservoir system behavior and power production. Average annual hydrogenation was decreased by 24%. Reduced streamflows and operating heads also resulted in substantial losses in dependable hydrosystem capacity, particularly during the summer months when power demands for residential and commercial cooling are high under current conditions.

Crissman (1988) assessed the impact of climate change on the power production capability of the New York Power Authority system on the Niagara River. Decreased flows from the Great Lakes drainage area, coupled with increased lake evaporation, could have dramatic negative effects on the scheduling of power based on historical records.

Warmer temperatures and hydrologic changes could also impact the operation of fossil and nuclear plants. Elevated water temperatures adversely impact thermal efficiency and the power output of steam turbines. Higher air and water temperatures also reduce cooling tower effectiveness. Elevated water temperatures, which can be exacerbated by high air temperatures and reduced streamflows, can increase the potential for violating environmentally-based thermal discharge standards in fossil and nuclear plants. Nuclear power plants also have limits on the maximum intake water temperatures for auxiliary safety systems. When this limit is reached, the Nuclear Regulatory Commission requires plants to shut down for safety reasons. Increased water temperatures, therefore, can constrain nuclear power plant operations (Miller et al. 1992).

In coastal areas, the prospects of rising sea levels, increased storm surges, and salt water intrusion could also impact power production. Power plant siting, the location of cooling water intakes, transmission efficiency, and system reliability become important issues (EPRI 1989). In inland areas, water intakes may have to be relocated during extremely dry conditions.

See Chapters 7 and 8 for further discussion of energy.

5.4.5 Recreation

Reduced flows with lowered lake levels and releases will diminish recreational use of lakes and their tailwaters. In most cases, an increase in water supply will increase recreational use of waterways (Miller and Block 1988). However, the timing of these increases in water supply may conflict with recreational needs.

5.4.6 Water Quality

The warming of the earth's atmosphere will result in significant modifications in the hydrologic cycle and the quality of surface and ground waters. The water quality of surface water bodies, such as rivers and reservoirs, could be significantly stressed by temperature increases, reduced dissolved oxygen levels, and potentially higher nutrient loads. Presently stressed systems will be affected the most.

5.4.7 Habitat

Aquatic and riparian ecosystems will change and may be dramatically altered in response to climate change. First, as noted elsewhere in this chapter, climate change will likely result in a change in reservoir operating rules, which may result in either enlarged or decreased riparian areas around and downstream of the reservoir. A fundamental concern would be the stability of the "new" riparian areas. If climate change includes long periods of drought followed by long wet periods, the reservoir riparian areas may never become fully developed, from an ecologic viewpoint, or hydraulically connected to the stream in the downstream reaches. Second, significant changes in reservoir operating rules and hydrologic conditions could also result in discharge temperatures and temperature ranges that may not sustain downstream fisheries. Third, under some climate change scenarios, some perennial rivers will become ephemeral and vice versa. A closely related fourth issue concerns the fate of terminal lakes, which provide critical habitats in many locations under various climate change scenarios. It should be mentioned that riparian and aquatic habitat associated with many terminal lakes (e.g., Walker and Pyramid Lakes in Nevada; the Salton Sea in California; and the Dead Sea in Jordan and Israel) are in danger, even under the current climate, because of changing water demand in their tributary areas. In some cases, climate change will result in ecosystem destruction, and in others, enhancement. Fifth, aquatic ecosystems associated with coastal areas and estuaries also will be affected by climate change. It is hypothesized that estuaries associated with rivers having continental watersheds (e.g., the Mississippi, Colorado, Nile, and Mekong rivers) covering multiple climate zones will be less affected than estuaries associated with rivers emanating from regional watersheds within a localized climate zone (e.g., Sacramento River).

5.4.8 Secondary Impacts

Areas of increased rainfall will have increased flow downstream of water control structures such as dams. This may cause increased flooding and limit agriculture activities in low lying areas. In most areas, an increased water supply will increase agricultural production and/or decrease its cost. Other areas may experience decreased flows, directly impacting activities depending on surface water for irrigation, aquaculture, municipal, or industrial use. This could lead to ground water overdraft. In the case of long term navigation shutdown, in which goods are moved by alternate land transportation modes, a higher unit transportation cost will result.

5.4.9 Issues by Region

Some regions will experience different or exaggerated climatic impacts because they exhibit extremes of natural climate. These include regions of deserts, permafrost, glaciers, and tropical rain forests. The IPCC (2011) projects that runoff will increase by mid-century at higher latitudes and in some wet tropical areas, but decrease over some dry regions at mid-latitudes and dry tropics, due to decreases in rainfall and higher rates of evapotranspiration. Areas affected by drought are projected to increase in areal extent.

5.4.10 Policy, Planning, and Management

Extreme shifts in rainfall and temperature could force changes in public works policies and planning. Project management may require re-analysis and alteration to reflect different future conditions. Projects currently in the planning stage may be adversely affected by conditions existing upon their completion in 10–15 years and throughout their economic life expectancy of an additional 50–100 years.

Public works activities tend to be reactive, in that public funds are often unavailable until a disaster strikes. For example, federal flood control activities had been sought for the lower Mississippi valley for almost a century before the great 1927 flood. Two years later, a comprehensive project was formulated and approved for the valley. Public policy should be active, anticipating potential climate change and working to adapt to climate changes during the early stages of project planning/design, rather than reacting to a disaster.

Where public works missions are divided among agencies and political units, fragmentation of responsibility can lead to less effective use of water resources. In reacting to potential climate change, comprehensive planning that crosses political and institutional boundaries—an aquascape approach—will mitigate the impact of adverse changes.

5.5 Research and Development

While predictions of regional climate change remain poorly defined and uncertain, the key climate change issue becomes how to prepare in the intervening years. The planning and implementation time horizon for major water resources projects is on the order of 10–30 years, and operational lifetimes often exceed 50 years. It is wiser and, ultimately, more cost-effective to consider the prospects of climate change than to ignore its possibility. Thus, in the near-term, response efforts should be directed at education, assessment, and research, accompanied by the development of longer-term adaptation and mitigation strategies. It is important to note that there will be significant benefits from these efforts even if global climate change does not occur.

The most pressing needs for research in water resources relating to climate change revolve around predicting future conditions and their variability;

including understanding the uncertainty of the predictions, determining how these changes will impact issues of importance to humans and the environment (e.g., floods, droughts, water quality), and determining how to mitigate or cope with the impacts.

One area of needed research relates to improvements to downscaling models that take climate data at scales of 2–5 degrees of latitude and longitude from GCMs and provide the information at regional scales. The North American Regional Climate Change Assessment Program provides the highest-resolution dynamically downscaled data in the United States. There are issues with the downscaling model for this program that need to be resolved, including model boundary conditions, internal variability, and physics consistency. More importantly, this data set has not shown convincing skill in reproducing past events—an indication of problems with GCMs or the downscaling model. Moreover, the data are only for the A2 scenario of the Intergovernmental Panel on Climate Change (IPCC)—the scenario with the highest growth of carbon dioxide to the year 2100. Data are needed for the range of IPCC scenarios. In addition, spatial resolution is only 50 km, and research is needed to further downscale this data to typical regional watershed basins, including addressing issues relating to bias corrections.

A second downscaling approach uses the statistically downscaled World Climate Research Programme's CMIP3 Climate Projections, which provide statistical information on about a 12 km^2 resolution. The projections have fundamental issues, including their assumption that statistical relationships—methods used to relate large scale circulation conditions and local surface conditions, spatial interpolations methods, and handling of GCM bias—are invariant in a changing climate. Different statistically downscaling methods have yielded different results, thereby reducing confidence in statistically downscaled data. Downscaling extreme events also is highly problematic, given the low predictability of the events and unknowns of how climate change will affect the frequency and magnitude of these events. Further research is needed to prove the efficacy of methods used to develop statistically downscaled data.

Research is needed to develop models of the physical environment that can take downscaled climate information and perform calculations over large regional basins, but at small enough scales to resolve important features of water resource projects (e.g., widths of levees, dam flood gates, and water diversion channels). For many important water resource problems, the models must be three-dimensional and dynamically couple surface and ground water, since the depletion and recharging of aquifers is an important issue for future water availability.

Research is needed to improve the physics of models that determine surface runoff through vegetation and urban areas, percolation of water to groundwater, groundwater flows through different porous media, sediment transport on watersheds and in rivers, turbulence modeling, and other phenomena. Research is needed to understand how climate change impacts variables such as evapotranspiration, groundwater recharge, land cover changes that affect watershed sediment yield, and moisture storage in soils and snow packs.

Research is needed to determine how climate change will affect the frequency, magnitude, and duration of extreme events such as floods and droughts. It is believed that climate change will increase climate variability and extreme events might be more frequent and longer lasting. Improved predictive capabilities—from determining the heights of levee systems to needed capacities of water supply projects to handle droughts—will be critical for engineering projects. However, it is not clear how to estimate hydrologic events in a changing climate.

Climate change and variability may significantly affect the environment, and research and development (R&D) is needed to determine how climate changes will affect chemical and biological processes that affect humans and plants and animals, including endangered and invasive species. R&D will be necessary to develop robust models that can determine the interaction of physical, chemical, and biological processes that are affected by climate change and variability.

Research is needed on how water quality characteristics that depend on climatic variables may evolve in a changing climate and how man can cope with or mitigate changes. For example, high water temperatures can reduce dissolved oxygen levels and impact aquatic life. Changes in the timing, intensity, and duration of precipitation can significantly affect water quality. Less precipitation can reduce streamflow and cause lakes and reservoir levels to fall, producing less dilution of pollutants. The increased frequency and intensity of rainfall can increase the movement of pollutants and sediment into these bodies of water.

To cope with or mitigate the effects of climate change and variability, models are needed that can determine how mitigation activities affect the environment from the scale of single species to ecosystem-level scales. For example, if water is released from water-control structures because of downstream low water levels, will it benefit an endangered species and how does this benefit compare with the risk of having low water levels in reservoirs that may impact the subsequent need for the water by urban areas?

There are socioeconomic areas needing further research. For example, there are instances of the successful reuse of waste water. For example, Orange County, California, uses waste water to recharge depleting aquifers. In Northern Virginia, 1.4 million people receive about 20% of their drinking water (up to 90% during droughts) from treated sewage water. However, a plan similar to the Orange County plan to recharge depleting aquifers was defeated in the San Gabriel Valley, with opponents citing calling the project "Toilet to Tap." A plan to use waste water in San Diego County, California, was defeated partially because of perceptions on the safety of using wastewater. An Australian study concluded, "It is now generally accepted that social marketing or persuasion is ineffective in influencing people to use recycled water." It is not understood what convinces some of the need and safety of recharging depleting aquifers or supplementing drinking supplies and generates emotional reactions in others.

Research is needed to understand the uncertainties in future scenario projections. There are uncertainties in climate science, simulations of global climate, emissions scenarios, bias-correction and spatial downscaling methods of GCM data, the physics, chemistry, and biology of natural systems, and relationships of variables in ecological systems. All of these uncertainties have to be understood to have confidence in projections.

5.5.1 Assessment and Measurement

An increased understanding of the potential impacts of climate change and of climate sensitive activities can be used to improve our flexibility to deal with future change. Sensitivity analyses should be used in the near-term to determine the critical thresholds for individual components of water resource systems, while scenario analyses can be effective in identifying the vulnerability of the integrated system to potential climatic changes and to climatic variability.

Increasing competition for a less predictable and more variable water supply will require improved data collection capabilities to manage water resource systems. In addition, improved data sets are needed to provide complete data records for impact analyses. Actions should include the increased gathering of climate, streamflow, and water quality data, improved instrumentation and data transmission equipment, better methods of managing and analyzing data, increased use of remote sensing technology, improved real-time operations and flood warning systems. To increase the quantity and quality of collected data, it is essential to find less costly ways of collecting and analyzing data.

Impact assessments, including sensitivity studies as well as scenario analyses, should be initiated quickly. Sensitivity studies can determine the critical thresholds for individual components of the water resource systems by evaluating the impacts of incremental changes in temperature, flow, and other pertinent meteorological variables, such as humidity or wind speed. Scenario analyses can be used to identify the vulnerability of integrated systems to changes in climatic variables. Changes in seasonality, the frequency and magnitude of extreme events, and the magnitude and distribution of key hydrologic variables, should be explicitly addressed in the formulation of scenarios. The impacts on multiple system uses and purposes, including those associated with environmental and recreational impacts, should be assessed.

5.5.2 Adaptation to Change

Strategies should be developed to incorporate climate change uncertainty into water resources planning, with the ultimate aim of creating robust, flexible water resource systems. Such strategies include improved hydrologic analyses procedures; monitoring of current trends; basin-wide integrated water management; flexible institutions and enhanced inter-agency and international cooperation; improved mediation procedures to resolve competitive water use issues; improved water conservation strategies; realistic water pricing policies; and more responsive legal and institutional frameworks to deal with future change.

The fragmentation of water resource missions and budgets among various agencies can lead to decisions and plans that are optimum for one agency's mission, but not for the nation.

Centralized planning and control should not be the goal, but centralized policies emphasizing a systems approach that transcends agency mission boundaries should be. Similarly, water resources planning and development by one nation can adversely impact other nations and, in cases of very large-scale projects, the global community. Integrated resource management across national boundaries should be encouraged.

Given the uncertainty of existing GCMs and regional models, their best near-term use may be in defining potential limits of the important factors. This information could be used by project personnel and designers in bracketing possible changes in the future. Current procedures can then be reviewed to determine if changes are desirable. While this process specifies neither the magnitude nor direction of the climate changes that occur, it does allow more formal consideration of the impacts of existing climate variability that now takes place.

Adaption strategies should be geared toward the development of techniques that incorporate climate change uncertainty into long-range planning with the aim of creating robust, flexible water resource systems. As historical records may no longer adequately predict future trends, hydrologic analyses procedures, such as flood forecasting and water supply determinations, need to account for the possibility of increased variability and changes in the frequency and intensity of extreme events.

As water supply variability is reduced with increased storage in larger basins, the need for basin-wide integrated water management supported by flexible institutions will increase in significance. Similarly, enhanced institutional arrangements for inter-agency cooperation, as well as improved mediation procedures among water use stakeholders, may be needed to resolve competitive use conflicts. Improved water conservation strategies, combined with realistic legal and water pricing policies, will become important in areas where dry conditions are likely to prevail.

Prospects of climate change offer the opportunity to increase the resiliency, efficiency, and productivity of existing water resource systems. Risk analysis should be included in the project design and operation to account for variability. Projects that are influenced by possible radical climatologic changes may require design lives of shorter duration. More responsive and flexible operating plans for existing and planned reservoirs will be required, and drought contingency plans should be developed. Measures to improve and encourage water and energy conservation will require development. The procedures that national or international water resource agencies must go through to permit modification to projects will most likely require streamlining.

While a variety of meaningful studies can be conducted using existing technology in conjunction with sensitivity and scenario analyses, a more definitive quantification of potential global climate change impacts on water resources systems/infrastructure will require improved GCM predictions, increased understanding of the various hydrometeorologic and ecologic processes impacted by potential climatic changes, and improved analytical tools. These long-term research and development requirements are discussed below.

5.5.3 Mitigation by Water Resources Development

Partial mitigation of greenhouse forcing may be effected through accomplishment of two basic objectives presented below.

5.5.3.1 Increased Use of Alternative Water-Related Power Sources

Approaches to accomplish this objective include the development and use of innovative hydropower sources such as low-head (i.e., at navigation dams or irrigation projects) tidal power, and increased use of pumped storage capabilities or additional off-peak alternatives. Each of these are known to represent largely untapped sources of potential power that, heretofore, have not been deemed cost effective.

Within this same topic area, the heat exchange capabilities of various ground and surface water sources can be developed and used. As an example, the use of groundwater aquifer as heat sinks/sources can be explored. The potential of geothermal resources, including hot springs, geysers, and so on, can also be investigated.

5.5.3.2 Increased Efficiency and Conservation in Water Resources Activities

The accomplishment of this objective can be promoted through improvement of hydropower generator and turbine efficiencies; the application of petroleum industry extraction techniques to extend groundwater development; and the examination of wastewater treatment methodologies that minimize methane and carbon dioxide production. Additional conservation measures and more efficient water use practices, including the lining of irrigation canals, best management practices on agricultural lands, the minimization of evaporation from arid-region transmission channels and reservoirs, the reduction of water use in individual households, the use of wetlands to trap runoff and assimilate pollutants, and increased recycling in industrial processes, appear promising. These and other measures and practices should receive increased emphasis in the near future.

5.5.4 Education

The education of water resource managers and planners, as well as the public, to the long-term nature and seriousness of the climate change issue is essential if we are to be prepared for the future. The assumption of a stationary climate, or at least, the existence of a climate with predictable and limited variability, is ingrained in engineering training, design, and operation. Hydrologists and water resource planners may find it difficult to consider climatic shifts outside traditional expectations. Recent advances in the use of risk/reliability analysis should be endorsed and more widely taught. The utilization of multidisciplinary teams in problem solving, with all specialties involved adequately represented, will improve the transfer of knowledge among professionals from diverse disciplines and will improve problem solving capabilities.

Information and awareness programs should be instituted in both industrialized and developing countries to educate the public. The importance of water supply, wastewater treatment, water conservation, and the quality of the aquatic environment should also be taught in primary and secondary schools. Information on water and energy conservation measures imparted in such programs could result in substantial resource savings.

5.5.5 Future Needs

Climatic change has the potential to impact many interrelated water users, purposes of water resources systems, and the many technical disciplines associated with them. With regard to water resources systems/infrastructure, significant research and development investments will be required to quantify the impacts of climatic change and variability, and to crease methodologies/designs to mitigate those impacts. We believe that the two broad goals listed below must be accomplished if the impacts of climatic change and variability on water resources systems/infrastructures are to be dealt with effectively.

Goal 1: Robust, viable water resources systems that can accommodate present climate variability and potential climatic changes

Goal 2: Mitigation of greenhouse forcing by increased use of alternative water-related power sources and increased efficiency and conservation in water resources activities

The amplification of these goals, and the basic components of the research and development required to accomplish them, are discussed below.

In the context of this discussion, the terms robust, viable, flexible, and resilient have specific meanings as they relate to the water resource system. Robustness refers to the ability of a system to perform in a predictable, controlled fashion when conditions approach and exceed its design limits. Viable systems are those whose design, construction, and operation have adequate flexibility and resiliency to meet performance, economic, social, cultural, and environmental quality objectives in the face of climatic variability and potential climatic changes. Flexible systems can be used in ways not originally intended if conditions require a change. Resilient system maintain their flexibility and operating effectiveness without frequent repair and rehabilitation. Vulnerability refers to the degree to which a water resource system fails to provide robustness, flexibility, and resiliency.

Accomplishment of Goal 1, which would result in the design, construction/modification, and operation of water resources systems capable of adequately responding to climatic change and variability, will require the completion of four basic technical objectives associated with climatic variability and potential climatic changes as described below.

5.5.5.1 Creation of an Adequate Understanding of Hydrometeorologic and Environmental Responses to Climatic and Anthropogenic Influences

This will require the development of more refined global climate models and an increased use of mesoscale modeling in the development of regional climate change scenarios. To produce hydrologic and climatic data for use in water resources development and management, the data from these models would be provided at a spatial scale of 250 km^2 and a daily temporal scale. Improved data collection efforts, for the analysis of historical climatic trends and verification of models, should be initiated. The coupling of process-based ground and surface water models, with other models that focus on natural ecosystems, environmental quality, economics, climate, and land use, is required to provide the basic modeling framework to assess the impacts of potential climatic change and climate variability on water resources systems. However, several of the underlying descriptions of the physical, chemical, and biological processes within even the best of the current component models are known to require additional investigation. For example, many aspects of groundwater flow and transport, especially within the unsaturated zone, are poorly understood. Even more well-understood processes, such as watershed rainfall-runoff or turbulent three-dimensional hydrodynamic flow fields, have known inadequacies that will require significant research and development.

5.5.5.2 Assessment of the Resiliency and Vulnerability (Including from an Environmental Quality, Performance, and Structural Perspective) of Present Water Resources Systems and Infrastructures

This will require the development of engineering tools that provide the capabilities to perform integrated systems analyses. These analyses are essential because they provide a mechanism for assessing the river basin-wide, or even inter-basin, responses to climatic inputs. Without these tools, the potential for current water resources infrastructures to flex, but not break, in the face of climatic change cannot be assessed. Use of these tools, however, requires the establishment of evaluation criteria and indices of economic value that appropriately reflect the environmental, economic, social/cultural, and performance requirements and public perceptions. The development of these criteria and indices—in forms ranging from reservoir release temperature objectives to hydropower efficiency targets to measures of the relative worth of meeting one of these objectives at the expense of the other (as examples)—should be coupled with integrated systems analysis tools. The development of such criteria will allow for the evaluation and identification of the resiliency and vulnerability of existing systems and infrastructure to climatic changes. This assessment of resiliency and vulnerability must then be documented along with the prioritization of efforts to ameliorate areas with significant levels of vulnerability.

5.5.5.3 Development of Planning and Design Procedures for Water Resources Infrastructures That Meet Multiple Project Uses and Objectives

The development of water resources planning and design procedures that meet multiple uses and purposes requires the incorporation of risk/reliability and uncertainty analyses in those procedures. Current planning and design procedures used by many water resources development agencies worldwide, provide for limited consideration of potential climatic changes beyond those previously observed and recorded. The incorporation of risk and uncertainty concepts will allow for a more straightforward consideration of climatic change and variability in system design and operation.

Methods should also be developed for identifying new water supplies. Weather modification, increased use of desalination, improved conservation, and the reuse of storm and wastewater are among those concepts meriting investigation. The implementation of institutional changes that make water a commodity responding to market forces, while providing for basic needs, should be explored within various international socioeconomic and cultural frameworks as a potential means of improving water conservation.

5.5.5.4 Development, Modification, and Operation of Water Resources Projects That Meet Multiple Uses and Stated Objectives in the Face of an Uncertain Climatic Future

Having assessed the need for more robust water resources systems, and having developed the procedures required to plan and design these systems for the future, one would naturally implement these more "climate change-proof" systems. While construction activities are implicit in this statement, there are a number of additional themes that should be investigated. The modification of the legal and institutional constraints that limit water resources systems' flexibility should be considered. This may prove a monumental effort in that several decades of agency inertia and legislation will have to be overcome. In concert with this, the scope of participants who input decisions to water resources project operations should be broadened. Given the multiple, and often conflicting, uses these projects generally have, it is imperative that trust and cooperation between all interests be voiced openly and equally. This is extremely important in the design of a project, when changes in project features can often be easily made. Such coordination is equally important during the consideration of project operational changes, so that modifications to improve or enhance the accomplishment of certain project objectives are not done to the detriment of differing, and perhaps less obvious, concerns.

New construction materials and methods for rehabilitation/retrofitting of existing water resources infrastructure are also needed. Assuming that climatic changes and variability will require future modifications to water resources systems, the need for cost-effective rehabilitation measures will, undoubtedly, increase.

The development and use of knowledge-based tools for the implementation of flexible operations will improve the responses of water resources systems to climatic changes and variability. These tools, to be most effective, must be driven by real-time data.

5.6 Questions for Discussion

The prospect of climatic change poses serious challenges to water resources managers concerning the availability and quality of future water supplies. Understanding the implications of these changes and preparing for an uncertain future raises several important issues regarding the development, maintenance, and management of water resources in the coming years. These issues are posed as a series of questions that must be adequately addressed to effectively cope with a changing world.

Reliable predictions:

1. What are the prospects for cyclic and non-cyclic climate change in the near and long-term?
2. Can the uncertainties regarding current climate change predictions be narrowed?
3. Can global climate change projections be translated into reliable regional climate scenarios on temporal and spatial scales useful for water resources analysis and management?

Impacts and vulnerabilities:

4. Can more reliable regional climate scenarios be used effectively to assess the sensitivity, range of potential impacts, and critical vulnerabilities of water resources systems?

5. Are the analysis tools and data sets currently available adequate to conduct reliable impact analyses?
6. Can integrated systems analysis procedures be used effectively to incorporate interrelated disciplines, such as forestry, agriculture, and environmental sciences, into water resources analysis?

Flexibility to cope with an uncertain future:

7. Are current water resources systems designed for current estimated levels of variability sufficient to accommodate predicted changes in climate?
8. Can information regarding future predictions and impact assessments be effectively incorporated into water resources planning, design, construction, and operation such that systems are rendered more robust and flexible to cope with an uncertain future?
9. Can water resources planners and agencies move to a more proactive role in the recognition and resolution of potential problems?

Institutional and legal considerations:

10. Are current institutional and legal frameworks capable of effectively dealing with increased stresses on available water supplies?
11. Are procedures available to effectively address issues of equity, water allocation, and competitive uses if the relative balance between available supplies and demand is significantly changed?
12. How can mechanisms be developed to increase inter-agency and intergovernmental cooperation in the sustainable development of water resources?
13. Should water be treated as a commodity subject to market forces after basic needs have been provided for?

Mitigation:

14. How can water resource systems be used to help mitigate the production of greenhouse gases?

References

Abramopoulos, F., C. Rosenzweig, B. Choudhury. September 1988. Improved Ground Hydrology Calculations for Global Climate Models (GCMs): Soil Water Movement and Evapotranspiration. *J. Climate*, Vol. 1, pp. 921–941.

Adams, D. 1987. *Dirk Gently's Holistic Detective Agency*. Pocket Books, New York.

Akcakaya, H. R., V. C., Radeloff, D. J. Mladenoff, and H. S. He, 2004. Integrating Landscape and Metapopulation Modeling Approaches: Viability of the Sharp-Tailed Grouse in a Dynamic Landscape. *Conserv. Biol.*, Vol. 18, pp. 526–537. doi:10.1111/j.1523-1739.2004.00520.x.

AWRA. August 1996. American Water Works Association Research Foundation. Minutes of Workshop on Total Water Management, American Water Works Association, Seattle, WA and Denver, CO.

Bull, W. B. 1964. History and Causes of Channel Trenching in Western Fresno County, California. *Am J Sci.*, Vol. 262, pp. 249–258.

CCSP. 2008. Climate Change Science Program-Synthesis and Assessment Product 4.5 February 2008, http://www.climatescience.gov/Library/sap/sap4-5/final-report/sap4-5-final-all.pdf. Accessed March, 2012.

Convertino, M. 2011. Neutral Metacommunity Clustering and SAR: River Basin vs. 2-D Landscape Species Patterns, *Ecol Modell*, Vol. 222, pp. 1863–1879.

Convertino, M., G. A. Kiker, M. L. Chu-Agor, R. Munoz-Carpena, C. J. Martinez, M. Aiello-Lammens, R. H., Akcakaya, R. A. Fischer, I. Linkov. 2011. Integrated Modeling to Mitigate Climate Change Risk due to Sea-Level Rise of Imperiled Shorebirds on Florida Coastal Military Installations, In *Climate Change: Global Change and Local Adaptation*, I. Linkov, Ed. NATO Science for Peace and Security Series. Springer, Dordrecht, the Netherlands, pp. 765–798.

Convertino, M., J. B. Elsner, R. Munoz-Carpena, G. A. Kiker, C. J. Martinez, R. A. Fischer, I. Linkov. 2010. Do tropical cyclones shape shorebird habitat patterns? Biogeoclimatology of Snowy Plovers in Florida, *PLoS ONE*, Vol. 6, No. 1, p. e15683. DOI:10.1371/journal.pone.0015683.

Convertino, M., R. Muneepeerakul, A. Azaele, E. Bertuzzo, A. Rinaldo, I. Rodriguez-Iturbe. 2009. On Neutral Metacommunity Patterns of River Basins at Different Scales of Aggregation. *Water Resour Res*, Vol. 45, p. W08424, doi: 10.1029/2009WR007799.

Crissman, C. A. 1988. Impacts of Electricity Generation in New York, State, in Impacts of Climate Change in the Great Lakes Basin, Joint Report No. 1 of the U.S. National Climate Change Program Office and the Canadian Climate Centre, Sept. 27–29, Oak Brook, IL.

Electric Power Research Institute. 1989. Development of a Scoping Document to Assess the Potential Effects of Climate Change on Electric Utilities, prepared by ICF, Inc., for EPRI Risk Analysis Program, Palo Alto, CA.

Emanuel, K. A. 1987. The Dependence of Hurricane Intensity on Climate. *Nature*, Vol. 326, pp. 483–485.

Frederick, K. D., P. H. Gleick. 1988. Water Resources and Climate Change in Greenhouse Warming: Abatement and Adaption, In *Proceedings of a workshop held in Washington, D.C., June 1988*, Wagonner, et al., Eds. Resources for the Future, Washington DC.

French, R. H. 1987. *Hydraulic Processes on Alluvial Fans*, Elsevier Scientific Publishing Co, Amsterdam, The Netherlands.

French, R. H. 1989. Effect of the Length of Record on Estimates of Annual Precipitation in NV. *ASCE. J. Hydraulic Eng.*, Vol. 115, No. 4, pp. 493–506.

Gleick, P. H. August 1989. Climate Change, Hydrology, and Water Resources. *Rev. Geophysics*, Vol. 27, No. 3, pp. 329–344.

Gleick, P. H. 1989. Methods for Evaluating the Regional Hydrologic Impacts of Global Climatic Changes. *J. Hydrol.*, Vol. 88, pp. 97–117.

Hains, D. K., H. R. Henry. 1989. Issues of Detectability of Climate Change Impacts on the Runoff of a Southeastern River Basin, In *Proceedings of the 1989 National Conference on Hydraulic Engineering*, M. Ports, Ed. New Orleans, LA. American Society of Civil Engineers, Reston, VA.

IPCC. 1990, *Intergovernmental Panel on Climate Change, Climate Change: the IPCC Scientific Assessment*, J. T. Houghton, G. J. Jenkins, J. J. Ephramus, Ed. Cambridge University Press, Cambridge, UK.

IPCC. 2007, In *Climate Change 2007: Impacts, Adaptation, and Vulnerability. Contribution of Working Group II to the Fourth Assessment Report of the Intergovernmental Panel on Climate Change*, M. L. Parry, O. F. Canziani, J. P. Palutikof, P. J. van der Linden, C. E. Hanson, Eds. Cambridge University Press, Cambridge, United Kingdom, p. 1000.

IPCC. 2011. Climate change and water resources in systems and sectors, Chapter 4 IPCC. http://www.ipcc.ch/pdf/technical-papers/ccw/chapter4.pdf. Accessed March, 2012.

Khor, M. 2010. Lessons on Climate Change from the Pakistan Floods, http://triplecrisis.com/lessons-on-climate-change-from-the-pakistan-floods/. Accessed March, 2012.

Konar, M., R. Muneepeerakul, S. Azaele, E. Bertuzzo, A. Rinaldo, I. Rodriguez-Iturbe. 2010. Potential Impacts of Precipitation Change on Large-Scale Patterns of Tree Diversity. *Water Resour. Res.*, Vol. 46, p. W11515, doi:10.1029/2010WR009384.

Leopold, L. B. 1951. Rainfall Frequency: An Aspect of Climatic Variation. *Trans Am Geophys Union*, Vol. 32, No. 3, pp. 347–357.

Lettenmaier, D. P., T. Y. Gan, P. R. Dawdy. 1989. Interpretation of Hydrologic Effects of Climate Change in the Sacramento-San Joaquin River Basin, California, In *The Potential Effects of Global Climate Change, Appendix A – Water Resources*, J. B. Smith, D. A. Tirpak, Eds. EPA Office of Policy, Planning, Evaluation, Washington, DC.

Lozar, R. C., M. D. Hiett, J. D. Westervelt. 2010a. The Effects of Climate Change on Ecosystems, Erosion, TES and Invasive Species at Army Installations, a Preliminary Evaluation, ERDC Technical Report.

Lozar, R. C., M. D. Hiett, J. D. Westervelt. 2010b. Climate Change Impacts and Adaptation on Conus Military Installations. In *Change and Local Adaptation. Modelling*, I. Linkov, Ed. Global NATO.

MacCracken, M. E., F. M. Luther. December 1985. *Projecting the Climatic Effects of Increasing Carbon Dioxide*, DOE/ER-0237. U.S. Department of Energy, Washington, DC.

Martin, P., N. J. Rosenberg, M. S. McKenney. 1989. Sensitivity of Evapotranspiration in a Wheat Field, a Forest, and a Grassland to Changes in Climate and Direct Effects of Carbon Dioxide. *Climatic Change*, Vol. 14, pp. 117–151.

Mearns, L. O., R. W. Katz, S. H. Schnieder. 1984. Changes in the Probabilities of Extreme High Temperature Events with Changes in Global Mean Temperature. *J. Climate App. Meteorol.*, Vol. 23, pp. 1601–1613.

Michaels, P. J. 1989. *Observed and Projected Climate Change: Fact and Fiction*, Dept. of Environmental Sciences, University of Virginia.

Miller, B. A., V. A. Alavian, M. D. Bender, D. J. Benton, M. C. Shiao, P. Ostrowski, J. A. Parsly. 1992. Sensitivity of the TVA Reservoir and Power System to Extreme Meteorology, TVA Engineering Laboratory.

Miller, B. A., W. G. Brock. 1988. Sensitivity of the Tennessee Valley Authority Reservoir System to Global Climate Change, Report No. WR28-1-680-101, TVA Engineering Laboratory, Norris, TN.

Muneepeerakul, R., E. Bertuzzo, H. J. Lynch, W. F. Fagan, A. Rinaldo, I. Rodriguez-Iturbe. 2008. Neutral Metacommunity Models Predict Fish Diversity Patterns in Mississippi-Missouri Basin. *Nature*, Vol. 453, pp. 220–222, doi:10.1038/nature06813.

Nash, L. L., P. H. Gleick. 1990. Sensitivity of Streamflow in the Colorado Basin to Climatic Changes, Paper Presented at Workshop: Implications of Climate Change for the Colorado River Basin, May 17–18, 1990, Denver, CO, Sponsored by the Pacific Institute for Studies in Development, Environment, and Security and the Institute for Resource Management.

Nemec, J., J. Schaake. 1982. Sensitivity of Water Resource Systems to Climate Variation. *Hydrol. Sci.*, Vol. 27, No. 3, pp. 327–342.

Nicholls, R., S. Hanson. 2007. Climate change could triple population at risk from coastal flooding by 2070, finds OECD Organisation for Economic Co-operation and Development (OECD), http://www.oecd.org/document/4/0,3343,en_2649_201185_39727650_1_1_1_1,00.html. Accessed March 2012.

Paine, R. T. 1966. Food Web Complexity and Species Diversity. *American Naturalist*, Vol. 100, No. 910:65–75.

Peterson, D. F., A. A. Keller. March/April 1990. Effects of Climate Change on U.S. Irrigation. *J. Irrigation Drainage*, Vol. 1. No. 2, p. 116.

Revelle, R. R., P. E. Waggoner. 1983. *Effects of a Carbon Dioxide-Induced Climatic Change on Water Supplies in the Western United States, in Changing Climate*, National Academy Press, Washington, DC, pp. 419–432.

Rind, D., F. Lebedeff. April 1984. Potential Climatic Impacts of Increasing Atmospheric CO2 with Emphasis on Water Availability and Hydrology in the United States, Prepared for U.S. EPA by the NASA Goddard Institute for Space Studies.

Rind, D., R. Goldberg, R. Ruedy. 1989. Change in Climate Variability in the 21st Century. *Climatic Change*, Vol. 14, No. 1, pp. 5–37.

Rodriguez-Iturbe, I., A. Porporato, F. Laio, L. Ridolfi. 2001. Plants in Water-Controlled Ecosystems: Active Role in Hydrologic Processes and Response to Water Stress I. Scope and General Outline. *Adv. Water Resour.*, Vol. 24, pp. 695–705.

Rodríguez-Iturbe, I., R. Muneepeerakul, E. Bertuzzo, S. A. Levin, A. Rinaldo. 2009. River Networks as Ecological Corridors: A Complex Systems Perspective for Integrating Hydrologic, Geomorphologic, and Ecologic Dynamics. *Water Resour. Res*, Vol. 45, p. W01413, doi:10.1029/2008WR00712.

Rosenberg, N. J., M. S. McKenney, P. Martin. 1989. Evapotranspiration in a Greenhouse-Warmed World: A Review and Simulation. *Agri. Forest Meteorol.*, Vol. 47, pp. 303–320.

Savory, A. 1966. *Holistic Management: A New Framework for Decision Making*. Island Press, Washington, DC.

Schnieder, S. H. 1989. The Greenhouse Effect: Science and Policy. *Science*, Vol. 243, pp. 771–781.

Sheer, D. P., D. Randall. 1989. Methods for Evaluating the Potential Impacts of Global Climate Change, In *The Potential Effects of Global Climate Change, Appendix A Water Resources*, J. B. Smith, D. A. Tirpak, Eds. EPA Office of Policy, Planning and Evaluation, Washington, DC.

Shiklamanov, J. August 1987. Changes in Runoff in Soviet Rivers Due to Climate Change, Paper presented at the International Union of Geodesy and Geophysics Symposium, Vancouver, British Columbia.

Smith, J. B., D. A. Tirpak. October 1988. *The Potential Effects of Global Climate Change on the U.S., Draft Report to Congress, Executive Summary*, EPA Office of Policy, Planning and Evaluation, Washington, DC.

Smuts, J. C. 1926. *Holism and Evolution*, Reprinted in 1999 by Sierra Sunrise Publishing, Sherman Oaks, CA.

Solomon, S., D. Qin, M. Manning, Z. Chen, M. Marquis, K. B. Averyt, M. Tignor, H. L. Miller (eds.). 2007. *Contribution of Working Group I to the Fourth Assessment Report of the Intergovernmental Panel on Climate Change*, Cambridge University Press, Cambridge, United Kingdom and New York, NY, USA.

Storch, D., P. Marquet, J. Brown (eds.). 2007. *Scaling Biodiversity, Series: Ecological Reviews*. Cambridge University Press, Cambridge, UK.

Szesztay, K. 1970. The Hydrosphere and the Human Environment: Results of Research on Representative and Experimental Basins, Proceedings, Wellington Symposium, UNESCO Studies and Reports in Hydrology, No. 12.

U.N. 2000. Best Practices on Flood Prevention, Protection and Mitigation, United Nations and Economic Commission for Europe (U.N./ECE) 2000—Guidelines for sustainable flood prevention, The Hague, Netherlands, 23–25 March 2000, http://www.floods.org/PDF/Intl_BestPractices_EU_2004.pdf. Accessed March, 2012.

U.N. 2011. Climate Change Strategy 2010–2013 U.N. Habitat, http://www.unhabitat.org/pmss/listItemDetails.aspx?publicationID=2861. Accessed March 2012.

U.S. 1933. United States Code 16U.S.C. § 831, et seq. Tennessee Valley Authority Act of 1933.

UNESCO. 2010. A world of science—a biocultural alliance. Technical Report 3, UNESCO, http://unesdoc.unesco.org/images/0018/001884/188440e.pdf. Accessed March 2012.

USACE. 2000. Engineer Regulation 1105-1-100, Planning Guidance Notebook, 2000, http://140.194.76.129/publications/eng-regs/. Accessed March 2012.

Waggoner, P. E (ed.). 1990. Climate Change and U.S. Water Resources, Report of the AAAS Panel on Climatic Variability, Climate Change and the Planning and Management of U.S. Water Resources, John Wiley & Sons, New York.

Weins, K., and Roberts, T. 2003. Effects of Headcutting on the Bottomland Hardwood Wetlands Adjacent to the Wolf River, Tennessee. WRP Technical Notes Collection (ERDC TN-WRP-HS-CP-2.1), U.S. Army Engineer Research and Development Center, Vicksburg, MS. http://www.wes.army.mil/el/wrtc/wrp/tnotes/tnotes.html. Accessed November 2012.

Westervelt, J., W. Hargrove. 2010. Forecasting climate-induced ecosystem changes on Army Installations, ERDC Technical Report, in press.

World Meteorological Organization. 1988. Water Resources and Climatic Change: Sensitivity of Water Resource Systems to Climate Change and Variability, World Climate Applications Program-4, Geneva, Switzerland.

6

Energy Demand, Efficiency, and Conservation

Hadi Dowlatabadi and Maryam Rezaei

CONTENTS

"There's plenty of room at the bottom."

Richard P. Feyman

6.1 Introduction

The history of human societies can generally be characterized by an ever-increasing pattern of energy use. As societies transition from hunting and gathering organizations to primitive and advanced agriculture, and then on to industrial and technological ones, they tend to engage in more energy-intensive forms of production. However, even as our overall energy consumption grows, we have found ways of performing the same activities more efficiently (often, as far as the time input is concerned, but occasionally reducing energy inputs as well), thereby freeing up time and other resources needed to contemplate and devise more sophisticated and elaborate activities in which to engage.

Energy consumption patterns have changed dramatically as we have adopted different forms of social organization (see Figure 6.1). As an example, note that energy consumption in the production of food has, at most, tripled while we have changed our methods from gathering nuts and fruits, to rain-fed local cultivars, to hybridized cultivars needing fertilizers, herbicides, irrigation, and mechanized harvesting. Yet other defining characteristics of energy consumption have led to overall energy consumption rising 30-fold, over the centuries. This dramatic increase in our energy consumption per

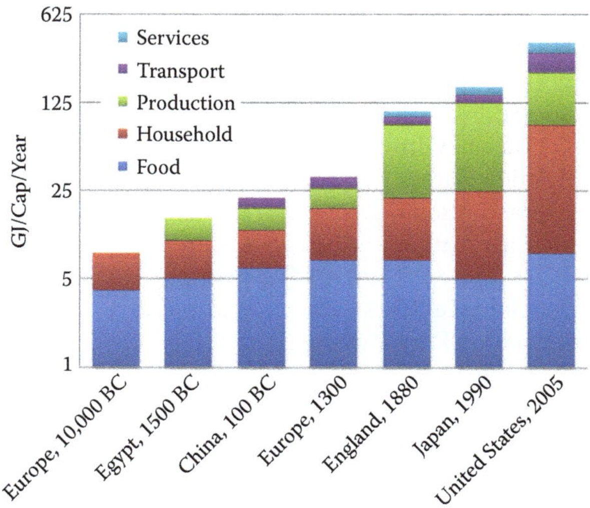

FIGURE 6.1
Per capita energy consumption patterns by social organization. (Based on Smil, V., *Energy in World History: Essays in World History*, Boulder, CO: Westview Press, 1994, pp. xviii–300.)

capita is a reflection of changes in behavioral norms, social organization, technological possibilities, economic power, and the energy forms utilized in our production processes. It also illustrates how energy consumption has enabled the social organizations that we now have.

URBANIZATION AND FUEL USE

Access to resources and norms shape fuel use in households as they migrate from rural to urban living. In 1993, 78% of rural Indian households and 30% of urban households used biomass for cooking. In 2005, the ratios were 75% and 28%, respectively (DeFries and Pandey 2010). This relatively static fuel share masks the growing prevalence of improved cook stoves, which have led to biomass use at far greater efficiency and rural forests being in much better health. Furthermore, the greater efficiency of such stoves is leading to a far smaller global climate change commitment from the products of incomplete combustion of the traditional biomass stove (Smith et al. 2000).

Much of this increase in our energy consumption has been made possible through the use of fossil fuels—the combustion of which has, incidentally, led to anthropogenic climate change. In tackling issues of climate change, it is, therefore, important to understand the various determinants of energy demand and consider the ways in which our demands for energy—especially those forms that contribute to greenhouse gas (GHG) emissions—can be curtailed or met more efficiently. The social and physical factors that lead to the different levels of energy demand, the resulting GHG emissions associated with these demands, and the efficiency with which energy is utilized in meeting our demands are, then, subjects with which we engage in this chapter: What will be the patterns of energy use in the future? And, how can we manage our energy resources most efficiently in response to these ever-changing demands? We shall return to these questions and discuss them with a focus on energy use in advanced technological societies, once we have clarified some definitions in the following section.

ENERGY LADDER

The "energy ladder" was first described by Kirk Smith and his colleagues (1994). It traced the changes in fuel use for cook stoves, starting with low efficiency biomass stoves, on to kerosene stoves, propane, and electricity. At the household level, these transitions represent significant improvements in the efficiency, rising from about 15% (three stone fire) to 84% (electric induction stove)—by delivering 40% of the fuel energy to the food being cooked.

6.2 Definitions

6.2.1 Energy Forms

Energy can be accessed in many forms. We have come to rely on electricity to run our electronics, just as we have come to rely on diesel to move our ships. Energy forms determine their ease of storage, conveyance, and use. Energy density, chemical and physical stability, toxicity, ease of handling, and ease of storage are among the factors that determine the desirability of different energy forms. Whether we can achieve greater energy efficiencies—and reduce GHG emissions—often depends on whether we can find appropriate pairings of energy forms and end uses.

6.2.2 Energy and Power

When we talk about energy, we are concerned about the absolute amounts of it contained in the desired form; but, when we talk about power, we are concerned about the speed with which that amount of energy can be accessed. In fact, power ratings indicate the amount of energy available for use per unit of time: a typical car has an engine capable of producing 100+kW of power; however, rarely do we use more than 30kW in everyday driving. It has a tank holding about 65 liters of gasoline (~2.25 GJ) and we use all of that capacity on long journeys. Similarly, when we talk about energy savings, we often talk about the absolute amount of energy saved through a specific action; but energy efficiency actions may also be motivated by the desire to reduce the amount of power required at a specific time (or by developing less generation capacity).

6.2.3 Energy End Uses and Services

From the production end, energy is defined as a commodity, that is, cubic feet of gas, barrels of oil, kilowatt-hours of electricity. However, to consumers, energy is a means to an end, which is more familiar in terms of end uses or services, such as air conditioning, lighting, internet, and air travel.

6.2.4 Energy Intensity

Energy intensity (1/efficacy) is defined as the energy input per unit of delivered output or service. The energy intensity of an economy is the energy use per unit of GDP. The energy intensity of a light bulb is the energy input per unit of lighting service provided and is expressed in Watts/lumen.

Energy intensity is sensitive to changes in both the numerator and the denominator. When both are changing, the ratio is a poor indicator of energy use. For example, take the reported reduction in the energy intensity of an economy: this can be due to a lower energy consumption due to efficiency gains or a change in the measured economic output. For example, all energy intensity studies of OECD countries lauded Ireland for its low energy intensity per GDP at the turn of the twenty-first century. However, by 2010, the measured GDP of Ireland was much lower and the energy intensity of the country was rising, despite the fact that due to economic hardship, the actual energy consumption was falling.

In general, energy intensity as an indicator of energy efficiency is most useful when talked about in sector-specific terms, such as the energy input associated with the production of, say, a barrel of oil, comparing conventional production to that of the synthetic crude from the Alberta oil sands. However, when the energy intensity of an economy is talked about, the term masks many differences in climate variations and economic organization across countries, and can lead to erroneous conclusions, as in the case of the Irish example.

6.2.5 Energy Efficiency and Conservation

The terms energy efficiency and energy conservation are often used interchangeably. However, they represent fundamentally different ways of thinking about the changing energy demand. Energy efficiency refers to reducing the amount of energy input required to achieve the same level of output or service. Energy conservation, on the other hand, can be achieved either through lowering the energy intensity or reducing the level of output or service. Typically, consumers associate energy conservation with a reduction in the level of service. For example, by using light bulbs with lower power ratings, or heating the house to a lower temperature, we can achieve energy conservation at the cost of darker and cooler rooms. However, improving energy efficiency can deliver both the same level of service and a reduction in total energy demand. For example, one can improve the energy efficiency of a home through upgrading its insulation and switching to more efficient lighting. These measures, if combined with the same level of heating and lighting demand, reduce energy consumption.

6.2.6 The Efficiency Gap

The efficiency gap is a phrase used to describe the difference between the amount of investment in energy efficiency anticipated by a techno-economic analysis of energy use and the actual amount of energy efficiency investment. In other words, the gap describes the difference between the potential, economically feasible energy savings and the actual level of savings in an economy.

6.3 Determinants of Demand

There are two approaches to considering energy demand. From a macroeconomic perspective, the resource endowments of a region determine the potential pathways for its economic activity and resource use. From a microeconomic perspective, lifestyles shape demand. In this section, we will discuss the various determinants of demand from both perspectives.

6.3.1 Economic Structure

Technological innovation, the price of resources, social organization, and barriers to trade have been key determinants of the patterns of economic activity and energy use since the Industrial Revolution. The following examples illustrate this point:

- *1750s–present*: Beginning with the Industrial Revolution, where we work and live continues to be redrawn in a most dramatic fashion. By 2007, there were more urban than rural dwellers. We used to farm, travel, and trade locally and use a small range of manufactured goods. In contrast, I am now using a laptop computer designed in the United States, assembled in China, using components from Taiwan and Japan, and a global flow of raw materials, including oil from the Islamic Republic of Iran and coltan from the Democratic Republic of Congo.
- *1973–1990*: In the wake of the oil crises, many established industries in Europe and North America lost market shares in energy-intensive, low-value products, such as steel, to Japan and Korea. This led to the development of economies that imported inputs to manufacturing, but continued to produce goods for domestic and export markets. In this example, importing steel from Japan represented a global improvement in energy and GHG intensity since the Japanese plants were more modern, and enjoyed better energy efficiency using the same fuels.
- *1990–present*: The increased access to cheaper labor in China and other less industrialized countries has led to the relocation of many manufacturing industries to these countries. This process of dematerialization gives the impression of post-industrial economies being more energy efficient and friendlier to the environment because their domestic energy use and GHG emissions are lower. However, once we take into account the embodied energy of imported and exported goods (Wyckoff and Roop 1994), we learn that their effective energy intensity may be much the same and GHG emissions higher than before. For example, importing a microchip from China has a higher GHG output because, while the same production process is used in China, the electricity production is more heavily reliant on coal.

The forces shaping these and other economic restructurings have created economies with varying degrees of reliance on energy-consuming activities. Furthermore, extensive trade and increased access to consumer and capital goods, themselves, have changed patterns of energy demand, and GHG emissions internationally, over the past few centuries.

6.3.2 Lifestyle

The external environment and individual factors determine the patterns of lifestyle, which in turn, affect the energy demand. For example, in a colder climate, households will demand more insulation and potentially more direct energy input to provide comfort. In such a setting, the household heating costs are such that they would consider a broader range of strategies and technologies for meeting their needs. It is very important to note that the interaction between patterns of activity, awareness of energy use, and the availability of alternative approaches shapes the ways in which we meet our energy needs and wants. In other words, our energy consumption patterns are in association with a lifestyle that reflects complex interactions between many external and internal factors (Shui and Dowlatabadi 2005) (see Figure 6.2).

It is also important to note that while households may be able to alter certain features of their lifestyle to affect their energy demand, many of the key features of a lifestyle are socially and environmentally determined. We may reason that walkable neighborhoods lead to healthier lives, with less time and

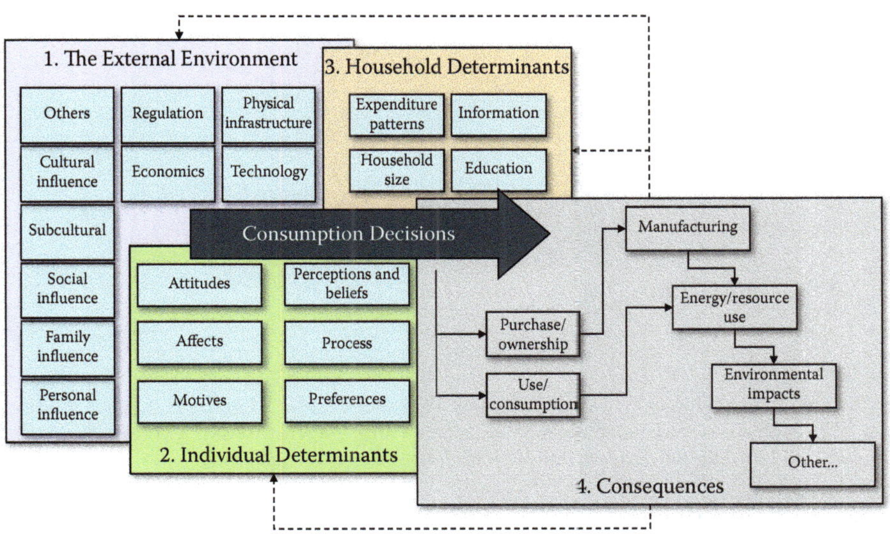

FIGURE 6.2
The factors shaping direct and indirect energy use at the household level. (Reprinted from *Energy Policy*, 33(2), Bin, S., and Dowlatabadi, H., Consumer lifestyle approach to U.S. energy use and the related CO_2 emissions, 197–208, Copyright 2005, with permission of Elsevier.)

energy lost in transit—this conforms to the Theory of Reasoned Action (Ajzen and Fishbein 1977). The focus of this theory is on the beliefs and attitudes of individuals. However, whether individuals act on this reasoning is also dictated by the price and availability of housing, quality of schools, perceptions of security, and so on. This is better described by the Theory of Planned Behavior (Ajzen 1991). This theory constrains individuals' reasoned actions by subjective norms, as shaped by their social environment, especially their chosen reference group. The gap between reasoned action and planned behavior is similar to the factors that have contributed to the so-called efficiency gap.

Besides attitudes, beliefs and social norms, other contextual factors and structural features, such as the arrangement of transportation networks, significantly impact lifestyles and their associated energy demands. Figure 6.3 displays the diffusion of new technologies in U.S. households from 1900–2005. In order to illustrate the interaction of technological progress, norms, social organization, and economic means, let us consider the energy use in transportation and automobile ownership. Personal automobiles were first invented in the 1700s. These steam machines were not robust and died out due to their fragility on poorly paved roads. They were reinvented with internal combustion engines in the late 1800s. This time, they were more robust. But this did not lead to a boom in automobile ownership. Initially, cars were a luxury item for the most affluent members of society. Only after Henry Ford introduced the Model T in 1907, with the expressed intent to market a vehicle that would suit the American family and its needs,* did

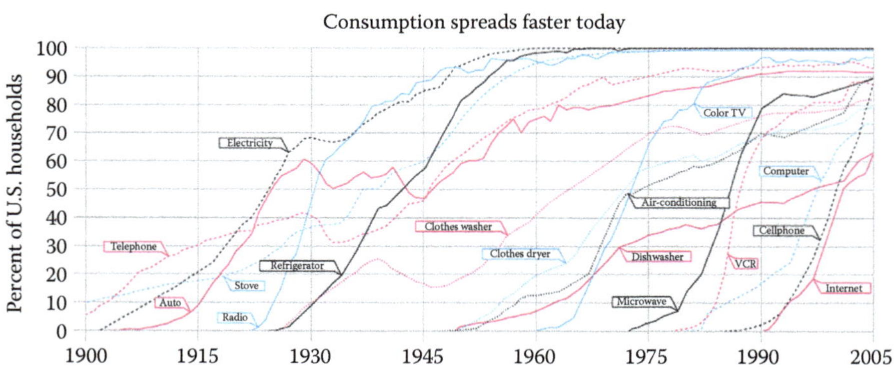

FIGURE 6.3
Diffusion of technologies in U.S. households 1900–2005. (From Thompson, D., *The Atlantic*, Apr. 12, 2012, Retrieved November 9, 2012, from http://www.theatlantic.com/technology/archive/2012/04/the-100-year-march-of-technology-in-1-graph/255573/.)

* "I will build a motor car for the great multitude. It will be large enough for the family but small enough for the individual to run and care for. It will be constructed of the best materials, by the best men to be hired, after the simplest designs that modern engineering can devise. But it will be so low in price that no man making a good salary will be unable to own one—and enjoy with his family the blessing of hours of pleasure in God's great open spaces" (Ford and Crowther 1922, Ch. IV).

auto ownership become a possibility for the masses. With more cars on the road, there followed improvements in road pavement, and the subsequent completion of the circle of interdependence between automobiles and roads. This newly mobile society and lifestyle, of course, demanded higher energy inputs for transportation such that, in 2010, the GHG emissions associated with the transportation sector in the United States were second only to that of electricity generation, at 1750 tera-gram CO_2 equivalent (EPA 2012).

While the pace of construction of enabling infrastructure determines the rate of expanding services, behavioral change allows significant response within that envelope of possibilities. The role of social norms in bringing about rapid change cannot be overstated. For example, Japanese consumers adopted air conditioning at a rapid rate through the last two decades of the twentieth century, from a virtual oddity to near universal presence. However, since the 2011 Tohoku earthquake and tsunami, Japan has, at least temporarily, foregone its dependence on nuclear power (30% of all production). This is being achieved by a massive campaign to change norms, including the dress code at workplaces. The circumstances and the determined response of the Japanese government and consumers have reversed the demand for air conditioning, built up over 20 years, in a matter of weeks.

The example above illustrates how social norms—in this case around acceptable workplace attire, and so on—can be changed with corresponding changes manifesting in aggregate energy demand. However, it is also important to note that many of the lifestyle factors and social norms that determine our energy demands, especially at the household level, fall in the realm of unconscious habits and routines (Stern 2000; Wilk 2002; Carlsson-Kanyama and Lindén 2007). Here, we are not only speaking of our differing and socially-determined notions of comfort, but also of engrained habits around how frequently we adjust our thermostat settings or open our windows for ventilation, how we do our laundry, and whether and how often we bathe or shower (Shove 2003).

6.3.3 Technological Change: New Demands

Technological change also has significant bearing on the changes in energy demand. As Figure 6.3 suggests, the availability of affordable personal transportation went hand-in-hand with a desire for personal mobility in the United States, and as long as the economic means for achieving that aspiration was available (until the Great Depression), auto sales flourished. While this trend did not survive the economic downturn of the 1930s, it would resume its previous ascent after the Second World War (WWII), when economic prosperity visited U.S. households again.

We should also not forget that the automobile is largely a technology to deliver pleasure and signify status—the personal transportation aspect is a happy coincidence. If automobiles' only function was to deliver passengers to their chosen destinations, we would all be driving nondescript boxes on wheels adequate to that task—much as standard light bulbs are uniform and deliver illumination. We shall return to this point later as it is a key aspect of understanding demand and the barriers to energy conservation.

In contrast, let us consider the diffusion of a product whose introduction filled a void in the household—the refrigerator. Prior to the availability of self-contained refrigerators in 1916, households relied on iceboxes, smokehouses, and salt or spices to preserve food. A careful review of Figure 6.3 shows that the pace of adoption of refrigerators by households was hardly affected by either the Great Depression or WWII. The need they filled was so significant that their adoption follows a classic diffusion curve, with early adopters hardly registering in 1925, reaching 10% of households by 1930, 40% by 1945, and reaching saturation by 1975.

These contrasting examples provide an opportunity to understand energy demand from the perspective of wants versus needs. Personal transportation can be considered a complex "want," where the resulting energy demand represents many factors, including the provision of social signaling (whether this be "I am macho and drive a Mustang Cobra" or "I drive a Prius and care for the environment"), pleasure (cruising the boulevard or heading out to the mountains), security (being able to avoid and if necessary protect its occupants in collisions), to simply taking you to work and back. The demand associated with refrigeration, on the other hand, represents something closer to a social necessity. But what is important to note is that technological change often creates new energy demands, by either satisfying a need or providing the opportunity for the realization of other desires, often improving the quality of life in the process. These factors are important to keep in mind as we think through questions of energy conservations and the best ways to achieve it.

6.3.4 Technological Change: Improved Efficiency

While technological inventions have whet our appetite for more goods and services, technological progress has, over time, also contributed to gains in energy efficiency. The efficiency of steam power generation, for example, has improved from about 1% in 1700 to about 50%—or nearly the theoretical maximum—today (Smil 2007). Another example of great efficiency gains can be found in the lighting industry, where efficiencies have improved by a factor of 1000 since the 1850s (see Figure 6.4).

Technological progress in energy efficiency takes two forms: one is the incremental improvement in a given technology; the other is the development of whole new approaches for the delivery of the same service. Take lighting as an example: fire, as the general class of lighting technology, has gradually moved from the use of candles to the Welsbach gas mantle over a 75-year span, improving in efficiency by a factor of 10. However, a totally different approach—the incandescent filament—was invented in 1880, gradually improving in efficiency by a factor of 10 over a similar span of time. The rapid pace of efficiency improvements for solid-state lights, emerging

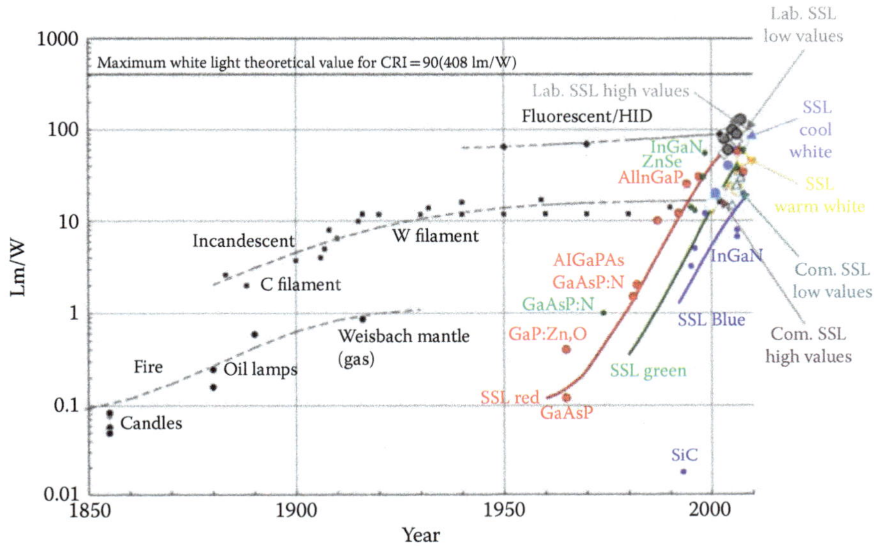

FIGURE 6.4
Improvements in lighting efficacy from 1850 to 2005. (From Azevedo, I. L., et al., *Proceedings of the IEEE*, 97(3), 481–510, 2009.)

since the 1960s, is promising the achievement of lighting efficiencies above 150 lm/W in commercial installations soon.*

It is important to notice that the pace of efficiency improvement within a technology family gradually trails off. It is also important to note that the successful adoption of a new family of technology (florescent, after incandescent, after fire) occurs when the victor offers clear advantages over the vanquished. Furthermore, the widespread presence of a winning technology does not rule out the continued use of an older technology if they offer qualitatively different services. For example, one rarely uses a florescent light to invoke a romantic atmosphere, even though it is 1000x more efficient than a candle.

6.4 "The Efficiency Gap"

In the aftermath of the oil shocks of the 1970s, many scholars turned to investigating the possibility of reducing the demand rather than increasing the energy supply as means of addressing energy scarcity and security

* The U.S. Department of Energy L-Prize was announced in August 2011 offering $10 million to the first manufacturer to offer true replacements for the 60W incandescent bulb and the PAR 38 halogen lamps with an efficacy exceeding 150 lm/W.

challenges. With rising concerns over climate change, curbing GHG emissions was added to the list of motivations for investment in energy conservation and efficiency. However, regardless of motivations, the literature identifies large energy saving potentials through cost-effective energy efficiency: Bevington and Rosenfeld (1990), for example, estimated energy saving potential of 50% in the building sector. Lovins and Lovins (1997) go a step further, arguing that not only could such savings be achieved in a cost-effective manner, but that the elimination of inefficiencies in the American economy would save about $300 billion per year.

Personal discount rates are a reflection of how we weigh a benefit today against a benefit at a future time. This is rarely an explicit calculation. It is, however, measured by economists on the basis of the choices made by consumers. Many factors contribute to this weighting: If this is simply a financial decision, we would weigh the pleasure from having the money in hand against investing the money in the hope of having more at some later date. Of course, this means we face uncertainties about getting our money back, whether and how much it would have gained from being invested, whether we are alive to enjoy the added cash, and whether we would enjoy cash in the future in the same way as we do today. Each of these uncertainties drives up the discount rate, making us expect higher returns on the investment before we can be persuaded to take the risk of investing.

In contrast, some economists maintain that if such possibilities did, in fact, exist, they would have been realized. Assuming that market behavior is reflective of true economic potentials, these scholars question the very existence of an efficiency gap (see Sutherland 1991, 1996, for example). The debate about the efficiency gap reflects the difference between analysis from a top-down perspective of an ideal energy delivery system and from actual consumer energy use and investment behavior.

From a supply perspective, the total resource cost (TRC) for supplying energy must be carefully managed. Ideally, this means that the cost of energy efficiency investment is evaluated against the cost of supplying the same amount of energy as that saved through any efficiency investment. Extending this philosophy to consumers and how they use the energy they buy, there is a perceived efficiency gap when consumers can, but fail to, invest in energy efficient technologies that could result in savings on their energy bills and eventually recover their original efficiency investment.

On the other hand, from an empirical perspective, if an obvious energy savings opportunity is available to consumers, it will be taken. This tautological statement highlights the importance of delving into why an efficiency investment may be obvious to an energy specialist and not to the majority of consumers:* in the wake of the energy crisis, economists studied the actual pattern of energy efficiency investments and compared these to the ideal. They found that consumers are investing less in efficiency than what they should if they were using discount rates used in financial decision making only. The observed personal discount rates are only a shorthand for conveying the gap between consumer perceptions and purely financial market conditions. Repeated studies found that consumer investment patterns reflected personal discount rates of 20%–30% (see Train 1985 for a review of the literature), while financially sound decisions would only apply a discount rate of 5%–8%.

In explaining the energy efficiency gap, some scholars emphasize the effects of technological risk and uncertainty (Sutherland 1991), along with the irreversibility and illiquidity of efficiency investments. Pointing to fears that the technological performance is not guaranteed and that investments are not reversible and lead to a loss of option value (Hassett and Metcalf 1993), these scholars explain the higher discount rates in private, consumer decisions on energy efficiency investment. Others point to the importance of access to information and the ways in which market information is communicated as important factors affecting the consumers' decision-making process. It is argued that gathering and processing information requires time and presents high transactions costs compared to inaction (Joskow and Marron 1992). Others explain this gap by pointing out various other factors affecting the end users who make energy decisions. We will explore these factors next in our discussion of the energy efficiency decision environment.

6.4.1 Why Do We Not Invest in Energy Efficiency?

If we are to bridge the gap between what should be and what is the pattern of energy efficiency investments, we need to understand why the gap exists. A number of factors have been proposed as contributors to the gap. These can be characterized as a decision environment within which those who invest in energy efficiency get familiar with their needs and options. A favorable decision environment can be depicted as pushing good choices towards appropriately primed efficiency investors—as shown in Figure 6.5.

6.4.2 The Decision Environment

As discussed earlier, energy decisions reflect the constraints and opportunities of a richly structured decision environment. This environment encompasses the supply chain of potential efficiency options, from

* Notice that one of these groups makes efficiency decisions infrequently, and can be thought of as novice decision makers, and the other is practiced in analyzing energy efficiency decisions, and can be thought of as an experienced decision maker.

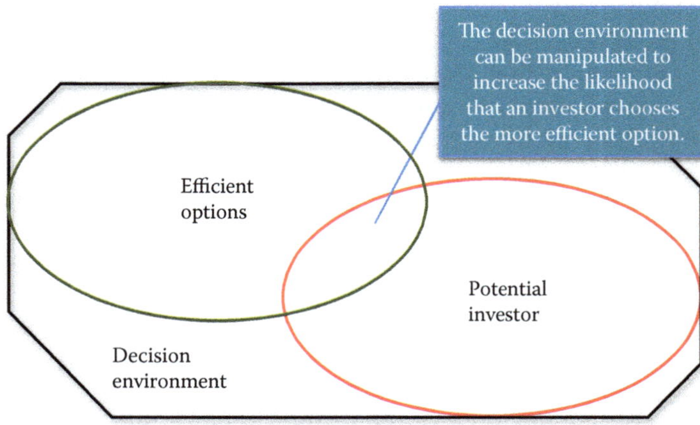

FIGURE 6.5
The decision-environment helps potential efficiency investors identify the appropriate efficiency options meeting their objectives.

hardware to behavioral patterns and decision makers. Here, we would like to stress the fact that when we speak of decisions and decision environments, we are using these words from the top-down perspective of an energy analyst and not that of the end users of energy services. To the energy analyst, a household's noninvestment in, say, a more efficient furnace is seen as a decision not to invest in energy efficiency. Household members that continue using their old furnace, however, will see this same outcome as a continuation of their daily lives and not a decision point. Keeping this in mind, you will note that, as we describe the energy efficiency decision environment from an analyst's point of view, we, in fact, highlight many factors that contribute to non-decisions—be it due to infrastructural, structural, or cultural factors, or simply a lack of experience or knowledge on the part of the end users. We cannot engineer our way past this difference in perspective between the top-down and bottom-up perceptions of engaging the public in energy efficiency investment (either through changes in the physical or behavioral elements of their lives). The only way we can see past this impasse is to redefine the decision-space, as exemplified in Section 6.6.

Here, we present five interacting factors that can be used to describe the decision environment and understand how it has taken shape:

1. Endowments set bounds on the possible development paths in any region. For example, abundant hydroelectric resources in British Columbia have led to the widespread use of electricity for space heating. The path to more efficient use of energy lies in ground-source heat pumps (GSHP) and central heat and power (CHP) fueled by biomass or more efficient housing envelopes. However,

the path-dependent nature of energy decisions hinders progress in that direction to a certain extent. Furthermore, if homes were made more efficient, the level of energy consumption would be too small to be able to recover the capital investments necessary to develop the network for distributed heating through energy savings.

2. Endowments also shape the local culture. The residents of British Columbia identify with limitless hydroelectric power and a long history of profitable exports of electricity to power-hungry California. The popular attitude to electricity pricing is one that is rooted in a culture of entitlement to this resource. However, very few realize that British Columbia is now a net importer of electricity and that the needs of its rapidly growing population cannot be met without significant price hikes permitting the expansion of generating capacity.
3. Expectations about energy prices are critical to whether energy expenditures enter the decision maker's worry budget. In the wake of the energy crises, tales of ever-dwindling energy resources and rising prices assailed the public. Even though the actual level of energy expenditures was still low compared to the pace at which income grew, the general public was persuaded that energy prices would only rise in real terms. This has certainly not been the case, and U.S. residential energy expenditures have only risen from 1.7% in 1980 to 2.5% of income in 2005.* Additionally, where the decision environment has shielded consumers from their true energy costs, consumers are less likely to care about energy efficiency options.†
4. Expectations about social stability and prosperity shape individual norms. The actual income level of Japanese households did not spike in the 1980s. Furthermore, they had just weathered a second energy crisis. However, their success at negotiating energy market challenges, which crippled their competitors, led them to feel safer in the persistence of their prosperity. One of the immediate consequences of this shift in attitudes was that many households chose to install two or more air conditioning units in their homes, beginning in the early 1980s, just as other countries who were not as successful in negotiating the second energy crisis were searching for new ways to save energy (Matsukawa and Ito 1998).

* Data sources: U.S. Census Bureau, Current Population Survey, Annual Social and Economic Supplements and Energy Information Administration, Office of Energy Markets and End Use, Forms EIA-457 A-G of the 1980–2005 Residential Energy Consumption Survey.

† U.S. Residential Energy Consumption data presents households in the Low-Income Household Energy Assistance Program as being a quarter as likely to use energy efficiency data as other households.

5. Learning environments provide a critical missing element for novice and infrequent decision makers.* The decision environment can enable access through shared trusted resources (e.g., consumer reviews), through transparency of decision making in a lead organization (e.g., the city decided to switch to LED lighting because…), or through a knowledgeable advisor (e.g., sales person at the store, general contractor for renovations at home, and consultant at the manufacturing plant).

6.4.3 The Options

The end use decision maker considers their options in meeting a need through the acquisition of an appliance and the energy to fuel it. Usually, the options have many ancillary attributes, some of which make the options more or less desirable to the decision maker. There are at least four factors that hinder consumer adoption of "the more efficient" alternative:

1. Full backward compatibility is a key challenge to the adoption of more efficient options. For example, florescent lights have been about five times more efficient than incandescent light bulbs and commercially available since 1938. Their superiority was not in question. By 1951, more lighting service was provided by florescent lights than by other technologies. However, a large fraction of residential fixtures continue to be compatible with the earlier incandescent technology. This led to the creation of the compact florescent lamp (CFL), designed specifically to be compatible with the standard lamp fixture. However, even this has not led to the old bulbs being replaced by the end-use decision makers. This failure to persuade consumers that CFLs (and lately solid state devices [SSDs]) are more efficient substitutes has forced regulators in many jurisdictions to phase out incandescent bulbs in their jurisdiction—changing the decision environment itself.† We will return to the role of regulations in Section 6.5.

* Homeowners are an exemplar of novice decision makers. Large energy consuming appliances (e.g., refrigerators and gas furnaces) in the last two decades or longer and home renovations are typically a decadal decision. Furthermore, most of the important energy-related features of our homes, such as the architectural design, or the choice of heating method and equipment, are already there when we purchase or rent our dwellings. Individual end users, therefore, do not have much opportunity to become experienced decision makers.

† Cuba led the world by banning incandescent bulbs in 2005. Other jurisdictions have followed with outright bans (Israel 2012), phase out of specific types (Switzerland) and exchange programs (India).

CFLs are far from a perfect substitute (Lefevre, de T'Serclaes, and Waide 2006):

- CFLs contain electronic circuitry sensitive to heat and requiring well ventilated fixtures. Many units continue to fail prematurely by being put into enclosed fixtures.
- Bulb bases flare out above the socket and fail to fit into all fixtures.
- CFLs are heavier, making them unsuitable for installation in lamp fixtures that can be easily toppled.
- CFLs contain mercury, posing a hazard if a bulb breaks, and require special care when being disposed.
- Early units were slow to turn on, had an unfamiliar spectrum, flickered, buzzed, and were not dimmable.

2. More efficient substitutes have a higher initial cost. There are three reasons for this. First, the manufacturer of the new product has to recover their R&D and manufacturing costs. Second, they also try to garner as large a share as possible from the consumer's potential energy savings. Finally, the manufacturer of the older product has already recouped their R&D and machining costs and can drop their prices and still stay profitable to protect their product's market share.
3. New options are often marketed with additional features to enhance their appeal. The added features strengthen the claim of manufacturers to higher prices. They may also cause the premature retirement of perfectly functional capital goods. Yet, the added features often involve sacrificing some of the available efficiency. A well-documented trend is that of the actual versus potential on-road fuel consumption of private automobiles. In Figure 6.6, the trend in automobile fuel consumption has been plotted for the period 1990–2003. This trend shows that cars in 2003 could be 29% less fuel-consuming than in 1990. However, they sacrificed 15% of this potential to more powerful engines and a further 7% to being heavier. Thus, actual fuel consumption only declined by 7%.
4. Limited availability is the final barrier to realizing the full potential of more efficient options. The higher price and slower sales of more efficient substitutes tend to limit local stocks and the immediate availability of more efficient capital goods, further delaying the diffusion of the technology.

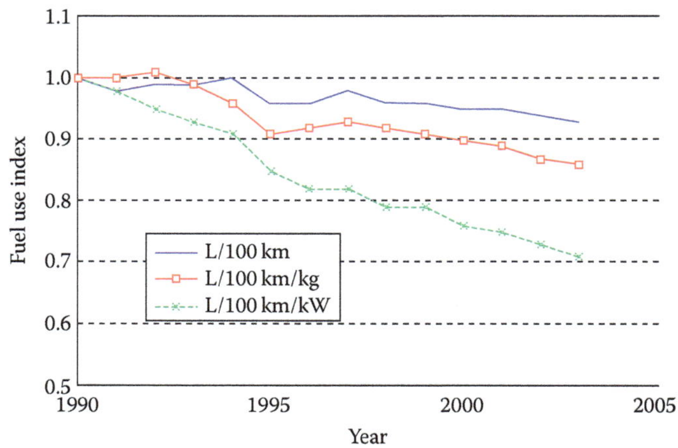

FIGURE 6.6
Automobile fuel consumption trends from 1990 to 2003, along with figures normalized for weight and horsepower. (Based on data from Natural Resources Canada, *The State of Energy Efficiency in Canada, Report 2006*. Ottawa: Natural Resources Canada's Office of Energy Efficiency, 2006.)

6.4.4 The Efficiency Investor

Five conditions must be met before the consumer or the end-use decision maker can make a wise energy efficiency investment:

1. The decision maker needs to be aware of their energy expenditures. There are a number of conditions that may prevent consumers from being aware of their energy costs. For example, living in housing where energy is included in the rent prevents consumers from being directly aware of their energy costs.
2. Consumers need to care about their energy expenditures. In high-income households, energy expenditures may be so small as to not command the attention of decision makers. Using the concept of worry budgets, this relative low cost of energy compared to other expenses leads to the assertion that rising U.S. incomes in the second half of the twentieth century have diminished the importance of energy expenditures and the attention they receive. Furthermore, decisions with energy use consequences are rarely framed as decisions regarding efficiency. Efficiency is, at best, only one of the factors considered while making decisions regarding the purchase of new energy consuming equipment or investing in home upgrades. Other attributes, such as the ease of use and aesthetic concerns, may be more prominent factors in the decision space than those relating to energy performance. The economic returns from greater energy efficiency, therefore, lacks salience in many decisions, especially given the incremental nature of the savings and the small gains in absolute dollar

terms associated with energy efficiency. Furthermore, recognizing the relative lack of salience of energy efficiency in a household or firm would suggest that energy efficiency investment decisions should be evaluated against all other possible decisions a firm or individual could make, both in terms of transaction costs and financial rewards.

3. Decision makers need to be able to access relevant information about their options. For example, a household may be interested in learning if they should invest in a more efficient appliance. Appliance efficiency labels and branding (e.g., EnergGuide) only go partway in providing information to the consumers. The labels show how a particular unit performs in their comparison group. This does not necessarily reflect how such units would perform in the decision maker's setting. In addition, there may be critical efficiency measures that could be taken that are not reflected in such labels. For example, washing clothes in cold water leads to very significant energy savings, as would the wearing of clothes more than once between washes.
4. Decision makers need to be able to process the information about their options. The information provided in Figure 6.7 demonstrates that the U.S. consumer is given information about estimated energy expenses while their Canadian equivalent needs to

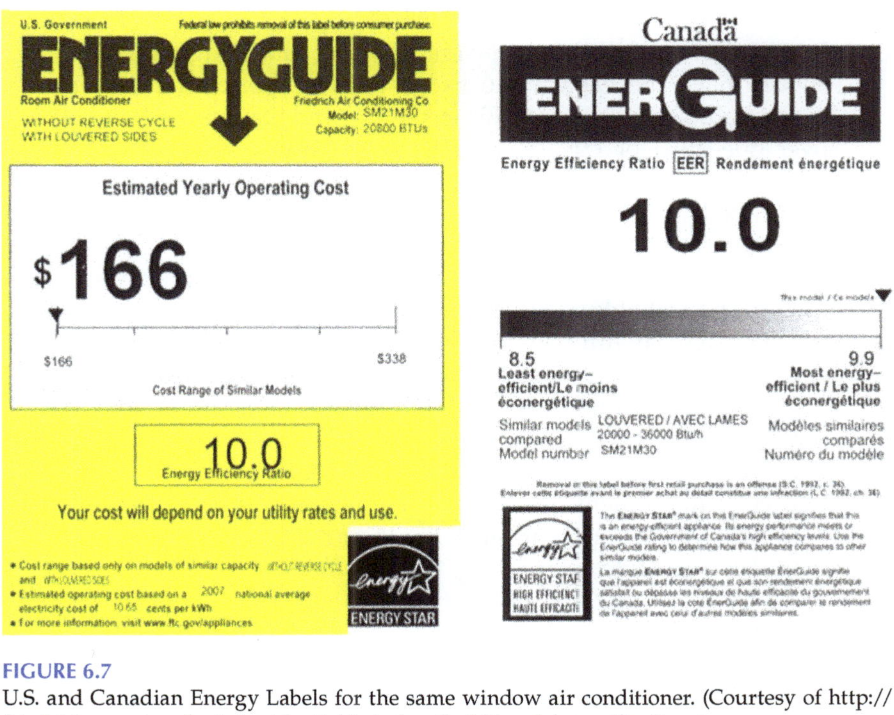

FIGURE 6.7
U.S. and Canadian Energy Labels for the same window air conditioner. (Courtesy of http://friedrich .com/products/residential/window/kuhl/model-specifications.)

translate the unit's energy efficiency ratio into some measure of running costs versus initial capital costs. As noted earlier, the label can inform consumers about how this unit fares compared to its peers. However, in order to make a financial decision, the energy costs of this unit and its comparison group over their lifetime must be combined with the capital costs in order to arrive at a financial decision about which offers the best balance between efficiency and investment. In order to make such a calculation, decision makers also need to place all the units on an equivalent footing (i.e., monetize their relative reliability, choice of features, and other qualitative factors), estimate their use patterns into the future, project energy prices into the future, and know how to perform net present value calculations. Such a careful process may be difficult or lack salience for most decision makers, and may represent a high transaction cost in energy efficiency investment (Jaffe and Stavins 1994). Product efficiency labels such as Energy Star partially remove the processing step for the efficiency investor. The label alone signifies that this product exceeds certain efficiency standards. This process simplifies the decision processing at the same time as providing an added incentive to manufacturers to reach beyond the minimum standard of performance in order to succeed in a market that is increasingly interested (but unable to process the information available) to achieve greater energy efficiency.

5. Finally, once the decision makers have the knowledge to act, they should have the capacity to exercise their choice. A number of agency problems can limit this step (Jaffe and Stavins 1994). In 2010, over one-third of U.S. (U.S. Census n.d.) and one-quarter of EU households lived in rental accommodations (Eurostat n.d.). Renters are unlikely to control the choice of appliances in their dwellings—a discount rate is hardly applicable as a descriptor when choices cannot be exercised. Furthermore, the typical U.S. home has an itinerant owner—households move every 7 years, on average. Expectations about moves, can limit the time horizon over which savings from an efficiency investment are taken into account—raising the equivalent discount rate dramatically. Last, but not least, the household may not be able to afford the energy efficiency investment. For example, the elderly in owner-occupied homes often could most benefit from efficiency investments. However, they lack the means to invest in renovating their homes or replacing their furnaces with new, much more efficient units.

6.4.5 Efficiency Gap or Noneconomic Motivations?

Much of the research reviewed in this section has been motivated by a concern about unutilized opportunities for investment in efficiencies with high rates of economic return. The corollary question of over-investment in efficiency has not received much serious attention. Consumers are not assumed to be half-witted if they buy hybrid cars, even though it can easily be demonstrated that there is no economic justification for such an investment. Interviews with consumers have repeatedly elucidated that noneconomic factors weighing in heavily on the side of their chosen alternative. If they sought to buy a small hybrid car, they are motivated by the need to signal their care of the environment. If they bought a sports utility vehicle, they are responding to their perceptions that occupants of heavier vehicles have lower odds of severe injury in two-car collisions. Likewise, the decision on whether or not to invest in economically beneficial energy efficiency is often motivated by noneconomic factors (Stern 1992).

> A number of studies have explored the motivations for buying hybrids. It is easy to show that hybrids are an excellent choice for taxi drivers in inner city traffic, but that they do not offer an economic return to the typical family. Yet, studies asking owners why they own a hybrid finds that they are making a subjectively more economic choice by buying a less expensive hybrid than a conventional car they would otherwise purchase (Klein 2007). The literature on the subject has found that the symbolic meanings of owning a hybrid vehicle, along with perceived economic benefits, tend to explain the decision investment. These symbolic meanings include the message that owning a hybrid car sends about being "green" (Heffner 2007), complying and belonging to a "green" community (Kahn 2007), as well as being on the forefront of the adoption of new technologies (Turrentine and Kurani 2007).

In short, both the intention of technological innovation and its patterns of adoption take direction from societal forces, beyond economic motivations. In fact, the so-called efficiency gap is only given such a label because the myriad of noneconomic factors shaping energy technology adoption patterns have been ignored. We will look next at conventional energy efficiency interventions that have mostly emerged in response to the barriers to energy efficiency investments identified above.

6.5 Efficiency Interventions

Energy efficiency objectives and interventions can be developed and implemented throughout the supply chain of energy commodities and capital goods, or target end users and their patterns of consumption. A range of policies have been developed for each of these intervention points, reflecting different philosophies and understandings about the context of the key decision makers. We will discuss the strengths and weaknesses of a number of these initiatives in this section.

6.5.1 Strategies Aimed at Appliance Manufacturers

At any given time, consumers face the choice of a range of appliances capable of meeting their core needs, but using very different amounts of energy. The horizontal line in the appliance efficiency labels presented in Figure 6.7 depicts this range of efficiencies. Policy makers have used energy efficiency standards to curtail the bottom end of the efficiency spectrum. They have also used incentives to challenge the industry to achieve greater efficiencies than the industry believes is economically within reach. Refrigerators offer an excellent example of these strategies at work.

The trends in energy efficiency and the internal volume of refrigerators have been plotted in Figure 6.8. The gray line presents the trend toward larger internal volumes, from 8 cubic feet in 1947 to 20 cubic feet by 1981. The black line depicts the energy consumption of the typical refrigerator. Prior to 1974, refrigerators were growing larger and less efficient—energy consumption quadrupled while usable volume increased by 150%. From that point onwards, energy use per unit has declined by more than 75%. This occurred because of mandated efficiency standards (led by the California

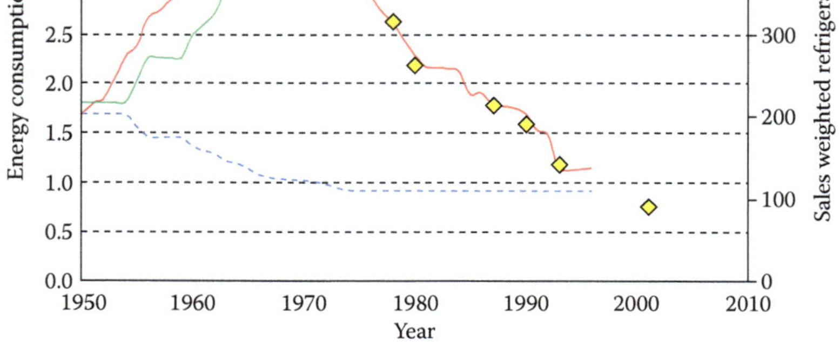

FIGURE 6.8
U.S. Household refrigerator volume and efficiency trends, 1950–2001. (Based on data from Rosenfeld, A., *Annual Review of Energy and the Environment*, 24(1), 33–82, 1999.)

Energy Commission) forcing manufacturers to curtail the sale of their less efficient units (Rosenfeld 1999).

As a complement to the removal of the least efficient units, we need to expand the offerings with significantly higher efficiency. Arguably, the most notable example of this approach is the Super Efficient Refrigerator Program Golden Carrot Award. This initiative pooled $30 million to motivate manufacturers to build an inexpensive, 50% more efficient CFC-free refrigerator. The $30 million was made available by shifting some of the funds allocated to DSM programs that attempted to entice consumers into buying more efficient refrigerators by offering rebates. The idea was conceptualized in 1990 and the prize was awarded to Whirlpool in 1993. The prize money offset the R&D and tooling cost of the new generation of refrigerators and Whirlpool marketed 250,000 units in 1994, with the runner up, Frigidaire, marketing their own units soon thereafter (Eckert 1995).

Three important lessons were learned from this initiative:

- Incentives to manufacturers were more effective than to consumers.
- What appeared to be a stretch target was closer at hand than anticipated.
- The prize is awarded to only one manufacturer, but engages many who follow up by offering their own more efficient models.

6.5.2 Policies Aimed at Energy Suppliers

Many believe that the profit motive would make it unlikely that energy suppliers would be wasteful. However, Herbert Simon's satisficing model (Simon 1979) is a better descriptor of how most organizations (including oil and gas companies and electric utilities) behave. The satisficing model suggests that, for complex decisions, when the full evaluation of options requires significant time investments, individuals and organizations rarely aim for maximizing profit or utility, but rather settle for a solution that *satisfies* the need and *suffices* under the circumstances. For example, the adoption of turbo generators to replace pressure release valves on gas delivery pipelines is still at its infancy even though the technology has been available for more than a decade. Typically, 1%–1.5% of the energy delivered as gas has been used to compress and move the gas along the transmission pipeline to the point of consumption. Using step-down turbogenerators to reduce the pressure of the gas can recover a substantial fraction of that energy as electricity. However, pressure release valves are still seen by many gas suppliers as adequate options that while not maximizing profits are perfectly capable of satisfying the need. Clearly, policies can be introduced to force technology adoption or to stipulate an upper bound on energy and gas lost in the transmission and delivery system.

Changing priorities have dictated the nature of policies aimed at energy suppliers. For example, nineteenth century electric utilities competed with one another for each customer, with each street crisscrossed by duplicated distributions lines. The high cost of such duplication led to policies that yielded monopoly supply territories, but regulated the price of electricity by capping the returns on investments in the supply infrastructure. In the wake of the energy crisis, the lock-step relationship between economic growth and the electricity demand was broken, and many utilities faced newly constructed power stations and no incremental demand. When utilities were denied the chance to charge their customers for these under-utilized plants, they had to absorb large capital losses and, subsequently, they shied away from building new plants ahead of demand growth. Of course, not long after this, utilities were facing capacity shortages and found that in many cases, it was less expensive to invest in demand side management (DSM) than to expand generation capacity.* As Laughran and Kulick (2004) report, between 1989 and 1999, utility companies in the United States spent $14.7 billion on DSM programs, which, among utility companies that continuously reported positive DSM activities, resulted in between 0.6 and 1.2% reduction in electricity sales. However, the success of DSM programs have been varied such that, on average, DSM spending in that same 10-year time span reduced electricity use by between 0.3% and 0.4%, at an average cost of $0.14–0.22kWh (Loughran and Kulick 2004). It is worth mentioning that the utility companies, themselves, often estimate this reduction in demand to be more significant.

More recently, regulators have forced utilities to present an integrated resource plan (IRP) for approval before new rates are introduced. IRPs go beyond traditional resource plans by forcing the consideration of energy conservation measures on an equal footing to capacity expansion in meeting growing demand. For example, the 2010 20-year demand forecast for BC Hydro projects a 40% increase in demand (BC Hydro 2010). The 2012 IRP aimed to meet half of this growth through energy conservation and the balance through new supply.

6.5.3 Financial Incentives

Energy suppliers and regulators use a number of strategies to improve energy efficiency and conservation by their clients—energy tariff structures, subsidies for energy efficiency investments, technical advice, and information programs are among these efforts.

* DSM captures both shifting of electricity demand to times of lower demand and overall energy conservation. Peak demand shifting can actually increase energy use, but is of tremendous value to utilities because of the difficulty in storing electricity and the cost of installing generation capacity to meet peak demands lasting a few hours per year.

1. Energy tariff structures aim to inform the consumers about the cost of the demand. Rising tariff structures penalize the consumers for using more electricity by increasing the unit price of consumption in excess of set thresholds. This approach to energy pricing aims to create an incentive for energy conservation, for those with higher consumption. Time-of-use pricing takes consumer engagement one step further, signaling consumers the time at which their demand costs more to meet—encouraging both demand shifting to times at which power is available at lower cost and/or the reduction of overall demand.
2. Efficiency investment subsidies: Many utilities (often in association with public authorities) provide incentives for consumers to buy more efficient products. For example, more efficient washing machines receive a subsidy from the energy utility and have a tax rebate in some jurisdictions.
3. Scrap-it programs: Utilities often also provide incentives for consumers who scrap their older, less efficient appliances. For example, the vast majority of homes with two refrigerators use an older unit to chill their beverages. Given the poor efficiency of older refrigerators, it is much to the benefit of energy suppliers to have these units scrapped; so they pay owners to take away their beer fridge.
4. Stocking subsidies: As noted earlier, the new efficient alternative is often in low demand and may not be stocked by local merchants. Both the utilities and regulators can help pay the carrying charge for more efficient alternatives locally in order to hasten the adoption of more efficient alternatives. Merchants are often further incentivized through a coordinated subsidy program aimed at consumers who are enticed to visit stores that stock the more efficient alternatives and become customers of the merchants participating in the program.

6.5.4 Information Programs

As noted in the discussion of the energy efficiency decision, a key prerequisite to more efficiency is an awareness of current consumption patterns (the bill) and awareness about alternatives. A number of initiatives have been undertaken in improving this awareness:

1. Outreach programs are a prominent feature of most energy bills. Energy suppliers try to alert their customers to their energy consumption patterns (often presenting present consumption rates vs. last year's or comparisons with an appropriate peer group). They also provide customers tips for realizing energy savings.

Products can earn the Energy Star label by meeting the energy efficiency requirements set forth in Energy Star product specifications. The EPA establishes these specifications based on the following set of key guiding principles:

- Product categories must contribute significant energy savings nationwide.
- Qualified products must deliver the features and performance demanded by consumers, in addition to increased energy efficiency.
- If the qualified product costs more than a conventional, less-efficient counterpart, purchasers will recover their investment in increased energy efficiency through utility bill savings, within a reasonable period of time.
- Energy efficiency can be achieved through broadly available, nonproprietary technologies offered by more than one manufacturer.
- Product energy consumption and performance can be measured and verified with testing.
- Labeling would effectively differentiate products and be visible for purchasers.

Source: http://www.energystar.gov

2. Real time feedback: Experts in education have long known that immediate feedback is a critical element of behavior change. Much of the discussion about efficiency has been about the efficiency of appliances, but behavior change has an even larger potential for energy conservation (see Lutzenhiser 1993)—that is, whole consumption categories could be eliminated. Automakers such as BMW have, for over 20 years, provided a real-time fuel efficiency indicator just below the speedometer—eliciting permanent annoyance or a lighter touch on the pedal from BMW owners. In a similar fashion, it is thought that real-time metering with prominent displays in the home (or office) will help consumers link their behavior to its energy consumption consequences and subjectively appropriate behavioral adjustments. Studies on feedback (real-time and otherwise) indicate that it can reduce residential energy use by about 1%–20%, with the majority of studies reporting values between 5%–12% (Fischer 2008).
3. Active guidance (utility): Some utilities have helped consumers make better choices for meeting their needs. In The Netherlands, local distribution companies identified customers who would

benefit from the cogeneration of power and heat instead of buying electricity from the utility and gas for heat separately. Their active support of these customers through transition to cogeneration has led to about 30% of all electricity demand in the Netherlands being met through decentralized cogeneration (Hekkert, Harmsen, and de Jong 2007). Suitable end users for such actions include hotels, hospitals, and greenhouses. In this transition, the utilization of primary energy has improved by over 50%.

4. Active guidance (merchant): While many programs continue to offer subsidies to consumers, it is well-known that this is not the best way to ensure against free riders and better decision making. As noted earlier, the infrequent nature of major decisions by most consumers makes it unlikely that they have the expertise or can process the information needed to purchase the alternative that is appropriate to their needs. In contrast, the intermediaries to such decisions are engaged in many such decisions as part of their function. The sales force in an appliance store should know the features of the units for sale and match customers to units that give them the most efficient alternative for meeting their needs. A program to help educate the sales force and provide them the incentive for selling more efficient appliances has already been proven more effective than the provision of incentives to end users.
5. Labeling: Efficiency branding, such as Energy Star, provides a shortcut to the process of option evaluations by many consumers. The label itself provides a trusted statement about the efficiency of a given product. However, products achieving higher efficiencies through proprietary technologies are not eligible for receiving the Energy Star label.

As discussed, despite this rich variety of intervention mechanisms, many scholars have identified vast untapped potential for further energy savings across all sectors of the economy. Yet others worry that further effort in gaining efficiencies will backfire through a rebound in demand.

6.5.5 The Jevons Paradox (a.k.a. The Rebound Effect)

The Jevons paradox, named after a nineteenth century British economist, stipulates that technical innovation leading to improved efficiency reduces energy expenditures and, by doing so, leads to additional demand for energy.

In simple economic terms, if the energy efficiency investment reduces overall energy expenditures, the investor has a budget surplus to redistribute. Unless there is no need at all for any additional energy consuming activities, some of this newfound budget will be spent on further energy purchases. This leads to an over-estimation of the final energy savings, also known as the direct rebound effect. For example, home owners often turn up their thermostats after investing in improving the insulation level of their homes. Clearly, they were not at their ideal comfort level before and part of the reason for their efficiency investment is to lower the cost of home heating so that they can approach their ideal comfort level without much higher energy expenditures.*

In a manufacturing setting, if the efficiency investment is such that a previously noncompetitive activity is now competitively priced, the economy-wide demand for this activity will rise—also a direct rebound effect. If this is a new product, it is creating a new demand category and the demand for it will mean that an energy efficiency investment has led to an overall increase in economywide energy consumption—also known as backfire.

The savings in energy expenditures may now be spent on other goods and services in the economy. Such incremental expenditures will have their own energy requirements and GHG emissions—termed an indirect rebound effect. Unless the part of the energy system being made more efficient is already more efficient than the whole economy, the additional expenditures in the economy will create a smaller indirect rebound than the efficiency gains achieved (i.e., no backfire).†

We explore this issue further using trends in efficiency improvements and the utilization rates of diesel trucks in Canada between 1990 and 2003, as depicted in Figure 6.9. The data show that fuel use per ton-km of freight

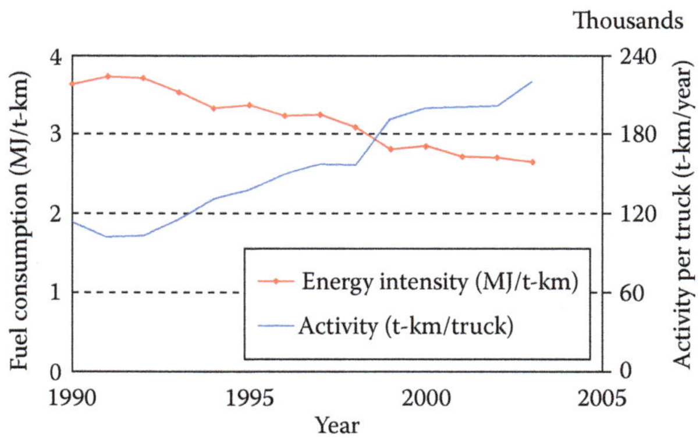

FIGURE 6.9
Trends in efficiency of truck transport and use (Based on data from Natural Resources Canada, *The State of Energy Efficiency in Canada, Report 2006*. Ottawa: Natural Resources Canada's Office of Energy Efficiency, 2006.)

* We know that consumers rarely calculate discounted costs of their energy efficiency and consumption expenditures. Therefore, one can also hypothesize that having invested in energy efficiency, homeowners feel virtuous and reward themselves with higher indoor temperatures and greater comfort.

† Of course, if the consumer chooses to spend their energy savings on an extremely energy intensive item, there would be backfire.

declined by 27% while the total freight moved per truck in each year increased by 94%. Is this evidence of backfire?

Increasing the utilization rate by almost 100% can only be achieved through five factors: driving faster, driving for more hours per day, increasing the hauling capacity per truck, increasing the ratio of moving to waiting at loading docks, or better capacity utilization. Driving faster would have decreased the fuel efficiency. Driving for longer hours is not permitted under Canadian law.* Increasing the capacity utilization and longer haul trips and shorter layovers are part and parcel of the drive to make trucking more competitive with rail and go hand in hand with better fuel efficiency. If long-haul trucking were to have become more efficient than rail due to the 27% fuel efficiency, trucks would have won contracts for long-haul container traffic away from rail, resulting in the pattern of change viewed here. However, it would be incorrect to then surmise that the overall energy use in the economy has risen, unless we can also calculate the associated changes in fuel use throughout the rail system and the rest of the economy. This example illustrates an indirect rebound effect, where the increase in demand for the service leads to a higher energy consumption in this particular mode of freight, but not necessarily economywide, as the patterns of reliance on rail versus trucks have changed.

The magnitude of the rebound effect is estimated variously, depending on the definition used. The literature on rebound reports values ranging from a few percent to over a 100% (backfire). However, the studies on rebound are often limited by the number of control variables, the availability of appropriate baseline data, and the limited time spans over which changes were monitored. Greening and Greene (2000) and Sorrell Dimitropoulos and Sommerville's (2009) judicious reviews of the empirical data for rebound estimates have, however, found the direct rebound associated with investment in home energy upgrades and more efficient personal transportation to be between 10% and 30% of the energy consumption savings, suggesting that investment in energy efficiency achieves between 70% and 90% of the predicted savings. They also found that across various end uses, and at the micro level, the rebound is well less than unity, suggesting that energy efficiency improvements do, in fact, lead to a reduction of energy use at the micro level and for the particular end use in which energy efficiency investment was made.

The economy-wide estimates of rebound are even more varied, and often less reliable. In fact, Greening et al.'s review of empirical evidence on rebound asserts that while it is conceptually possible to extend the definition of rebound to include transformational effects across an entire economy, it is analytically impractical since both the theory and data for such predictions are lacking. They further argue that "attempting to assign causal linkages between changes in society and changes in energy efficiency, without addressing all

* Canada limits truckers to 14 hours of duty per 24 hours and drivers rarely double up in the cab.

of the potential confounding factors, would likely lead to unsupported and incorrect conclusions" (Greening, Greene, and Difiglio 2000, p. 399).

In short, the empirical evidence suggests that the magnitude of the direct rebound associated with energy efficiency is quite small, and certainly not cause for detraction from energy efficiency efforts. It is also important to point out that there are circumstances where seeing a rebound effect or increases in energy use resulting from energy efficiency upgrades is actually desirable, if it leads to better standards of life or improves comfort. In the case of underserved communities, whose comfort is compromised by their ability to afford energy services, for example, moderate increases in energy consumption because the household is able to meet their basic energy needs as a result of improvements in their energy efficiency should, in fact, be welcomed. Furthermore, as some environmental economists argue, the rebound effects may be mitigated by the introduction of taxes and price increases in conjunction with energy efficiency measures so that the consumer cannot invest the savings in energy consuming activities (see Wackernagel and Rees 1997 for example).

6.6 Re-Imagining the Energy System

As noted earlier, from a techno-economic perspective, most modern economies can be much more energy efficient. The specific efficiency investments and their net savings have also been studied and depicted in summary presentations, such as Figure 6.10. However, the failure to be more efficient despite the rich panoply of possible interventions serves to reiterate that perhaps we are tackling this challenge using a wrong approach.

Energy can be thought of in two, somewhat distinct, ways. Conventionally, energy is talked about—and sold—as a commodity, in energy units. Utility companies charge customers per units of energy consumed: Electricity is billed by the kilowatt-hour, natural gas by the gigajoule. The idea is to charge the customers for what they use, letting them manage how they use the energy purchased. Thinking of energy as a commodity assumes that what interests customers is the purchase of energy as an end in itself, and that they can, and will, manage it in their best economic interest, such that if they feel their energy bills are too high, they will take appropriate actions to consume less by modifying their needs or investing in energy saving measures.

Alternatively, energy may be thought of as the prerequisite to the production of a service, rather than a commodity. After all, what people are interested in when purchasing energy is the utility that they get out of converting the energy they purchase into a service. If they are buying natural gas to heat their homes, what they care about is a warm house. If they are consuming electricity to light their homes, they are interested in proper lighting. Talking

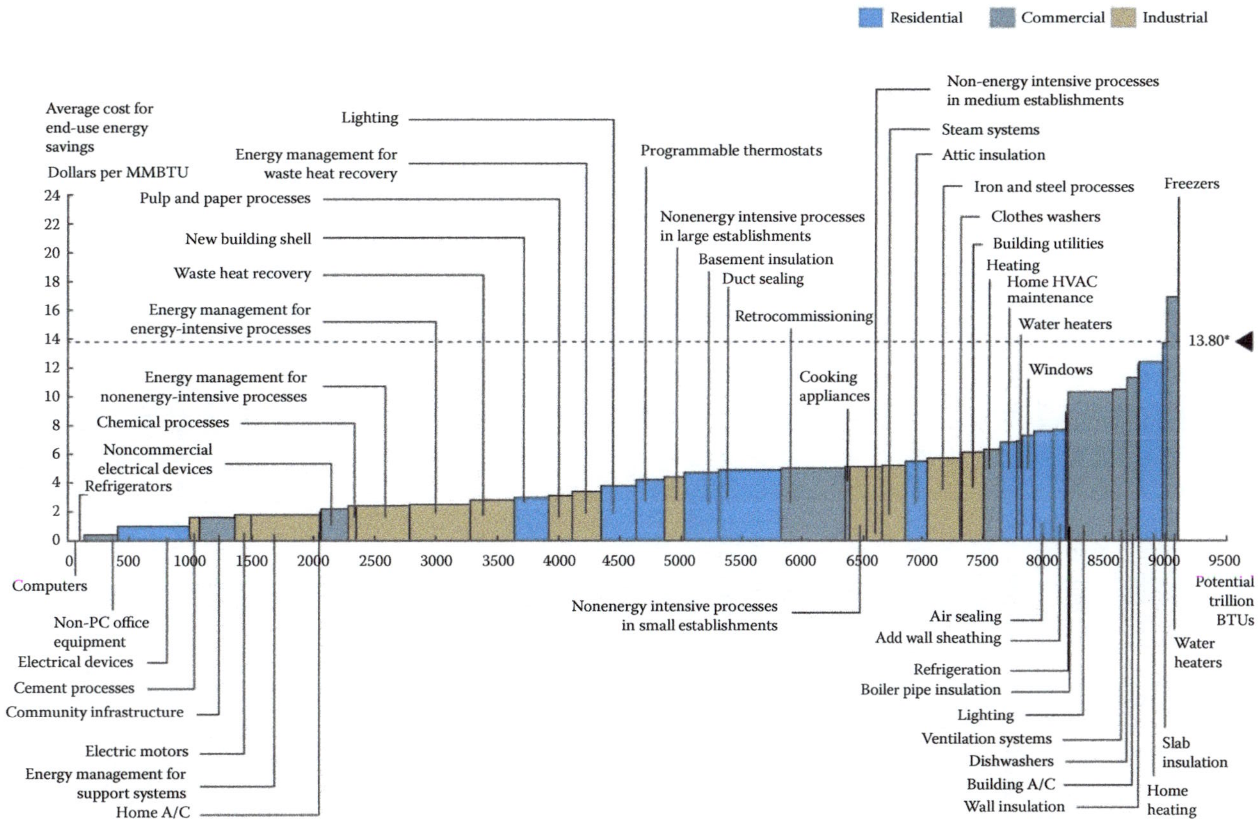

FIGURE 6.10
The U.S. Energy conservation supply curve to 2020. (From Granade, H. C. et al., *Unlocking energy efficiency in the U.S. economy*, 2009, p. 165.)

about energy services rather than the units of energy consumed shifts the focus from the rather abstract concept of energy as a commodity to the more tangible ways in which it is used. Those advocating this way of thinking (and selling) energy also argue that by selling a service rather than a commodity, utility companies or service providers will have more incentive to deliver this service in the most economic (and efficient) way, essentially shifting the burden of responsibility for efficiency away from the customer and to service providers.

More generally speaking, re-imagining energy systems to deliver services more efficiently relies on a shift in the sites of decision making, from infrequent, novice decision makers (often end-use customers) to experienced decision makers who, through sustained feedback and motivated by profit, become familiar with the more efficient ways of delivering energy services and identify and remove barriers to their access. Many successful examples of using this approach to increasing energy efficiency already exist; others are yet to be imagined. In the remainder of this section, we will review a number of these strategies.

An example of an existing approach to shifting the sites of decision making from the end user to a more experienced efficiency decision maker can be found in programs that aim to provide efficiency incentives and information to contractors—rather than to homeowners—who can create energy efficiency opportunities in home renovation projects (Wilson 2008).

Another example of shifting these sites of decision making to more experienced entities is the provision of hot water services. This kind of a service attempts to replace a world in which consumers have to assess their hot water needs, choice of fuel, and choice of water heater individually, with one where they can choose a hot water service provider. This kind of an approach to meeting hot water demand is just one step removed from hot water heater rentals, currently available in Ontario, Canada (for an example, see Direct Energy Home Services 2012). The current practice of hot water rentals has led to the installation of high quality and high reliability units, replacing the need for frequent and costly maintenance and repairs of the less reliable units that consumers independently installed before they could rent the service. This higher reliability is also a positive incentive for consumers who hate to wake up to cold showers and flooded basements—due to failed water heaters. However, the hot water rental contract does not entice the vendor to ensure that the most efficient units are being installed. If the contract was per volume of hot water used, then consumers could be enticed to conserve their usage, and hot water service providers would financially benefit from having the most efficient units possible (so long as they are not an arm of the energy supply company).

Central- or district-heating networks are common in many colder and older European cities, providing heating services, and allowing for the possibility of competition among different energy forms and conversion technologies to achieve a more efficient use of energy (Grohnheit and Gram Mortensen 2003).

Solid waste incineration, biomass, and natural gas are common sources of energy in these systems. They often supply heat obtained through the cogeneration of electricity and heat and have been recognized as a major opportunity for the reduction of GHG emissions. These networks are excellent examples of providing the end user with efficient energy services, while simultaneously simplifying the decision space.

Moreover, service provision, rather than commodity provision, has already been available in the transport sector for some time. We buy transportation services from buses, taxis, and airplanes. With community car-sharing programs, like Zip, we have access to a range of vehicles, each suitable for different transport tasks—matching our vehicles to our needs.

Many of the same factors that have influenced our energy consumption behavior shape the uptake of energy services. In addition to addressing those challenges, we need to transform the system so that current energy utilities profit from the sales of services, not commodities, providing them with the financial incentive to deliver energy in the most efficient way. Furthermore, we need to realize the potential of systems, such as municipal scale ground-source heat pumps, with larger energy efficiency gains. One possible model, allowing for this kind of a transition is the energy service company (ESCO) model, which we will discuss next.

ESPCS AND ESCOS

The idea of selling energy services and not energy as a commodity was first embodied in energy service providing companies (ESPCs). These companies provide and maintain the equipment necessary to supply their services. First Energy, for example, has provided hot water services for 50 years. Energy service companies (ESCOs) go one step further, guaranteeing energy cost savings and earning their keep from sharing the savings. The ESCO concept has met with varying degrees of success in proliferation and value in different countries, from $6/capita in the United States to 3¢/capita in China (Ürge-Vorsatz et al. 2007). Some of that variance stems from the initial conditions in each country: by comparison, the United States is far more energy profligate than China, and, as such, there is far greater scope for energy efficiency projects in the United States.

ESCOs face different barriers in various sectors: In the public sector, their clients fear loss of jobs or control. In the industrial sector, their clients have far shorter time horizons than the norm in ESCO payback for efficiency investments. In the residential sector, their individual projects are too small unless they are easily replicable (Ürge-Vorsatz et al. 2007).

6.6.1 Energy Service Companies (ESCOs)

The Energy Service Company model arose in the 1970s out of a marketing strategy by Time Energy, a Texas company. Time Energy was failing to convince customers to buy their automatic switch to turn off lights outside office hours. Customers were suspicious of the promised energy savings. So, Time Energy installed the switches for customers at no cost, and on the basis of recovering a percentage of their realized savings. Energy crises and the deregulation of the utilities, both served to promote the growth and proliferation of ESCOs. They now constitute a multibillion industry. However, ESCOs have made greater inroads with large institutional customers than other customer classes. In the United States, for example, ESCOs have the highest penetration among municipal and state governments, universities and colleges, schools, and hospitals (Hopper, Goldman, Gilligan, Singer, and Birr 2007).

This approach only requires the end user to identify their energy service needs (e.g., lighting, IT, comfort, cooking, transport). Consumers then enter the market with the aim of procuring energy services—not appliances and energy as a commodity. This changes the nature of the decision faced by the end user, from one of having to evaluate different energy vendors, their rate structures, and appliance performance information, to one of having to evaluate service providers, their cost structure and deciding which to trust. This approach has a number of very desirable characteristics:

1. ESCOs are dedicated to creating economically efficient opportunities for technology selection suitable to the reliable delivery of energy services.
2. System consumers are no longer burdened by the challenge of deciding which appliance to purchase, which energy commodity to use, and so on.
3. Service providers face strong incentives to strike an economic balance between the investment in energy efficient appliances and the payment for energy as a commodity.
4. Service providers have concentrated purchasing power, enabling them to procure efficient appliances at a more favorable pricing, driving down their costs and nudging the capital expenditures further towards energy efficiency.
5. Service providers have an incentive to keep a close eye on the net present value of their capital stock, given the advances in technology and energy efficiency. They are likely to invest in new energy efficiency equipment as soon as it is economically advisable.
6. The end user capital constraints are no longer a barrier to investment in appropriate energy efficiency measures.
7. If ESCO projects are funded using bonds financed through real estate taxes, the barrier of the itinerant homeowner with a short time horizon and the renter without an agency is also removed.

8. By focusing on a geographic region, ESCOs can initiate community heating and cooling systems and install cogeneration plants, realizing economies of scale unavailable to their clients individually.

6.6.2 The Softer Energy Path

When established independently of a utility, ESCOs may also help address the issue of choosing the right energy form for each specific end use. This approach is consistent with Lovins' conceptualization of soft energy paths (Lovins 1978), which pair end uses with the appropriate modes of delivering them. For example, we can use solar thermal collectors, or conventional fossil fuel, or electric water heaters for hot water. Electric units are the least expensive to install. However, solar thermal units are more cost effective in almost all climates and lead to net savings in five years or less. Furthermore, solar thermal units are more suitable for harnessing solar energy than the current generation of photovoltaic systems.

Lovins' 1978 analysis of energy demand suggests that only about 8% of the energy used in the United States requires high quality energy in the form of expensive electricity. The single largest energy end use, is, in fact, space heating. For example, over 40% of the energy used in the U.S. residential sector in 2005 was used for space heating, while another 20% was used in heating water (U.S. EIA 2005). Arguing that space heating needs are more efficiently met at the local scale, Lovins suggests that the more expensive and high quality forms of energy, such as electricity, are best reserved for applications that truly demand them, such as powering IT.

A good contrast can be drawn in relation to the generation of electricity: Electricity is typically generated in large, centralized power plants, with efficiencies of close to 35%. A further 10%–12% of the electricity generated is lost in transmission and distribution to consumers. Thus, about 30% of the primary energy content is delivered as electricity while the rest is lost as heat to the environment. In contrast, a small distributed cogeneration plant can muster a 30% efficiency rating in electricity production, but about 80% of the remaining thermal energy is available to meet low-grade heating (and cooling) requirements locally, leading to an overall primary energy savings of close to 50%. As discussed above, this more appropriate pairing of energy forms and uses can be realized in district heating networks, and through energy service companies.

6.7 Concluding Remarks

In considering energy conservation as means of reducing energy demand and mitigating GHG emissions, we have discussed some of the social, cultural, behavioral, technological, and environmental determinants of demand, exploring energy efficiency decision spaces and end users that are typically responsible for making these decisions. We have also presented two distinct ways of looking at securing one's energy needs. First, the conventional paradigm, which forces the end user to purchase energy as a commodity and secure appropriate conversion technologies to achieve the desired service. We have presented examples and discussions of how this saddles the end user with the burden of a complex decision—that is, if it even registers for them as a decision point—so that issues such as risk and unfamiliarity with the new, more efficient technology or the upfront cost of this new technology become a hindrance to energy efficiency investment.

Alternatively, as we have pointed out, meeting one's energy needs may be accomplished within an energy services paradigm, where the end user hires a service provider to supply the required service, such as space heating or lighting. Approaching the question in this way displaces the responsibility for making decisions about the most appropriate form of energy and the most efficient way of converting that energy to the desired service onto the service provider, who stands to gain financially from more efficient investments and through repeated experience in making efficiency-related decisions, becomes an expert in the field, navigating riskier and unfamiliar decisions more easily.

In other words, achieving a higher energy efficiency and conservation as a means of mitigating GHG emissions will require a transition from our current energy system organization: under the current energy system, consumers face too high a burden to identify and implement the full complement of energy efficiency measures that offer reasonable economic returns. In addition, the energy system itself is poorly structured to facilitate the use of appropriate energy forms for specific energy services. A transition to an energy services paradigm, as we have argued and demonstrated in numerous examples, has the potential for sidestepping many of the barriers that currently hinder investment in energy efficiency.

6.8 Questions for Discussion

1. How is energy different as a service versus a commodity?
2. How should we calculate the footprint of post-industrial societies?
3. How useful an indicator are economywide energy intensity values? What are their limitations?
4. Why don't consumers buy more efficient appliances? Discuss barriers to energy efficiency investment.

5. What is the difference between the uptake of automobiles and refrigerators in the twentieth century, and what factors led to these differences?
6. Explain why fluorescent lights have not completely replaced the incandescent light bulb.
7. What five factors shape the energy efficiency decision environment?
8. Why are energy efficiency labels difficult to interpret by consumers?
9. Where should financial incentives for energy efficiency be directed and why?
10. What are the different types of barriers ESCOs face in delivering services to different end users?
11. What pairings of energy form and service would you recommend for: room heating, lighting, refrigeration, and telecommunications?

References

Ajzen, I. (1991). The theory of planned behavior. *Organizational Behavior and Human Decision Processes, 50*(2), 179–211.

Ajzen, I., and Fishbein, M. (1977). Attitude-behavior relations: A theoretical analysis and review of empirical research. *Psychological Bulletin, 84*(5), 888–918.

Azevedo, I. L., Morgan, M. G., and Morgan, F. (2009). The Transition to Solid-State Lighting. *Proceedings of the IEEE, 97*(3), 481–510.

Bevington, R., and Rosenfeld, A. (1990). Energy for buildings and homes. *Scientific American, 263*(3) 76–80, 82, 84, 86.

Bin, S., and Dowlatabadi, H. (2005). Consumer lifestyle approach to U.S. energy use and the related CO2 emissions. *Energy Policy, 33*(2), 197–208.

Carlsson-Kanyama, A., and Lindén, A.-L. (2007). Energy efficiency in residences—Challenges for women and men in the North. *Energy Policy, 35*(4), 2163–2172.

DeFries, R., and Pandey, D. (2010). Urbanization, the energy ladder and forest transitions in India's emerging economy. *Land Use Policy, 27*(2), 130–138.

Direct Energy Home Services. (2012). Worry Free Water Heater Rental from Direct Energy Home Services. Retrieved April 28, 2012, from http://residential.directenergy.com/EN/Home-Services/ontario/Pages/Rental-Water-Heaters/worry-free-rental.aspx.

Eckert, J. B. (2007). The Super Efficient Refrigerator Program (No. NREL/TP-461-7281) (pp. 1–55). U.S. Department of Energy.

EPA. (2012). *2012 U.S. Greenhouse Gas Inventory Report*. Inventory of U.S. Greenhouse Gas Emissions And Sinks: 1990–2010. U.S. Environmental Protection Agency. Retrieved November 9, 2012, from http://epa.gov/climatechange/emissions/usinventoryreport.html.

Eurostat, (n.d.). Housing statistics—Statistics Explained. Retrieved April 29, 2012, from http://epp.eurostat.ec.europa.eu/statistics_explained/index.php/Housing_statistics#Tenure_status.

Fischer, C. (2008). Feedback on household electricity consumption: a tool for saving energy? *Energy Efficiency, 1*(1), 79–104.

Ford, H., and Crowther, S. (1922). *My Life and Work.* Book Jungle, Published by Gutenberg.org.

Granade, H. C., Creyts, J., Derkach, A., Farese, P., Nyquist, S., and Ostrowski, K. (2009). *Unlocking energy efficiency in the U.S. economy* (p. 165). McKinsey & Company. Retrieved November 9, 2012, from http://www.mckinsey.com/USenergyefficiency.

Greening, L. A., & Greene, D. (2000). Energy efficiency and consumption—the rebound effect—a survey. *Energy Policy, 28*(6), 389–401.

Grohnheit, P. E., and Gram Mortensen, B. O. (2003). Competition in the market for space heating. District heating as the infrastructure for competition among fuels and technologies. *Energy Policy, 31*(9), 817–826.

Hassett, K. A., and Metcalf, G. E. (1993). Energy conservation investment* 1: Do consumers discount the future correctly? *Energy Policy, 21*(6), 710–716.

Heffner, R. R. (2007). *Semiotics and Advanced Vehicles: What Hybrid Electric Vehicles (Hevs) Mean and Why it Matters to Consumers.* Research Report UCD-ITS-RR- 07-30, Institute of Transportation Studies, University of California Davis.

Hekkert, M. P., Harmsen, R., and de Jong, A. (2007). Explaining the rapid diffusion of Dutch cogeneration by innovation system functioning. *Energy Policy, 35*(9), 4677–4687.

Hopper, N., Goldman, C., Gilligan, D., Singer, T. E., and Birr, D. (2007). *A Survey of the U.S. ESCO Industry: Market Growth and Development from 2000 to 2006* (No. LBNL-62679). Berkeley: Lawrence Berkeley National Laboratory. Retrieved November 9, 2012, from http://eetd.lbl.gov/ea/EMS/EMS_pubs.html.

Hydro, B. C. (2010). *Electric Load Forecast*, 2010/11 to 2030/31. Retrieved November 9, 2012, from http://www.bchydro.com/energy_in_bc/irp/document_centre/reports.html.

Jaffe, A. B., and Stavins, R. N. (1994). The energy paradox and the diffusion of conservation technology. *Resource and Energy Economics, 16*(2), 91–122.

Joskow, P. L., and Marron, D. B. (1992). What Does a Negawatt Really Cost? Evidence from Utility Conservation Programs. *The Energy Journal, 13*(4), 41–74.

Kahn, M. E. (2007). Do greens drive Hummers or hybrids? Environmental ideology as a determinant of consumer choice. *Journal of Environmental Economics and Management, 54*(2), 129–145.

Klein, J. (2007). *Why People Really Buy Hybrids. Topline Strategy Group.* Topline Strategy Group, Retrieved November 9, 2012, from http://www.cleanenergycouncil.org/files/Topline_Strategy_Report_Why_People_Really_Buy_Hybrids.pdf.

Lefevre, N., de T'Serclaes, P., and Waide, P. (2006). Barriers to technology diffusion: the case of compact fluorescent lamps. *Joint OECD and IEA Paper, Paris,* (No. JT03216791) 1–35.

Loughran, D. S., and Kulick, J. (2004). Demand-side management and energy efficiency in the United States. *The Energy Journal, 25*(1), 19–43.

Lovins, A. B. (1978). Soft energy technologies. *Annual Review of Energy, 3*(1), 477–517.

Lovins, A. B., and Lovins, L. H. (1997). *Climate: Making Sense and Making Money.* Snowmass, CO: Rocky Mountain Institute.

Lutzenhiser, L. (1993). Social and behavioral aspects of energy use. *Annual Review of Energy and the Environment, 18*(1), 247–289.

Matsukawa, I., and Ito, N. (1998). Household ownership of electric room air conditioners. *Energy Economics*, *20*(4), 375–387.

Mulbrandon, C. (n.d.). Adoption of New Technology since 1900. visualizingeconomics.com. Retrieved July 22, 2012, from http://visualizingeconomics.com/2008/02/18/adoption-of-new-technology-since-1900/#.UAxblo5pvgr.

Natural Resources Canada. (2006). *The State of Energy Efficiency in Canada, Report 2006.* Ottawa: Natural Resources Canada's Office of Energy Efficiency.

Rosenfeld, A. (1999). The Art of Energy Efficiency: Protecting the Environment with Better Technology. *Annual Review of Energy and the Environment*, *24*(1), 33–82.

Shove, E. (2003). *Comfort, Cleanliness and Convenience: The Social Organization of Normality* (p. 221), London: Berg Publishers.

Simon, H. A. (1979). Rational decision making in business organizations. *The American Economic Review*, *69*(4), 493–513.

Smil, V. (1994). *Energy in World History. Essays in World History* (pp. xviii–300). Boulder, CO: Westview Press.

Smil, V. (2007). *Energy in Nature and Society: General Energetics of Complex Systems* (p. 496). Cambridge, MA: The MIT Press.

Smith, K. R., Apte, M. G., Yuqing, M., Wongsekiarttirat, W., and Kulkarni, A. (1994). Air pollution and the energy ladder in Asian cities. *Energy*, *19*(5), 587–600.

Smith, K. R., Uma, R., Kishore, V., Zhang, J., Joshi, V., and Khalil, M. (2000). Greenhouse implications of household stoves: an analysis for India. *Annual Review of Energy and the Environment*, *25*(1), 741–763.

Sorrell, S., Dimitropoulos, J., and Sommerville, M. (2009). Empirical estimates of the direct rebound effect: A review. *Energy Policy*, *37*(4), 1356–1371.

Stern, P. C. (1992). What psychology knows about energy conservation. *American Psychologist*, *47*(10), 1224–1232.

Stern, P. C. (2000). New environmental theories: Toward a coherent theory of environmentally significant behavior. *Journal of Social Issues*, *56*(3), 407–424.

Sutherland, R. J. (1991). Market Barriers to Energy-Efficiency Investments. *The Energy Journal*, *12*(3), 15–34.

Sutherland, R. J. (1996). The economics of energy conservation policy. *Energy Policy*, *24*(4), 361–370.

Thompson, D. (2012). The 100-year march of technology in 1 graph. *The Atlantic*. Retrieved November 9, 2012, from http://www.theatlantic.com/technology/archive/2012/04/the-100-year-march-of-technology-in-1-graph/255573/.

Train, K. (1985). Discount rates in consumers' energy-related decisions: a review of the literature. *Energy*, *10*(12), 1243–1253.

Turrentine, T. S., and Kurani, K. S. (2007). Car buyers and fuel economy? *Energy Policy*, *35*(2), 1213–1223.

Ürge-Vorsatz, D., Köppel, S., Liang, C., Kiss, B., Nair, G. G., and Celikyilmaz, G. (2007). *An assessment of energy service companies (ESCOs) worldwide*. World Energy Council, Retrieved from www.worldenergy.org/documents/esco_synthesis.pdf.

U.S. Census: 2010. (n.d.). Profile of General Population and Housing Characteristics: 2010. Retrieved November 9, 2012, from http://factfinder2.census.gov/faces/tableservices/jsf/pages/productview.xhtml?pid=DEC_10_DP_DPDP1&prodType=table.

U.S. EIA. (2005). 2005 Residential Energy Consumption Survey. Energy Consumption and Expenditure Tables. Retrieved April 28, 2012, from http://205.254.135.7/consumption/residential/data/2005/#undefined.

Wackernagel, M., and Rees, W. E. (1997). Perceptual and structural barriers to investing in natural capital: Economics from an ecological footprint perspective. *Ecological Economics*, *20*(1), 3–24.

Wilk, R. (2002). Consumption, human needs, and global environmental change. *Global Environmental Change*, *12*(1), 5–13.

Wilson, C. (2008). *Understanding and Influencing Energy Efficient Renovation Decisions*. (H. Dowlatabadi, Ed.) (pp. 1–313). Vancouver: UBC.

Wyckoff, A. W., and Roop, J. M. (1994). The embodiment of carbon in imports of manufactured products: Implications for international agreements on greenhouse gas emissions. *Energy Policy*, *22*(3), 187–194.

7

Renewable Electricity

Walter Short

CONTENTS

7.1 Introduction and Scope

Today, combustion of fossil fuels is the world's primary source of energy, and it is also the primary source of greenhouse gas (GHG) emissions to the atmosphere. While the world economies clearly need energy to sustain them, there are alternative energy sources available that do not emit significant GHGs. The two primary sources are nuclear energy and renewable energy. Renewable energy includes many different forms of energy, such as solar, wind, geothermal, biomass, and ocean. These renewable energies are used

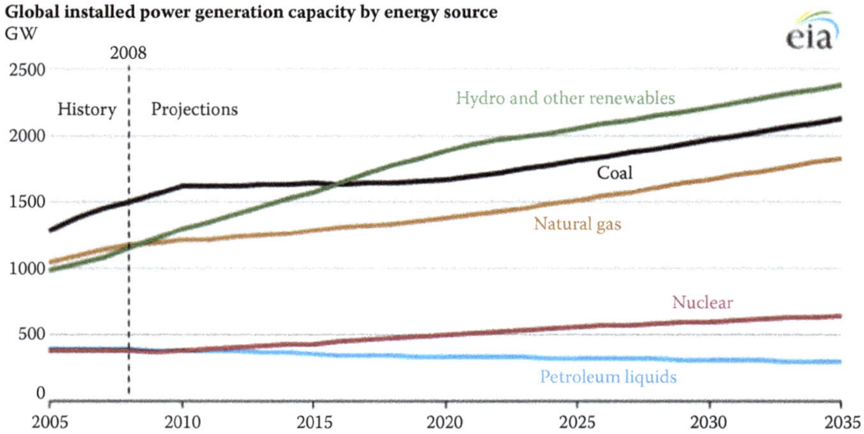

FIGURE 7.1
Global power generation from RETs. (From Energy Information Administration [EIA], International Energy Outlook 2012, Report No. DOE/EIA-0484, http://www.eia.gov/forecasts/ieo/pdf/0484(2011).pdf [accessed February 19, 2012], 2011.)

today in all energy markets from residential to electric power production. The generation of power from renewable electric technologies (RETs) has been growing worldwide and shows every prospect of continuing to grow in the future. In its 2011 World Energy Outlook (International Energy Agency [IEA] 2011), the IEA estimates global use of nonhydro renewables for power generation will increase from 3% in 2009 to 15% in 2035. Similarly, in its 2012 International Energy Outlook (Energy Information Administration [EIA] 2011), the U.S. EIA projects that by 2020 the electric-generating capacity of all RETs together will be larger than that of any single fossil fuel (see Figure 7.1). This projected growth of RETs, combined with the fact that the U.S. electric power sector is the primary source of U.S. carbon emissions (see Figure 7.2), provides a substantial opportunity to reduce carbon emissions. This chapter addresses this opportunity.

7.2 RET and Resources

Renewable resources vary substantially around the world, as shown in Figure 7.3 for total solar irradiance. In this section, we address the magnitude and spatial variability of these resources. Not all resources are equal; some are more difficult (i.e., expensive) to access; some vary temporally as well as spatially and show various degrees of correlation with load. These quality aspects of renewable resources will be addressed in subsequent sections.

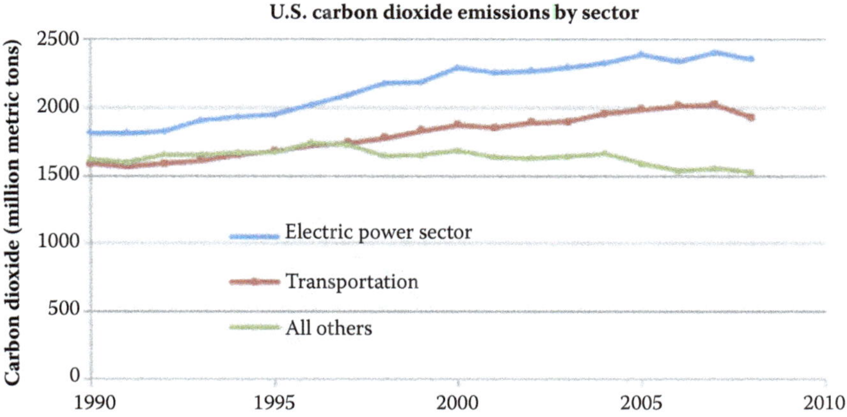

FIGURE 7.2
U.S. carbon dioxide emissions by sector. (From Energy Information Administration [EIA], Emissions of Greenhouse Gases Report, Report No. DOE/EIA-0573, http://www.eia.gov/oiaf/1605/ggrpt/carbon.html [accessed February 4, 2012], 2009.)

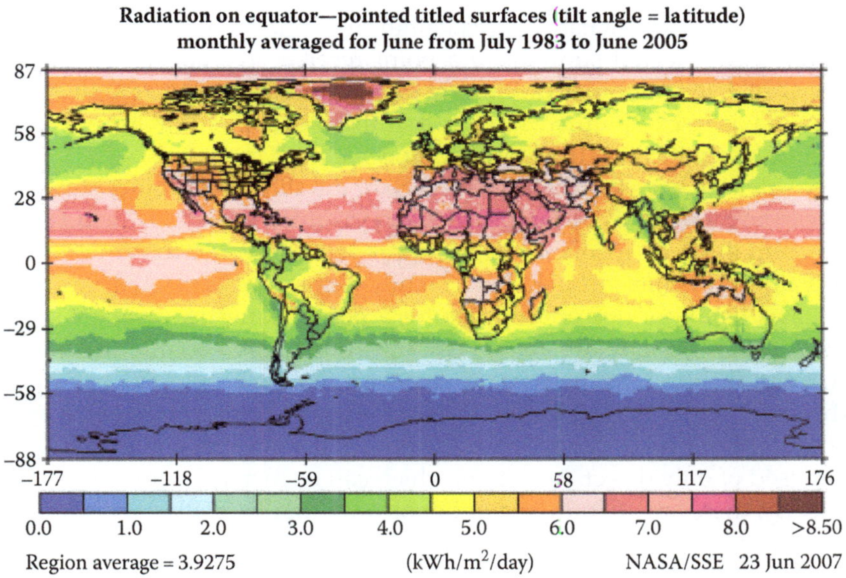

FIGURE 7.3
Worldwide total solar radiation. (U.S. Department of Energy [U.S. DOE], Report to Congress on Renewable Energy Resource Assessment for the United States, Draft Report, Office of Energy Efficiency and Renewable Energy, January 28, 2011.)

7.2.1 Solar Energy

There are several solar generation technologies. The most commonly known are photovoltaic (PV) panels that one can observe on the roofs of many buildings and elsewhere. These fixed PV panels use both direct (directly from the sun) and indirect solar irradiance to generate electricity. To maximize annual generation, they are generally installed facing south and tilted at an angle from the horizontal approximately equal to the latitude of the installation. Some flat panel systems are installed on structures that track the sun to increase the incident insolation; their performance is better than fixed systems, but the cost of the tracking apparatus increases their cost. Generally, these flat-plate tracking PV systems track the sun along one axis only (generally east to west), but higher performance (and higher cost) is available with two-axis tracking systems. Two-axis tracking systems are also used for concentrating PV systems that use lenses or reflectors to concentrate the insolation on a smaller PV cell area.

There is a whole other class of solar generation technologies referred to as concentrating solar power (CSP) technologies that use tracking technologies and reflectors to heat a transfer fluid that provides the energy for steam generation for power production through a steam turbine. Three primary configurations of CSP plants exist today. The more common are parabolic troughs that use one-axis tracking to focus direct insolation on a long receiver tube that contains the heat transfer fluid. The second CSP technology is a parabolic dish, somewhat like a satellite dish but with two-axis tracking, that focuses reflected sunlight to a point receiver just above the center of each dish. A less modular CSP technology are central receiver systems that use two-axis tracking mirrors to focus sunlight onto a central tower receiver through which the heat transfer fluid flows.

The solar resource associated with each of the above solar power technologies can be different. For example, global horizontal insolation (includes both direct and indirect* insolation) is the appropriate resource for a nontracking flat-plate PV system installed horizontally. But global horizontal insolation estimates would not be appropriate for a nontracking flat-plate PV system installed facing south at an angle from the horizontal equal to the installation latitude. Similarly, a one-axis, east-west, flat-plate tracking system has a different resource available to it than does a fixed PV system. And again, a concentrating solar system has yet another resource available to it that is composed of only direct insolation, which can be focused onto a receiver.

Clearly, there are an infinite number of ways that fixed PV panels could be oriented or that concentrating solar systems could track the sun, and thus an infinite number of resource estimates. We show here some of the more common estimates. Figure 7.4 shows the variability in the United States for the annual solar energy available to a fixed, tilted-at-latitude, flat-plate PV array (nonconcentrating). Figure 7.5 shows that similar data for a concentrating solar collector varies by more than a factor of two just across the United States due to latitude, clouds, atmospheric differences (e.g., airborne particulates, humidity), and so on. Clearly, the sites in the southwestern

* Often referred to as diffuse insolation because it is refracted and reflected before arriving at the collector.

United States with the better solar resources, shown in Figure 7.5, for concentrating solar collectors will produce more annual energy and be more cost competitive than a similar array placed in New England. Fortunately, as intimated by Figures 7.3 through 7.5, there are ample solar resources; it has been estimated (University of Michigan 2011) that the nation's electric needs could be met by moderate-efficiency PV panels covering only 0.4% of the land area of the United States.

7.2.2 Wind Energy

While many different wind turbine configurations have been proposed and developed over the years, large horizontal axis turbines have become the standard for central utility applications. Today, the state-of-the-art large onshore wind turbine may have a rotor diameter as large as 125 m with three blades attached to a turbine in a nacelle on top of a tower with a hub height of 80–100 m. The size of such turbines has been gradually increasing with

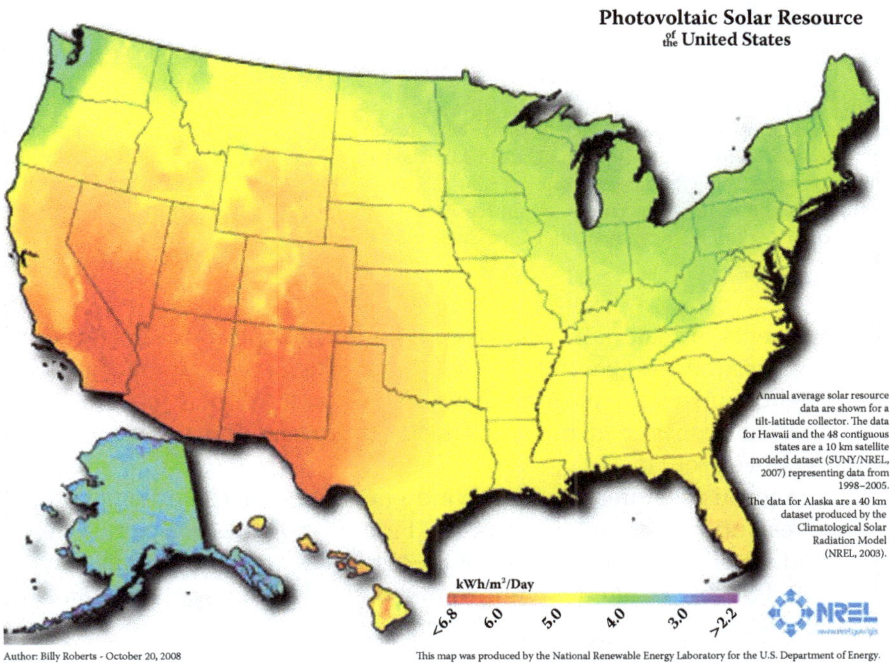

FIGURE 7.4
Estimated annual average solar resource for a flat-plate collector oriented with a south-facing tilt equal to site latitude based on satellite model results for 1998–2005 with 10 km resolution. Data for Alaska are from the 40 km estimates by the Climatological Solar Radiation model. (From U.S. Department of Energy [U.S. DOE], Report to Congress on Renewable Energy Resource Assessment for the United States, Draft Report, Office of Energy Efficiency and Renewable Energy, January 28, 2011.)

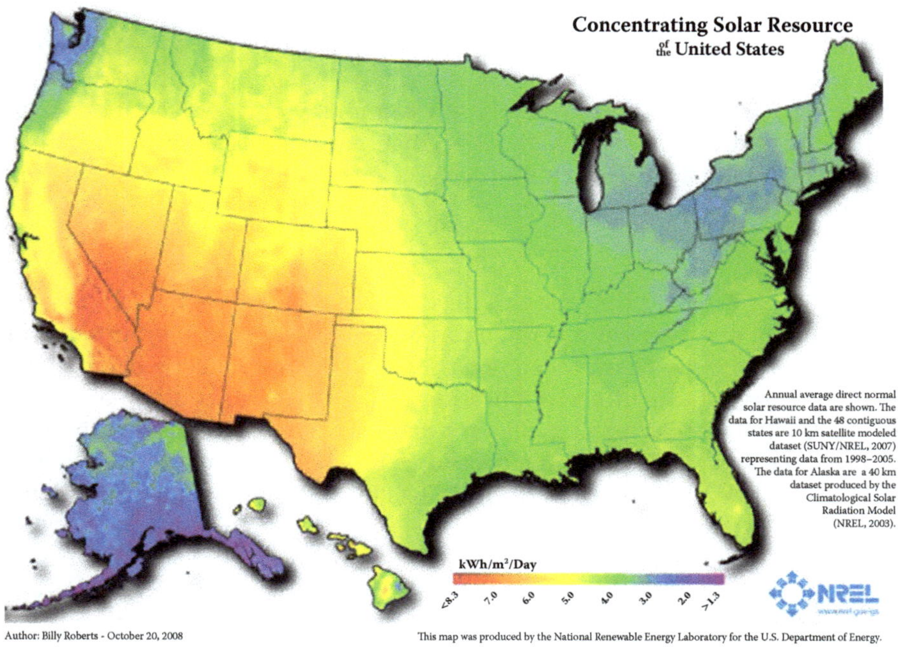

FIGURE 7.5
Estimated annual average direct normal solar resource based on satellite model results for 1998–2005 with 10 km resolution. Data for Alaska are from the 40 km estimates by the Climatological Solar Radiation model. (From U.S. Department of Energy [U.S. DOE], Report to Congress on Renewable Energy Resource Assessment for the United States, Draft Report, Office of Energy Efficiency and Renewable Energy, January 28, 2011.)

improvements in materials, design, and manufacturing to reach elevations above the ground where wind resources are greater. As shown in Figure 7.6, these higher hub height machines have opened up many new resource areas that were not considered in the past due to lower wind speeds at lower heights above ground.

Due to terrain features, onshore wind energy resources vary spatially at a finer resolution than solar resources. Figure 7.6 shows the annual average wind speed at the hub height of 80 m for new onshore turbines across the United States. While the wind resource is not as smoothly available as the solar resources of Figure 7.3 through 7.5, onshore wind resources are clearly concentrated in the center of the country, a portion that is not heavily populated and therefore has lesser need for electric power. Offshore wind resources also exist off all coasts, including the Great Lakes. However, as can be seen from Figure 7.7, the best resources occur in the northeast and the least in the southeast and eastern Gulf of Mexico. The U.S. Department of Energy (U.S. DOE) has recently estimated that wind energy technologies could provide 35,000 TWh of generation each year, almost nine times the current total

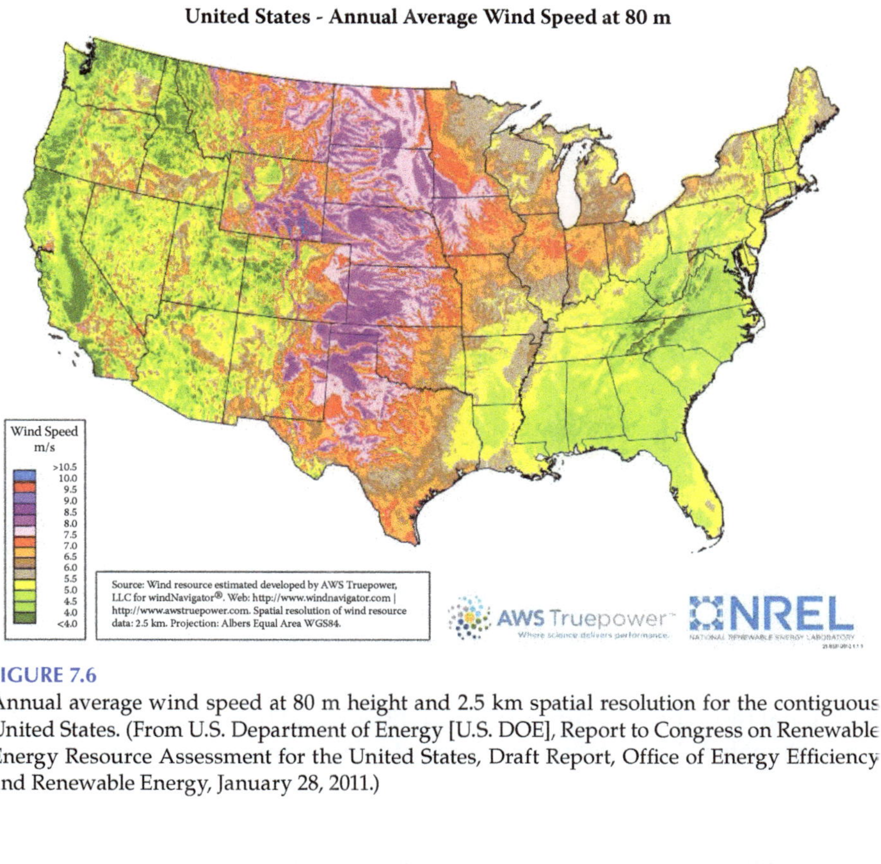

FIGURE 7.6
Annual average wind speed at 80 m height and 2.5 km spatial resolution for the contiguous United States. (From U.S. Department of Energy [U.S. DOE], Report to Congress on Renewable Energy Resource Assessment for the United States, Draft Report, Office of Energy Efficiency and Renewable Energy, January 28, 2011.)

U.S. power generation of 4,000 TWh (National Renewable Energy Laboratory [NREL] 2010a).

7.2.3 Hydroelectric Power

Hydroelectric power is generated using water from rivers. There are two basic forms of hydropower technology—reservoir hydro that stores water in a reservoir behind a dam, and run-of-river hydro that places turbines within a river to generate power as the river water flows through the turbine blades. Of the 78 GW of existing hydroelectric capacity, 66 GW are reservoir-based systems with capacities of 30 MW or more. These hydro facilities have provided 6%–7% of U.S. generation over the last 10 years, depending on annual precipitation (Center for Climate and Energy Solutions [CCES] 2012). Today, this generation comprises about 70% of all power generation from renewable sources.

There are several potential sources of new hydropower. Only about 3% of existing dams have hydro generators; the efficiency of existing hydro plants could be significantly improved to produce more power; there are sites where

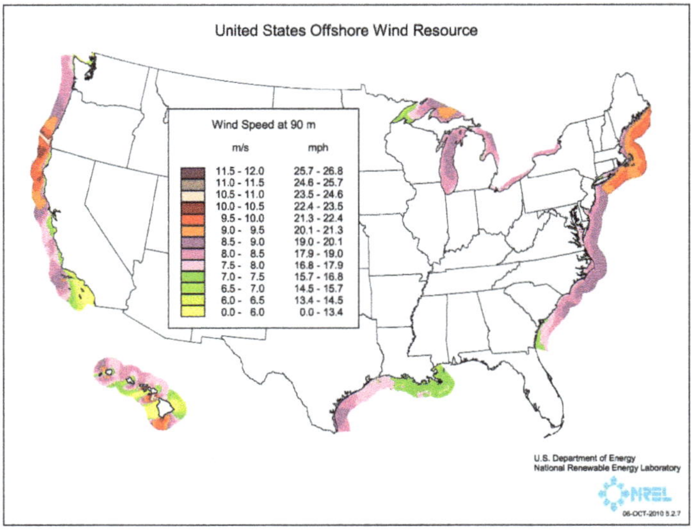

FIGURE 7.7
Offshore annual average wind speed at 90 m height and 0.2 km spatial resolution for the contiguous United States and Hawaii. (From U.S. Department of Energy [U.S. DOE], Report to Congress on Renewable Energy Resource Assessment for the United States, Draft Report, Office of Energy Efficiency and Renewable Energy, January 28, 2011.)

additional reservoir and run-of-river facilities could be built. Hall et al. (2004) have estimated this potential in terms of GW of annual mean power* (see Figure 7.8) to be approximately 170 GW.

7.2.4 Other Renewable Energy Resources for Power Generation

Although relatively smaller in size, other renewable resources for power—biomass, geothermal, and ocean—are adequate to meet significant fractions of current demands for power. There are three primary technologies for power production from biomass energy. The most conventional is direct combustion of biomass to produce steam to power a steam turbine. A more advanced alternative that could facilitate the capture of any carbon emissions emitted during the combustion process is a combined-cycle† biomass gasification system. A third hybrid option is the cofiring of coal-fired power plants, wherein biomass is substituted for 5%–15% of the coal normally used by the power plant.

* A GW of annual mean power is the capacity equivalent of a plant operating around the clock that produces the same amount of annual generation as a variable generation source. Inasmuch as most hydro plants are constrained by water availability, they operate approximately 50% of the time so that 100 GW of plant capacity would equal 50 GW of annual mean power capacity.

† A combined-cycle power plant generally refers to a plant that combines one or more Brayton cycle combustion turbines with a Rankine cycle steam turbine. The waste heat from the combustion turbines is fed to a heat recovery steam generator, where it is used to produce steam for the Rankine cycle.

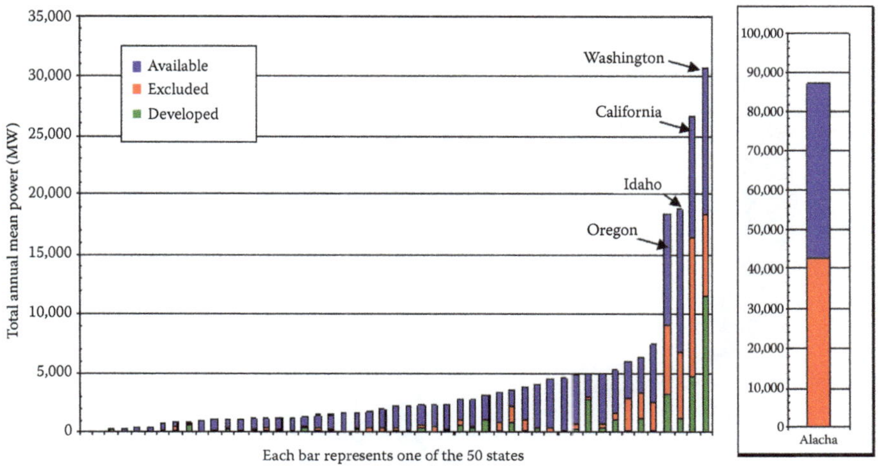

FIGURE 7.8
Total power potential of water energy resources in the 50 states of the United States divided into developed, excluded, and net available constituents. (From Hall, D., et al., Water Energy Resources of the United States with Emphasis on Low Head/Low Power Resources, Report No. DOE/ID-11111, Idaho National Engineering and Environmental Laboratory, ID, April 2004.)

The biomass resources used to fire these technologies can take many forms from waste materials—agricultural, industrial, municipal, forest—to dedicated energy crops grown specifically for energy. These same biomass resources are more often used today to produce fuels or in some cases, for space heating for buildings. Similarly, dedicated energy crops must compete with other uses of land, fertilizer, and water for the growth of crops for food, fabrics, biochemicals, and so on. Thus, resource estimates for biopower are dependent on markets and competition between resources and their use. Most estimates of the resource attempt to separate out the conventional use of land for food and other conventional uses, from the use for energy, but make no attempt to estimate the resource available for power versus fuels. There is also the complexity introduced by increasing yields in crop production as new agricultural technologies and practices are introduced. Thus, it is important to distinguish the year for which a biomass resource is estimated.

For example, Figure 7.9 shows the total biomass resource that could be available today accounting for conventional land use. It illustrates that like wind and solar resources, biomass resources are not uniformly available across the country. As might be expected, they are concentrated in the Corn Belt in the middle of the country with significant additional resources on the west coast, the mid–south central region, and Maine.

Estimates of biomass resource potential vary depending on the market, competition, and time frame assumptions mentioned above. Ranges from

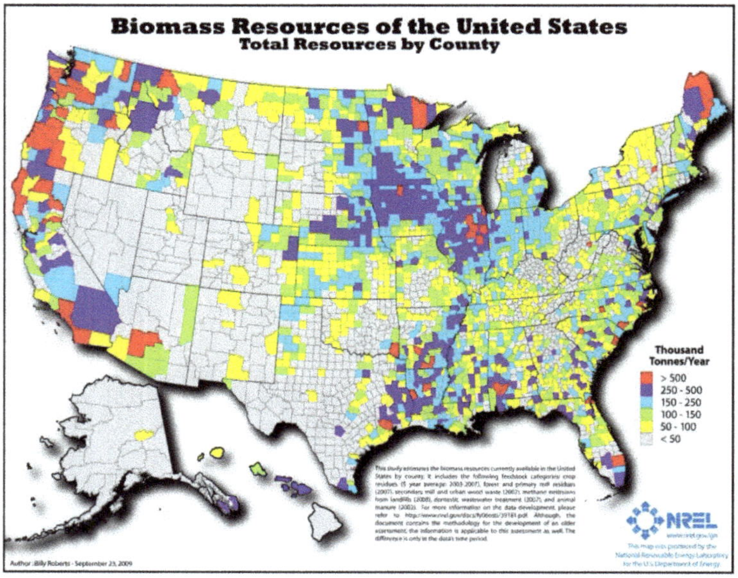

FIGURE 7.9
Total current biomass resources by county, including agricultural residues, wood residues, and municipal discards. (From U.S. Department of Energy [U.S. DOE], Report to Congress on Renewable Energy Resource Assessment for the United States, Draft Report, Office of Energy Efficiency and Renewable Energy, January 28, 2011.)

400 to 1300 million metric dry tons/year are cited by DOE (U.S. DOE 2011) from different sources. Translating these estimates to power generation is complicated by the fact that the heat content of the biomass can vary from 12 to 20 million British thermal unit (MBtu)/dry ton (Oak Ridge National Laboratory [ORNL] 2012), and the heat rate* of the power generation equipment will also vary depending on the technology and its state of operation. Assuming a heat rate of 10 MBtu/MWh and a heat content of 15 MBtu/dry ton, the ends of our resource range (400–1300 million dry tons) yield a range for power generation from 600 to 2000 TWh. Today's biopower production is on the order of 56 TWh of total U.S. generation of 4000 TWh (EIA 2012).

There are also several different geothermal power generation technologies. The largest geothermal power facility in the United States today is a unique direct-steam resource at the Geysers in northern California; all other existing U.S. geothermal power plants are hydrothermal power plants in which hot water is taken from underground reservoirs and either flashed to steam or used to generate steam in a secondary fluid by means of a heat exchanger.

* The heat rate of a power generation technology is the ratio of the input thermal energy per unit of output electric energy. It is usually expressed in the United States in terms of British thermal unit of input per kilowatt hour of output, but other units can also be used as we have done here.

Estimates of the U.S. hydrothermal resource are difficult to estimate precisely due to geology, hydrology, and other uncertainties. The DOE has estimated known resources at 9 GW and undiscovered resources at 30 GW (U.S. DOE 2011). Although these are relatively small quantities, a much larger resource is available in what is termed "enhanced geothermal systems" (EGSs). EGSs require that hot underground rock structures be fractured and water circulated through and recovered from those fractures to produce steam-driven power. The DOE (2011) estimates these resources could be as large as 518 GW of electric capacity.

As shown by Figure 7.10, geothermal energy resources in the United States are concentrated in the west, with known hydrothermal deposits concentrated in Nevada, Oregon, and Idaho with some significant sites in Colorado, Utah, Montana, and New Mexico. To date, less than 3 GW of the 40 GW of hydrothermal and direct steam sites have been developed

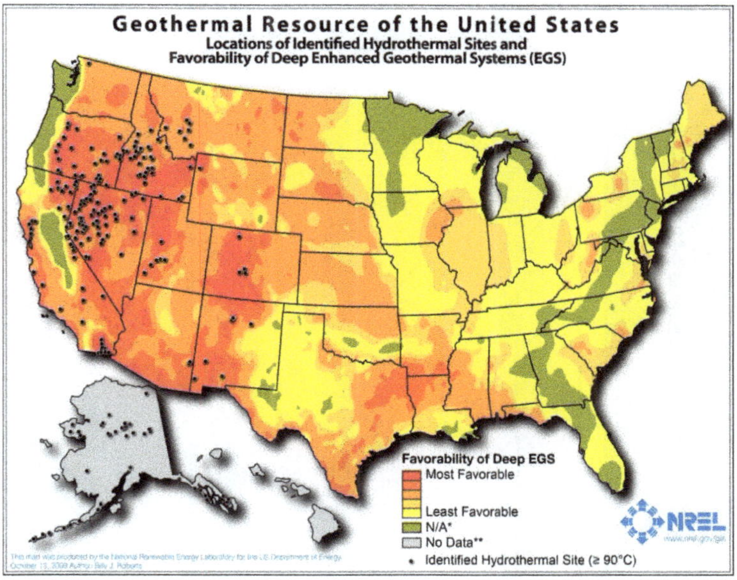

FIGURE 7.10
Geothermal resource map showing locations of identified hydrothermal sites and favorability of deep EGSs. Map does not include shallow EGS resources located near hydrothermal sites or U.S. Geological Survey assessment of undiscovered hydrothermal resources. Source data for deep EGS includes temperature at depth from 3 to 10 km, provided by Southern Methodist University Geothermal Laboratory and analyses (for regions with temperatures ≥150°C) performed by NREL in 2010. *"N/A" regions have temperatures less than 150°C at 10 km depth and were not assessed for deep EGS potential. **Temperature-at-depth data for deep EGS in Alaska and Hawaii not available. (Data from U.S. Department of Energy [U.S. DOE], Report to Congress on Renewable Energy Resource Assessment for the United States, Draft Report, Office of Energy Efficiency and Renewable Energy, January 28, 2011.)

for power production. However, as shown by Figure 7.10, there are ample additional resources available at enhanced geothermal sites. Unfortunately, these resources and others to include geopressurized brines have not been widely exploited because technical challenges remain and costs are relatively high.

The primary hydrokinetic ocean energy technologies include wave energy, ocean currents, and tidal energy. Generation of power from waves uses the surface movement of water, which is generally driven by winds over the water surface. Wave resources are the most promising of the hydrokinetic technologies, but the total potential for the contiguous United States is less than 40 GW (Previsic et al. 2011) and is focused on the West Coast (see Figure 7.11). Power can be derived from ocean currents through underwater turbines that convert the kinetic energy of the current into rotational energy and then to electricity. Ocean currents are created by winds, ocean salinity gradients, rotation of the earth, and other factors. The only ocean currents close enough to the United States to access for power are off the east coast of

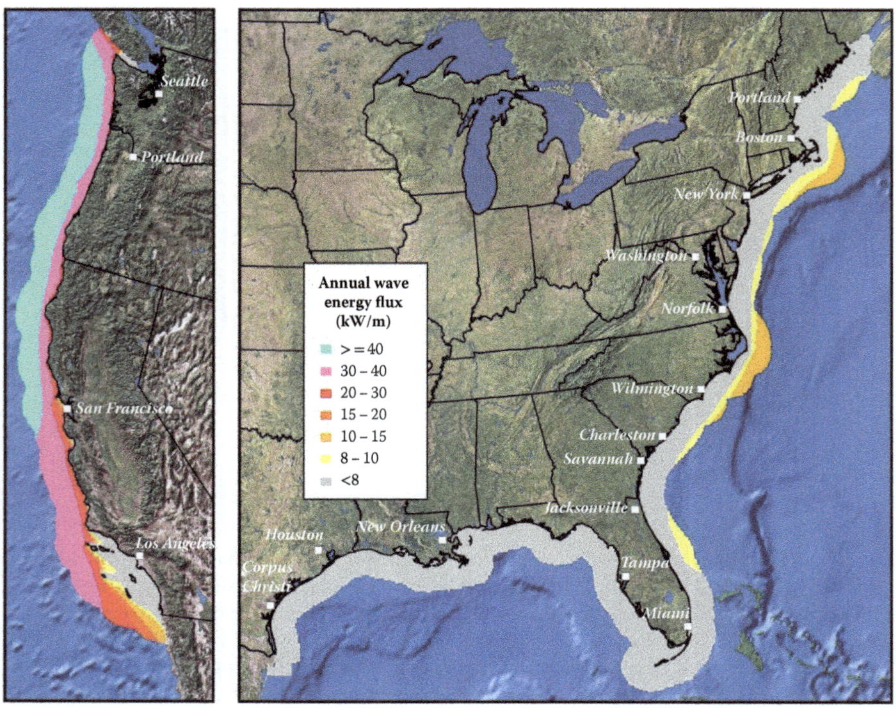

FIGURE 7.11
Annual average wave flux for the contiguous United States. (From Previsic, M., et al., Lifecycle Cost Assessment of Hydrokinetic Technologies in the U.S. over Time, Draft Report, RE Vision Consulting, December 2011.)

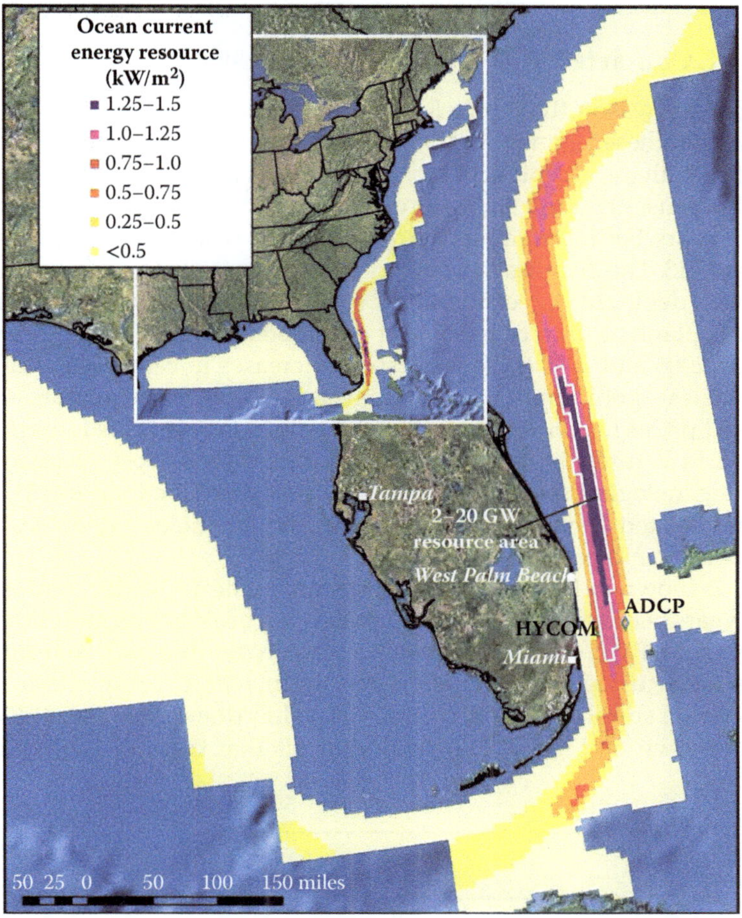

FIGURE 7.12
Ocean current power density. (From Previsic, M., et al., Lifecycle Cost Assessment of Hydrokinetic Technologies in the U.S. over Time, Draft Report, RE Vision Consulting, December 2011.)

Florida (see Figure 7.12). While there may be as much as 25 GW of potential near Florida, concerns about the effect of energy turbines on current flow may limit the practical potential to only 5 GW (Previsic et al. 2011). Ocean tides are the result of the gravitational pull of the sun and the moon combined with the rotation of the earth. Tidal energy can be converted to power either through entrapment of high-tide waters through dams or barrages with subsequent release through conventional hydropower generators, or through strategically placed underwater turbines. Expert estimates of the tidal energy available to the contiguous United States are limited to less than a GW (Previsic et al. 2011).

7.3 RET Characteristics

The primary reason that more RETs have not been deployed is the cost of energy from these technologies. Their costs are composed primarily of capital costs. (With the exception of biomass, these technologies do not have fuel-related expenses.) By and large, the capital costs for RETs have been declining over the years. See for example, the decline in the cost of wind and PV shown in Figure 7.13. The declines can be attributed to learning-by-doing as systems have been deployed, to economies of scale in both system size and manufacturing plants, and to research and development (R&D) efforts. However, between 2006 and 2010 there were some increases in capital costs of many technologies—both renewable and conventional power generators—due to increased prices for commodities like steel and concrete. This is exemplified by wind in Figure 7.13. Wind technology prices (and those of many other power technologies) seem to have leveled off or declined since 2011 due to recession-related decreases in commodity and technology demand, and to continued learning and R&D.

As also shown in Figure 7.13, PV costs have followed a different trend in recent years because PV is not as dependent on bulk commodities like steel and concrete. PV costs have continued to decrease due both to technology and manufacturing improvements, and to increased competition among international suppliers. In particular, China has dominated recent PV markets, as shown by Figure 7.14, reminding us that there is a difference, at

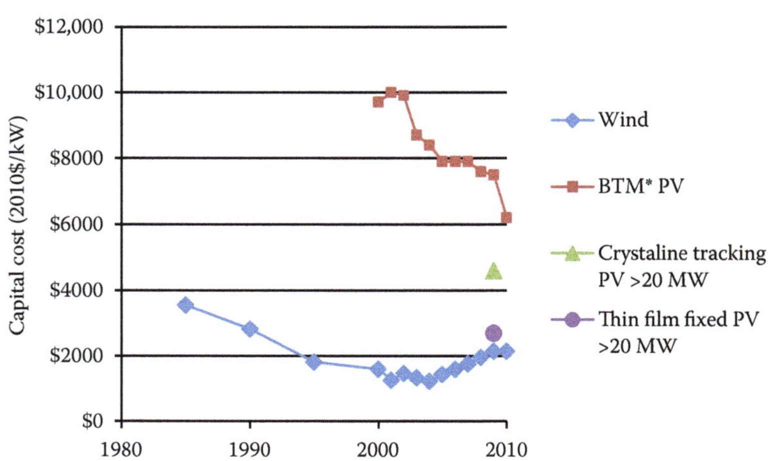

FIGURE 7.13
Wind and small PV capital cost trends. *BTM = behind the meter, that is, on the customer side of the electric meter. (Data from Wiser, R., and M. Bolinger, 2010 Wind Technologies Market Report, U.S. Department of Energy, June 2011, 2011 and from Barbose, G., et al., Tracking the Sun IV: An Historical Summary of the Installed Cost of Photovoltaics in the United States from 1998 to 2010, Report prepared by the Lawrence Berkeley National Laboratory, CA, September 2011.)

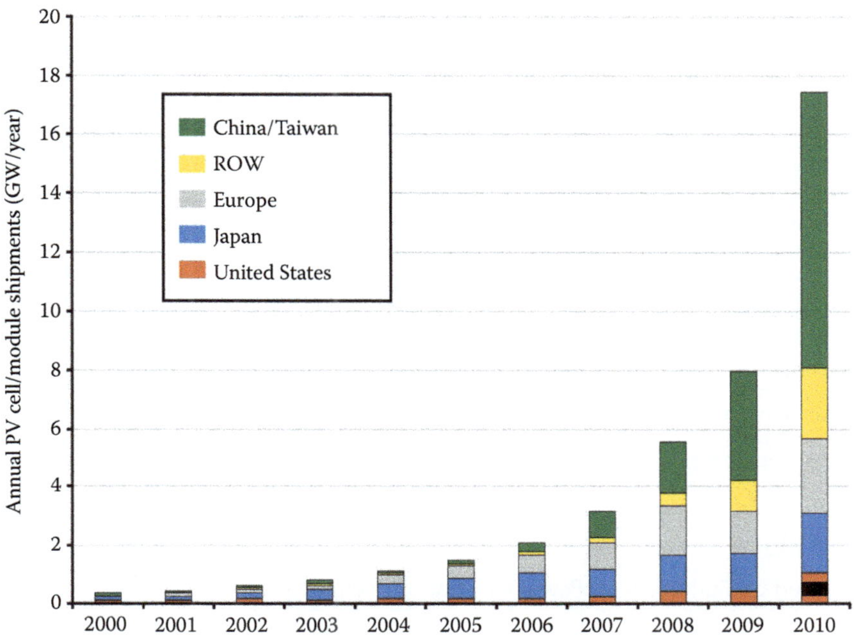

FIGURE 7.14
Regional PV cell and modular shipments, 2000–2010. (From U.S. Department of Energy [U.S. DOE], SunShot Vision Study, February 2012.)

least in the short term, between cost and price. PV costs in Figure 7.13 are also differentiated by small PV systems that are "behind the meter" and larger systems that may be owned by utilities or independent power producers. Note the smaller costs of new thin film utility installations in Figure 7.13.

Table 7.1 presents additional cost data for wind, PV, CSP, and geothermal 2010 installations that is largely consistent with the wind and PV costs of Figure 7.13. The Environmental Protection Agency (EPA) data reminds us that even within a given year, RET capital costs can vary significantly and that RET generation costs vary even more widely due to resource quality, operations and maintenance (O&M) costs, and capital costs.

Table 7.2 presents technology cost data from a different source. We include it because

- It has data from conventional and renewable generation technologies.
- It has data for multiple future years.
- Its costs are generally higher than those of Table 7.1 because the costs were estimated in 2009 before commodity prices softened, highlighting the fact that it is important today to consider the time at which technology cost estimates were made.

TABLE 7.1

2010 Costs of RETs

	Capital Costs ($/kW)	Fixed O&M ($/kW-Year)	Variable O&M ($/MWh)	LCOE ($/MWh)
Wind minimum	1938	12	8	84
Wind maximum	2145	31	32	142
PV minimum	4978	10	—	132
PV maximum	9322	25	—	298
CSP minimum	2024	61	101	109
CSP maximum	6071	71		335
Geothermal minimum	3491	170	25	59
Geothermal maximum	4655	—	30	94

Source: Environmental Protection Agency (EPA), Renewable Energy Cost Database, http://www.epa.gov/cleanenergy/energy-resources/renewabledatabase.html (accessed February 6, 2012), 2012.

TABLE 7.2

Projected Capital Costs for Power Generation Technologies

	Capital Cost (2010 $/KW)			Heat Rate (MMBtu/MWh)		
	2010	2025	2050	2010	2025	2050
Nuclear	6,283	6,283	6,283	9.7	9.7	9.7
Gas Combustion turbine	671	671	671	10.4	10.4	10.4
Gas Combined cycle	1,267	1,267	1,267	6.7	6.7	6.7
Gas Combined cycle sequestration	NA	3,863	3,863	10.08	10.08	10.08
Coal pulverization	2,977	2,977	2,977	9.37	9	9
Integrated gasification combined cycle	4,130	4,130	4,130	9.03	7.95	7.95
Integrated gasification combined cycle sequestration		6,798	6,798		10.38	10.38
Hydro	3,605	3,605	3,605			
PV	6,129	3,564	3,018			
PV tracking	NA	2,482	2,091			
CSP storage	7,272	6,098	4,841			
Wind onshore	2,039	2,039	2,039	41[a]	42[a]	43[a]
Wind offshore	3,409	3,162	3,080	45[a]	45[a]	45[a]
Bio-cofire	1,020	1,020	1,020	10	10	10
Biopower	3,945	3,945	3,945	14.5	13.8	12.5
Hydrothermal	6,118	6,118	6,118			
EGS	10,197	9,721	8,673			

Source: Black and Veatch Corporation, Cost and Performance Data for Power Generation Technologies, Report No. NREL-SR-48-595-201202-0, Draft Report prepared for the National Renewable Energy Laboratory, Golden Colorado, February 2012 (costs estimated in 2009, but published in 2012).

[a] Annual capacity factor of class 5 wind.

What really matters in terms of the market adoption of a technology is the cost of generation relative to the cost of competing technologies.* Of course, the capital costs given in Tables 7.1 and 7.2 cannot be directly compared as there are other cost and performance data associated with each technology, most importantly fuel costs for conventional technologies. We do not try to estimate those here because the cost of natural gas has been so volatile over the years and is currently low due to the recent production increases in shale gas. Marginal costs of electricity from new generation investments are also heavily influenced by the cost of money used to amortize the capital costs given in Tables 7.1 and 7.2. These costs of money (i.e., interest rates) have also been volatile in recent years and are at an all-time low, making any comparison of the cost of electricity from future installations of these technologies tenuous at best.

Notwithstanding the above concerns regarding future prices, there is a reasonable expectation that the costs of renewable technologies that are still in their market infancy—like PV, CSP with thermal storage, and offshore wind—will continue to decline in the future due to learning, R&D, and competition. Thus, in the remainder of the chapter we focus on the market prospects for those solar and wind technologies that have large resource bases and potential for low cost.

7.4 Variability of Wind and Solar Resources

Clearly, cost is not the only driver of market deployment of RETs, especially the more cost-competitive technologies of wind and PV. The competitiveness of these two technologies is strongly affected by the fact that their renewable resources are variable and uncertain, as shown for wind by Figure 7.15. This variability introduces several difficulties in the planning and operation of the grid. It is these difficulties that can make wind or solar less competitive even if their base costs of energy are less than that of competing technologies.

As shown in Figure 7.15, the introduction of variable RETs (VRETs) into the grid exacerbates variability and uncertainty (VU) faced by grid planners and operators. In particular, Figure 7.15 shows that

- The net load (original load minus the generation provided by VRETs, that is, the load that remains to be met by non-VRETs) varies more from its low to its peak than the load itself. Witness April 14 when the load ranged from about 27 GW overnight to about 47 GW in mid-afternoon. With VRETs, the net load ranges from 18 to about 43 GW. This means that with VRETs in the system, non-VRETs will have to

* Later, we will show that cost is a strong, but not the only, critical factor in market adoption, especially for variable renewable generators.

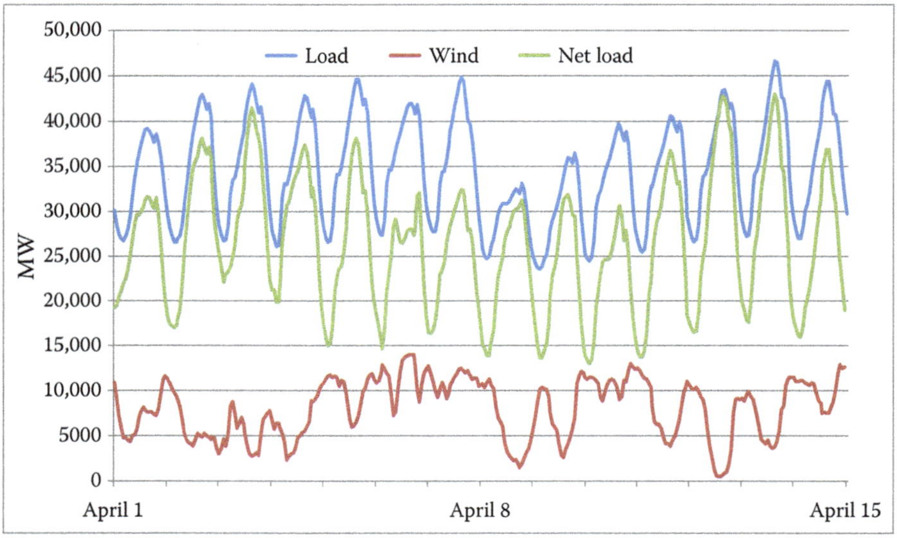

FIGURE 7.15
Load and wind variability. (From Denholm, P., et al., The Role of Energy Storage with Renewable Electricity Generation, Technical Report No. NREL/TP-6A2-47187, January 2010.)

ramp up to 25 GW between morning and afternoon, whereas without VRETs, the ramp range would have been only about 20 GW. This increase in ramp range can impose additional demands (e.g., thermal stress) on the non-VRETs that must meet the net load.

- The net load descends to new lows; for example, for the period shown in Figure 7.15, the lowest load point was about 24 GW on April 9, but the lowest net load is only 13 GW on the night of April 10. Clearly, this has the advantage that it requires less power from non-VRETs at those times, but that may not be entirely a good thing. Before the VRETs were added, base load plants like nuclear could have been providing all this minimum load level. However, with the addition of VRETs, these base load plants will either have to ramp down to 13 GW or the generation from the VRETs will have to be curtailed. Neither is really desirable. The owners of the VRETs want to be compensated for this power that they could have generated at essentially no cost to them, and the owners of the nuclear plant do not want to cycle their plant to follow the net load because it causes thermal stress on their generator.
- The maximum wind contribution during this 2-week period is almost 15 GW; however, the net peak load (see April 13) is only 3 GW below the peak load of 46 GW on April 14. This is a simple illustration of the fact that the capacity value of VRET can be considerably lower than the nameplate capacity of the VRET system.

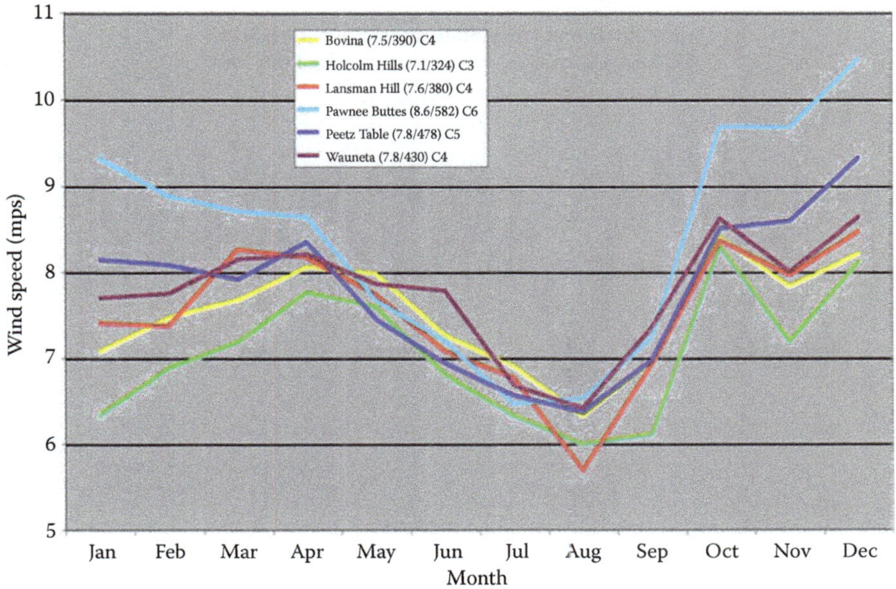

FIGURE 7.16
Seasonal variation in wind speed measured at 40 m above ground at six locations in northeastern Colorado. (From U.S. Department of Energy [U.S. DOE], Report to Congress on Renewable Energy Resource Assessment for the United States, Draft Report, Office of Energy Efficiency and Renewable Energy, January 28, 2011.)

Figure 7.16 presents monthly average generation from six wind sites in Colorado that further illustrate why the capacity value of wind may be considerably less than the nameplate capacity, or even the capacity factor,* of the wind system. The Colorado wind farms all show a distinct downturn in the generation in summer months when air-conditioning loads generally result in peak annual loads for most utilities within the contiguous United States. This summer downturn in wind generation is common for most wind sites in the United States.

The profile of generation from PV is distinct from that of wind, as shown by Figure 7.17. Whereas wind generation generally is less during the summer and during the daytime, solar generation peaks in the middle of summer days. As shown by Figure 7.18, this peak is largely, though not entirely, coincident with peak summer loads.

Variability of wind and other variable renewable resources occurs not only from hour-to-hour, as shown in Figure 7.15, but also from month-to-month and year-to-year, as shown for wind in Figure 7.16 and solar in Figure 7.17.

* Annual capacity factor for wind is the total annual wind generation divided by the generation that could have been produced by the wind system, had the wind blown at full strength all year-round (i.e., divided by nameplate capacity times 8760 h/year).

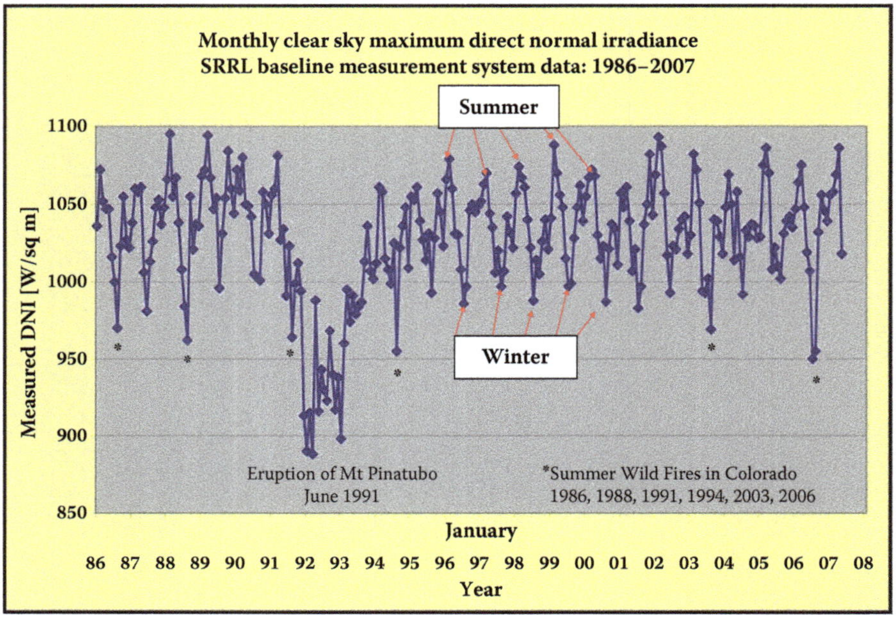

FIGURE 7.17
Inter- and intra-annual variation in direct normal irradiance. (From U.S. Department of Energy [U.S. DOE], Report to Congress on Renewable Energy Resource Assessment for the United States, Draft Report, Office of Energy Efficiency and Renewable Energy, January 28, 2011.)

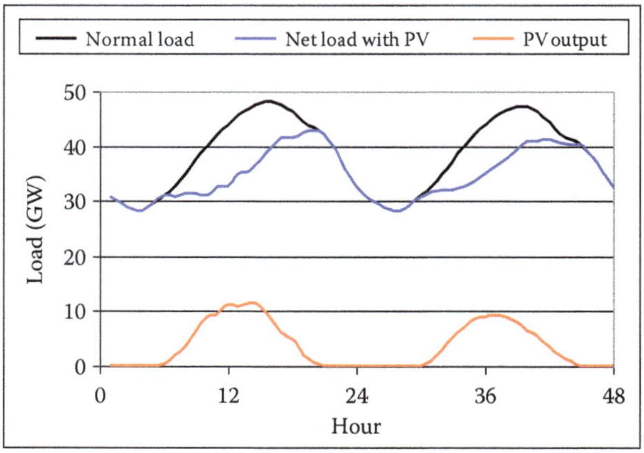

FIGURE 7.18
PV coincidence with summer loads.

Fortunately, much of the impact of wind variability can be mitigated through geographic diversity. Wind installations that are far apart or separated by significant topological features will be less correlated with each other so that many times when the wind is not blowing at one site, it will be blowing at the other site. Figure 7.19 shows that the correlation between the output of two sites decreases with the distance between them, but increases with the time span considered. Thus, by using wind from multiple geographically dispersed sites, a utility can easily reduce the short-term fluctuations in wind generation, but will find it more difficult to smooth out the daily wind generation. Figure 7.20 shows how the coefficient of

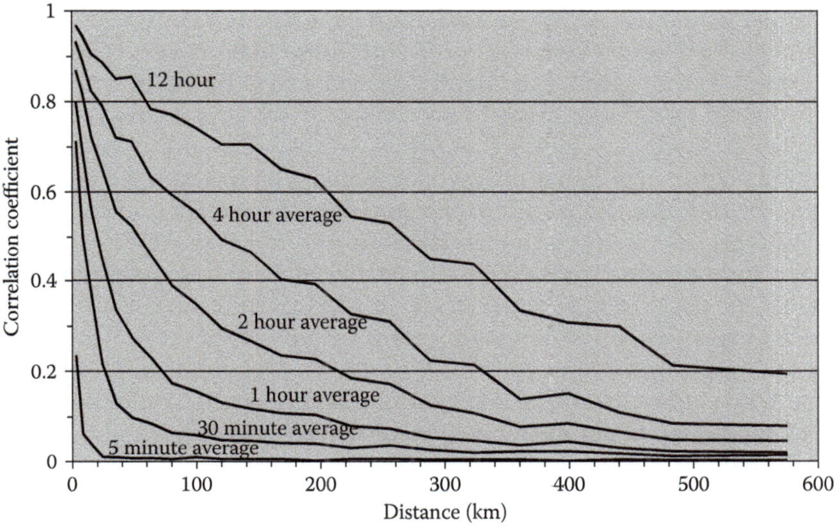

FIGURE 7.19
Impact of geographic diversity on variability in wind generation. (From Ernst, B., et al., Short-Term Power Fluctuation of Wind Turbines: Analyzing Data from the German 250-MW Measurement Program from the Ancillary Services Viewpoint, Report No. NREL/CP-500-26722, National Renewable Energy Laboratory, Golden Colorado, July 1999.)

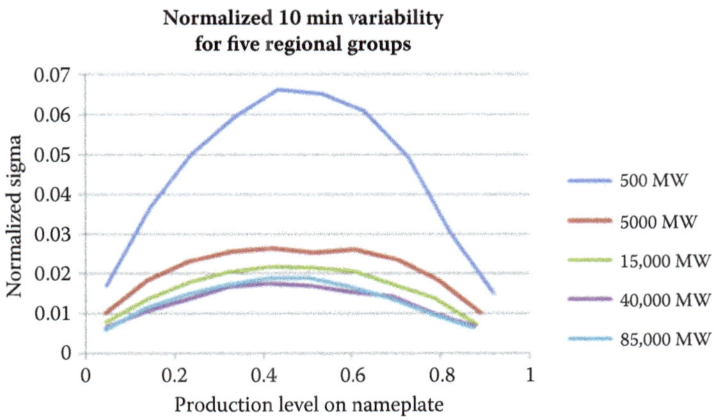

FIGURE 7.20
Variability of wind from multiple sites. (From NREL, Eastern Wind Integration and Transmission Study [EWITS], Report No. NREL/SR-550-47078, Report prepared for NREL by Enernex, TN, January 2010.)

variation (standard deviation divided by the mean) of short-term (10 min) generation decreases significantly as wind farms are added from multiple dispersed sites.

Figures 7.15 through 7.18 would seem to suggest that were the costs of energy the same from wind and PV, PV would always be a better option than wind because it matches load better. However, reality is more dynamic. While the correlation of PV generation and peak loads does give PV an initial advantage, two issues begin to occur with more market penetration. First, with about 15% of generation met by PV, there may be times in the spring when the sun is shining brightly yet there are not large loads and the PV must be curtailed, especially if there are base load generators that cannot be operated below some minimum level or there is significant wind generation during the day. Second, once PV has penetrated to a certain level (perhaps 20%–25% of generation, far higher than we find in today's U.S. system), its strong afternoon generation could reduce the net load in summer afternoons to below the net load in summer evenings, that is, shift the net peak load to a point in time when PV cannot contribute much. With such a shift in the net peak load, the capacity value of PV would fall dramatically. This is one reason why studies that include high renewable electricity penetration often show PV reaching about 15%–25% penetration by generation and then leveling off.

The value of variable generation is reduced not only by its variability but also by the uncertainty of that variability. If a grid dispatcher is aware that variable generation is about to ramp up or down, adjustments in the output of non-VRET generators can be made. If accurate forecasts are available hours or days in advance, needed adjustments to planned generation from non-VRET units can be replanned in advance and take advantage of less costly units. Forecasting generation from variable renewable energy sources is an expanding science and a growing business for which utilities are willing to pay handsomely to reduce their planning costs. These and other costs required to ensure grid adequacy and security are discussed below.

7.5 Grid Adequacy and Security

Grid planners and operators are entrusted by the public to ensure what the planners refer to as system adequacy and security. Essentially this means that they are responsible for ensuring that there is adequate capability to provide electricity to customers and that the grid can securely provide power even if some untoward event happens, for example, the failure of a large power plant or transmission line. Grid operators have been dealing with these issues long before RETs entered the market; they have to confront the VU of loads that ebb and surge from day-to-day and season-to-season, the downing of transmission lines in storms, and the unexpected failure of power plants.

To cope with these VUs, the grid planners and operators have instituted a large number of protocols, metrics, and technological safeguards to ensure system adequacy and security. Many books and articles have been written on this subject, and there is no way this chapter can fully capture all the issues and the safeguards employed. However, to help the reader understand the issues involved with the additional VU introduced to the system by the adoption of VRETs, a basic introduction is presented in the following paragraphs.

Electric power must be available to meet loads with exceptionally high reliability. This is currently accomplished by generating sufficient power at all times and by ensuring reserve generation is available. These reserves guarantee that sufficient power can be generated at all times even with frequent variations in loads, unplanned outages of generators and/or transmission lines, and unforeseen generation variations from variable generation sources like combined heat and power, wind, and solar.

Reserves are ensured when planning and operating. To accommodate maintenance and breakdowns of generators, utilities have typically planned reserves of generation capacity 15%–20% larger than the highest peak load during the year. While utilities count the full nameplate capacity of thermal power plants toward their capacity reserve needs, they cannot do the same for VRET capacity due to its variable output. Nonetheless, they do usually give VRET capacity some credit toward meeting the planning reserve requirement because VRET capacity does increase the probability that sufficient capacity will be available at the time of peak load. This capacity value of VRET is difficult to calculate because it depends on

- The type of VRET (e.g., PV output is often coincident with the peak loads that are usually due to air-conditioning loads, while wind does not usually blow much during hot summer afternoons.)
- The amount of VRET installed relative to the size of the load (e.g., the first PV system installed will have a higher capacity value than one installed when PV is already generating 20% of load because at that level, the summer afternoon air-conditioning loads may have been met and the remaining peak load will have shifted to late summer evenings when the intensity of the sunlight has decreased significantly.)
- The geographic diversity of the VRET systems (e.g., 500 MW of wind capacity spread over five sites will have more capacity value than 500 MW at one site because the probability that there will be some wind from the five sites during the peak load moment is higher than that for the 500-MW-at-one-site case.)

Operating reserves are even more complicated. They largely address the fact that although sufficient capacity may be available, it cannot necessarily turn on (or off) fast enough to provide the additional (or reduced) generation needed. One way to classify* operating reserves is by how large and unexpected the need is for changes in generation.

- "Regulation" reserves provide a near-instantaneous response to meet small, short-lived (less than 1 min) random variations in loads and generation from VRETs (see Figure 7.21). They are usually automatically dispatched through centralized control.
- "Following" reserves provide a response to anticipated increases or decreases in generation needed over a sustained period; for example, increasing loads as a workday gets underway (see Figure 7.21), which can require further reserves if VRET output is expected to decline at the same time. Units providing these reserves can be scheduled for as much as a day or more ahead of time.
- "Contingency" reserves are usually substantial generation needed instantaneously and for a sustained period to make up for the unexpected loss of a generator and/or transmission line. Contingency reserves are provided by an assortment of generation technologies

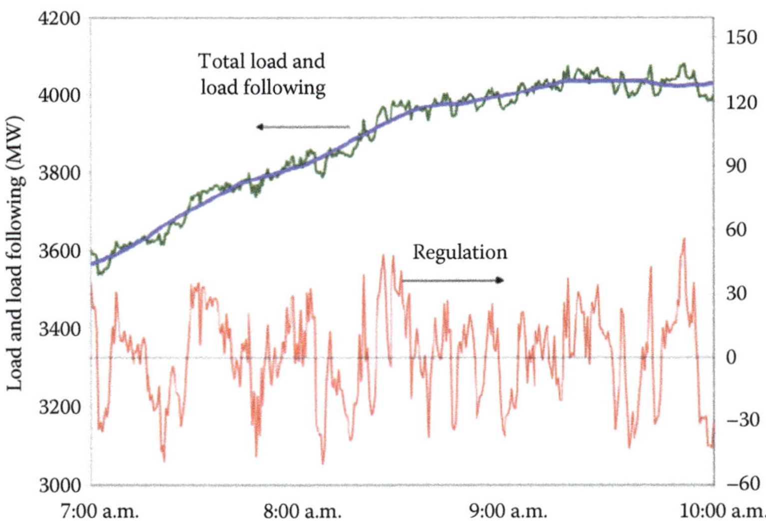

FIGURE 7.21
Time scales for regulation and following reserves. (From Kirby, B., Frequency Regulation Basics and Trends, Report No. ORNL/TM 2004/291, Oak Ridge National Laboratory, December 2004.)

* There is no universally accepted classification of operating reserves. Regulation and contingency reserves are commonly used terms, while the term "following reserves" is not as widely used and is often restricted to reserves for the following of loads only (i.e., not the following of net loads).

determined by the time it takes for each technology to respond. As shown in Figure 7.22, at the instant of loss, the frequency of AC power begins to drop immediately. The inertia of rotating generators provides the first level of contingency reserves; they simply slow the rate at which the frequency of AC power declines. Second, automatic generation control (automatic governors on power plants, also used to control regulation reserves) further slows and maybe arrests the decline in frequency. Next, spinning reserves are ramped up and

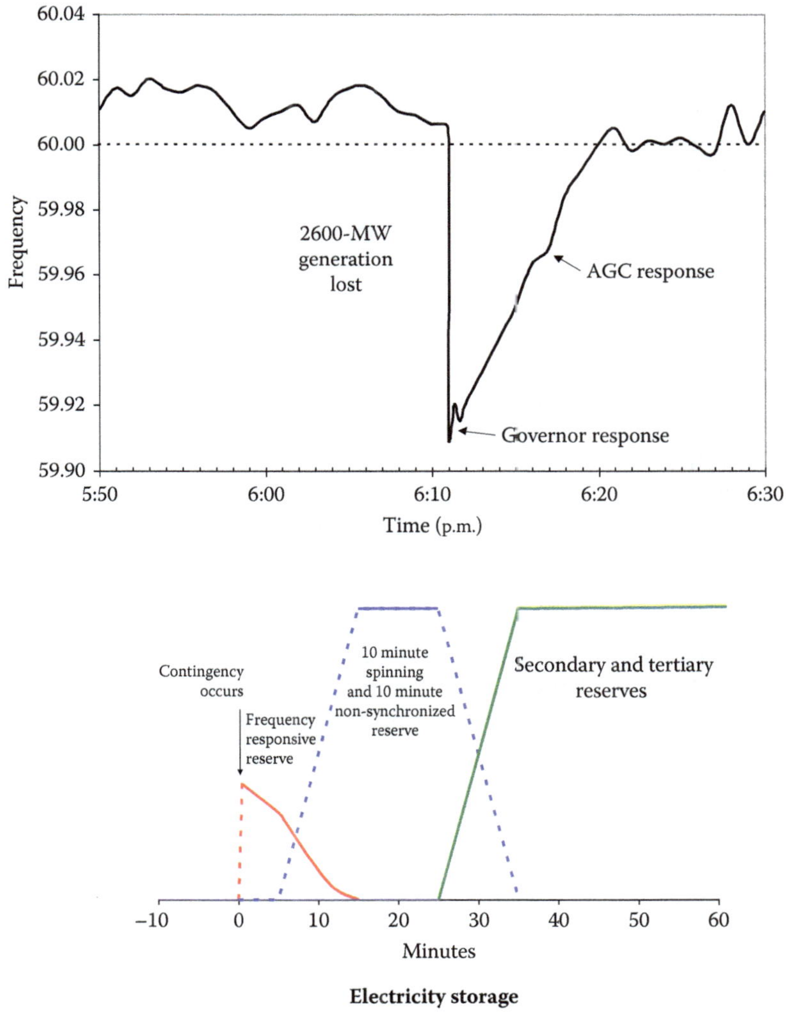

FIGURE 7.22
A series of contingency reserves are available to respond to the unexpected loss of a generator or transmission line. (From Kirby, B., et al., Cost-Causation-Based Tariffs for Wind Ancillary Service Impacts, Report No. NREL/CP-500-40073, Preprint of presentation for WindPower 2006, Pittsburgh, PA, June 2006.)

dispatched to increase the frequency. Spinning reserves are provided by rotating generators that are already providing power to the grid, but not at their maximum power level; thus, they can ramp their power output up in a matter of minutes. Spinning reserves may be supplemented by nonspinning reserves that can be started and synchronized to the grid within 10 min. To bring the frequency back to 60 Hz, slower starting secondary reserves are manually dispatched once it is determined that a contingency has occurred. With frequency restored and stabilized at 60 Hz, tertiary reserves are dispatched to displace these primary and secondary reserves so that they can be available to address another contingency, should one occur.

Renewable VRETs increase the need for following reserves because their output can gradually ramp up or down over hours. Investigations continue on the additional regulation reserves required with VRETs; much of the second-by-second variation in the output of a single VRET generator is offset by variations in the output of other VRET generators. Renewable VRETs do not require an additional contingency reserve because they are typically modular (e.g., many wind turbines make up a wind farm), and, therefore, an entire installation does not fail at once. As with planning reserves, the additional operating reserves required by renewable VRETs depends on the type of VRET, the amount of VRETs in the system, and the geographic diversity of the VRETs. They also depend on the ability to forecast the variation in output from the VRET.

The variability of generation from renewable VRETs creates another issue in addition to reserve requirements. At significant levels of VRET penetration, there can be times when more VRET generation is available than can be used. In this case, the VRET generation must be curtailed and the VRET owner may not be paid for that generation. This impacts the economic viability of a new VRET investment. Once again, the frequency and level of curtailments depend on the type of VRET, the amount of VRETs in the system, the accuracy of forecasts regarding the VRET output, and the geographic diversity of the VRETs. Curtailments will also increase if there are base load generators that cannot be easily ramped down further to make room for the VRET generation. Finally, there are curtailments at today's relatively low level of VRET penetration due to a lack of adequate transmission to move wind from source to load in some locations (see Section 7.7 on transmission issues).

7.6 Ensuring System Adequacy and Security with VRETs

Ensuring system adequacy and security is clearly made more difficult with the introduction of VRETs into a system. However, there are a number of ways grid planners and operators are mitigating their impact. The most obvious mitigation strategy is electricity storage. Unfortunately, electricity storage can be an expensive proposition due to equipment costs and energy losses as the power is stored and later extracted for use. We will explore storage options later after we have exhausted the potential of more cost-effective options.

7.6.1 Supply-Side Options

We have already introduced one of the most cost-effective options to reduce the VU associated with the RETs. This is to take advantage of diversity, both geographic and technological. Where transmission is adequate, geographic diversity can be achieved at essentially no cost. One bounding analysis by Short and Diakov (2012) has shown that geographic diversity could allow as much as 50% more generation to be supplied by wind. Short and Diakov have also shown that by adding PV for technology diversity, the contribution from VRETs can be increased by another 10%.

Figure 7.23 shows a set of additional options on both the supply and demand sides for further mitigation of the VU of RETs. We begin with the supply-side options. One that is already taking place is the expansion of balancing areas to larger areas with more loads and more generation and transmission capacity. These areas are exemplified by the formation of large regional

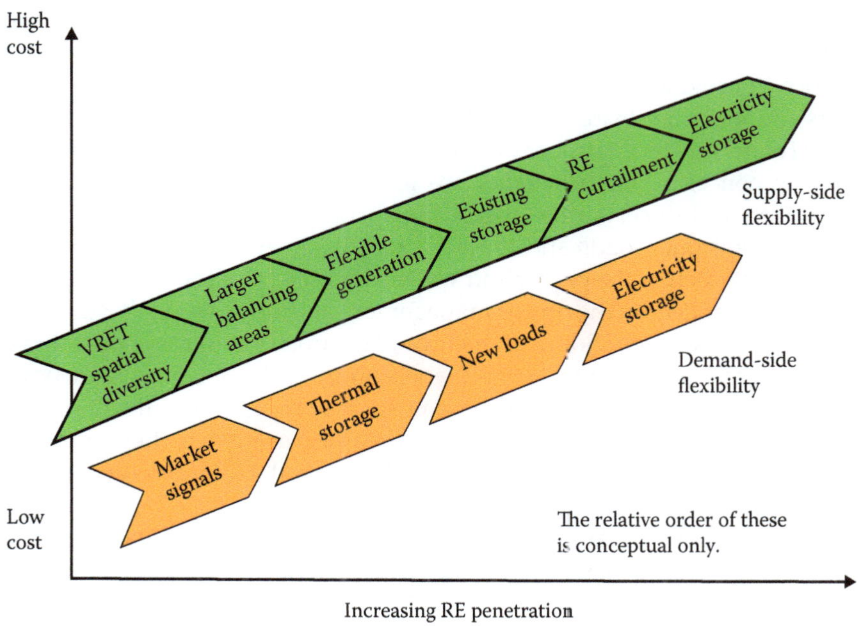

FIGURE 7.23
Mitigation options for variability of renewable electricity. (Modified from Denholm, P., The Role of Energy Storage in the Modern Low-Carbon Grid, http://.nrel.gov/wind/systemsintegration/energy_storage.html [accessed February 7, 2012], 2008.)

transmission organizations (RTOs) like PJM in the East or independent system operators such as Midwest Independent System Operator (MISO) or California Independent System Operator. These larger areas are able to better balance variability from renewables.

The next option is to simply have flexible dispatchable generators that can easily follow net loads as they increase and decrease. Existing deployments of wind and solar resources already take advantage of this option because many such flexible generators already exist within the U.S. grid to handle variations in loads and contingencies as described above. Primary among these are combustion turbines and internal combustion units that can ramp up quickly from a cold start to provide following reserves. Since these units do not run that often due to their high fuel and operating costs, they generally are not the most cost-effective source of spinning reserves, which are commonly provided by base and intermediate load* gas combined-cycle units and coal-fired generators.

Hydroelectric units can also start and vary their output relatively quickly. However, hydro plants have several constraints on their operation that in actual practice have historically limited their ability to mitigate the VU of wind and solar resources. The primary constraint on the ability of hydro plants to further mitigate the impact of new wind and solar resources is that no major (>500 MW) hydro plant has been built in 15 years largely due to environmental concerns and a scarcity of promising sites. Second, the use of existing hydro plants for unplanned following reserves is often limited by current scheduling procedures wherein water flows are scheduled far in advance to provide irrigation needs, leaving little room for real-time adjustments to follow net loads. Other flow constraints include the mitigation of downstream erosion with heavy flows, fish migration issues, and reservoir overflow concerns. Hydro plants are also often limited by overall water availability and uncertainties regarding spring rains and runoff as well as the size of summer loads and the need for power in the late summer/early fall. Finally, much of the power and water available from the extensive hydro facilities controlled by the power management authorities (e.g., Western Area Power Administration, Bonneville Power Administration) are committed by grandfathered contracts to farmers and communities for base load power and irrigation needs. Although these current barriers to the flexible use of hydro in support of wind and solar resources are extensive, they present an opportunity for change that could make the overall grid more secure in the presence of expanded use of wind and solar resources.

Another source of supply-side flexibility comes from base load units that are often operated at less than full power so that they can be ramped up

* Generating units are often classified by the types of loads they meet best. Base load generators are those whose economics lend themselves to continuous operation and therefore to meeting base loads. For example, coal-fired power plants have low fuel costs and high capital costs, making them ideal candidates for base load generation.

and down quickly to provide regulation and spinning or following reserves. The rotating mass associated with these base load turbines also provides primary regulation reserves that simply limit the rate of change of frequency as net loads change or a contingency occurs. Currently there are adequate base load units in operation around the clock to provide VRETs with the backup needed. In a scenario with high RET penetration, there may be many fewer base-load plants in operation and, therefore, less spinning reserves available.

Fortunately, market structures exist to ensure these ancillary services are available. In regulated markets, such services are simply a part of an integrated utility's operation. Deregulated markets vary in their approach to ensuring reserves are available; some have explicit markets for capacity, some for other forms of reserves, and some rely principally on energy markets with real-time adjustments.

Since flexible generators were part of the system before wind and solar resources were introduced into U.S. markets in significant quantities, they are the primary means today of ensuring grid adequacy and security with today's variable generators. And with more variable generation added in the future, more flexible generation will need to be added to the grid. However, adding flexible generators simply to provide reserves can be an expensive proposition and if they are not in operation, they may not be able to provide the spinning reserves when needed. Thus, for the reasons summarized below, flexible generators will not be able to provide the full solution to the VU of wind and solar resources as their penetration into the grid increases.

- As the fraction of load met by wind and solar resources increases, there may be times when there are not enough dispatchable units in operation to provide adequate regulation and spinning reserves to cover the variations in net load.
- There may be times when the generation from variable generators together with that of base load plants that cannot be ramped all the way down to zero without turning them off completely (which can imply a multi-day cool-down and warm-up process to get them back into service) exceeds the system load. During these times, the variable generators must either be curtailed or some other use found for their generation.
- Increasing VRET penetration will increase the rate at which net load changes, as was shown in Figure 7.15. This means more flexible generators are needed for ramping at the same time as less dispatchable generators are needed to provide energy to meet loads.

Fortunately, there are other options beyond flexible generators as we saw in Figure 7.23. Another option in use today is existing electricity storage facilities. There are currently 21 GW of pumped hydroelectric storage (PHS) spread around the country. These PHS units were built largely in the 1970s and 1980s to provide flexibility to the grid as large, relatively inflexible nuclear plants were being added. Today, they pump water to an upper reservoir when inexpensive power is available for the pumping and send the water back through the hydro turbines to produce power when loads are larger and more power is needed. For the integration of VRETs, these plants have two advantages over the flexible generators discussed above. In particular, some of them can be ramped up very quickly to provide following reserves. Second, they can absorb the output of the VRET when it is greater than needed (i.e., they can be used to reduce curtailments of the output of VRETs). They are often well-suited for this role when they have large impoundment reservoirs capable of storing large amounts of energy. Probably, their biggest limitation is that there are not a lot of ideal sites for the construction of new PHS facilities. Alternative configurations using underground storage reservoirs are being considered.

This ability to store the surplus output of VRETs is common to all forms of storage—both electrical and thermal. However, those with large inexpensive reservoirs for stored energy are generally more cost-effective in this role. For example, the most familiar form of electric storage, batteries, are not usually used to support VRETs because they are expensive per kilowatt hour of energy stored. Their primary use today is more for regulation and following reserves because they can be ramped up almost instantaneously to provide short bursts of power. Like batteries, flywheels can provide short bursts of power, but cannot store large amounts of energy cost-effectively. Thus, like batteries, they are not ideal for the integration of VRETs. On the other hand, storage technologies like compressed air energy storage (CAES) with their ability to store large amounts of energy in the form of compressed air in large underground geologic formations are better suited for use with VRETs.

So far we have not seen many storage systems being installed such as CAES and PHS with the large energy reservoirs necessary to accommodate VRETs. This is largely due to the costs—equipment and operating costs—associated with storage and the losses incurred during storage operation. All storage technologies have the drawback that they are not 100% efficient, for example, batteries may be 80%–90% efficient, while CAES may be only 75% efficient.

The storage systems being installed in the grid today are designed to provide other benefits—regulation reserves, peak capacity, distributed capacity, transmission and distribution deferrals, and so on. These values can be achieved without the large energy reservoirs needed to store surplus VRET output. Nonetheless, the cost-effectiveness of storage is improving due to technology improvements and due to the increased value incurred as more variable generation is introduced into the system.

Because of their multiple value streams and the need for general arbitrage capabilities within the grid, storage is generally not associated with a particular generation technology. There are two exceptions. In some restructured markets, the market rules may provide such a large advantage to firm bids that storage must be associated with a particular VRET installation. A second exception exists with CSP plants that use a heat transfer fluid to produce steam. That fluid can be stored as thermal energy for later use; the storage must be on-site as the fluid cannot be transported long distances without excessive heat loss. While this is a relatively inexpensive form of storage, it is associated exclusively with CSP technologies and, therefore, will not enjoy widespread use until such point in time that CSP technologies are widely economically competitive.

7.6.2 Demand-Side Options

In addition to these supply-side options, as shown in Figure 7.23, there are additional options on the demand side for mitigating the VU associated with the VRETs. The least expensive options are those associated with the market structure itself. There are numerous ways that markets can be structured to allow the VRETs to compete on a level playing field with other generator types, without disadvantaging one technology relative to another. The most influential of these will be those that elicit the most economic market response from producers and consumers alike. A key measure that is being tried and considered in many parts of the country is electric rates that vary with the cost of electricity generation. The most common approach is some form of time-of-day rates that reflect the higher cost of generation during peak usage periods like summer afternoons. There can be a handful of hours each year when high air-conditioning loads can require that old inefficient generators be kept in operational order to provide power for just those hours. Another option that is possible with today's "smart grid" is a rate structure that changes rates from one moment to the next, and smart appliances and machinery that automatically suspend or initiate operation in response to those rates. Such rate structures will elicit demand response that automatically curtails demand when rates are too high or that accomplishes marginal tasks when rates are low. Such rates will allow the market to respond appropriately to variations in the availability of generation from VRETs; if VRET generation drops off, real-time rates would increase and smart appliances would shut off in response as preprogrammed by their owners.

Another market innovation that will accelerate the use of VRETs, and that has been recently adopted in some RTOs is subhourly bidding for generation, ancillary services, and transmission. Until recently, the common practice in restructured markets has been to employ day-ahead and hour-ahead bidding for generation and transmission. Given the uncertainty associated with day-ahead or even hour-ahead forecasts for VRET generation, subhourly bidding allows VRET generator owners to better contribute their output to the market. Recent market adoption of subhourly bidding has demonstrated that today's computer networks and smart grid are up to the challenges imposed by such an interactive system.

Another recent introduction to market practices has been the testing of nonfirm transmission tariffs that allow a VRET generator to pay to use a transmission line that has available capacity even though others may have tied that transmission capacity up through contracts. Clearly, this increases the potential contribution of VRETs. Finally, FERC order 888 guaranteeing transmission access to all bidders has ensured VRET power equal access to existing transmission lines.

There are also technological innovations on the demand side that could mitigate the impacts of VRET variability. The development and penetration of these innovations will depend on the appropriate market structure being in place (e.g., real-time pricing). For example, with a large and frequent enough price disparity between on-peak electricity prices and off-peak prices, customers might elect to install thermal storage devices such as ice-storage (for cooling) or other forms of storage, including on-site batteries, flywheels, and so on.

An even cheaper option than customer storage might be the development of alternative uses of electricity so that when VRET output is high, the power can be used and useful. The defining characteristics of such alternatives are that they can be used when the VRET generation is available; that their initial cost is low enough that one can afford to have them idle much of the time (i.e., when surplus VRET generation is not available); and that the savings associated with their use is significant. A good example of such additional demand could be the use of baseboard-resistance space heaters; they are cheap; they could be used only when VRET generation is overabundant, and they displace the use of natural gas or oil for space heating.

Studies have been conducted to assess the integration costs of VRETs at low to moderate levels of VRET market penetration. For the most part, these studies do not consider the VU mitigation options just discussed. Nonetheless, as shown in Table 7.3, they show relatively little cost for the integration of wind into the different systems analyzed. (These costs do not consider transmission costs that are discussed in Section 7.7.) There is still significant uncertainty and debate over the integration costs that might be expected were VRET penetration to reach levels as high as 50% or more of generation. To some extent, the uncertainty resolves around the efficacy of the mitigation measures portrayed in Figure 7.23.

7.7 Transmission Issues Associated with VRETs

Inasmuch as the most widely deployed VRET in the United States today is wind turbine technology, the transmission issues associated with wind are critical to the expanded use of VRETs. Transmission issues specific to wind include the following:

- Wind resources are frequently remote from load centers because people do not like to live where it is extremely windy.

TABLE 7.3

Summary of Findings from Wind Energy Integration Cost Studies

				Integration Cost ($/MWh)			
Year	**Study**	**Wind Capacity Penetration (%)**	**Regulation**	**Load Following**	**Unit Commit**	**Gas Supply**	**Total**
2003	Xcel-UWIG	3.5	0	0.41	1.44	—	1.85
2003	We Energies	29	1.02	0.15	1.75	—	2.92
2004	Xcel-MNDOC	15	0.23	—	4.37	—	4.60
2005	PacifiCorp-2004	11	0	1.48	3.16	—	4.64
2006	California (multi-year)	4	0.45	trace	trace	—	0.45
2006	Xcel-PSCo	15	0.20	—	3.32	1.45	4.97
2006	MN/MISO	31	—	—	—	—	4.41
2007	Puget Sound Energy	12	—	—	—	—	6.94
2007	Arizona Public Service	15	0.37	2.65	1.06	—	4.08
2007	Avista Utilities	30	1.43	4.40	3.00	—	8.84
2007	Idaho Power	20	—	—	—	—	7.92
2007	PacifiCorp-2007	18	—	1.10	4.00	—	5.10
2008	Xcel–PSCo	20	—	—	—	—	8.56
2009	Bonneville Power Administration	36	0.22	1.14	—	—	5.70
2010	EWITS	48	—	—	1.61	—	4.54
2010	Nebraska	63	—	—	—	—	1.75

Source: U.S. Department of Energy (U.S. DOE), 2009 Wind Technologies Market Report, Prepared by the Lawrence Berkeley National Laboratory, CA, http://www1.eere.energy.gov/wind/pdfs/2009_wind_technologies_market_report.pdf (accessed August 2010), 2010.

- Transmission lines from the strongest VRET resource areas to load areas do not generally already exist.
- Any radial transmission lines built to move wind to loads will be used only when the wind is blowing, decreasing their cost-effectiveness.

As with almost all generalities, there are exceptions to each of the above issues. These exceptions often represent the "low-hanging" fruit. For example, there are transmission lines from some remote mine-mouth coal plants to load centers. Some of those mine-mouth coal plants are in rich wind resource areas (e.g., in South Dakota), and transmission lines already exist that could be used if the wind displaced the coal generation. Similarly, the third issue does not always apply as there are many wind resources in areas with strong transmission systems that would not require extended radial lines to connect the wind resources to the grid; any line added might strengthen the overall grid in addition to allowing the transmission of the wind generation to load. Furthermore, there may be multiple wind sites along the length of any new transmission line addition that would add diversity to the wind resources using the line, increasing the load factor of the line and improving its economics.

Nonetheless, there are and will be transmission issues with wind that will need to be addressed to take further advantage of good wind resources. For example, in Texas, almost 17% of the wind that could have been generated in 2009 had to be curtailed because adequate transmission did not exist to move the wind energy from western Texas to demand centers like the Dallas-Fort Worth area. Similarly, there were congestion events in 2010 in Texas (8% wind curtailments), MISO (5%), and other locations during the periods of high wind generation. Nationwide almost 5% of wind generation was curtailed in 2010, considerably higher than the average level of curtailments in Europe where wind penetrations are as high, but the grid may be stronger (Porter et al. 2011). Curtailments in the United States are very region specific with the highest fraction occurring in Texas, where the fraction appears to be declining (2010 curtailments were about half of 2009 curtailments) while they are increasing in other regions. To some extent, there is a "chicken and egg" problem as to whether to build transmission in anticipation of wind farm development or to build the wind farms and suffer congestion until the transmission system catches up. The economic optimum lies somewhere in the middle and will also be a system in which there is always the possibility of some curtailments during high wind periods.

Transmission issues are less of an issue with respect to PV because total insolation does not vary so strongly from one location to another. Thus, PV can be deployed near local loads. However, at least one generator manufacturer has recently begun to promote hybrid PV/wind systems that can take more full advantage of transmission because solar generates during the daytime and wind output is generally larger during winter nights.

CSP plants that focus direct insolation are by far more economic in the desert Southwest where direct insolation can be several times stronger than in the more humid air of the east coast. Thus, there could be future scenarios with very high penetrations of renewable energy that would call for the transmission of CSP-generated power from the southwest to other portions of the country.

In addition to the economic issues associated with transmission, there is the more concerning issue of public resistance to the siting of transmission lines. This can be an even larger issue when the lines are carrying power from VRET resource areas to distant load centers with no apparent benefit to the regions crossed by the lines. This issue is further accentuated for high-voltage direct current transmission with its better economics for long distances, but its inability to cost-effectively connect to generators and loads along its length.

Many of the integration and transmission issues discussed above for VRETs like wind and solar resources do not exist for non-VRETs like biopower and geothermal electricity. Biopower is largely dispatchable and can be considered a base load technology given adequate biomass resource (there can be seasonal variations in resource availability if proper biomass storage is not available). Geothermal has no seasonal resource issues and is generally considered a base load technology. Geothermal may have more transmission issues as the resource location is fixed, unlike the biomass collection point.

7.8 Carbon Reductions Associated with RETs

With the exception of biopower, RETs emit only insignificant amounts of GHGs at the point of generation. Although as shown in Figure 7.24, there are GHG emissions associated with the manufacturing, installation, and decommissioning of RETs. However, these are usually very small compared to those associated with the combustion of fossil fuels for power, as also shown in Figure 7.24.

While biopower does produce carbon emissions at the point of combustion of the same order of magnitude as the emissions of coal plants, the growth of the biomass itself absorbs an equal amount of carbon from the atmosphere. However, biopower produces more net carbon emissions than other RETs because there can be significant carbon emissions associated with the cultivation, harvesting, and collection of biomass. The range of carbon emissions shown in Figure 7.24 largely reflects the variability in these biomass production emissions, which depend strongly on the type of biomass used; for example, switchgrass planted for energy purposes will have emissions associated with cultivation while forest residues will not.

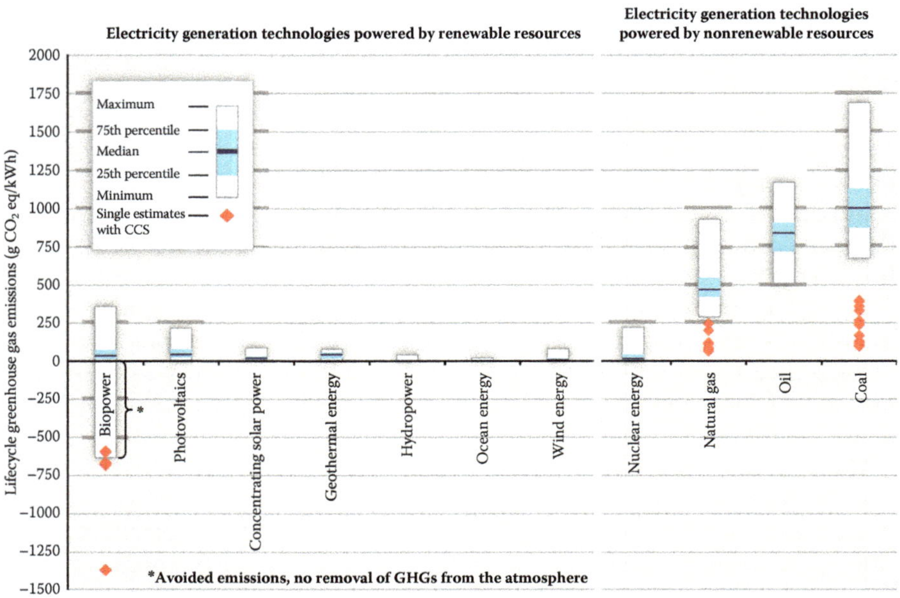

Count of estimates	222(+4)	124	42	8	28	10	126	125	83(+7)	24	169(+12)
Count of references	52(+0)	26	13	6	11	5	49	32	36(+4)	10	50(+10)

FIGURE 7.24
Estimates of lifecycle GHG emissions (g CO_2 eq/kWh) for broad categories of electricity generation technologies, plus some technologies integrated with carbon capture and sequestration (CCS). Land-use-related net changes in carbon stocks (mainly applicable to biopower and hydropower from reservoirs) and land management impacts are excluded; negative estimates1 for biopower are based on the assumptions about avoided emissions from residues and wastes in landfill disposals and co-products. The number of estimates is greater than the number of references because many studies considered multiple scenarios. Numbers reported in parentheses pertain to additional references and estimates that evaluated technologies with CCS. Distributional information relates to estimates currently available in lifecycle analysis literature, not necessarily to underlying theoretical or practical extrema, or the true central tendency when considering all deployment conditions. (From IPCC, IPCC Special Report on Renewable Energy Sources and Climate Change Mitigation, Prepared by Working Group III of the Intergovernmental Panel on Climate Change, Cambridge University Press, Cambridge, U.K., and New York, NY, p. 1075, 2011.)

Figure 7.24 also shows that coal-fired power plants emit almost twice as much carbon dioxide per kilowatt hour as natural gas combined-cycle plants. And of course, there is even a bigger difference in carbon emissions from coal plants compared to hydroelectric and nuclear plants, which like renewables, are largely carbon free, even on a lifecycle basis.

Thus, the reduction in carbon emissions that is associated with a given amount of RET generation clearly depends on which source of power is displaced by the RET generation. As can be seen from Figure 7.24, if coal-fired generation is displaced, the reduction in carbon emissions is on average 1000 g of CO_2 per kWh of RET generation. On the other hand, if nuclear power is displaced by the RET generation, the change in carbon emissions will be negligible. Identifying which power source (or combination of power sources) is displaced by RET generation is not a trivial exercise. One common practice is to assume that the RET generation will displace a mix that is proportional to the overall generation mix in the region. However, this can be a very misleading result because it is the marginal contributor's emissions that count and the RET generation is not uniform over time, especially for the more cost-effective VRETs of wind and PV. The high correlation of the time of generation from PV with the time of peak loads in the summer means that PV is more likely to displace peaking units, which are probably fired by natural gas. But there are complications even in this simple case. For example, the natural gas peaking unit displaced could be an older combustion turbine with a heat rate of 13,000 Btu/kWh or an advanced unit with a heat rate of only 9,000 Btu/kWh. The difference in carbon emissions between two such units would be proportionately as large. Another example of the complexity is that PV generation in the winter is more likely to displace base load generation, which could be either coal with its high carbon emissions or nuclear with essentially no carbon emissions.

Even if the generation mix is known for each point in time, there are other complications in computing the carbon reduction impacts of VRETs. For example, the variability in the generation from VRETs necessitates that dispatchable units fired by gas, oil, or coal cycle up and down to follow net load (load minus the VRET generation). When such units are started up, cycled, and/or run at other than their design point, their performance can degrade and consequently, their carbon emissions per kilowatt hour generated can increase.

Due to such complications, the preferred method of estimating the carbon reductions due to RETs is to simulate the production of the electric system using a production cost model that is capable of assessing which generator is operating and its emissions at each point in time. To estimate the RET impact on carbon emissions, such a model can be run with and without the RETs.

Because their structure and data are usually based on hour-long time steps, models can have difficulty in capturing these off-design conditions, which vary subhourly in step with the subhourly variations in VRET generation. Some production cost models are now being modified to handle subhourly variation, but they can still be limited in their ability to optimize across startup costs/emissions and the nonlinear performance curves associated with off-design operation.

These difficulties in simulating the impact of variability in generation from VRETs on system-wide carbon emissions have led to controversy over the level of carbon emission mitigation that can be provided by wind and solar generation. However, recent ongoing analysis of 20%–30% wind penetration in the Western Electric Coordinating Council area by the NREL has shown that for this region startup, cycling and part-load operation of dispatchable units reduces the carbon emission reduction impacts of the wind generation by less than 3% (Brinkman 2011).

7.9 Estimates of U.S. RET Potential and Carbon Mitigation Impacts

One of the more interesting and difficult questions to answer today is how far can we go with RETs in the electric sector. In one sense, the answer to that question is that we can go all the way—100% RETs if we also buy enough storage capacity. With enough batteries, we could store all the surplus VRET generation and use it essentially instantaneously when we need it. However, given the cost of batteries and other forms of electric storage, this would be a very expensive proposition. So, the more appropriate question is what would it cost us to meet different fractions of our generation needs with RETs? And, how would those costs compare to the cost associated with a business-as-usual scenario going forward? Even that last question is somewhat of an apples-to-oranges comparison because the RET scenario will have less pollution, less impact on climate, and more price stability—factors that are not really quantifiable in terms of cost. Nonetheless, it is a useful comparison and often addressed by the studies of RET potential.

NREL, in conjunction with utility, RTO, industry, and consultant partners, has conducted several studies in recent years to assess the feasibility and cost of meeting various fractions of load with RETs at the national and interconnect levels. The first of these was the 20% Wind Energy by 2020 study conducted jointly by the American Wind Energy Association, NREL, and DOE (U.S. DOE 2008). This study used a detailed model of capacity expansion for the United States—the wind deployment system (WinDS) model (Short 2003)—to estimate the sites that would be best, the transmission required, and the cost were it required that 20% of all U.S. load be met by wind energy. At the time of its undertaking in 2006–2007, this study was considered highly controversial and overly aggressive. But it found no major "show stoppers" with respect to this level of deployment, even though storage and the other mitigation measures described above were not considered. Furthermore, as shown in Figure 7.25, the overall direct system cost was not much higher than that of a business-as-usual scenario. And, it found that the overall cost to the economy as a whole would be lower than that of a business-as-usual scenario if one considered the value in other market sectors (buildings, industry) of reduced natural gas prices from reduced electric demand for gas. The study also estimated these cost results would be even better if the study had valued the reductions in carbon emissions associated

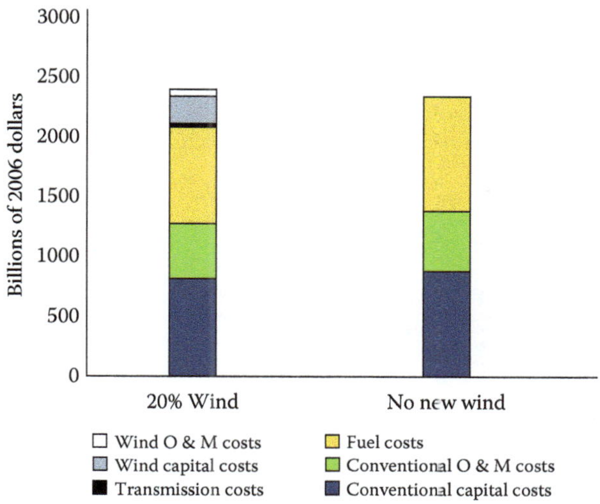

FIGURE 7.25
Present value of 20-year system costs for the 20% wind energy scenario. (From U.S. Department of Energy [U.S. DOE], 20% Wind Energy by 2030, Report No. DOE/GO-102008-2567, May 2008.)

with the 20% scenario. Due to coal and gas being the usual marginal generators, the carbon reductions were actually slightly larger than 20% of the business-as-usual carbon emission from the electric sector in 2030. Finally, the study showed that the optimum 20% scenario would need a fair amount of new transmission to move the wind from the "wind belt" in the center of the country to the load centers on the coasts (see Figure 7.26), but that the cost of this new transmission was modest relative to overall system costs.

As with all studies, the 20% study was not able to consider all possible issues, costs, and solutions. Some of the more important considerations that required further examination were

- Transmission power flow (the WinDS model used a simplified "transportation" model of transmission)
- Local transmission congestion within the regions of the WinDS model
- Transmission contingencies (the 20% study increased the cost of transmission to account for redundant capacity, but did not actually build the redundant lines)
- Reactive power flow
- Subhourly/hourly treatment of wind variability (the WinDS model used a statistical treatment as opposed to a discrete moment-by-moment treatment)
- Consideration of VU mitigation measures like those described earlier (e.g., demand response, storage)

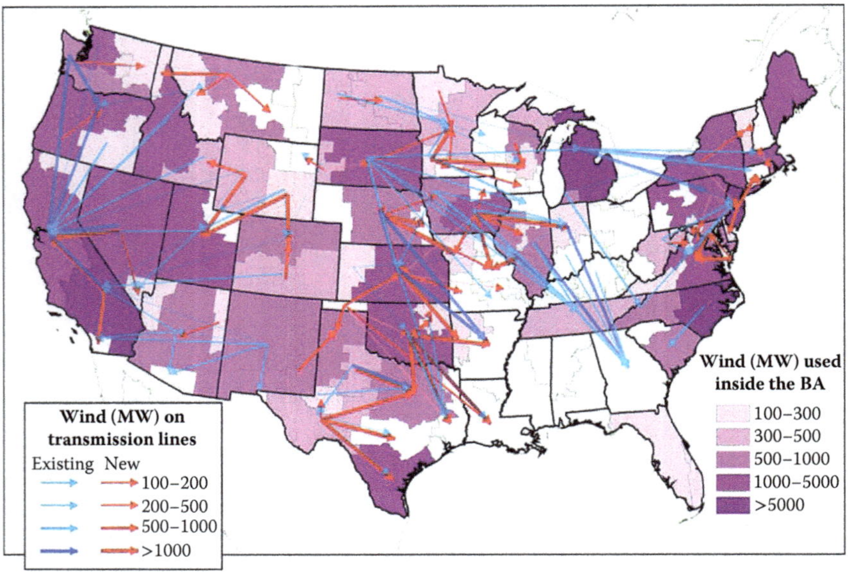

FIGURE 7.26
Transmission requirements of the 20% wind energy by 2030 study. (From U.S. Department of Energy [U.S. DOE], 20% Wind Energy by 2030, Report No. DOE/GO-102008-2567, May 2008.)

- Optimization over the sensitivity of natural gas prices to demand for gas
- Interactions with other RETs
- Cycling/startup/part-load impacts on thermal generators, and so on

As can be seen from this list, there are a myriad of factors to consider, and this list is not complete, as it does not include all the issues that are considered by the WinDS model, nor secondary issues that were not considered.

In spite of the limitations, the 20% study demonstrated that the central issues of cost and wind variability may not be as constraining as once thought. Thus, the 20% study along with the results from the more local studies shown in Table 7.3 led to other studies that expanded on the issues considered as well as on the fraction of load met by RETs.

Two noteworthy studies, also from NREL, were the studies of wind and solar resources in the two major U.S. interconnects—the Eastern Wind Integration and Transmission Study (EWITS) (NREL 2010b) and the Western Wind and Solar Integration Study (WWSIS) (NREL 2010c). Both studies examined scenarios with wind penetration as high as 30% of generation, while the WWSIS effort also examined concurrent PV penetration as high as 5%. These two studies also extended beyond the 20% study by considering transmission congestion and contingencies, hourly treatment of wind variability, and interactions between wind and PV (in the WWSIS). Like the 20% study, they found nothing to indicate that these levels of VRETs would necessarily jeopardize security or degrade reliability. To accomplish this, they relied on more aggregated/larger balancing areas, extensive transmission build-outs, and new market structures with shortened bidding and response times. Their results did vary from those of the 20% study in that they did hypothesize (capacity expansion was based on expert judgment, not on the modeling-based approach of the 20% study) a different build-out of wind and transmission (see Figure 7.27). The EWITS study did find that the cost to build and operate the system increased by as much as 37%. However, both studies found that locational marginal prices (LMPs) that represent the marginal cost of power in a particular location at a particular time decrease in the VRET scenarios relative to the reference cases. This is not surprising given that VRETs will often be the marginal generation source that sets the LMP and given that VRETs have marginal operating costs close to zero.

The carbon reductions due to the wind in these studies are not as clearly presented. For the 30% wind penetration case by 2024, the EWITS report shows a 19% reduction compared to the emissions in 2008. It is not clear from the report what the carbon emissions are in the reference case in 2024, but the load assumed in 2024 appears to be about 30% larger than that of 2008.

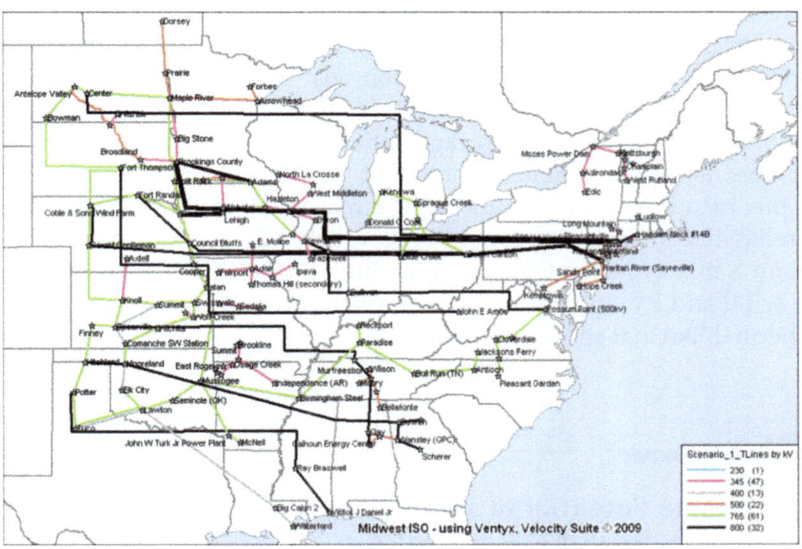

FIGURE 7.27
EWITS scenario 1 (20% wind) transmission build-out. (From NREL, Eastern Wind Integration and Transmission Study [EWITS], Report No. NREL/SR-550-47078, Report prepared for NREL by Enernex, TN, January 2010.)

Thus, the carbon emission reduction appears to be greater than 30% overall relative to the reference case emissions in 2024. For the WWSIS study, the carbon emission reductions within the study area were 23% from the base case when 30% of the load was met by wind.

All three of these NREL studies were based on natural gas price forecasts for the future that exceed today's natural gas price. They did not foresee the decrease in gas prices to under \$3/MMBtu that has occurred with the success of shale gas development. Such lower gas prices effectively increase the relative cost of these renewable energy scenarios, making RETs less economically attractive.

While these interconnect level studies have increased the confidence of market decision makers with respect to the potential of renewables, there remain issues that need to be addressed. In fact, a second version of each of these interconnect level studies is underway to expand the analysis to include such things as subhourly modeling of the variability of wind and solar resources, cycling/startup/part-load operation impacts of conventional generators, lower natural gas prices, and so on.

Similarly, NREL is expanding on the 20% study to examine scenarios in which RETs meet as much as 90% of load nationwide. To do this, NREL is using a new capacity expansion model—regional energy deployment system (ReEDS)—evolved from the WinDS model to include all major RETs (PV, CSP, geothermal, and biopower), the leading forms of storage (PHS, CAES, batteries, thermal with CSP, and thermal end-use storage), demand response, and endogenous natural gas price response to gas demand (Short et al. 2011). This new study further improves on the 20% study by examining the ReEDS results for the capacity build-out with GridView, a commercially available, hourly, optimal power flow, production cost model. While the results of this study have yet to be published, it is known that by considering these additional RET technologies, storage options, demand response, and market structures, the study is able to show that the cost of achieving 80%–90% of load met by all renewables could still be reasonable without sacrificing system reliability and security. These high levels of RET generation are achieved through a mix of technologies with about half the RET generation coming from wind and PV, and the other half from hydropower, biopower, CSP and geothermal like that shown in Figure 7.28.

7.10 Ultimate Potential of Wind and PV in the U.S. Central Electric Market

The above studies examined the transition scenarios from the current U.S. electric network to a grid with high levels of RETs. NREL has also examined how far the country might be able to go with wind- and PV-generated

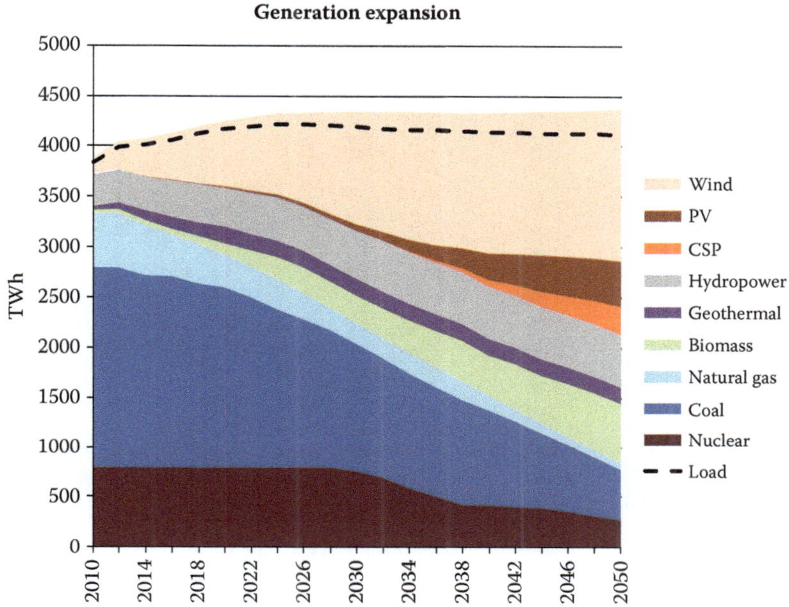

FIGURE 7.28
Generation in an 80% renewable electric case study. (From Short, W., Renewable Electricity Futures, Presentation of Internal NREL, November 16, 2010.)

electricity if the electric system were designed from the ground up (Short and Diakov 2012). Short and Diakov used the RELM (Renewable Energy Load Matcher) model to select the best sites for wind and PV in the western United States, not on the basis of cost, but so as to minimize the sum of the generation needed from other dispatchable sources and the VRET electricity that is lost as surplus. This capacity expansion modeling has only recently become possible with the availability of detailed resource data sets for wind and PV that include hourly data for 30,000 potential wind and PV sites along with hourly load data for 3 years (2004 through 2006). The study found that for the Western Interconnect, wind and PV sites could be selected that meet 80% of the load with about 10% of the VRET generation being curtailed, mostly in the spring (hours in which the blue VRET generation dots of Figure 7.29 are above the pink load dots). The dispatchable generators filled in behind the VRET generation to meet the other 20% of the load.

This was accomplished by taking advantage of both technology and geographic diversity available in wind and solar resources across the west. Normally, one would expect that the mix of wind and PV would be heavily dominated by wind because it is the least expensive of the two technologies. But the RELM optimization does not consider costs, yet it still produced a mix with 60% of the VRET capacity being wind and the remaining 40% PV.

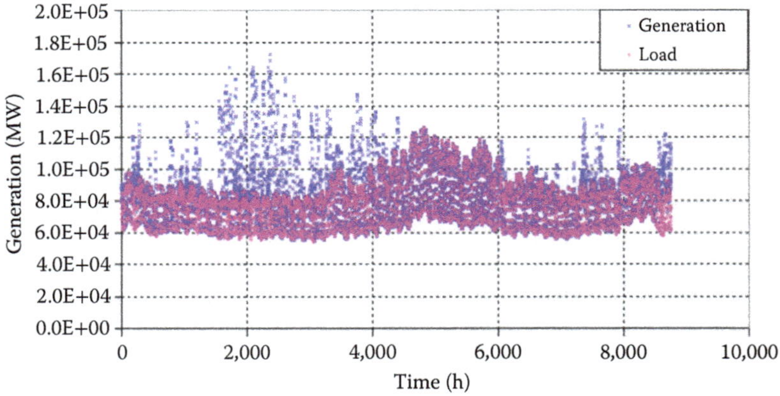

FIGURE 7.29
RELM results for 2005 WECC optimization with wind and PV. (From Short, W., and V. Diakov, Matching Western U.S. Electricity Consumption with Wind and Solar Resources, Presented at AWEA WindPower 2011 and draft submission to *Wind Energy*, 2012.)

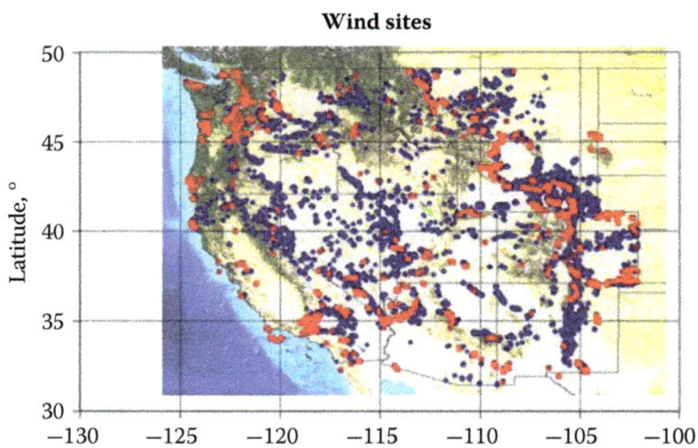

FIGURE 7.30
Wind sites selected (red dots) from the potential sites (blue dots). (From Short, W., and V. Diakov, Matching Western U.S. Electricity Consumption with Wind and Solar Resources, Presented at AWEA WindPower 2011 and draft submission to *Wind Energy*, 2012.)

Even though the PV generation profile matches well with the summer afternoon peak loads, the advantage of wind is simply that it can provide power both day and night. These results required the deployment of 175 and 113 GW of wind and PV, respectively, in a Western Electricity Coordinating Council (WECC) system with only 125 GW of peak load. Even at these levels of deployment, curtailments amounted to only 10% of the output of wind and solar resources. That is possible primarily because the wind and PV sites selected were spread around the west (see Figure 7.30) to take advantage of the fact that variable resources not only vary temporally but also spatially. The benefits of such geographic diversity are illustrated in Figure 7.20, wherein the variability of output reduces with the combination of wind deployments in multiple regions.

While these results are encouraging, they are optimistic in that they do not consider several critical features of the electric system—transmission constraints, reactive power, contingencies, cycling/part-load operation of dispatchable generators, and so on. They are simply meant to be illustrative of the future potential of VRETs (NREL is expanding their analysis to address some of these issues).

7.11 Summary and Conclusions

Renewable resources for power generation are widespread across the country, but not uniformly available. The levelized cost of energy (LCOE) from most RET technologies has been declining for years, and wind is now competitive with new coal and nuclear generators. However, recent declines in the price of natural gas have increased the competitiveness of gas-fired generators. The two principal additional issues with the widespread introduction of RETs are the integration issues associated with the VRETs and the transmission issues associated with accessing remote renewable resources. While storage is generally thought of as the primary solution for introducing more VRETs into the grid, there are other less expensive measures that can be taken—for example, flexible dispatchable generators, resource diversity, demand response, and market design. Recent studies at the national and interconnect levels have not found the integration and transmission issues associated with high levels of RET penetration to be insurmountable or even that costly. However, regional issues may present themselves due to the regional availability of existing infrastructure and resources. Regional differences will also occur with respect to the conventional fuels displaced and thus the amount of carbon dioxide emissions reduced through generation from RETs. However, studies suggest that at the national level, carbon emission reductions will be at least proportional to the fraction of load met by RETs.

Questions for Discussion

1. Which renewable electric technologies have the most potential?

   (*Answer:* Solar and wind have the most potential due to their low cost relative to other RETs and their larger resource base.)

2. What are the biggest issues with the market penetration of wind and solar?

   (*Answer:* Cost, transmission requirements, and resource variability and uncertainty are the biggest issues.)

3. What kind of electric system reserves must increase with the addition of variable renewable electric technologies?

   (*Answer:* Contingency reserves, net load following reserves, and regulation reserves will need to increase with the addition of VRETs.)

4. Why does the marginal capacity value of wind decrease as more wind is added to the system?

   (*Answer:* The marginal capacity value of the next unit of wind will be less because its output will be positively correlated with existing wind farms. For example, if the new wind turbines were installed at the same location as the existing wind farm (i.e., 100% correlated) and if the existing farm output completely met the original peak load and contributed nothing to the new residual peak load, then this marginal increment of wind capacity would have no capacity value.)

5. Why do curtailments of PV increase as more PV is added to the system?

   (*Answer:* Adding more PV to the system increases the probability that there will be a point in time at which there is more PV than can be used. This probability is essentially zero at lower levels of penetration but can become significant when generation from PV reaches the levels of 15% or more.)

6. What are five ways in which the variability of wind and solar can be mitigated?

   (*Answer:* See Figure 7.23 for the different ways that wind and solar variability can be mitigated.)

7. Give two reasons that transmission is more of an issue for wind than for other conventional generation technologies.

   (*Answer:* Transmission is an issue with wind generation because most load centers are not in strong wind areas. Transmission is also an issue for VRETs because they have lower capacity factors and therefore the cost of transmission can be spread over fewer kilowatt hours of generation.)

8. How can the carbon emission reductions that are possible with variable renewable electric technologies be estimated?

   (*Answer:* A production cost model can be used to estimate which existing generators will be displaced by the addition of VRETs. The reduced generation can be translated into emission reductions based on the type of generation displaced and the pattern of its operation.)

References

Barbose, Galen, N. Darghouth, R. Wiser, and J. Seel. 2011. Tracking the Sun IV: An Historical Summary of the Installed Cost of Photovoltaics in the United States from 1998 to 2010. Report prepared by the Lawrence Berkeley National Laboratory, CA, September 2011.

Black and Veatch Corporation. 2012. Cost and Performance Data for Power Generation Technologies. Report No. NREL-SR-48-595-201202-0. Draft Report Prepared for the National Renewable Energy Laboratory, Golden Colorado, February 2012.

Brinkman, Greg. 2011. Emissions Impacts of Fossil Fuel Unit Cycling. Presentation at the Utility Wind Integration Group User Meeting, October 12, 2011.

Center for Climate and Energy Solutions (CCES). 2012. Hydropower. Available at http://www.c2es.org/technology/factsheet/hydropower (accessed February 5, 2012).

Denholm, Paul. 2008. The Role of Energy Storage in the Modern Low-Carbon Grid. Available at http://.nrel.gov/wind/systemsintegration/energy_storage.html (accessed February 7, 2012).

Denholm, Paul, E. Ela, B. Kirby, and M. Milligan. 2010. The Role of Energy Storage with Renewable Electricity Generation. Technical Report No. NREL/TP-6A2-47187, January 2010.

Energy Information Administration (EIA). 2009. Emissions of Greenhouse Gases Report. Report No. DOE/EIA-0573. Available at http://www.eia.gov/oiaf/1605/ggrpt/carbon.html (accessed February 4, 2012).

Energy Information Administration (EIA). 2011. International Energy Outlook 2012. Report No. DOE/EIA-0484. Available at http://www.eia.gov/forecasts/ieo/pdf/0484(2011).pdf (accessed February 19, 2012)

Energy Information Administration (EIA). 2012. Annual Energy Outlook 2012. Available at http://www.eia.gov/oiaf/aeo/tablebrowser/#release=EARLY2012&subject=0-EARLY2012&table=8-EARLY2012&region=0-0&cases=full2011-d020911a,early2012-d121011b (accessed February 5, 2012).

Environmental Protection Agency (EPA). 2012. Renewable Energy Cost Database. Available at http://www.epa.gov/cleanenergy/energy-resources/renewable-database.html (accessed February 6, 2012).

Ernst, Bernhard, Y. Wan, and B. Kirby. 1999. Short-Term Power Fluctuation of Wind Turbines: Analyzing Data from the German 250-MW Measurement Program from the Ancillary Services Viewpoint. Report No. NREL/CP-500-26722. National Renewable Energy Laboratory, Golden Colorado, July 1999.

Hall, Douglas, S. J. Sherry, K. S. Reeves, R. D. Lee, G. R. Carroll, G. L. Sommers, and K. L. Verdin. 2004. Water Energy Resources of the United States with Emphasis on Low Head/Low Power Resources. Report No. DOE/ID-11111. Idaho National Engineering and Environmental Laboratory, ID, April 2004.

International Energy Agency (IEA). 2011. World Energy Outlook 2011. Available at http://www.iea.org/Textbase/npsum/weo2011sum.pdf (accessed February 4, 2012).

IPCC. 2011. IPCC Special Report on Renewable Energy Sources and Climate Change Mitigation. Prepared by Working Group III of the Intergovernmental Panel on Climate Change (O. Edenhofer, R. Pichs-Madruga, Y. Sokona, K. Seyboth, P. Matschoss, S. Kadner, T. Zwickel, P. Eickemeier, G. Hansen, S. Schlömer, and C. von Stechow [eds]). Cambridge University Press, Cambridge, UK, and New York, NY, p. 1075.

Kirby, Brendan. 2004. Frequency Regulation Basics and Trends. Report No. ORNL/TM 2004/291. Oak Ridge National Laboratory, December 2004.

Kirby, Brendan, M. Milligan, and Y. H. Wan. 2006. Cost-Causation-Based Tariffs for Wind Ancillary Service Impacts. Report No. NREL/CP-500-40073. Preprint of presentation for WindPower 2006, Pittsuburgh, PA, June 2006.

NREL. 2010a. Department of Energy Releases New Estimates of Nation's Wind Energy Potential. Available at http://www.nrel.gov/wind/news/2010/816.html (accessed February 4, 2012).

NREL. 2010b. Eastern Wind Integration and Transmission Study (EWITS). Report No. NREL/SR-550-47078. Report prepared for NREL by Enernex, TN, January 2010.

NREL. 2010c. Western Wind and Solar Integration Study (WWSIS). Report No. NREL/SR-550-47434. Report prepared for NREL by General Electric, May 2010, p. 536.

Oak Ridge National Laboratory (ORNL). 2012. Heat Content Ranges for Various Biomass Fuels. Available at http://cta.ornl.gov/bedb/appendix_a/Heat_Content_Ranges_for_Various_Biomass_Fuels.xls (accessed February 5, 2012).

Porter, Kevin, J. Rogers, and R. Wiser. 2011. Update on Wind Curtailment in Europe and North America. Presentation of the Center for Resource Solutions. Available at http://www.efchina.org/csepupfiles/report/201163041046377.1820833464392.pdf/Update%20on%20Wind%20Curtailment.pdf (accessed February 12, 2012).

Previsic, Mirko, J. Epler, M. Hand, D. Heimiller, W. Short, and K. Eurek. 2011. Lifecycle Cost Assessment of Hydrokinetic Technologies in the U.S. over Time. Draft Report. RE Vision Consulting. December 2011.

Short, Walter. 2003. Modeling the Long-Term Penetration of Wind in the United States. In *Proceedings of WindPower 2003,* May 2003, Austin, TX.

Short, Walter. 2010. Renewable Electricity Futures. Presentation of Internal NREL, November 16, 2010.

Short, Walter, P. Sullivan, T. Mai, M. Mowers, C. Uriarte, N. Blair, D. Heimiller, and A. Martinez. 2011. Regional Energy Deployment System (ReEDS). Report No. NREL/TP-6A20-46534. National Renewable Energy Laboratory, Golden Colorado, December 2011.

Short, Walter, and V. Diakov. 2012. Matching Western U.S. Electricity Consumption with Wind and Solar Resources. Presented at AWEA WindPower 2011 and draft submission to *Wind Energy*.

University of Michigan. 2011. Renewable Energy Fact Sheet. Center for Sustainable Solutions. Available at http://css.snre.umich.edu/css_doc/CSS03-12.pdf (accessed February 4, 2012).

U.S. Department of Energy (U.S. DOE). 2008. 20% Wind Energy by 2030. Report No. DOE/GO-102008-2567, May 2008.

U.S. Department of Energy. 2010. 2009 Wind Technologies Market Report. Prepared by the Lawrence Berkeley National Laboratory, CA. Available at http://www1.eere.energy.gov/wind/pdfs/2009_wind_technologies_market_report.pdf (accessed August 2010).

U.S. Department of Energy. 2011. Report to Congress on Renewable Energy Resource Assessment for the United States. Draft Report. Office of Energy Efficiency and Renewable Energy, January 28, 2011.

U.S. Department of Energy. 2012. SunShot Vision Study. February 2012.

Wiser, Ryan, and M. Bolinger. 2011. 2010 Wind Technologies Market Report. U.S. Department of Energy, June 2011.

8

The Future of Energy from Nuclear Fission

Son H. Kim and Temitope Taiwo

CONTENTS

8.1 Introduction

Nuclear energy is an important part of our current global energy system, and contributes to supplying a significant percentage of the electricity for many nations around the world. There are 433 commercial nuclear power reactors operating in 30 countries, with an installed capacity of 367 GWe as of October 2011 (IAEA PRIS 2011). Nuclear electricity generation totaled 2630 TWh in 2010, representing 14% the world's electricity generation. The top five countries of total installed nuclear capacity are the United States, France, Japan, Russia, and South Korea at 102, 63, 45, 24, and 21 GWe, respectively (WNA 2012a). The nuclear capacity of these five countries represents two-thirds of the total global nuclear capacity.

The role of nuclear power in the global energy system today has been motivated by several factors, including the growing demand for electric power, the regional availability of fossil-fuel resources and energy security concerns, and the relative competitiveness of nuclear power as a source of base-load electricity. There is additional motivation for the use of nuclear power because it does not produce greenhouse gas (GHG) emissions or local air pollutants during its operation and contributes to low levels of emissions throughout the lifecycle of the nuclear energy system (Beerten et al. 2009).

Energy from nuclear fission—primarily in the form of electric power and potentially as a source of industrial heat—could play a greater role for meeting the long-term growing demand for energy worldwide while addressing the concern for climate change from rising GHG emissions. However, the nature of nuclear fission as a tremendously compact and dense form of energy production with associated high concentrations of radioactive materials has particular and unique challenges as well as benefits. These challenges include not only the safety and cost of nuclear reactors, but proliferation concerns, the safeguard and storage of nuclear materials associated with nuclear fuel cycles.

In March 2011, an unprecedented earthquake of magnitude 9 on the Richter scale and the ensuing tsunami off the east coast of Japan caused a severe nuclear accident in Fukushima, Japan (Prime Minister of Japan and His Cabinet 2011). The severity of the nuclear accident in Japan has brought about a reinvestigation of nuclear energy policy and deployment activities for many nations around the world, most notably in Japan and Germany (BBC 2011; Reuters 2011). The response to the accident has been mixed and its full impact may not be realized for many years to come.

The nuclear accident in Fukushima, Japan has not directly affected the significant on-going nuclear deployment activities in many countries. China, Russia, India, and South Korea, as well as others, are continuing with their deployment plans. As of October 2011, China had the most reactors under construction, at 27, while Russia, India, and South Korea had 11, 6, and 5 reactors under construction, respectively (IAEA PRIS 2011). Ten other nations have one or two reactors currently under construction. Many more reactors are planned for future deployment in China, Russia, and India, as well as in the United States. Based on the World Nuclear Association's data, the realization of China's deployment plan implies that China will surpass the United States in total nuclear capacity sometime in the future (WNA 2012a).

8.2 Energy from Nuclear Fission

Energy is liberated when a heavy nucleus is split into two parts, otherwise known as nuclear fission. The kinetic energy of the split atoms or fission fragments and neutrons, is converted to useful energy, such as heat and electricity. When fission occurs, two or more neutrons are released, along with fission fragments. These neutrons initiate additional nuclear fissions, releasing yet more neutrons and a chain reaction occurs.

Some background on nuclear materials, provided here, is useful for understanding the current issues related to nuclear energy use. The most important naturally occurring elements for generating energy from nuclear fission are uranium and thorium. Of the two, only natural uranium contains an isotope that readily undergoes fission, namely uranium-235 (U-235). Isotopes that readily undergo fission, such as U-235, are referred to as fissile material. The bulk of natural uranium, however, is composed of uranium-238 (U-238) that can be induced to fission, and are said to be fissionable, but are not fissile. The fissile U-235 content of natural uranium is relatively small at 0.7%. In a nuclear reactor with enriched uranium fuel, the neutron capture of U-238 produces plutonium-239 (Pu-239), a fissile material. Natural thorium consists of only one isotope, thorium-232 (Th-232), a fissionable but not a fissile material. For thorium to be useful in typical nuclear fuel cycles, it must initially be used with fissile material.

Nuclear reactors for power generation are devices for harnessing the energy from nuclear fission by controlling the rate of nuclear chain reactions derived from heavy elements such as uranium, plutonium, and thorium and transferring the kinetic energy of the fission byproducts into useful heat and electric power. In most operating reactors, water is used as a moderator for slowing down the neutrons and as the coolant for transferring the heat from nuclear reactions for conversion to power generation. The majority of commercial reactors operating around the world rely on water as the moderator and coolant and are referred to as light-water reactors (LWRs).

In light-water reactors, fresh nuclear fuel is composed of uranium, with a fissile U-235 enrichment of up to 4%–5%. The fission of U-235 is the primary source of energy from current light-water reactors, with contributions from the conversion of U-238 to Pu-239 and the subsequent fission of Pu-239. Once fresh fuel has been spent, the composition of the spent nuclear fuel is approximately 93% uranium with 1% U-235 content, 5% fission fragments, 1% plutonium, and traces of minor actinides. Very little of the initial uranium content is consumed and the bulk of uranium remains in the spent fuel. The compositions of fresh and spent fuels are provided here so as to highlight several important issues for the current nuclear energy system, namely the sustainability of uranium resources, proliferation concerns, and waste disposal requirements.

Reliance on U-235, a relatively scarce uranium isotope, as the primary source of nuclear fission with the bulk of fissionable U-238 relegated to the waste stream means that the current nuclear fuel cycle does not effectively utilize the available uranium resource. While the ultimate availability of natural uranium resources is uncertain, the poor utilization of existing uranium resources implies quicker transition to ores grade of lower uranium concentration and higher uranium cost (Schneider and Sailor 2008).

Additionally, the necessity of U-235 enrichment for light-water reactor fuels and the very fact that the enriched uranium fuel cycle leads to the production and presence of fissile plutonium in the spent fuel are the primary proliferation concerns. There are differing national policies for the use or storage of fissile plutonium in the spent fuel, however, with some nations electing to recycle plutonium for use in new fuels and others electing to leave it intact within the spent fuel.

The presence of plutonium, as well as minor actinides, in the spent fuel leads to greater waste disposal challenges as well. Heavy isotopes, such as plutonium and minor actinides, have very long half-lives, as high as tens to hundreds of thousands of years, which require final waste disposal strategies to address the safety of waste disposal on such great timescales. Fission fragments, the inevitable byproduct of any fission reaction, have significantly shorter half-lives. Waste disposal challenges for fission fragments are in the order of several hundreds of years. Thus, strategies to isolate and dispose of only the fission fragments in spent fuel could have significant beneficial impact on waste disposal requirements (Wigeland et al. 2006).

Mixed oxide (MOX) fuel, which replaces U-235 in enriched uranium fuel with recycled or excess plutonium, has been in greater use in Europe, and more recently in Japan (WNA 2011a). The reprocessing of plutonium from spent fuel and its use as new fuel improves the utilization of uranium resources and reduces the mass of high-level wastes requiring disposal. There is debate as to whether the separation and use of plutonium in MOX fuel reduces or raises proliferation concerns, however (Becker and Broad 2011).

Use of excess plutonium from military stockpiles, originally intended for weapons application, as a source of fuel for nuclear power plants has clear proliferation benefits of reducing the amount of surplus plutonium. However, widespread global reprocessing of spent LWR fuels and the separation of plutonium from spent fuel raises the concern for increased access to plutonium. Those in favor of MOX fuel use see it as an economically attractive approach to extracting more energy from existing fuels and extending limited fissile resources while minimizing nuclear wastes. However, those opposed to MOX fuel use are concerned with the potential increased access and vulnerability to theft of plutonium from the expansion of reprocessing infrastructures that involves the isolation of plutonium, the transport of MOX fuels, and their storage at commercial sites. Currently, MOX fuel use constitutes a small portion, about 2%, of the new nuclear fuel used (WNA 2011a).

8.3 Global Nuclear Energy System Today

Energy from nuclear fission contributes significantly to the world's electricity needs today. It contributes to the stability of electricity supply and prices by providing a steady source of base-load electricity. Nuclear power provides energy security to many nations that lack sufficient domestic fossil or renewable energy resources as it generates electricity uninterrupted throughout the year, with minimal downtimes for maintenance and refueling. Furthermore, nuclear power contributes to reducing local air pollution and GHG emissions compared to fossil-fuel power plants.

The majority of the world's reactors are based on light-water technology of similar concept, design, and fuel cycle. Of the 433 reactors worldwide, more than 80% or 353 of the reactors are light-water reactors, of which 268 are pressurized water reactors (PWR) and 84 are boiling water reactors (BWR) (IAEA PRIS 2011). The remaining commercial reactor types consist mostly of pressurized heavy-water reactors (PHWR), and a few gas-cooled reactors (GCR) and graphite moderated reactors (RBMK/LWGR).

The historical deployment of commercial nuclear power plants began in earnest in the late 1950s and the early 1960s in several countries around the world. In 1960, the first BWR as well as the first fully commercial PWR started operation in the United States (DOE/NE 1987; WNA 2010). Alternatives to light-water reactor designs were being deployed at the same time in Great Britain, France, Canada, and the Soviet Union. Great Britain and France developed and deployed the gas-cooled graphite moderated reactors in the late 1950s. France has eventually transitioned to PWRs. Great Britain has deployed a PWR and is considering more such reactors, but the majority of nuclear plants remain gas-cooled reactors. Canada developed a unique reactor design that utilizes natural uranium fuel without enrichment and heavy water as the coolant and moderator in the early 1960s. This design, called the CANDU reactor, continues to be refined and operated today. This reactor system or similar designs are in use in South Korea, China, Romania, and India. However, the predominant reactors in South Korea and China today are PWRs. Also in the early 1960s, the Soviet Union developed yet a different design, a boiling water graphite channel reactor. This reactor design evolved to the larger RBMK reactors that are still in operation today. The predominant reactor system in the Russian Federation today is the PWR.

Nuclear power quickly expanded in the United States throughout the 1970s and the 1980s, along with significant deployments in the former Soviet Union, France, Japan, Germany, and Great Britain. In 1973, U.S. utilities ordered a record 41 nuclear power plants in one year (DOE/NE 1987). Nuclear power continued to be deployed at a modest rate outside of these early adopters, with notable deployments in Canada, Sweden, Belgium, Spain, and Finland. In the 1980s and the 1990s, a rapid deployment of nuclear power occurred in South Korea. The deployment of reactors in China and India occurred in earnest in the twenty-first century, although India has had a reactor program since the 1960s.

The brief history of nuclear reactor deployment reveals that nearly 30–40 years have passed since the bulk of the world's reactors have come online. Twenty-four of the currently operating reactors are more than 40 years old, and an additional 138 reactors are between 30 and 40 years old

(IAEA PRIS 2011). In total, 37% of the currently operating reactors are more than 30 years old. License renewal of older reactors has become more common. In the United States, where the majority of the older reactors reside, the original operating licenses of 40 years have been, or are in the process of being, extended to 60 years (U.S. NRC 2011a; U.S. NRC/DOE 2008). Life extensions of older reactors are also occurring in France and Russia, with license extension periods dependent on specific reactor types and national regulatory policies (WNA 2012b; WNA 2012c).

Regardless of the ultimate operating lifetimes of legacy reactors, replacing the power capacity lost from aging reactors are potentially more challenging for the United States and France, considering that the bulk of their historical deployment occurred within a relatively short time period of 10–15 years. For some countries, such as Germany and Switzerland, the nuclear accident in Fukushima, Japan has accelerated the phase-out of existing reactors and no further life extensions are proposed. Worldwide responses to life extension and the retirement of legacy reactors are likely to have important consequences for regional economic growth, nuclear safety, and carbon mitigation efforts (WEO 2011).

8.4 Motivations for Expanded Nuclear Energy Use

There are 65 nuclear reactors, representing 63 GWe of capacity, currently under construction in 14 countries (IAEA PRIS 2011). Forty-nine of the reactors under construction are located in only four countries: China, Russia, India, and South Korea. China, with 27 reactors under construction, has the most active nuclear reactor deployment program of any nation. Beyond those under construction, there are future plans or proposals for the deployment of nearly 500 additional reactors worldwide (WNA 2012a). The deployment plans of China, India, Russia, United States, and Ukraine represent nearly 70% of the potential future growth in nuclear power. Although smaller countries have fewer nuclear deployment plans, the realization of such efforts could constitute a significant portion of their total electric power capacities.

Motivations for new nuclear deployment are many, including energy security, economic competitiveness, local air pollution, and climate change concerns. For most nations, the current and planned deployment of nuclear power is strongly motivated by the rapid growth in the demand for electricity and the desire for an increased diversification of power supplies. For instance, China and South Korea's demand for electricity has grown at a rapid rate in the last few decades, exceeding 10% per annum for many years (IEA 2010). More recently, India, Russia, Ukraine, and others have had high growth rates in electricity demand (IEA 2010). Other contributing factors vary and are specific to each region. China and India's motivation is to reduce their reliance on coal (CAEA 1999; Government of India 2006), Russia's motivation is to minimize the use of natural gas for power (Ministry of Energy of the Russian Federation 2010), and South Korea's motivation is to ensure security of power supplies due to the lack of domestic energy resources (MKE 2012).

Steady-state base-load electricity comprises the bulk of total electricity generation. Coal, nuclear and, to a lesser degree, natural gas are the primary sources of base-load electricity. Improvements to the operation of nuclear power plants in the last several decades have increased the average annual capacity factor, ratio of the energy produced to the reference energy generation, of nuclear plants worldwide to over 80% (IAEA PRIS 2011). Several countries, such as Finland, Romania, Slovenia, Netherlands, South Korea, Switzerland, and the United States, have capacity factors greater than 90% (IAEA PRIS 2011). The ability for steady-state continuous generation of electricity throughout the year with minimal downtime for refueling and maintenance has increased the economic competitiveness of currently operating nuclear reactors worldwide.

There is considerable interest in the potential of nuclear energy for replacing fossil power options for mitigating CO_2 emissions beyond nuclear power's current contribution. Nuclear power does not produce CO_2 during its operation, while the estimate of lifecycle emissions ranges from 8–58 g CO_2/kWh (Beerten et al. 2009). Lifecycle CO_2 emissions from nuclear power are significantly lower than that from coal or natural gas-fired power plants, while the estimated range of nuclear lifecycle CO_2 emissions is comparable to that of solar and wind technologies (Lenzen 2008; Fthenakis and Kim 2007; Voorspools et al. 2000).

There are twelve nations in Europe, Eastern Europe, and Asia that have more than 30% of their electricity coming from nuclear power. The most notable is France, with the largest nuclear share of electricity generation of any nation, at 74% (WNA 2012b). The significant reliance on nuclear power for electricity generation has contributed to low CO_2 emissions in France. For the last three decades, annual CO_2 emissions in France have not exceeded the 1979 peak level of 144 MTC (WDI 2011; CDIAC 2011). The 2008 CO_2 emission of 103 MTC is nearly 30% below the 1979 peak. The per capita CO_2 emissions rate in France has fallen from the 1979 peak of 2.7 tons C/capita to the 2008 rate of 1.7 tons C/capita (CDIAC 2011). Several countries, Japan, South Korea, United States, China, India, and others, have stated the reduction of GHG emissions as a motivating factor in their interest in the expansion of nuclear power.

8.5 Limitations and Concerns of Nuclear Deployment

The resolution of many issues remains for the continued use and further expansion of nuclear energy worldwide. While nuclear plants continue to be deployed and become more widely distributed throughout the world, several factors limit its global acceptance: concern over the safety of existing and new reactors, the cost of new reactors, the availability of uranium resources, the proliferation of nuclear materials, and the challenge of waste disposal.

8.5.1 Safety

The safety of nuclear power has been thrust to the forefront of discussions on global energy policy with the nuclear accident in Fukushima, Japan in 2011. The earthquake and ensuing tsunami on the east coast of Japan caused a major nuclear accident at the Fukushima Dai-ichi nuclear facility. The significant release of radioactivity from the Fukushima accident is rated at a Level 7, the maximum level of the International Nuclear and Radiological Event Scale (INES) for nuclear accidents, and the same level as the 1986 Chernobyl nuclear accident in Ukraine (IAEA 2008).

In the Fukushima accident, operating reactors at the Units 1-3 of the Fukushima Dai-ichi facility shutdown automatically and nuclear reactions stopped in response to the earthquake. The ensuing tsunami, however, destroyed the plant's backup power system, leaving the reactors without the ability to operate their cooling systems. Although nuclear reactions were halted, the residual decay heat of nuclear fuels in the reactor cores, as well as in spent fuel pools, could not be dissipated without power. Increasing reactor core temperatures led to fuel rod damage, the production of hydrogen gas and hydrogen gas explosions, and the subsequent release of radioactivity to a wide area (Prime Minister of Japan and His Cabinet 2011).

Twenty-five years prior to the Fukushima accident, another major nuclear accident occurred at Chernobyl Unit 4. The Chernobyl accident was caused by operator error, but the design features of the RBMK reactor had a significant influence on the course of the accident and it consequences (IAEA 1992). The accident was triggered during experiments to test the design operation of the independent power supply in the event of external power source disruptions (SCSSINP 1991). The inability to manage and control unstable reactor conditions initiated by the planned test led to a runaway increase in reactor power. The sudden power surge led to explosions and fire in the reactor building and the significant release and dispersion of radioactivity to a wide area.

Another significant accident occurred at the Three Mile Island nuclear reactor in the United States in 1979. The Three Mile Island accident was rated at INES Level 5 and was not as severe as the accidents at Chernobyl and Fukushima. A valve failure and operator error led to the partial core meltdown at Three Mile Island Unit 2 that was operational prior to the accident. Although the automatic response had halted nuclear reactions with the insertion of control rods, cooling water could not be circulated through the core to remove the decay heat of nuclear fuels. The rising reactor core temperatures and a faulty valve led to core damage and the release of radioactive coolant from the primary containment. The Three Mile Island accident did not cause any deaths and very small releases of radioactivity to the environment were reported (U.S. President's Commission 1979; U.S. NRC 2011b).

The series of nuclear accidents at Fukushima, Chernobyl, and Three Mile Island occurring within several decades of each other has brought about calls for a reevaluation of nuclear energy policies worldwide. The debate over the acceptable levels of safety for nuclear power varies from region to region and public sentiments of nuclear power and its safety continue to evolve.

The nature of nuclear fission is that an enormous amount of energy can be liberated from a small amount of material. This fact translates to the extremely high power density of the current generation of nuclear power plants. For safety reasons, the majority of the operating reactors around the world have inherent designs in which the reactor power decreases with increases in reactor temperature. This design feature ensures the stability and passive safety of nuclear reactors. Another commonality of all nuclear power plants that are in operation today is that the reactor cooling system requires active management. It is, therefore, crucial that cooling systems and redundant backup power systems work as intended in the event of reactor power loss.

It is important to understand the root causes of each specific accident to improve the safety of nuclear energy use. Differences in the nature of nuclear accidents at Fukushima, Chernobyl, and Three Mile Island are many, and involve differing factors, such as inherent nuclear plant designs (BWR, RMBK and PWR), backup and safety systems, age, initiating events, incidence response and operating procedures, and siting and environmental factors. The inherent designs of new nuclear reactors, direct responses from knowledge gained from previous incidents, reevaluation and modifications to operational procedures and safety systems of existing reactors, and improvements to and transition to new reactors with improved passive and active safety features are likely to reduce the potential risk of future accidents.

8.5.2 Nuclear Costs

The economics of new nuclear power is dominated by the capital cost. Most of the difference in capital cost between nuclear and conventional power is due to the high cost of the reactor and associated equipment and the provisions for safety. Except for the reactor and associated equipment, cost items for nuclear plants are similar to those of conventional power plants. Fuel costs for nuclear plants represent only a small portion of the cost of power. For a prospective new nuclear plant, the front-end fuel cost is typically only 15%–20% of the total (WNA 2011b). Other contributing costs specific to nuclear power are decommissioning and waste disposal costs. Provisions for these costs are, however, typically incorporated into the generation cost.

It is important to distinguish the cost of electricity generation of a prospective new nuclear plant from an existing nuclear plant with sunk or depreciated capital costs. For the latter, the cost of electricity is extremely low and is relatively insensitive to the price of uranium ore. Comparisons of recent historical electricity production costs in the United States show that nuclear has the lowest cost compared to electricity from coal, natural gas, and oil (NEI 2010).

Overnight capital costs of new nuclear power plants have increased in recent years. However, there is a broad range of reported capital costs of new nuclear plants contributing to the uncertainty in the economic competiveness of nuclear power. Some of the uncertainty, however, stems from inconsistencies in the reported costs that may or may not include the bare plant cost (engineering-procurement-construction EPC cost), owner's cost (land, cooling infrastructure, administration and associated buildings, site works, switchyards, project management, licenses, etc.) and financing cost.

Each cost component varies further, contributing to the difficulty in comparing costs among nuclear and competing technologies. The overnight capital costs vary by vendor, model and location, ranging from as low as $1556/kW for APR-1400 in South Korea to as high as $5863/kW for EPR in Switzerland (OECD NEA/IEA 2010). The U.S. AP1000 ranges from $2444 to $3582/kW and the U.S. ABWR is reported at $2900/kW (WNA 2005; AEO 2011). The owner's cost, included in the overnight capital cost, depends on the existing infrastructure and whether the new plant is Greenfield or at an established site. The financing cost varies by interest rate and the duration of the construction period that has varied in the past, depending on the specific site and project. Financing cost can add an additional 30%–40% to the overall cost of a new nuclear plant (WNA 2011b; OECD NEA/IEA 2010).

Although nuclear plant costs have increased, it is important to recognize that construction costs of all types of large-scale engineering projects have escalated and that capital costs of other power technologies, such as coal, natural gas, and wind, have increased as well (MIT 2009; AEO 2011; WNA 2005). In general, acquiring financing for any large infrastructure project with high capital cost is likely to be difficult without government support. Comparisons of capital costs among power plant options reveal that the overnight capital costs of nuclear, coal, natural gas, wind, and others have changed less significantly on a relative basis within the last decade (MIT 2009; AEO 2011; WNA 2011b). Relative differences in the levelized electricity costs of new power options must be considered in the assessment of the economic competitiveness of nuclear power.

8.5.3 Uranium and Thorium Resources

The current worldwide nuclear fuel cycle is predominantly based on natural uranium. Natural uranium is a relatively common metal and the important factor in the consideration of uranium resource availability is the economic recoverability of uranium ore and not its ultimate availability, which could be significant.

An authoritative estimate of recoverable uranium is provided by the OECD NEA/IAEA Uranium Resource, Production and Demand, also referred to as the "Red Book" (OECD NEA/IAEA 2009). Estimates of recoverable conventional uranium resources have increased steadily within the last several decades even as consumption has increased. The 2009 Red Book estimate of Identified Conventional Resources recoverable at less than $260/kgU is 6.3 million metric tons of uranium (MtU) (OECD NEA/IAEA 2009). At current usage of 68,000 tons of U per year, global identified resources for conventional nuclear power are sufficient for 90 years (WNA 2011c). The demand for new uranium resources has been reduced to some degree from the use of excess military uranium and plutonium. The Red Book further estimates an additional 10.4 MtU of undiscovered resources that can extend the supply of uranium for an additional 150 years at the current consumption rate. Beyond these conventional resources, an additional 22 MtU is available in phosphate deposits, and up to 4000 MtU is available in seawater at a significantly higher cost.

Thorium is widely distributed and considered to be at least three times more abundant than uranium. Although thorium is not commercially utilized for nuclear power today, thorium can be used as a fuel for nuclear power. The thorium cycle, however, cannot be initiated without an external source of fissile material. Once initiated, the breeding and recycling of fissile U-233 from natural thorium can sustain the thorium cycle. The thorium fuel cycle has been studied for 40 years and experiments conducted worldwide have confirmed its potential as a nuclear fuel (Hargraves and Moir 2010). The Red Book's current thorium estimates are 3.6 Mt of known and estimated resources, and 6 Mt of total resources, including undiscovered resources. Due to limited commercial demand, these estimates are not complete and are based on data that do not include the contributions of thorium resource estimates from many of the world's regions. Thorium's abundance, however, implies significantly greater availability relative to uranium at an equivalent extraction cost.

There are potentially multiple fuel cycles and reactor designs for generating energy from nuclear fission that can affect the utilization of natural uranium. Heavy-water reactors, such as Canada's CANDU reactor, do not require enriched uranium fuel, and so, the total demand for natural uranium is significantly less. The reprocessing of plutonium from spent fuel for use as MOX fuel has reduced natural uranium requirements as well. The trend towards extracting greater energy from the initial mass of nuclear fuel (higher burnup) could further reduce natural uranium requirements. Even efforts to improve the efficiency of uranium enrichment processes and the thermal efficiency of nuclear electricity generation can reduce the need for natural uranium. Although current nuclear electricity generation is based predominantly on the once-through uranium fuel cycle, multiple options are available for more effective utilization of natural uranium as a response to resource economics.

Ultimately, concern over the scarcity of uranium resources is unwarranted since breeding and recycling of fissile materials from either uranium or thorium in a fully closed fuel cycle can increase the production of energy from nuclear resources by more than two orders of magnitude relative to the current once-through fuel cycle. The material losses from reprocessing would reduce this magnitude to a small degree. A closed fuel cycle would effectively extend the use of nuclear energy for thousands of years. Breeding and recycling of fissile material with fast reactors are proven concepts. However, present cost estimates for reprocessed fuels and fast reactors are higher than that for enriched natural uranium fuels and current light-water reactors. On the other hand, high-level waste reductions from fast reactors with fuel recycling could have significant economic and environmental value, and lower the total cost of nuclear energy use.

8.5.4 Waste and Proliferation

Five decades of electricity generation from nuclear energy has resulted in the accumulation of spent nuclear fuel worldwide. The total historical amount of spent fuel cumulatively generated worldwide by the beginning of 2003 was approximately 255,000 metric tons (Fukuda et al. 2003). About 27 metric tons[2] of spent or used fuel with a discharged volume of 20 m^3 is generated from a typical 1000 MWe light-water reactor in a year (WNA 2012d). The term used fuel in reference to LWR spent fuel is in recognition of the remaining uranium and plutonium in spent fuel that could be used as new fuel. Relative to other base-load electricity generation such as coal, the quantity of waste generated from nuclear power is very small. For instance, in one year alone, the same sized coal-fired power plant produces 400,000 metric tons of ash and 5.8 million metric tons of CO_2.[3]

Although the sheer amount of waste is small, the radiotoxicity of nuclear wastes requires safe methods for its storage and ultimate disposal. Waste in the form of fission fragments is an inevitable byproduct of nuclear fission, with no additional value as a fuel. Used nuclear fuel, however, contains significant quantities of unused uranium and other transuranic elements (elements heavier than uranium) that can be recycled as new fuel. The policy to leave spent fuel intact for direct disposal or recycling as new fuel varies across regions. Reprocessing occurs or is the waste management policy in several nations such as Belgium, China, France, India, Japan, Russia, and United Kingdom, while others such as Canada, Finland, South Korea, Spain, Sweden, and the United States have elected direct disposal as their policy. Alternative disposal options are currently under consideration in the United States, however (BRC 2011).

The reprocessing of used fuel affects resource sustainability and waste management, as well as proliferation concerns. Recycling uranium and fissile materials in used fuel reduces the demand for natural uranium in fabricating new fuel, while reducing the mass of high-level wastes that require disposal. Recycling transuranics can also significantly reduce the challenge of isolating high-level waste from the biosphere in any geologic disposal option from tens of thousands to hundreds of years. High-level waste containing only fission fragments would require about 300 years for the radiotoxicity to drop below the level of natural uranium ore used for making the initial fresh fuel. However, proliferation concerns are heightened from the potential access to fissile materials from the worldwide application of reprocessing technologies and policies.

There is not a commonly accepted single worldwide approach to dealing with the long-term and permanent disposal of high-level waste. Regional differences in the availability of uranium ore and land resources, technical infrastructure and capability, nuclear fuel cost, and societal acceptance of waste disposal have resulted in alternative approaches to waste storage and disposal. Regardless of these differences and the fuel cycle ultimately chosen, some form of long-term storage and permanent disposal, whether surface or geologic (subsurface), is required. Finland and Sweden are the furthest along in selecting a site for the geologic disposal of their high-level waste. Other countries, particularly in Europe, have chosen to recycle used fuels as part of their waste disposal strategy. Yet others, such as South Korea, are pursuing a synergistic application of light and heavy water reactors to reduce the total waste by extracting more energy from used fuels. In the United States, waste disposal options are currently under review with the termination of the Yucca Mountain facility in Nevada as a candidate site for waste disposal. The Yucca Mountain facility, originally approved in 2002 as a geologic repository for spent nuclear fuel and other high-level waste, was cancelled in 2009. Indefinite dry cask storage of high-level waste at interim storage facilities (mostly at reactor sites) is to be pursued until decisions on waste disposal are resolved.

8.6 Future Nuclear Energy Systems

Nuclear power has been around for many decades, but efforts to improve the safety, economics, resource sustainability, waste management, and proliferation concerns of nuclear power use continue. Significant potential exists for making progress in all of these areas, and both private industries and national governments are developing new reactor and fuel cycle technologies through their research efforts.

The Three Mile Island and Chernobyl accidents emphasized the importance of the human factor, operational procedures and man-machine interface, and the value of well designed and constructed containment structures of nuclear power plants. The Fukushima accident will further emphasize these lessons learned, as well as the need to consider a greater range of conceivable events when evaluating safety designs and possible accident consequences of nuclear reactors.

Light-water reactors continue to evolve with improved passive and active safety features. Evolutionary reactors currently available for commercial deployment, such as the European Pressurized-Water Reactor (EPR France), Advanced Passive 1000 Reactor (AP 1000, USA-Japan), Water-Water Energetic Reactor-1200 (VVER-1200, Russia), and Advanced Pressurized-Water Reactor-1400 (APR-1400, South Korea), all have improved safety features over the previous generation of light-water reactors.

Other more revolutionary light-water and gas reactors with even greater passive safety features are near commercial status. These small modular reactors (SMR) are typically less than 300 MWe and much smaller than the 1000 MWe size of current reactors. SMRs with lower power density, large heat capacity, and heat removal through natural means are designed to reduce the consequences of human errors and unforeseen events, and contribute to their improved safety. SMRs based on light-water designs from South Korea, Russia, and the United States are near commercial status (KAERI 2012; WNA 2012e; NuScale 2012). In addition to their passive safety features, smaller gas-cooled reactors from South Africa, China, United Kingdom, and United States are designed to operate at higher temperatures for improved electricity generation efficiencies over LWRs and potential industrial applications as a source of process heat. Moreover, smaller reactors that can be constructed in a factory setting with modular construction techniques and flexibility for incremental additions to total power capacity could shorten the duration of construction periods and improve the quality and economics of new nuclear plants.

Alternative nuclear fuel cycles, beyond the once-through uranium cycle, and related reactor technologies are also under investigation. The partial recycling of used fuels, such as the use of MOX fuels, is already being employed. Dramatic improvements to nuclear resource utilization and waste management, and non-proliferation can be achieved from the realization of full recycling options based on either uranium or thorium that is combined with alternative reactor designs where only fission fragments are relegated as waste. Multiple reprocessing and recycling options, along with higher burn up fuels, are currently being investigated to determine the merits of each approach.

The research and development of next generation nuclear energy systems is being undertaken internationally through such collaborations as the Generation IV International Forum (GIF). The GIF consists of 13 members, including Argentina, Brazil, Canada, China, Euratom (the European Atomic Energy Community), France, Japan, the Republic of Korea, the Republic of South Africa, Russian Federation, Switzerland, the United Kingdom, and the United States (OECD GIF 2001). The GIF has as its goal improving sustainability, economics, safety and reliability, and proliferation resistance and the physical protection of nuclear energy use.

Various reactors and fuel cycle technologies have been proposed for achieving the objectives of advanced nuclear energy systems. Simplifying reactor designs, improving passive safety features, increasing the temperature of working fluids, reducing reactor pressures, more efficient uranium use, and burning transuranic isotopes are some commonalities of advanced nuclear energy systems. To meet these objectives, Generation IV (Gen IV) reactors using sodium, lead, molten-salt, supercritical-water, or very-high temperature gas coolants are being studied and developed in various countries (OECD GIF 2007). Multiple strategies and approaches for used fuel processing are being considered to identify approaches that are acceptable from non-proliferation and economic viability perspectives.

8.7 Transition to Future Nuclear Energy Systems

The transition from the currently operating reactors to advanced fuel cycle and nuclear reactor systems will not occur immediately and is expected to take several decades. According to the GIF, the deployment of Gen IV reactors is not envisioned until after 2030 (see Figure 8.1). The commercial availability

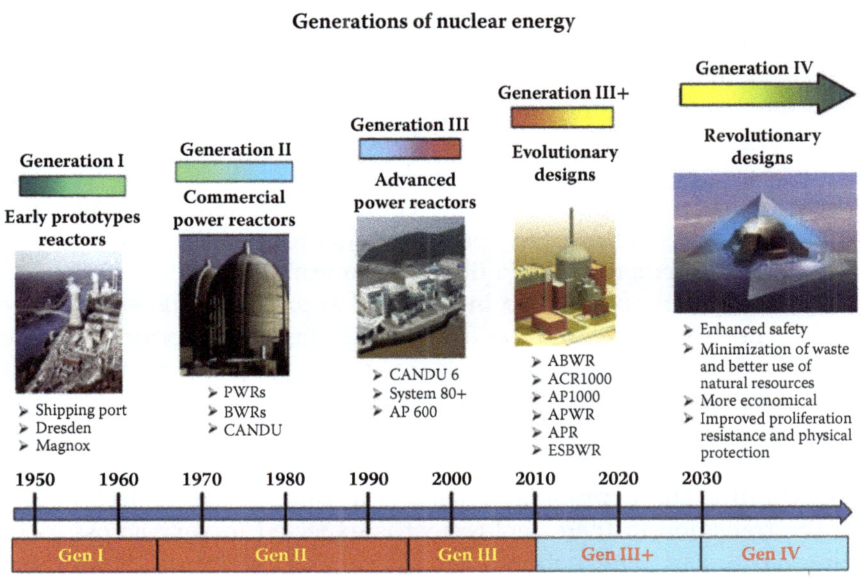

FIGURE 8.1
Generation IV International Forum evolution of nuclear power. (Courtesy of http://www.gen-4.org/Technology/evolution.htm).

of new fuel cycles and reactor technologies requires demonstration of new reactor concepts and their economic viability, as well as regulatory reviews and approvals. The realization of these efforts will require long-term commitments and time.

For the most part, only advanced LWRs are being considered for deployment in the near term. Considering that LWRs have a lifetime of 60 years, and potentially up to 80 years (NRC/DOE 2008), they will remain as the bulk of the global nuclear energy system well into the twenty-first century. While some nations are phasing-out or are not further extending existing licenses, they represent only a small portion of the total global nuclear capacity. For this reason, even with future availability of Gen IV reactors, full replacement of LWRs cannot be expected to occur until the end of the twenty-first century or into the twenty-second century. If advanced reactors are to have a significant impact in the future, they must demonstrate economic competitiveness relative to the mainstream LWR or contribute to the synergistic use with LWRs in a way as to improve the nuclear energy system as a whole. Such improvements could be in extending nuclear resources and reducing nuclear fuel costs, minimizing waste generation, and strengthening proliferation resistance.

The rapid economic growth and high demand for energy from non-OECD countries could encourage the transition to alternative nuclear technologies from the current LWRs and uranium fuel cycle. Along with the construction of LWRs, China and India have shown significant interest in the development of advanced nuclear reactors. China has made rapid progress on the high-temperature gas-cooled pebble-bed reactor design, which has significantly greater inherent safety features and higher thermal efficiency than an LWR. The Chinese HTR-PM (high-temperature-reactor pebble-bed module), due to be completed by 2013, is a demonstration of the Gen IV class of reactors (Zhang et al. 2009). Fast reactors have been demonstrated previously in countries such as France, Germany, Japan, Russia, the United Kingdom, and the United States. India is the latest entrant to this group, with the near completion of a 500 MWe prototype fast breeder reactor (PFBR) (Chetal et al. 2009). The development of the PFBR is a commitment to nuclear energy use in India and an important step in a phased approach to the expansion of domestic nuclear power and the use of their abundant thorium resources (Jain 2004). Interest in advanced nuclear power technologies necessitated by the growing demand for energy could encourage a more rapid deployment of advanced nuclear reactors from countries outside of the OECD, such as China and India.

Determining the appropriate choice of advanced nuclear technologies and the transition to their deployment is expected to vary across global regions in accord with national assessments of energy needs, domestic energy resources, industrial capabilities, and technology readiness (NEA 2009). Countries with a small nuclear fleet may find that multilateral arrangements for managing nuclear resources and used nuclear fuels are more attractive than embarking on a full-fledged national advanced fuel cycle program. Countries with existing advanced nuclear capabilities and a larger nuclear fleet may find the converse to be true. The globalization of nuclear energy technologies and the transition to advanced nuclear energy systems will require sustained efforts and long-term planning by national governments to demonstrate the capabilities of advanced fuel cycle and reactor concepts. Along with the realization of advanced nuclear energy systems, international cooperation for ensuring the safety of global nuclear energy systems and safeguard of nuclear materials will continue to remain important.

The limits to harnessing energy from nuclear fission have not been fully explored. While five decades of worldwide experience with nuclear power have led to improvements in the current nuclear energy technology, significant research and demonstration efforts are underway to develop new fuel cycles and reactor technologies that address many concerns of global nuclear energy use.

Questions for Discussion

1. What are the factors that have contributed to the deployment of nuclear power?
2. What socioeconomic factors have to be considered in developing nuclear power as an energy source?
3. What are the benefits and concerns of nuclear energy use? Are they the same across all countries currently utilizing or planning to utilize nuclear power?
4. What are the potential impacts of a major accident on nuclear power utilization? Under what conditions would you propose to scrap the technology?
5. Discuss any potential role for nuclear power in a national economy determined to reduce GHGs as a response to climate change. How, and to what degree, can nuclear power contribute to the climate change problem? Are there any compelling reasons for using nuclear power if climate change is not a concern?
6. Is nuclear power destined to be a major source of energy or a bridging technology to those based on renewables?
7. What types of nuclear power systems are currently being used internationally? Compare the relative availability (capacity factor) of these power generation systems to those of other sources of energy.

8. Compare the energy density in a nuclear fuel to those of other sources of energy (e.g., coal and gas). Compare the type and amount of waste generated from a nuclear power plant to conventional fossil power plants.
9. What are the differences and similarities between a nuclear and coal power plant?
10. What is the composition of fresh and spent nuclear fuel (LWR) and how does it affect resource utilization, proliferation, and waste disposal issues?
11. Is nuclear power a mature or an immature technology?

End Notes

1. Based on summation of Operable, Under Construction, Planned, and Proposed reactor categories from World Nuclear Association; does not include capacity reductions from future decommissioned reactors.
2. Mass of used fuel from 1000 MWe LWR reactor with burnup of 50 GWd/tHM, 90% annual capacity factor and thermal efficiency of 33%.
3. Quantity of coal ash from WNA, 2011 (http://www.world-nuclear.org/info/inf04.html). CO_2 emissions for a coal-fired power plant based on the following calculation: 2.1 lbs CO_2/kWhr * 1000 MWe * 8760 hrs/yr * 0.7 (CF) * metric ton/2204.6 lbs = 5.84 million metric tons of CO_2. 2.1 lbs CO_2/kWhr emissions factor from U.S. EIA. http://www.eia.gov/cneaf/electricity/page/co2_report/co2report.html#table_2

References

AEO. 2011. *Annual Energy Outlook 2011*. U.S. Department of Energy, Energy Information Agency, Washington, DC.

BBC. 30 May 2011. Germany: Nuclear power plants to close by 2022. British Broadcasting Corporation, London.

Becker, J., and William J. Broad. 2011. New Doubts About Turning Plutonium Into a Fuel. *New York Times*. April 10, 2011. http://www.nytimes.com/2011/04/11/us/11mox.html.

Beerten, J., E. Laes, G. Meskens, and W. D'haeseleer. 2009. Greenhouse gas emissions in the nuclear life cycle: A balanced appraisal. *Energy Policy*, 37 (2009) 5056–5068.

BRC. July, 2011. Draft Report to the Secretary of Energy, The Blue Ribbon Commission on America's Nuclear Future (BRC). U.S. Department of Energy, Washington, DC.

CAEA. 1999. Nuclear Power Development and Related Policies in China. China Atomic Energy Authority, People's Republic of China. http://www.caea.gov.cn/n602670/n621894/n621898/32153.html.

CDIAC. 2011. Carbon Dioxide Information Analysis Center, Oak Ridge National Laboratory, Oak Ridge, Tennessee. http://cdiac.ornl.gov/trends/emis/tre_fra.html.

Chetal, S. C., P. Chellapandi, P. Puthiyavinayagam, S. Raghupathy, V. Balasubramaniyan, P. Selvaraj, P. Mohanakrishnan, and B. Raj. 2009. A perspective on the future development of FBRs in India, in *Fast Reactors and Related Fuel Cycles: Challenges and Opportunities* (FR09). Proceedings of an International Conference. International Atomic Energy Agency. Vienna, 2012, 91–107.

Commission to the USSR State Committee for the Supervision of Safety in Industry and Nuclear Power. 1991. Causes and Circumstances of the Accident at Unit 4 of the Chernobyl Nuclear Power Plant on April 26, 1986. Moscow.

DOE/NE. 1987. *The History of Nuclear Energy*. U.S. Department of Energy, Office of Nuclear Energy, Sciences, and Technology, Washington, DC. DOE/NE-0088.

Fthenakis, V. M., and H. C. Kim. 2007. Greenhouse-gas emissions from solar electric and nuclear power: A life-cycle study. *Energy Policy*, 35 (2007) 2549–2557, Elsevier Ltd.

Fukuda, K., W. Danker, J. S. Lee, A. Bonne, and M. J. Crijns. 2003. IAEA Overview of global spent fuel storage, Department of Nuclear Energy, IAEA, Vienna Austria (IAEA-CN-102/60). https://iaea.org/NewsCenter/Features/.../Grimsel/storageoverview.pdf.

Government of India, Planning Commission. 2006. Integrated Energy Policy, Report of the Expert Committee. Government of India, Planning Commission, New Delhi. August 2006.

Hargraves, R., and R. Moir. 2010. An old idea in nuclear power gets reexamined. *American Scientist*, 98 (2010) 304–313.

IAEA. 1992. The Chernobyl Accident: Updating of INSAG-1. A report by the International Nuclear Safety Advisory Group. International Atomic Energy Agency, Vienna.

IAEA. 2008. International Nuclear and Radiological Event Scale (INES). Information Series, Division of Public Information, 08-26941/E. International Atomic Energy Agency, Vienna. www.iaea.org/Publications/Factsheets/English/ines.pdf.

IAEA PRIS. 2011. Power Reactor Information System. International Atomic Energy Association, Vienna. http://www.iaea.org.

IEA. 2010. Energy Balances of non-OECD Countries and Energy Balances of OECD Countries. International Energy Agency, Paris.

Jain, S. K. 2004. Nuclear power in India—the fourth revolution. *An International Journal of Nuclear Power*, 18 (2–3) 2004.

KAERI. 2012. System-integrated Modular Advanced ReacTor (SMART). Korea Atomic Energy Research Institute, Daejeon, South Korea. www.smart.kaeri.re.kr.

Lenzen, M. 2008. Life cycle energy and greenhouse gas emissions of nuclear energy: A review. *Energy Conversion and Management*, 49 (2008) 2178–2199, Elsevier Ltd.

Ministry of Energy of the Russian Federation. 2010. Energy Strategy of Russia for the period up to 2030. Ministry of Energy of the Russian Federation, Moscow.

MIT. 2009. *Update of the MIT 2003 Future of Nuclear Power*. Massachusetts Institute of Technology, Cambridge, MA.

MKE. 2012. Ministry of Knowledge Economy, Republic of Korea. http://www.mke.go.kr/language/eng/policy/Epolicies.jsp.

NEA. 2009. Strategic and Policy Issues Raised by the Transition from Thermal to Fast Nuclear Systems. OECD Nuclear Energy Agency, Paris, France.

NEI. 2010. U.S. Electricity Production Costs (1995–2010). Nuclear Energy Institute, Washington, D.C. http://www.nei.org/resourcesandstats/documentlibrary/.reliableandaffordableenrgy/graphicsandcharts/uselectricityproductioncosts/

NuScale. 2012. NuScale Power, Inc. Corvallis, Oregon. www.nuscalepower.com.

OECD GIF. 2001. Generation IV International Forum, GIF Charter. OECD, Paris, France. http://gen-4.org/GIF/Governance/charter.htm

OECD GIF. 2007. Generation IV International Forum (GIF), Annual Report 2007. Organization for Economic Co-operation and Development, Paris, France.

OECD NEA/IAEA. 2009. *Uranium 2009: Resources, Production and Demand*. A Joint Report by the OECD Nuclear Energy Agency and the International Atomic Energy Agency, Paris.

OECD NEA/IEA. 2010. Projected Costs of Generating Electricity 2010. OECD/IEA, Paris.

Prime Minister of Japan and His Cabinet. 2011. Report of Japanese Government to the IAEA Ministerial Conference on Nuclear Safety—The Accident at TEPCO's Fukushima Nuclear Power Stations. http://www.kantei.go.jp/foreign/kan/topics/201106/iaea_houkokusho_e.html.

Reuters. 10 May, 2011. Japan says nuclear policy must be reviewed from scratch. Thomson Reuters, New York.

Schneider, E. and W. Sailor. June 2008. Long-term uranium supply estimates. *Nuclear Technology*, 162, (June 2008): 379–387.

SCSSINP. 1991. Report by a Commission to the USSR State Committee for the Supervision of Safety in Industry and Nuclear Power (SCSSINP). Causes and Circumstances of the Accident at Unit 4 of the Chernobyl Nuclear Power Plant on 26 April 1986. Moscow.

U.S. NRC. 2011a. U.S. Nuclear Regulatory Commission. http://www.nrc.gov/reactors/operating/licensing/renewal/applications.html.

U.S. NRC. 2011b. Three Mile Island Accident. U.S. Nuclear Regulatory Commission. http://www.nrc.gov/reading-rm/doc-collections/fact-sheets/3mile-isle.html.

U.S. NRC/DOE. 2008. Life Beyond 60 Workshop Summary Report. NRC/DOE Workshop. U.S. Nuclear Regulatory Commission and the U.S. Department of Energy, Bethesda, Maryland.

U.S. President's Commission. October 1979. *The Report of the President's Commission on The Accident at Three Mile Island*. Washington, DC.

Voorspools, K. R., E. A. Brouwers, and W. D. D'haeseleer. 2000. Energy content and indirect greenhouse gas emissions embedded in 'emission-free' power plants: results for the Low Countries. *Applied Energy*, 67 (2000) 307–330, Elsevier Science Ltd.

WDI. 2011. World Bank Development Indicators. Washington, DC. http://data.worldbank.org/country/france.

WEO. 2011. "Lower-Nuclear Case." World Energy Outlook 2011. International Energy Agency, Paris.

Wigeland, R. A., T. H. Bauer, T. H. Fanning, and E. E. Morris. April 2006. Separations and transmutation criteria to improve utilization of a geologic repository. *Nuclear Technology*, 154, no. 1 (2006): 95–106.
WNA. 2005. *The New Economics of Nuclear Power*. A Report by the World Nuclear Association. London.
WNA. 2010. Outline History of Nuclear Energy. World Nuclear Association, London. http://www.world-nuclear.org/info/inf54.html.
WNA. 2011a. Mixed Oxide (MOX) Fuel. World Nuclear Association, London. http://www.world-nuclear.org/info/inf29.html.
WNA. 2011b. The Economics of Nuclear Power. World Nuclear Association, London. http://www.world-nuclear.org/info/inf02.html.
WNA. 2011c. Supply of Uranium. World Nuclear Association, London. http://www.world-nuclear.org/info/inf75.html.
WNA. 2012a. World Nuclear Power Reactors & Uranium Requirements. World Nuclear Association, London. http://world-nuclear.org/info/reactors.html.
WNA. 2012b. Nuclear Power in France. World Nuclear Association, London. http://www.world-nuclear.org/info/inf40.html.
WNA. 2012c. Nuclear Power in Russia. World Nuclear Association, London. http://www.world-nuclear.org/info/inf45.html.
WNA. 2012d. Radioactive Waste Management. World Nuclear Association, London. http://www.world-nuclear.org/info/inf04.html.
WNA. 2012e. Small Nuclear Power Reactors. World Nuclear Association, London. www.world-nuclear.org/info/inf33.html.
Zhang, Z., Z. We, D. Wang, Y. Xu, Y. Sun, F. Li, and Y. Dong. 2009. Current status and technical description of Chinese 2x250 MWth HTR-PM demonstration plant. *Nuclear Engineering and Design*, 239 (2009) 1212–1219, Elsevier.

9

Energy from Nuclear Fusion

Arthur W. Molvik

CONTENTS

9.1 Introduction to Fusion Energy

In this chapter, I present my vision of fusion's potential along with the difficulties that have prevented us from already achieving fusion energy. Fusion should become a major source of energy in the latter half of the twenty-first century—it is a long-term, inexhaustible energy option that offers the potential of generating power in an economically and environmentally attractive system, which is compact relative to renewable energy power plants, and does not emit carbon dioxide (CO_2) or other greenhouse gases (GHGs). The progress made in both magnetic and inertial fusion energy provides confidence that fusion power will be achieved; however, problems remain to be solved, and I suggest pursuing some promising developments; these and other innovations could revolutionize the field.

I see fusion energy as a challenging and rewarding field for students beginning their careers today, as it has been for me. The viewpoints of the students and researchers at the Massachusetts Institute of Technology on this topic and on technical issues for fusion can be found on their website (MIT Issues). While this chapter contains my personal opinions, I make an effort to reference differing opinions too. In the interests of full disclosure, I worked in two areas of fusion energy for at least a decade: magnetic mirrors and heavy-ion inertial fusion; although neither is currently well funded, I remain enthusiastic about their potential as attractive low-activation fusion power plants. I also worked on tokamaks and spheromaks for shorter periods.

In one sense, fusion is the only long-term option other than fission, since the sun, which is the source of all forms of renewable energy, is powered by fusion reactions. Fusion is simple and reliable in star-sized power plants; however, the sun's output is 100 million billion times greater than is convenient for a power plant on earth. (This number can be written more compactly and conveniently as 1×10^{17}. This notation, called scientific notation, is convenient for very large or very small numbers. In this case, the exponent "17" means 17 zeros between the decimal point and the number "1." It could be written out in decimal notation as 100,000,000,000,000,000. To give other examples: if the number were 1.6×10^{-3}, in standard decimal notation it would be 0.0016; or if it were 1.6×10^4, it would be 16,000.)

The small loss of mass when light nuclei fuse into heavier nuclei provides the source of energy. Fusion nuclear reactions occur when two positively charged light nuclei approach closely enough for the attractive short-range nuclear forces to overcome the repulsive electrostatic force between two particles with the same sign of charge. Getting two nuclei this close requires that they have a high energy corresponding to a temperature of about 10,000 electron volts (10 keV $\approx$ 100 million Kelvin = 1.6×10^{-15} J/particle). At temperatures above 0.01 keV, the atoms have such a high velocity that electrons are knocked off by collisions. An atom that becomes charged because of missing, or extra, electrons is called an ion. A mixture of equal numbers of positively charged ions and negatively charged electrons is a fluid, known as a plasma. In addition to needing a high temperature for fusion, the hot ions (and electrons) must be held at a high enough density for a long enough time so that they release more energy through fusion reactions than went into heating them and that they lose by radiation and plasma-particle loss. At densities far below those required for fusion, particles with a single sign of charge (+ or −) develop impractically high electrical potentials (voltages); therefore fusion must employ plasmas for which nearly equal numbers of positive and negative charges result in quasi neutrality and low electrical potentials. Fusion power has been difficult to demonstrate in the laboratory because it is hard to satisfy these requirements of density, confinement time, and temperature.

The difficulty of fusion has a positive side: inherent-passive safety—reactions will not, and cannot, run away. (Passive safety means that an off-normal condition will correct itself, rather than requiring positive action, as in "active safety," to prevent the escalation of an accident.) The fuel in the core of a fusion power plant is sufficient, at most, for only a few seconds of operation, and needs to be continually replenished. Fusion reaction products are nonradioactive, a significant environmental advantage. This distinguishes fusion energy from fission, in which a heavy nuclei splits or fissions into two or more lighter nuclei, that are frequently radioactive. Fission releases energy when a critical mass is gathered into a volume that is small enough so that a neutron released by a fission reaction will cause another fission reaction before escaping from the fissionable material in a chain reaction. (Fission energy is discussed in more detail in Chapter 8.) On the one hand, passive safety is much more difficult to achieve with fission reactors because the energy in the fuel of a fission core is sufficient for about two years of operation. On the other hand, a fission core is conceptually simple, which translates to a low cost of electricity (COE) that will be difficult for fusion to equal.

Several possible fusion fuels exist among light nuclei. These are listed in Table 9.1, along with the reactions. The deuterium–tritium (DT) reaction has a larger cross section (i.e., larger reactivity) at a lower energy than the others do, so it is the nearest-term fusion fuel. (Deuterium is heavy hydrogen that occurs naturally. It has a nucleus consisting of one proton and one neutron; therefore it has a mass twice that of ordinary hydrogen, which has one proton in its nucleus. Tritium has three times the mass of ordinary hydrogen, with one proton and two neutrons in its nucleus. It is radioactive, decaying with a half-life of 12.3 years, so it does not occur naturally and must be produced.) A problem with DT fuel is that 80% of the energy, produced by the fusion of deuterium and tritium, is carried by 14.1 MeV neutrons (we will round this to 14 MeV in subsequent discussions), which can damage the power plant structure and make it radioactive. The remaining 3.5 MeV is

TABLE 9.1

Candidate Fusion Reactions

Fuel	Reaction
D–T	$^2H + ^3H \rightarrow ^1n + ^4He + 17.6$ MeV
D–D	$^2H + ^2H \rightarrow ^1H + ^3H + 4.0$ MeV
	$^2H + ^2H \rightarrow ^1n + ^3He + 3.3$ MeV
D–^{3}He	$^2H + ^3He \rightarrow ^1H + ^4He + 18.7$ MeV
P–^{6}Li	$^1H + ^6Li \rightarrow ^3He + ^4He + 3.9$ MeV
D–^{6}Li	$^2H + ^6Li \rightarrow ^1H + ^7Li + 4.9$ MeV
P–^{11}B	$^1H + ^{11}B \rightarrow 3^4He + 8.7$ MeV

carried by a helium nucleus, ^4He. This is called an alpha particle when it is emitted by a nuclear reaction.

The 14 MeV neutrons damage materials near the reacting plasma, with additional damage to the first wall (the surface nearest the plasma) from the 3.5 MeV alpha particles and plasma losses. The major materials damage effect is analogous to billiard ball collisions—energetic neutrons knock atoms out of their position in the lattice, causing damage to the microstructure. One displacement per atom (dpa) is defined as when, on the average, every atom in a material has been shifted from its position once. Neutrons, at 14 MeV, have a typical range between collisions with material of 0.05 m (in high-density materials such as tungsten) to 0.1 m (in lower-density materials such as water); this results in a shield thickness of 0.5–1 m to protect the components outside a shield from neutron damage. To achieve a several-year lifetime, unshielded materials near the reacting plasma must survive 100–200 dpa.

Neutron reactions with material also generate helium and hydrogen atoms within the material; in certain temperature ranges, these gas atoms can migrate. If they migrate and accumulate at trapping sites, such as grain boundaries; they can weaken the material. Similarly, alpha particles striking the first wall can be implanted in the surface, causing the flaking away of surface layers. Radiation damage to materials can significantly alter materials' properties. Materials issues include hardening, embrittlement, and swelling and are reviewed by Zinkle (2005).

Modeling of materials radiation damage is particularly challenging, covering 10–18 orders of magnitude (each order of magnitude is a factor of 10, so 9 orders of magnitude is a factor of a billion) in length and time scales: from atomic dimensions to structural component sizes, and from damage cascades on 0.1 ns (a tenth of a billionth of a second) time scale to several-year lifetimes. An intense neutron source that can test materials to end of life, and provide experimental benchmarking of computer codes is needed to develop materials for fusion. This will be discussed further in Section 9.1.3.

Anneutronic fusion, using advanced fuels that do not generate neutrons (see the three reactions at the bottom of Table 9.1), would eliminate most materials issues except for energetic ion implantation and flaking of surfaces, and the activation of the thin surface layers near the plasma. This eliminates damage and activation of the power plant structure beyond the first wall. Because all the reaction products are charged particles, this also offers the possibility of directly converting the reaction product energy to electricity by the interaction of the energetic ion reaction products with electric or magnetic fields. Direct conversion can, in principle, achieve efficiency levels greater than those obtained with a thermal cycle. The anneutronic ions, however, have a higher charge, and hence require not only a higher plasma temperature to overcome the electrical repulsive force, but also a greater density-confinement time product to overcome smaller cross sections; both factors increase the difficulty of achieving adequate plasma confinement. Considering the effort that it has taken to approach sufficiently good confinement for DT fusion, it appears prudent to stick with DT and work on the materials issues as well as to look for ways to bypass these issues.

Reactions using D-^{3}He can be anneutronic if the deuterium ions can be kept cold so that they do not react with other deuterium ions. (This notation indicates helium (He) with a mass of 3 atomic mass units. Ordinary helium has a mass of 4.) The deuterium ions naturally collide with hot ^{3}He ions and electrons, which heat them. If confined long enough, all species of ions and electrons would equilibrate to a common temperature. Hot deuterium ions will also fuse with other deuterium ions, producing 2.5 MeV neutrons. These are of lower energy, so are less damaging than 14 MeV DT neutrons, but they still produce neutron damage.

9.1.1 Fuel Resources and Cost

The attractiveness of any energy source is determined initially by its potential, and finally by its performance, in the areas of fuel resources, cost, plant size, safety, environmental effects, land usage, and output products. These areas are discussed below for fusion energy.

Fuel resources of deuterium are effectively unlimited. Water, by weight, contains 75 times as much energy as gasoline for the deuterium–deuterium reaction, even though there is only one deuterium atom for every 6400 hydrogen atoms. (This corresponds to 10 billion years of primary energy at 1990 "burn rate" of 11 TW [1 TW = 10^{12} W or a thousand billion watts].) We expect that power plants using deuterium–deuterium reactions will be feasible in the future; but the first fusion power plants will use DT fuel. For DT, assuming that half of the 17.6 MeV from the reaction is due to deuterium, water, by weight, contains 330 times as much fusion energy as the chemical energy in gasoline. Lithium, the source of tritium for first-generation fusion power plants, is significantly more abundant in the earth's crust than are the fission fuels, uranium, or thorium. However, the isotope, ^{6}Li, which is used to produce tritium, must be separated from the 13 times more abundant ^{7}Li isotope, both being nonradioactive. Lithium is also about 50 times more abundant than uranium in seawater. Developing low-cost methods of extracting materials from seawater would be beneficial to both fission and fusion.

Costs of fusion power are estimated to be in the \$0.04–\$0.10/kWh for 1000 MW_e size plants (Delene et al. 2001; Najmabadi et al. 2006). (These estimates are based on conceptual designs, not detailed engineering. As designs become more detailed, and the cost-estimating algorithms are validated by building near-power-plant size facilities, the cost estimates will become more accurate and certain.) Doubling the plant size to 2000 MW_e can reduce the COE further by about 25% due to economy-of-scale (Logan et al. 1995). Large metropolitan areas could use several power plants of this size range, yielding sufficient redundancy to deal with plant outages with minimal long-distance power transmission. The lower cost range is competitive with other advanced forms of energy, and with coal. However, at present, the lower cost range is populated only by more innovative concepts, ranging from advanced tokamaks with advanced materials to other confinement concepts that involve more unknowns than conventional (well-demonstrated) tokamaks, which are predicted to occupy the higher cost range. Researchers are achieving some success in reducing the costs with innovative concepts, by simplifying old concepts or devising cheaper manufacturing techniques to reduce capital costs, and by introducing concepts that reduce the downtime for periodic maintenance.

However, we note that estimating the future costs of electricity is fraught with uncertainties, even for well-established technologies where we know current costs accurately. This is because we do not know when, or if, carbon sequestration will be required for fossil fuel plants, whether nuclear (and fossil fuel) waste disposal/recycling costs will be included, whether power plant decommissioning costs will be included, and whether infrastructure costs—such as natural gas pipelines—will be included.

9.1.2 Safety and Environment

Fusion power plant concepts are passively safe. Since the amount of fuel within a fusion power plant will not sustain operation for more than a few seconds at most, there is minimal stored nuclear energy available to trigger severe accidents. If we show that they meet "no-evacuation in case of an accident" requirements, then we could site them in highly populated areas and tap the market for heat to drive industrial processes or even sell low-grade heat for space-heating applications. We expect that these stringent safety constraints can be met by fusion power plants; however, it is not a no-brainer; as we point out below, there are several issues to consider that will require innovative and thorough engineering.

To achieve these strong safety characteristics and ensure passive safety, all sources of stored energy (chemical, magnetic, thermal, etc.) must be contained so that they cannot drive an accident; and materials need to be chosen that minimize activation of the power plant and its subsequent afterheat. It is particularly important to minimize the volatile hazardous materials that could be released in an accident, especially gases containing tritium. The tritium handling systems must be carefully engineered to keep tritium loss to extremely low levels in either routine operation or under any conceivable accident scenario. Furthermore, as emphasized by the effects of the 2011 tsunami in Japan, we must be imaginative in conceiving of all possible accidents. While first-generation power plants using DT must contain the tritium, the reaction products are not radioactive. The materials forming the containment vessels must themselves be shielded from radiation to prevent the degradation of their properties. With careful and thorough design and construction, probably including secondary containment structures surrounding the first, radioactive releases are not expected to be a problem. If low activation can be achieved, this also minimizes after-heat issues and eliminates the possibility of needing emergency cooling systems. These issues address local safety, in the vicinity of a power plant.

Fusion is also attractive on the global safety issue: nuclear weapons proliferation—for pure fusion, this can be minimized by careful controls on the tritium inventory, and by ensuring that fissionable materials are not being exposed to the neutron flux. If, however, the world is using a significant amount of fission energy, then a fission–fusion hybrid power plant could be a highly effective fission breeder (MIT Fusion–Fission Research Workshop 2009). A hybrid would consist of a fissionable blanket surrounding a fusion core, to take advantage of the copious neutron production from fusion. Hybrids have been discussed for three purposes: burning fission waste, to reduce the amount that needs geological long-term storage; electrical power production; and fissionable fuel production. Each hybrid fuel-production plant could supply fuel for several nonbreeder fission power plants, as well as generating power. (Fission proponents claim to be able to accomplish similar goals, except for the high rate of breeding, if allowed to reprocess fuel [MIT Fusion-Fission Research Workshop 2009]; to date, no detailed study of this has been done with both fission and fusion proponents.) Preventing nuclear weapons proliferation becomes more difficult than in a pure fusion scenario, but is similar to the safeguards needed for fission power plants.

Reprocessing of the fissionable materials is likely to be required if either fission or fusion power plants are used to burn radioactive waste. Reprocessing of fissionable fuel is currently forbidden by regulations in the United States. If reprocessing were allowed, then further care in design and operation would be required to minimize proliferation risk. The current U.S. policy of passing fissionable fuel once through a light-water reactor, then disposing of it was initiated to minimize the proliferation risk (for a defense of the current U.S. policy, and issues that must be considered before changing it, see Fetter and von Hippel 2005); however, the rest of the world has not followed the United States on this. The result is that the United States is losing influence in the fission energy field as other countries are selling more reactors. This policy eventually becomes even more counterproductive: the highly radioactive fission products, which self-protect burnt fuel, decay to low levels after a few centuries. Then, the separation of plutonium becomes relatively safe and simple. The current policy for fission fuel, one-pass-through a light-water reactor burns less than 1% of the energy contained in the fuel, the remainder is waste—adding to waste disposal problems that are not yet solved, rather than generating more power from it. The current U.S. policy passes these problems along to future generations.

Fusion, as well as fission, power plants should be constructed from a few standardized designs that are engineered to make diversion of radioactive, especially fissionable, material very difficult; and all nuclear power plants and reprocessing plants should be operated under international supervision to further minimize proliferation risks.

9.1.3 Radioactive Waste

Fusion is environmentally attractive due to no CO_2 or other GHG production, small land use compared with renewable energy, and nonradioactive reaction products. The major waste issues to be addressed are the activation (radioactivity) level, volume, and lifetime of the power plant chamber components under bombardment by 14 MeV primary neutrons from DT fusion, plus bombardment by scattered neutrons that are degraded in energy. These issues are interconnected: reducing the volume exposed to neutrons will reduce the amount of radioactive waste, doubling the lifetime of a component will halve the waste production (Fetter 1990). Conceptual studies of tokamak-based fusion power plants have emphasized reduced fusion waste production, and succeeded in reducing it by a factor of 2 (El-Guebaly et al. 2008). (Waste and proliferation issues for fission energy were discussed in Section 8.5.4 of Chapter 8.)

There are some ways to minimize fusion waste and its effects:

- First, much of the waste is only slightly activated, with radioactivity levels so low that these materials can be released for unrestricted use in consumer products or disposal in nonnuclear landfill; this is referred to as "clearance." Approximately 80% of fusion waste and near 95% of fission waste qualify for clearance (El-Guebaly et al. 2008).
- Second, most of the radioactivity is in structural materials, which can be chosen for lower activation and shorter half-lives, among the other important qualities. Biohazards in particular can be minimized; fusion can avoid materials such as strontium, cesium, and iodine that are unavoidable fission products.

In addition, other innovations are being investigated to significantly reduce the disposal of radioactive waste from fusion energy:

- First, the industrial ecology concept of recycling materials and components, rather than disposing of them (Frosch 1992), is beginning to be applied to fusion power plants, and promises further reduction in waste volumes (El-Guebaly et al. 2008). Extensive recycling requires the use of relatively low-activation materials; "low-activation" in this context means that remote handling and accident containment and cleanup are feasible during reprocessing; or better yet, that activation is so low that hands-on manipulations are feasible during reprocessing. It is not necessary to separate all activated materials during reprocessing; for example, replacement components for use in radioactive environments could use activated materials. Wherever possible, highly activated materials with inconveniently long half-lives would be separated for specialized uses or disposal. Recycling has the potential of diverting and re-using much of the waste stream, greatly reducing the amount of waste that requires disposal.

- Second, it is important to develop the facilities to irradiate materials with a fusion spectrum of neutrons (a continuous distribution of neutron energies up to, but not exceeding, 14 MeV). Such facilities are needed to measure the damage and activation of candidate materials and to develop long-lifetime low-activation materials and concepts (ReNew Report 2009, Thrust 14). Neutron sources and fusion materials science and development in the future will need to be funded at a level comparable to the confinement program, to develop and qualify materials for fusion power plants within the next few decades. The first materials to be qualified may not be low activation, but developing and qualifying low activation materials will be a long-term goal.

The International Fusion Materials Irradiation Facility (IFMIF; also see the website) is an accelerator-driven neutron source; with intentions, but no commitment to construct it. The IFMIF will provide power plant levels of neutron flux at an average energy of 14 MeV, but with some neutrons with energy up to 30 MeV. Plasma-based neutron sources can provide similar neutron flux with a neutron energy spectrum that closely approximates that of a fusion power plant (Fischer et al. 2000). Plasma-based neutron sources provide a high flux of neutrons over a larger area, so in addition to testing materials, they can also test subcomponents of tritium-breeding blankets in the case of a magnetic mirror neutron source (Fischer et al. 2000), or they can be predominantly designed for testing components in the case of a tokamak-based Fusion Nuclear Science Facility (FNSF), called for in the ReNew Report (2009), Thrust 14, two candidates for which are described by Peng and Stambaugh (2012) or Stambaugh (2009).

- Third, an innovative method, proposed to reduce materials damage and radioactive waste, is the use of thick (0.5–1 m) liquid walls or jets of a low-atomic-number material (Moir and Rognlien 2007; Moir 1997). This is not compatible with all fusion power plant concepts. It will be discussed further in Sections 9.2.4 and 9.3.5.

9.1.4 Land Use and Fusion Products

The high power density of fusion requires little land for the plant, compared with the requirements for the various forms of renewable energy. The space occupied can be made even smaller if we bury the larger elements deep enough for roads and agriculture to exist above it. This has been done at major high-energy physics accelerators, such as the Stanford Linear Accelerator (SLAC) near Palo Alto, California, a 3-km-long linear accelerator. It currently drives light sources for studies in life sciences, materials, environmental science, and accelerator physics. Similarly, the European Laboratory for Particle Physics (CERN) accelerator complexes, including the Large Hadron Collider (LHC), with a circumference of 27 km, near Geneva, Switzerland, are underground. Such dual use of land will become increasingly important as population growth results in an increased need to use all the available land for growing food and minimizing that devoted to energy generation. (For that reason, I believe that most forms of renewable energy can produce a significant fraction of our energy needs for only a few decades, unless population growth stops or reverses. Photovoltaic cells are one exception; they can be used to cover roofs that do not have any food-growing applications. Other views on land use issues can be found in Section 3.5.3 of Chapter 3 and Chapter 7.)

Electricity and heat are the primary products of fusion power plants. The heat can have direct value in industrial processes, or space heating for cities—if safety constraints can be met, as we think is possible. Locating fusion power plants near the largest electricity and heat consumers will reduce costs by minimizing electrical transmission costs in addition to increasing the income through the sale of heat. Other products are also possible. The power can be applied to generating hydrogen or other synthetic fuels for efficient long-distance power transmission through pipelines and for portable needs such as transportation, but competing in this area requires very low COE (Logan et al. 1995). Like fission, fusion is most cost effective when operated at a constant power level, although power variation within a limited range is possible with many fusion concepts. Fusion is appropriate for recharging hydropower or other energy storage during off-peak demand periods. Radioisotope production and the transmutation of fission waste would take advantage of the intense flux of neutrons available, but would also raise issues such as nuclear proliferation and possible additional activation of the fusion plant.

9.1.5 Approaches to Fusion Energy

A generic fusion power plant, Figure 9.1a, consists of a central reaction chamber (the core), surrounded by an approximately 1-m-thick blanket that contains lithium. The blanket performs three functions: it generates the tritium fuel from neutron interactions with the lithium; it converts the neutron energy to heat that generates electrical power through a heat cycle, such as a steam turbine; and it shields the surrounding structure from neutron bombardment.

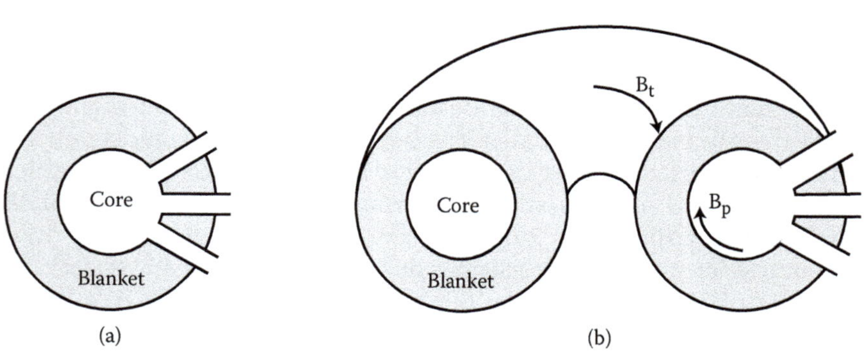

FIGURE 9.1
(a) Generic fusion core with blanket/shield. (b) Toroidal fusion device showing direction of toroidal and poloidal magnetic fields, B_t and B_p, respectively.

Penetrations through the blanket must be provided for injecting fuel and energy with either magnetic fusion energy (MFE) or inertial fusion energy (IFE). These penetrations may be from one side as shown, but more generally are from opposite sides or even uniformly distributed around the periphery. Outside the blanket are superconducting magnets for magnetic fusion, and beam lines for inertial fusion. The power plant in Figure 9.1a could be spherical, cylindrical, or wrapped into a toroid, as shown in Figure 9.1b.

Historically, two approaches to harnessing fusion power for energy production have been followed. The first is MFE, for which we create "magnetic bottles" that need to hold the plasma for a time of order 1 s (Chen 1974; Sheffield 1994). The best-known example is the tokamak, a toroidal geometry. (A toroid, or torus, is a donut-shaped configuration, which has come to be preferred for most magnetic fusion applications, because magnetic field lines can encircle the toroid nearly endlessly without intercepting a wall. Plasma ions and electrons flow along magnetic field lines much more easily than they flow across them, so plasma loss rates to walls are reduced in toroids.) The second approach is IFE, which today uses powerful lasers, but may use particle beams or x-rays in the future, to compress a millimeter-sized capsule of fusion fuel to the high densities and temperatures needed for fusion to occur (Lindl 1998). For IFE, the confinement time is the time for the capsule to blow itself apart, of order nanoseconds (10^{-9} s or one billionth of a second), hence the name "inertial" fusion.

The progress in MFE by the 1980s was sufficient to initiate the design of the International Thermonuclear Experimental Reactor (ITER), shown in Figure 9.2. It is a joint project between the United States, Russia, Europe, Japan, China, India, and South Korea. ("Iter" in Latin means "the way," see the ITER website. For more extensive information on MFE programs, see the FIRE website.*) ITER, a large tokamak, is intended both to study plasmas producing fusion power greater than the external power used to heat the plasma, with the fusion gain predicted to reach about 10, and to begin studying power plant engineering. Construction is underway in southern France, with the first plasma operation scheduled for 2019, and the DT operation beginning in 2028.

Ignition in MFE means that heating of the plasma by alpha particles is able to completely replace heating by other means that we will discuss. We may choose to operate MFE power plants slightly below ignition, effectively as power amplifiers with gains in the range of 20, so that we can vary the input heating power to control the plasma parameter profiles and to regulate the output fusion power.

Experiments to achieve inertial fusion ignition and energy gain of 1–10 are in progress at the National Ignition Facility (NIF), where up to 1.8 MJ of 0.35 μm wavelength laser light from 192 beams are focused into a target, as shown in Figure 9.3. Ignition is the scientific demonstration of inertial fusion feasibility.

* The Fire-Place website, in addition to providing information on the FIRE burning plasma concept, is perhaps the most complete single website with information on magnetic fusion energy progress and issues, http://fire.pppl .gov/.

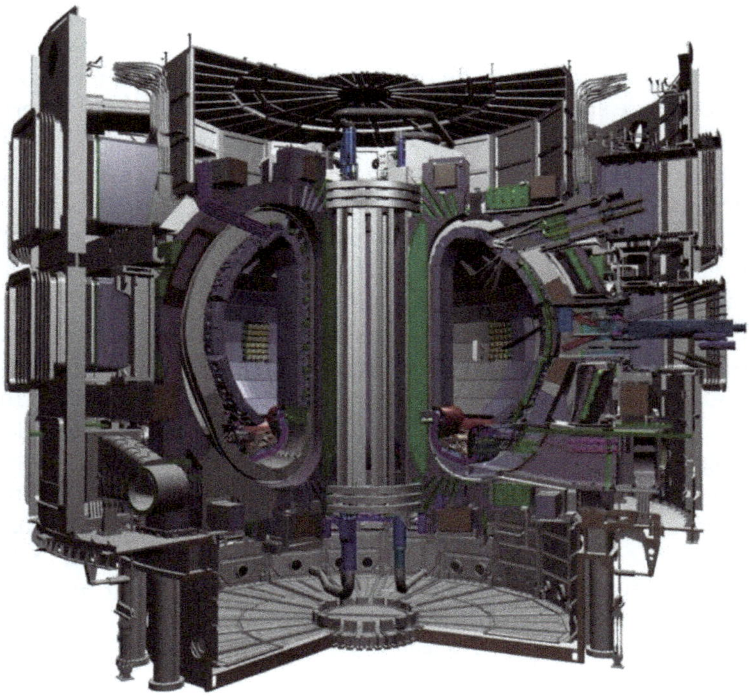

FIGURE 9.2
ITER is shown with its current design (as of 2011).

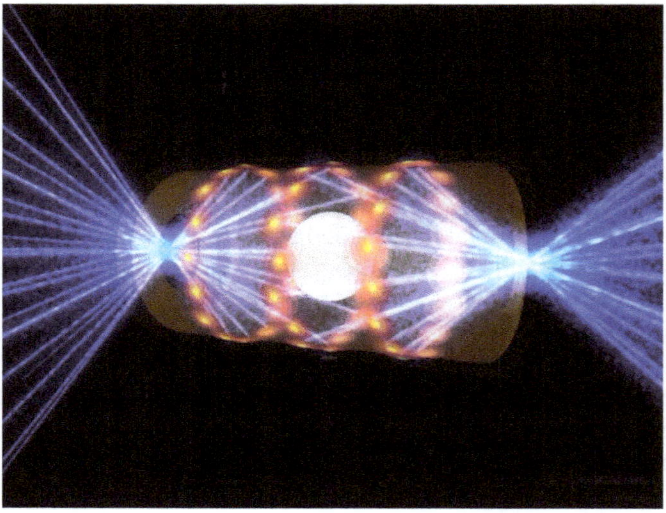

FIGURE 9.3
Artist's conception of an NIF ("Hohlraum"). The scale can be obtained from the white, 2-mm-diameter spherical pellet in the center. Laser beams enter from either end. (Courtesy of Lawrence Livermore National Laboratory.)

Ignition in IFE means that the 3.5 MeV alpha particles from fusion reactions deposit enough energy in nearby fuel to heat it to fusion energies, analogous to heating a small portion of firewood hot enough to ignite and burn. Energy gain is the ratio of fusion energy out to laser energy in. Studies are in progress to exploit success in NIF, termed LIFE, for laser inertial fusion energy (see the LIFE website* for LIFE and NIF information). Inertial fusion power plants typically pulse about 10 times per second; significant engineering development is required to get there from NIF's few shots per day. This development includes not only repetition-rated lasers, with efficiency approaching 10% or better rather than NIF's 1% and high-gain targets that yield more than 100 MJ per pulse rather than the approximately 10 MJ predicted for NIF's first ignition experiments, but also new technologies—for example, clearing shot-debris from the fusion chamber within 0.1 s so it does not interfere with the next shot.

These approaches are at roughly equivalent levels of development: Although IFE ignition experiments have begun about 17 years ahead of the MFE schedule, the engineering and technology of present magnetic fusion experiments is much closer to that needed for power production than is the case for IFE, and ITER is much closer to a prototype fusion power plant than is NIF, so the time to develop MFE and IFE power plants is likely not that different (see the LIFE website for a more optimistic viewpoint of IFE [LIFE Website entry for December 2011 lists 7E14 neutrons from a 1.4 MJ laser shot in 12/11. With 14 MeV neutrons, this corresponds to fusion energy of about 1E-3 of laser energy.]). NIF reached fusion energy gain of 10^{-3} at the end of 2011 (see the LIFE website), and hopes to reach an energy gain of 10 within a few years. Magnetic fusion has demonstrated approximately as much fusion power produced as heating power injected into the plasma in the Joint European Torus (JET) tokamak (Watkins et al. 1999), and equivalent confinement in the JT-60 tokamak in Japan (Kishimoto et al. 2005).

9.2 Magnetic Fusion Energy

9.2.1 Basic Principles

Magnetic fusion energy is based on confining hot, reacting plasmas in magnetic bottles. In this section, we will outline the basic physics of magnetic confinement to give a feeling for some of the significant phenomena. (Students without a physical science or engineering background can skim or skip this section.) Equations that quantify these principles will be presented and discussed, but will not be derived here; derivations can be found in a plasma physics textbook (e.g., Chen 1984). We use SI units in all equations; however, results will frequently be quoted with other units that are convenient for

* The LIFE website has tabs for NIF laser facility and for LIFE inertial fusion energy, https://lasers.llnl.gov/about/missions/energy_for_the_future/life/.

fusion energy purposes, and which will be defined, with appropriate conversion factors to SI and other familiar units. Vectors are indicated by bold face; they are defined to have both magnitude and direction. Non-bold face representation of a vector quantity indicates that only its magnitude, not its direction, are used in that equation.

Magnetic bottles to confine hot plasmas are possible because charged particles are bent by magnetic fields; with sufficiently high magnetic fields, charged particles spiral around field lines. The force on a charged particle in electric (**E**) and magnetic (**B**) fields is given by

$$F = q(\mathbf{E} + \mathbf{v} \times \mathbf{B})$$

That is, a particle, with charge (q), is accelerated in a direction that is parallel to an electric field and mutually perpendicular to its velocity (**v**) and the magnetic field direction (which is indicated by the "cross-product" symbol "×"). This results in particles gyrating around magnetic field lines, with a gyroradius (ρ) given by

$$\rho = \frac{mv}{(qB)}$$

where m is the particle mass. Ion masses greatly exceed electron masses (by 1837 for hydrogen and by near-integral multiples of this for heavier hydrogen isotopes and other ions), therefore the ion gyroradius is much larger than that for electrons. For good confinement, the radius of the magnetic bottle should be many ion gyroradii across.

In a straight, uniform magnetic field, with no electric field, ions and electrons have a uniform velocity along the magnetic field. If the magnetic field strength varies, we find that the magnetic moment μ is a conserved quantity ("conserved" means it remains constant for a given particle),

$$\mu \equiv \frac{1/2\ mv_{\perp}{}^{2}}{B}$$

where the total particle energy (another conserved quantity) is the sum of the perpendicular and parallel (to the magnetic field) energies,

$$\frac{1}{2}mv^2 = \frac{1}{2}mv_{\perp}^2 + \frac{1}{2}mv_{\parallel}^2$$

As the magnetic field B increases along a particle trajectory, conservation of μ requires that the particle energy, perpendicular to the magnetic field, increases linearly with B; for sufficient increase in B, the perpendicular energy increases enough to become equal to the total energy of the particle, then the parallel energy goes to zero and the particle reflects. This provides what is known as a magnetic mirror, which can be used as a confinement mechanism; it also separates particles into trapped and passing particles in toroidal devices where the magnetic field is bent into a circle. Trapped particles bounce back and forth between mirrors, within a fixed azimuthal range, and passing particles encircle the torus. Toroidal configurations were motivated by the goal of eliminating plasma loss of passing particles along field lines, by keeping the magnetic field lines within the device.

So far, our discussion dealt with the behavior of individual charged particles. Now we begin to consider plasma effects, in which the positively and negatively charged particles in the plasma can act together like a fluid. Bending the magnetic field to create a toroidal configuration causes the magnetic field magnitude to vary with $1/R$, where R is the major radius of the toroid. This variation, or gradient, in a magnetic field causes particles to drift along constant-B surfaces, with the positive and negative particles drifting in opposite directions. This partially separates the ion and electron charges at the top and bottom of the plasma, producing a vertical electric field, **E**, across the plasma, which causes the entire plasma to drift outward at a velocity $\mathbf{v}_{\mathbf{E}\times\mathbf{B}}$ given by

$$\mathbf{v}_{\mathbf{E}\times\mathbf{B}} = \frac{\mathbf{E}\times\mathbf{B}}{B^2}$$

This was understood early in the fusion program, and various methods were invented to short-out this charge separation; most methods cause magnetic field lines to follow a spiral path, which then connect top to bottom. In other words, the field lines wrap around the minor radius in the poloidal direction (B_p direction in Figure 9.1b) while wrapping around the major radius in the toroidal direction (B_t direction in Figure 9.1b). This spiral path is termed "rotational transform." A rotational transform can be accomplished by driving current around the toroidal direction; this current produces a poloidal magnetic field; such a device with a high toroidal field, in which the field makes less than one poloidal rotation for each toroidal rotation, is known as a tokamak. Integer rotational transforms occur when field lines close on themselves by making one complete poloidal rotation after one or more toroidal rotations. Instabilities reduce performance for integer rotational transforms of one to three; field line closure prevents field lines from covering an entire shell and shorting out all charge separations. The other extreme, has a much lower toroidal magnetic field so that the magnetic field makes multiple poloidal rotations for each toroidal rotation. This is known as a reversed-field pinch (RFP) and has produced impressive results that are discussed by Chapman. A rotational transform can also be accomplished with external coils to produce the poloidal magnetic field; this is known as a stellarator or heliotron (Sagaro et al. 2010).

The radial pressure gradient of the plasma in a magnetic field drives the drift current in the poloidal direction, which provides radial confinement from the $\mathbf{j} \times \mathbf{B}$ force:

$$\mathbf{F}_{\mathbf{j}\times\mathbf{B}} = \mathbf{j}\times\mathbf{B}$$

A density decreasing outward from the minor axis generates a current in a direction to apply an inward force that restrains the plasma from expanding in the radius; thereby confining it. It is still subject to other processes that can cause smaller regions of plasma, or individual particles, to move outward.

9.2.2 Equilibrium, Stability, and Confinement

Plasma equilibrium and stability was an early challenge to confining plasma. Toroidal devices, with few exceptions, achieved MHD equilibrium by the rotational transform where magnetic field lines spiral around the minor radius, and achieved MHD stability by shear where magnetic field lines spiral at different rates at adjacent radii. This results in nonparallel adjacent magnetic field lines, which prevent flux tubes from interchanging because they would become tangled. Early axisymmetric magnetic mirrors were plagued by plasma moving radially nearly as rapidly as it could escape out the ends. Theory identified this as an MHD instability caused by the magnetic field decreasing radially from the axis; magnetic flux tubes filled with plasma could find a lower energy state by interchanging with empty flux tubes at a larger radius. This is analogous to the instability of a ball on the top of a round hill—with any small disturbance, the ball will roll off. This stability problem was solved by the invention of minimum-B magnets, which produced magnetic fields that increased radially as well as axially. This was stable because the lowest energy state was on the axis, analogous to a ball at the bottom of a bowl. In the last two decades, a number of methods have been devised to achieve stability with axisymmetric magnetic fields (Ryutov et al. 2011). This provides advantages that will be discussed further in Section 9.2.5.

Before we can have stability, we must have equilibrium. Equilibrium in nature generally requires uniform density and temperature and isotropic velocities, without gradients in any of the parameters; a plasma confinement device must violate some of these equilibrium constraints to isolate hot, dense plasma away from the much colder walls. These violations cause problems. For example, pressure gradients lead to drift waves, which can be unstable under certain conditions. Gradients in the electron and ion temperatures lead to instabilities that have been receiving considerable attention in recent years. Tokamaks are subject to current-driven instabilities. Magnetic mirrors are subject to loss-cone instabilities, driven by the anisotropy of ions: few ions have most of their velocity parallel to the magnetic field because such ions are unconfined. None of these instabilities are disastrous, but keeping them sufficiently suppressed places restrictions on the operating parameters, generally reducing performance.

The fusion power P_F in watts is given by

$$P_F(\mathrm{W}) = n_D n_T < \sigma_{DT} v > (17.6 \times 10^6\ \mathrm{eV/fusion})(1.6 \times 10^{-19}\ \mathrm{J/eV})\ V(\mathrm{m}^3)$$

and the power plant quality factor (or power gain) Q is given by

$$Q = \frac{P_F}{P_{\text{wall plug}}}$$

where the density per cubic meter of deuterium is given by n_D and of tritium by n_T; the effective area in square meters of deuterium and tritium ions for collisions that are close enough to fuse (termed the cross section) by σ_{DT}; the ion velocity by v; constants give the energy per fusion event in electron volts, then convert that to Joules to give the fusion power in watts, and the plasma volume. We see that the fusion power increases with the product of the ion densities. The wall-plug power is the power that is used to operate and heat the fusion power plant; this is known as the engineering Q, the value that an electrical utility would want to know. It should have a value near 20, or more. The scientific Q is obtained by replacing $P_{\text{wall plug}}$ with the power to heat the plasma, this quantity becomes infinite in an ignited MFE plasma.

Impurity ions must be kept to low levels, to prevent dilution of the DT fuel, which reduces fusion power. High-atomic-number elements are especially undesirable: if they are dislodged from the walls and enter the hot plasma, not only do they dilute the ions more severely by becoming multiply ionized, but if not completely ionized with all electrons removed from orbiting the nucleus, the electrons still trapped by the nucleus radiate energy copiously, which has a strong cooling effect on the free plasma electrons.

The increase of the fusion power P_F with ion density suggests increasing the ion density to increase the fusion power per unit volume. However, as we increase the ion density, at the required temperature of at least 10,000 eV, the pressure of the plasma increases. The plasma pressure is the sum of the pressures of deuterium, tritium, and impurity ions plus the electron pressure.

$$P_P = \frac{3}{2}\,(n_D k\,T_D + n_T k\,T_T + n_{im} k\,T_{im} + n_e k\,T_e)$$

where P_P is the plasma pressure, k the Boltzmann constant, and T the temperature. (The subscripts indicate the particle species. T as subscript denotes tritium.) We define beta, β, as the ratio of the plasma pressure, P_P, to the magnetic field pressure:

$$\beta = \frac{P_P}{(B^2/2\mu_0)}$$

where the magnetic field strength B is measured in Tesla, and μ_0 is a constant $\mu_0 = 1.26 \times 10^{-6}$ T m/A. Magnetic bottles confine plasma only up to a limiting β, before they "rupture"; like glass or plastic bottles, they work only below a limiting pressure. The magnitude of β_{lim} is usually in the range of 0.01–1.0, depending upon the details of the magnetic field configuration. If we double the magnetic field for fixed β_{lim}, we can increase the plasma pressure, and therefore the plasma density, by a factor of 4 (because β scales inversely with the square of the magnetic field), and the fusion power increases by a factor of 16.

Magnetic bottles to produce 1000 MW (a typical conventional power plant size) of fusion power are large, with a volume of approximately 500 m^3. If this were in the shape of a cube, it would be quite large—7.9 m (26 ft) on a side. The size is set by the allowable wall loading at the first wall and at the tritium-generating blankets. The wall loading is the fusion power per square meter (typically a few MW/m^2), which depends on the materials chosen, and how they are cooled. Other limits to power production must be adjusted to reach the wall-loading limit, if possible. These include plasma pressure limits, which are adjusted by the details of the magnetic field shape, as well as by the field strength (within limits of superconducting materials and the strength of materials that constrain the conductor winding to staying in the proper place). Other limits also need to be considered, such as the total power density to any plasma strike points, as well as any other structure exposed to high neutron flux.

Maximizing the effectiveness of magnetic bottles, that is, achieving good confinement of plasmas, requires minimizing the transport of plasma to the walls. The minimum radial transport rates arise from classical diffusion, which is caused by ion–electron collisions that produce step sizes of a gyroradius. Faster diffusive processes will dominate over classical diffusion in all configurations, with the possible exception of two highly symmetric long-cylinder configurations: axisymmetric mirrors with axial magnetic field predominating, and field-reversed configurations (FRCs) with poloidal field predominating. (Both configurations will be discussed in more detail in Section 9.2.5.) In toroids—or minimum-B linear—geometries, trapped particles drift on different surfaces that are separated by many gyroradii between clockwise- and counterclockwise-drift directions. If we project the particle positions onto the plasma cross section at one toroidal angle, we find that it maps out a banana-like shape; hence, this is frequently termed "banana-diffusion," or more generally, neoclassical diffusion. Neoclassical diffusion occurs at the same collision frequency as classical diffusion, but the radial step size is much larger, resulting in more rapid plasma loss. Even so, it is rare to obtain plasma confinement that is limited by neoclassical diffusion; other effects such as electron temperature gradient or ion-temperature gradient instabilities frequently increase the plasma radial transport rates above that of neoclassical (or classical in axisymmetric linear configurations).

Understanding and reliably predicting plasma confinement is a long-term goal in the fusion program. Until that is achieved, empirical scaling laws from a variety of existing facilities provide guides to plasma confinement in planned facilities. Computer simulations are providing a second method of projecting the confinement of future facilities. Due to the large number of interacting processes in plasmas, this is a very complex undertaking, which is making progress as computational power is increasing and as the detailed physical understanding of individual processes is being developed in current experiments to benchmark the simulations.

9.2.3 Fueling and Heating

Plasmas must be fueled and heated to reach and be sustained at power plant level parameters. Fueling can be as simple as injecting gas at the plasma edge; as gas penetrates the plasma, it becomes ionized. The major fueling issue is to penetrate deeply enough to fuel the plasma core; gas fueling has proven inadequate on large tokamaks. Pellets of frozen hydrogen isotopes can be accelerated to hundreds of m/s with gas guns or centrifuges, and can penetrate much further than gas, so have become the standard fueling technique for present and future large tokamaks.

To attain steady-state plasma parameters with steady-state heating and fueling of plasmas, one must also remove power and particles, usually from the edge, at the same rate that they are injected. In particular, it is desirable to remove helium ash; otherwise it will dilute the DT fuel, reducing the fusion power. But it would be preferable, if possible, to remove helium ash from fusion reactions only after it has cooled—giving up its 3.5 MeV of energy per alpha particle to heating the plasma. Pumped diverters enable particles to be removed without an equal number of particles returning to the plasma, which is otherwise typical of losses to walls or unpumped diverters. However, the diverter area is a small fraction of the wall area, so the power density there reaches levels that can be difficult to handle.

Heating is done by a number of techniques. Tokamaks usually start up by inductively driving the toroidal voltage (with a transformer), which breaks down the background gas and drives a toroidal current. This ohmically heats the electrons to a few keV where the heating plateaus, because the plasma resistivity decreases with increasing electron temperatures. Other heating techniques are necessary to reach the required 10 keV range for ion and electron temperatures.

The most reliable heating technique is neutral beams: megawatt-level high-current (tens of amperes) ion beams are accelerated to hundreds of keV, passed through a gas cell where a significant fraction of the ions charge exchange with gas (i.e., pick up electrons from the gas) to become neutrals. These energetic neutrals freely cross the magnetic field until they reach the hot plasma, where collisions with ions and electrons ionize the neutrals. The resulting hot ions are then trapped by the magnetic field and heat the plasma through collisions with cooler ions and electrons. Positive ions are used for energies up to 150 keV (with deuterium or tritium). Beyond that energy, charge exchange cross sections become too low for a gas cell to efficiently neutralize a positive ion beam; but negative ions can be neutralized with efficiencies exceeding 50%, at the expense of more complex and lower current-density ion sources. Negative ion neutral beams are planned for future tokamaks, to inject at 1 MeV.

Neutral beam technologies are complex and expensive, and require a line of sight to the plasma, but they have become highly reliable. The physics of trapping and heating is simple and can be reliably computed. Besides heating, neutral beams can drive current in a tokamak, and can also provide fueling.

Radiofrequency, rf, waves are also used for heating and driving current in plasmas. The most common frequencies correspond to the ion and electron cyclotron frequencies, ω_{ic} and ω_{ec}

$$\omega_{ic} = \frac{qB}{m_i} \quad \text{and} \quad \omega_{ec} = \frac{qB}{m_e}$$

respectively, and their harmonics, and lower hybrid, ω_{lh}, which occurs at $\omega_{lh} = (\omega_{ic}\omega_{ec})^{0.5}$. The frequency in the more familiar unit of hertz, which is an rf cycle per second, can be obtained from ω by dividing by 2π. Ion cyclotron heating typically involves frequencies in the range of tens to hundreds of megahertz, lower hybrid 0.1–10 gigahertz, and electron cyclotron heating tens to hundreds of gigahertz, where the prefix giga represents a billion. The physics of rf-heating is more complex: an antenna must be able to launch, that is, couple to waves that can propagate in plasma near the antenna, the waves must propagate to the region where heating is desired, and there they must couple to cyclotron or other resonances of particles, or at some location, they must mode-convert to another type of wave that will heat ions or electrons at the desired location. RF can also be used to drive the current. RF technology is somewhat less complex than neutral beam technology, and does not require line-of-sight ports; however, the physics is more complex, and rf has not always yielded the anticipated results. Also, except for electron cyclotron heating, antennas must be near the plasma and are subject to damage.

9.2.4 Magnetic Confinement Concepts—Tokamaks

Many concepts in magnetic fusion have been studied during the last six decades, with a strong emphasis on the tokamak during the last four decades. The tokamak consists of an externally applied toroidal magnetic field B_t, and a toroidal plasma current that creates a poloidal magnetic field, B_p; see Figure 9.1b. A transformer initially drives the plasma current in nearly all tokamaks, with auxiliary current drive by radiofrequency waves or neutral-atom beams on many tokamaks. Additional externally applied magnetic fields provide vertical fields for plasma-position control and shaping, and can produce diverter geometries, where the outermost magnetic field lines strike a diverter plate to allow the removal of power and impurities from the plasma edge.

Within the past two decades, innovations in tokamaks have dramatically increased their ability to confine plasma and to make them more desirable as potential power plants, as summarized below.

1. Until the 1980s, tokamaks were inherently pulsed devices. Now, potentially steady-state operation has been demonstrated with the current driven by neutral beams or rf. Current drive efficiencies are adequate when most (perhaps 80%–90%) of the current is supplied by a bootstrap effect in which the outward diffusion of plasma density, peaked on the axis, drives the required currents.
2. Higher-β operation has been achieved by using vertically elongated plasmas, with sharper bends at upper and lower corners.
3. Confinement and β have been increased by discovering methods of producing transport barriers, which reduce plasma particle or energy radial losses, thus enhancing confinement. Transport barriers typically enable a large increase in density or temperature across the barrier; the latter is analogous to insulation in a thermos bottle.
4. About 16 MW of fusion power has been achieved in the European JET facility (Watkins et al. 1999), almost equal to the 18 MW of heating power.
5. The density–time–temperature product has increased by more than 100 million times in 30 years to 1.5×10^{21} $m^{-3}s^{-1}$keV in the JT-60U in Japan (Kishimoto et al. 2005). If the JT-60U operated with DT, this confinement would be sufficient to produce fusion power equal to 1.25 times the heating power.
6. Studies have identified conditions leading to disruptions. In a disruption, the plasma current decays much faster than usual, and exerts large sudden forces on the walls, which could result in structural damage. The disruption energy comes from that stored in the poloidal magnetic field and the plasma thermal energy. Understanding developed through experiments and theory has enabled reducing the occurrence rate of disruptions. To mitigate against residual disruptions, and to reduce the generation of runaway (high-energy) electrons associated with disruptions, techniques such as massive gas injection have been developed.
7. The Tore Supra facility has maintained hot plasma discharges for 6.5 min, and has the potential of operating for 1000 s. Pumped diverters have removed power and particles from the core plasma without refluxing impurities back into the core plasma.

Conceptual fusion power plant studies, such as the ARIES-AT, given in Figure 9.4, apply these innovations to improving the power plant core

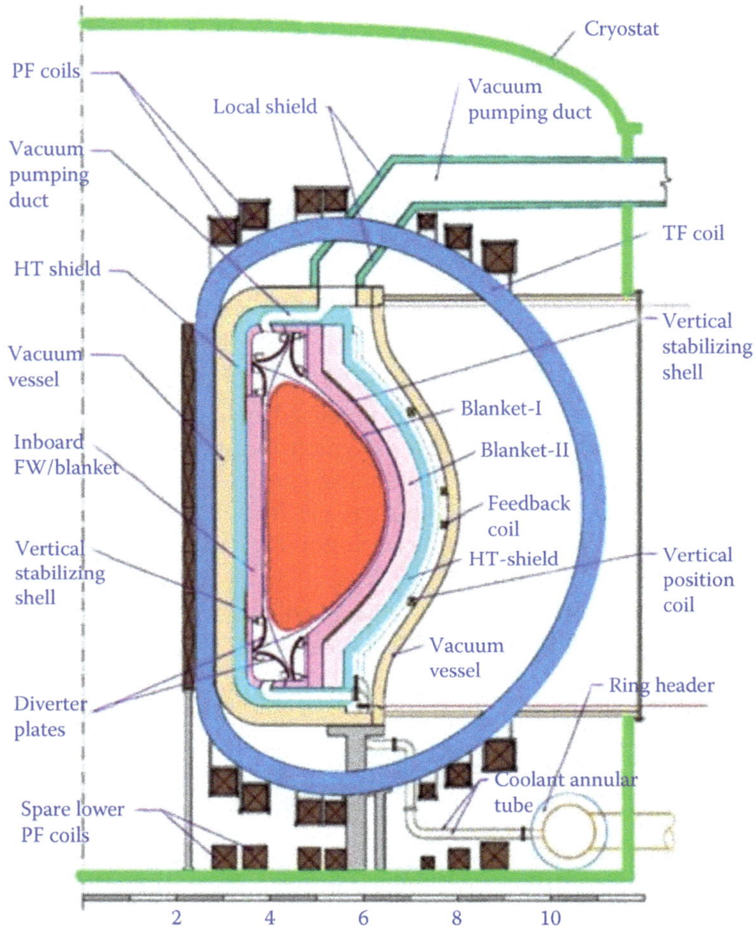

FIGURE 9.4
ARIES-AT advanced tokamak power plant cross section. Toroidal and poloidal magnetic field coils are TF and PF, respectively. The first wall is labeled FW; high- and low-temperature shields are HT and LT, respectively; and a feedback controlled coil to control resistive wall mode instabilities is RWM.

(Najmabadi et al. 2006), which lowers the estimated COE to $0.046/kWh. In addition to capitalizing on the plasma-confinement developments above, ARIES-AT is innovative in its use of silicon carbide as a low-activation structural material. This is a brittle, ceramic-like material, for which methods of joining silicon carbide and surviving, or avoiding, mechanical shock need development. Tests with a fusion-neutron source are also needed to determine longevity under neutron bombardment.

Many of the innovations discussed above have been demonstrated only for short periods during a tokamak discharge. The initial phase of ITER will provide an opportunity to further develop the physics and technology of extending operation toward steady state, for as long as 1000 s.

The second phase of ITER operation will operate with DT, to study the physics and engineering of burning plasmas (see the FIRE and ITER websites for more detail). A burning plasma is one in which the fusion power in alpha particles heats the plasma at a rate exceeding the external heating power. ITER will enable the study of many of the remaining physics and engineering issues for a magnetic fusion power plant.

9.2.5 Magnetic Confinement—Concept Improvements

Concept improvement is an ongoing effort even in established fields. For power plants, generic goals include lower capital costs, lower operating costs, easier maintenance, increased reliability, increased safety, and increasingly in the future—reduced waste generation.

There are two main strategies that can be followed: first, improve the mainline concept, which as described in the previous section, has proven extremely productive over the last few decades, and is rarely a controversial approach; second, develop alternatives that offer significant advantages over the mainline in some characteristic(s).

Looking at the mainline concept—current tokamak performance is close to that needed for power plants. The major issue, beyond those discussed earlier, with applying the tokamak to power plant applications at low cost is that much of the complexity is packed into the first 1–2 m radius outside of the plasma. This includes a first wall (the surface nearest to the plasma); blankets to convert lithium to tritium, heating a fluid for a thermodynamic cycle to generate electrical power, and shielding structures beyond from neutrons; and magnets outside the shield. Penetrations through the first wall and blankets are needed for neutral-beam injection, rf antennas/launchers, pellet injectors for fueling, and perhaps disruption mitigation, vacuum pumping, and diagnostics. These components are exposed to neutron bombardment, or must be shielded from neutrons, as in the case of superconducting magnets. The space constraints complicate the design, construction, and maintenance, adding to the costs of construction and operation. Progress can, and is, being made in the advanced physics developments described above that get more performance out of the same apparatus. Engineering developments can lower the costs of construction and maintenance, and can find new ways to replace complex structures with simpler ones. IFE avoids this tokamak issue; most of its complexity is in the driver (laser, heavy-ion accelerator, or pulsed power) and the target fabrication, not in the fusion chamber. Some alternative concepts for MFE also provide much less complexity near the fusion plasma, as we will discuss below.

A technical analogy to improving the mainline approach is the internal combustion engine, which still dominates automobile power systems in 2011. One hundred years ago, engines were likely to break down on a trip of a 1000 km. Today, after 100 years of engineering development, they frequently operate reliably for a few hundred thousand kilometers with few or no breakdowns, and cost less in inflation-adjusted currency, while producing much more power.

A common issue for all fusion concepts is that materials have not been developed for components exposed to neutrons, some of which need to survive for 100–200 dpa, let alone materials that can survive without becoming significantly activated. Adequate solid materials cannot be developed, demonstrated, and qualified for power plant licensing until a high flux 14 MeV neutron source is constructed to test them. The lack of qualified fusion materials is common to all fusion concepts except, perhaps, those that have simpler structure or that can use thick-liquid walls.

The international fusion materials program has never reached a funding level exceeding 10% of the confinement program, and usually has been closer to a few percent of confinement funding. First-wall materials have been found to have a significant influence on plasma confinement, yet we do not know what materials will survive in a fusion environment. Eventually, the materials program will require a budget similar to that of the plasma confinement program; we suggest a strategy that sooner would be better, both to give more time to develop materials and to give nearer-term experiments the benefit of more appropriate materials choices. This would include one, or more, neutron sources to irradiate materials, and allow for the correction and validation of computational studies. This should be an international program; the accelerator-based IFMIF is likely to be built in Japan. The rest of the world should participate in that and also consider participating in the construction of a plasma-based neutron source to speed the material development, and to provide a source with a neutron energy spectrum very close to that expected in ITER (Fischer et al. 2000). Appropriate neutron spectrum irradiations will be more convincing to licensing authorities. Materials program results will be likely to change the content or course of fusion program elements. A significant materials program would ideally bring in additional funding as the fusion program is seen to make a serious effort to address all the major obstacles to fusion energy production; if not, the confinement budget would need to be cut to a painful degree.

A fusion materials program should be pursued, even if the ideas suggested under the secondary strategy below prove successful. There are three reasons for this: (1) It will be difficult (but perhaps possible) to engineer systems that completely avoid having some solid materials exposed to neutron bombardment, either because the materials penetrate the liquid walls or because they are visible through gaps in the liquid walls; (2) thick-liquid walls have not been demonstrated, the issues discussed below may prove too challenging, and unexpected issues are frequently encountered in research and development; (3) liquid walls are unlikely to be compatible with all fusion concepts.

The second concept-improvement strategy is to develop alternatives that offer significant advantages over the mainline in some characteristic(s) that were listed at the beginning of Section 9.1.1. Optimizing this strategy would require approximately 10% of the fusion confinement budget. One way to approach this is to ask—Is there an idea whose success would dramatically change the prospects for fusion energy, or that would eliminate a major research and development area? Here, we give an example of one such concept—thick-liquid walls that are 0.5–1.0 m thick.

Why do we think that thick-liquid walls are attractive for fusion? Thick (0.5–1 m) liquid walls or jets of a low-atomic-number material would use a material like flibe. Flibe is salt that is composed of the low-atomic-number elements fluorine, lithium, and beryllium. The low atomic number restricts activation to low levels that may, or may not, be obtainable with solid walls. The liquid state eliminates the damage suffered by solid materials from atomic displacements because in a liquid, displacements of atoms or molecules do not change the properties. Helium and hydrogen generation under neutron bombardment are also less problematic as gases are more easily separated from liquids than diffused out of solids, and liquid properties are not degraded by bubbles to the degree of solid properties, such as tensile strength and ductility. Liquid walls may also allow further significant reductions in the volume of activated waste if they can operate at a higher power density, thereby reducing the wall area (Moir 1997), although higher power density has an issue in MFE— can the higher evaporation rate from the walls be handled without eroding the plasma edge?

If thick-liquid walls can be successfully integrated with a fusion core, much of the need for a fusion materials development program would be eliminated. This could accelerate the introduction of fusion power plants by a decade or two, and by providing continuous change-out of tritium-generating blankets, would eliminate one reason for regular shutdowns of a power plant, which could result in higher availability that would reduce the cost of electricity (Moir 1997)—these are exciting prospects!

Liquid walls are not the current baseline in mainline fusion planning because they are not readily usable in the toroidal geometries of tokamaks, nor are they compatible with the extreme cleanliness required by laser inertial fusion energy, which allows no droplets or fogging on laser optics, or the optics will be destroyed by a subsequent laser pulse. Thick liquid walls are the baseline chamber concept for heavy-ion inertial fusion and pulsed-power inertial fusion, which do not have the final optics issues of lasers.

Thick-liquid walls bring their own challenges. Careful, innovative engineering will be required to prevent or mitigate these issues (e.g., Peterson 2001):

1. Flibe will be dissociated into its component atoms, and some of it ionized by interactions with fusion products such as neutrons and alpha particles. The individual atoms and ions of fluorine and lithium are very active chemically; so mitigating corrosion will be an issue.
2. All of the components are toxic. Releases to the atmosphere must be prevented in both operation and accident scenarios.

3. If the liquid varies in temperature where it strikes solid materials, temperature cycling and thermal expansion will disintegrate the materials over time.
4. Tritium and other materials will be dissolved, or otherwise present in the many cubic meters of liquid. Tritium inventories must be limited, especially to minimize any release to the atmosphere in an accident. Heavy elements could become highly radioactive, or eventually change the flibe characteristics. All of these impurities need to be removed from the flibe, down to the level that is determined to be allowable.
5. Liquid walls are probably not compatible with wall stabilization, because low-frequency perturbations could move the walls. Furthermore, many wall stabilization mechanisms require conductive walls to carry currents, and flibe is a poor electrical conductor, similar to semiconductors.
6. Achieving thick-flowing liquids is most readily done in simple geometries, especially cylinders, and perhaps spheres, for which inlet and outlet ducts are less challenging.

If thick-liquid walls are to accelerate fusion development, an obvious Achilles' heel is a compatible fusion core that will not take as long to develop a suitable product as the fusion materials program is likely to. Below, we discuss one concept that shows considerable promise and two others that the author regards as "dark-horse" candidates.

Axisymmetric magnetic mirrors, Figure 9.5, have demonstrated significant progress in the past two decades. Long ignored because of MHD instability, more recent work has shown that they can be stabilized by a number of techniques while maintaining axisymmetry (Ryutov et al. 2011). Most of these stabilization techniques do not depend on close-fitting conductive

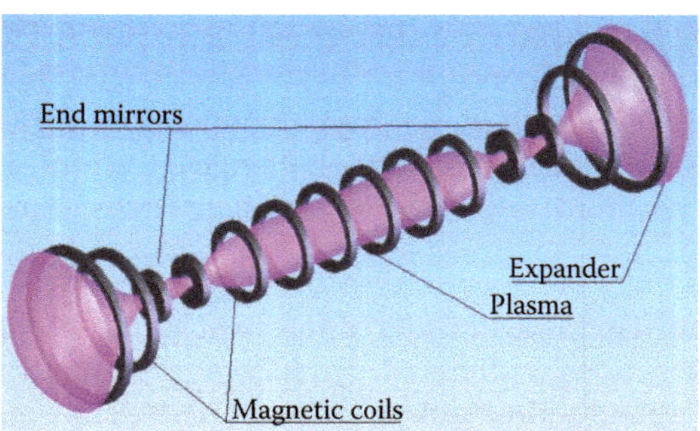

FIGURE 9.5
Axisymmetric tandem mirror coils are simple and accessible. Other structures, such as first walls, blankets, and shields, are not shown.

walls, so axisymmetric mirrors should be compatible with thick-liquid walls. Axisymmetry in magnetic mirrors has significant advantages:

1. Higher-field magnets are feasible with circular windings, which reduce the volume of tandem mirror end cells, consequently reducing the power to drive them. This could enable a tandem mirror to provide adequate confinement for a fusion power plant, without the complexity of thermal barriers.
2. Radial transport is reduced, with no neoclassical or resonant transport.
3. Power losses can be spread over large areas in end tanks.
4. No disruptions.
5. Axisymmetry with a cylindrical geometry makes thick liquid walls much easier to implement.
6. Construction and maintenance are simplified.

Axisymmetric mirrors could build on the achievements of minimum-B single-cell and tandem mirrors, which received significant development in the 1970–1980s, with the potential of significantly higher performance. Mirrors are inherently steady-state, driven by steady-state neutral beams and steady-state fueling. Existing medium-scale facilities in Russia and Japan enable new concepts to be tested expeditiously. Axisymmetric stability has been experimentally demonstrated by using the pressure of the end-loss plasma in the gas dynamic trap (GDT) in Russia (Ivanov et al. 2010). The major issues to resolve for power plant quality confinement are (1) to demonstrate one, or more, methods of stabilization that do not depend on plasma losses (which must be very small in a power plant) and (2) to show that the electron temperature can be considerably higher than about 1% of the ion temperature, which has characterized mirror confined plasmas since the 1970s. A few percent of the international fusion budget, for example, would be sufficient to resolve these issues, in collaboration with the GDT program.

The two "dark-horse" concepts use toroidal magnetic fields, but can fit into spherical or short-cylindrical enclosures. These are the spheromak, and the field-reversed configuration, FRC. Their simple boundary shape makes the formation of liquid walls less difficult, and their high plasma density is more likely to withstand erosion by the high vapor pressure of the liquids.

Perhaps the simplest magnetic configuration is the FRC, which generates a poloidal field with plasma currents, but has little or no toroidal field, and is immersed in an axial magnetic field, for example, a magnetic mirror field, which is analogous to the vertical field of a tokamak, but simpler. FRC shapes range from cigar-shaped to nearly spherical, and have been recently reviewed by Steinhauer. FRC experiments have demonstrated surprising stability, but scaling to larger sizes is not understood. Unlike most plasma configurations, theoretical studies that treat plasma as a fluid fail to explain experimental observations, so more complex theoretical treatments are necessary. FRCs can be isolated from walls by an externally applied magnetic field; such a configuration could be surrounded by thick-liquid walls and probably remain stable, possibly the only toroidal configuration that can make that claim. FRCs have been formed by a number of different techniques, including successful formation by spheromak collision, which implies that an FRC configuration is robust. A large uncertainty is whether confinement can be improved to a level that is adequate for a power plant.

The spheromak has a more complex magnetic geometry with both toroidal and poloidal currents to create both the poloidal and toroidal magnetic fields, respectively. It looks similar to a spherical tokamak (a compact tokamak in which the donut hole of the toroid is shrunk to a thin center leg that is part of the toroidal-field coils), but with no center conductor. Spheromak performance and physics have been reviewed by Jarboe (1994). More recently, a small spheromak, SSPX, achieved confinement with electron temperatures exceeding 500 eV (Hudson et al. 2008). This is more promising than the confinement results of FRCs to date, but, like FRCs, the scaling to power plant levels is unknown.

Counter-intuitively, spheromaks do not appear to be subject to disruptions, even though plasma currents produce all the magnetic fields. This is attributed to their existing in a minimum-energy Taylor state that produces a robust stability in the magnetic configuration because no rapid release of energy from the reconfiguration of the magnetic field (disruption) is possible. A disruption from a minimum-energy state would be analogous to falling out of a basement window—it cannot happen. A spheromak plasma will persist for some time when placed inside a can with conducting walls that the magnetic field will slowly soak into; therefore, the structure near the plasma is potentially much simpler than with a tokamak. The short-cylindrical geometry is compatible with thick-liquid wall formation, but spheromaks gain stability from the plasma pressure against the walls; whether liquid walls are too easily pushed aside by the plasma or can provide stability without high-electrical conductivity remains to be determined.

Steady-state fueling and heating of spheromaks and FRCs, which, like the early tokamaks, are pulsed devices, could build on the array of techniques developed for tokamaks, as well as other techniques like rotating magnetic fields that have had some success in driving current in FRCs and RFPs (discussed under Section 9.3.1). Even if steady-state solutions do not exist for FRCs and spheromaks, pulsed configurations may be attractive if their repetition rate is high enough to minimize thermal cycling in the liquid that operates the heat cycle.

Critics of alternative concepts argue that tokamak performance surpassed that of other concepts when tokamaks were poorly funded, and that little promise has been demonstrated by alternative concepts in subsequent decades. I will answer with the following comments: The amazing array of diagnostics and the incredible finesse in fueling and heating developed for tokamaks can, to a large degree, be applied to these other concepts. In addition, if these concepts had more and stable funding, we could expect further innovations to improve these concepts, just as the tokamak concept has evolved. I am confident that the current poor confinement of these concepts, relative to tokamaks, will evolve significantly toward (how far is unknown) power plant levels if given consistent funding. This confidence is supported by the achievement of 500 eV electron temperatures in the SSPX Spheromak (Hudson et al. 2008), which was funded for nearly 8 years, and the RFP experiments reported by Chapman, which have achieved ion temperatures exceeding 1 keV and electron temperatures of 2 keV, after being funded for more than 20 years.

Finally, I list one additional area of fusion research as a bridge topic to discussing IFE. The extreme pulsed limit of magnetic fusion is magnetized target fusion, MTF (Siemon et al. 1999), which fits a largely unexplored parameter space between MFE and IFE. MTF employs magnetic fields to reduce electron thermal conduction, thereby allowing plasma densities a few orders of magnitude lower than with IFE, but with plasma density and magnetic fields exerting pressures much too large to be restrained by the strength of materials. As a result, some of the structure blows apart and must be replaced each shot. The structure to hold the plasma is small and relatively inexpensive, allowing multiple confinement geometries to be evaluated in one facility. The development path to ignition may be the least expensive of any fusion concept. The path to a power plant is less clear, but might involve replaceable structures of frozen liquid and/or closely coupled liquid walls, fusion yields exceeding a gigajoule, and repetition rates of slower than one per second (to provide about 1000 MW of power), to allow time for reloading, similar to the concepts for pulsed-power inertial-fusion energy discussed by Cuneo et al. (2010).

9.3 Inertial Fusion Energy

9.3.1 Basic Principles

Inertial fusion requires depositing energy in a short time (less than 10 ns—the time for light to travel 3 m or 10 ft) on the outside of a millimeter-sized capsule. The outside of the capsule blows off at high velocity, causing the rest of the capsule to rocket inward. This reduces the capsule radius by a factor of 10–30, which compresses the fuel to high density and temperature, with the goal of reaching sufficiently high values to ignite a fusion burn. Several issues arise in trying to do this:

1. The compression must be spherically symmetric, compressing the sphere as a sphere rather than becoming pancake or sausage shaped.

On NIF, this is accomplished by controlling the laser energy illuminating each of the rings inside the target Hohlraum (German for "radiation room"), as shown in Figure 9.3.

2. The time-dependence of the laser energy must be carefully controlled, timing several shock impulses to minimize heating of the compressing fuel, which can limit compression. Ideally, the fuel is compressed before it is heated.
3. The energy deposition to the outside of the capsule must be spatially uniform; small variations can cause portions of the capsule to compress faster than neighboring regions, leading to large amplitude waves, termed the Rayleigh–Taylor instability, which can prevent the capsule from compressing as a sphere by 10- to 30-fold in radius that is required to obtain fusion energy, and can mix the fuel with surrounding capsule material, reducing its reactivity. This requirement implies that each laser beam must have uniform power across the beam (for direct drive, described later), and that different laser beams all deliver the same power. Lasers are not naturally uniform across the beam, however, various techniques have been developed that smooth out nonuniformities. These are now standard practice on all high-power lasers for IFE.

9.3.2 Inertial Fusion Targets

Target design is usually the starting point for any inertial fusion energy concept because the requirements for igniting a target design set the requirements for the driver.

Inertial fusion is characterized as direct or indirect drive. For direct drive, the energy to compress the target, a spherical capsule of fusion fuel, is incident directly on the outside of the capsule. Experiments with lasers have demonstrated that the energy must be extremely uniform over the surface of the capsule to have the entire surface rocket inward at the same velocity, a requirement if the capsule is to remain spherical during compression without Rayleigh–Taylor instabilities.

Indirect drive uses a target similar to that shown in Figure 9.3. The driver (e.g., laser) energy is deposited on the inside of a can or ("Hohlraum"), which contains the capsule. The energy creates soft x-rays of a few hundred electron volts (few million Kelvin) energy that reflect off the walls of the can and uniformly illuminate the capsule. This is analogous to indirect lighting that creates a relatively uniform, diffuse illumination, as contrasted with direct lighting, which is more efficient, but creates higher contrast lighting. Direct drive, for which the laser beams impinge directly on the outside of the spherical capsule, is similarly more efficient since energy is not expended in heating the inside of the ("Hohlraum"). Indirect drive is more compatible with the liquid-wall fusion chamber design because the beams can be clustered to come in from one or two sides, rather than distributed uniformly over the surface of a spherical chamber.

9.3.3 Inertial Fusion Drivers

Inertial fusion can also be categorized by the type of driver it uses. Experiments to date have been driven by lasers (see the LIFE website) or pulsed power (Cuneo et al. 2010). Most of the lasers are flashlamp-pumped solid-state lasers that have low electrical efficiencies (of order 10^{-3} to 10^{-2}) and low repetition rates of a few shots per day. Lasers are ideal for IFE ignition experiments: photons are electrically neutral, so can be focused to very high power densities with no limits from space charge effects; the time dependence of laser beam power can be precisely controlled; and the reproducibility is excellent. A few experiments have also used z-pinch radiation sources or light ions driven by pulsed power.

Other drivers are being developed. Diode-pumped solid-state lasers deliver efficiencies near 10% and repetition rates that can exceed 10 Hz. Electron-beam-pumped gas lasers, such as KrF (krypton fluoride), have nearly as high efficiency. Heavy-ion accelerators are attractive driver candidates for inertial-fusion energy, building on extensive experience with high-energy and nuclear physics accelerators. They promise high efficiency (10%–50%), high repetition rate, long life, and magnetic final optics to focus the beam onto the target (Yu et al. 2003). Compared with the mirrors or lenses that focus laser beams, magnetic optics are immune to the effects of target explosions because neutrons, x-rays, and debris can pass through the aperture of the magnet, while the magnet windings can be shielded from neutrons. The high beam currents required with heavy-ion fusion are a new and challenging element for both accelerators and focusing systems.

Target designs, similar to Figure 9.3, have been developed for both laser and ion drivers. These designs are predicted with two-dimensional codes to have reasonable gains (near 100) for a power plant. A recent X-Target design is driven only by heavy-ion beams, because of their ability to deposit energy inside matter rather than just at the surface, as lasers do (Henestrosa et al. 2011).

9.3.4 Inertial Fusion Power Plants and Issues

Review panels evaluating IFE have concluded that heavy-ion fusion has the most promise for power plants—see, for example, the 1996 FESAC Report (DOE Archives). A heavy-ion driven, inertial fusion power plant is shown in Figure 9.6. Like all inertial-fusion concepts, it is characterized by modularity, where the driver, chamber, target fabrication plant, and target injector are all separated. Thick-liquid walls within the chamber are composed of jets of flibe, some of them oscillating in position. These enclose the reaction region

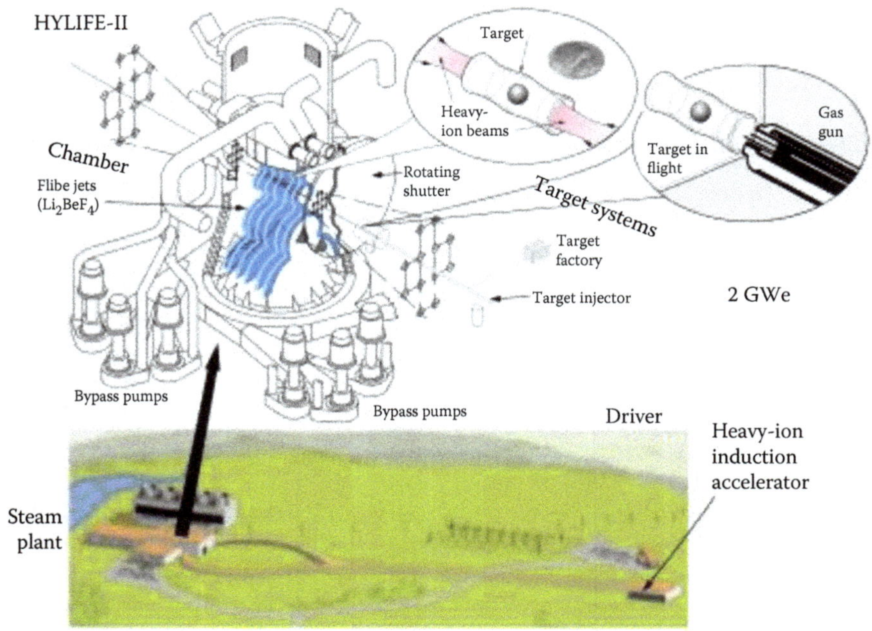

FIGURE 9.6
Inertial fusion power plant driven by a heavy-ion accelerator with a HYLIFE-II thick-liquid-wall fusion chamber.

and protect the solid chamber walls from neutrons and shock waves. A conservative conceptual design for a heavy-ion inertial-fusion power plant was described by Yu et al. (2003), in which the authors satisfied all known scientific and technological constraints (a significant accomplishment), but with the disadvantage of high capital cost ($5 billion) and high COE ($0.07/kWh). These costs are expected to decrease with optimization and further innovation.

Critics point to several difficult issues with IFE, which proponents believe can be handled. These have been studied through experiment, theory, simulations, and conceptual designs with the following conclusions:

1. Targets that currently cost a few thousand dollars each need to be about 10,000 times less expensive for economical power production. When the target shooting rate increases from a few per day to about 10 s^{-1}, the manufacturing will have to shift from a few at a time to mass production. This has been evaluated by conceptual engineering of a manufacturing facility, with the conclusion that targets with the required precision can be achieved at a tolerable cost, given a supporting development program (Goodin et al. 2006).
2. Target must be made to extremely precise requirements. Surfaces must be smooth to avoid seeding Rayleigh–Taylor instabilities. NIF

has stringent requirements, smoothness ranging from the order of nanometers (beryllium capsule smoothness) to about a micrometer (smoothness of the DT fuel layer). These requirements have been met (see the LIFE website). Serendipity has helped here: if the frozen DT fuel layer varies in thickness, the beta decay of radioactive tritium warms the thicker regions more than thin regions. The warmer thick regions evaporate and deposit on thinner, cooler, regions. A recent target design for heavy-ion fusion is likely to require much less precision than do NIF targets (Henestrosa et al. 2011).

3. Target injection must place targets in position with an accuracy of 20–200 μm (μm means a micrometer, 1000 μm = 1 mm), depending on the target design, over a distance of about 7 m, or more. The first experimental gas gun injector succeeded in placing surrogate targets within an angular spread of 1.4 mrad (corresponding to a 1 cm circle at a range of 7 m) (Petzoldt 1998), which is more than a factor of 10 worse than the accuracy of competition-rifle sharpshooters who can achieve 0.05–0.1 mrad spread over multiple shots, corresponding to less than 0.07 cm at a range of 7 m. Electrostatic steering of targets is near the threshold of feasibility with a 1 cm circle; if the injection uncertainty were reduced a factor of 2–3 (still several times worse than a sharpshooter achieves), electrostatic steering would become feasible, and this has been demonstrated with a second-generation target injector (Petzoldt 2007).
4. The final optics will be exposed to neutrons, x-rays, and target debris. This will cause damage to mirrors and lenses, resulting in the destruction of the optic on a subsequent laser shot. Techniques such as annealing are being studied to mitigate damage (see the LIFE website). This issue is eliminated for heavy-ion beams, which use electromagnets for focusing the beams to the target; there is no damage from neutrons and debris passing through the winding aperture. Lifetimes of magnet windings to neutron damage have been engineered to exceed 30 years (Yu et al. 2003).
5. Laser or ion beams can interact with plasma near the target, or in the ("Hohlraum"), absorbing or deflecting the beam energy from illuminating the target as intended. This is an area that continues to receive attention, but is beyond the scope of this chapter. In the case of laser-plasma interactions, an authoritative book is devoted to the subject (Kruer 2003). Similar to MFE, none of these instabilities are disastrous, but keeping them sufficiently suppressed places restrictions on operating parameters, generally reducing performance.

IFE provides a paradigm shift from MFE tokamaks in two areas. First, it places most of its complexity in the driver, which is decoupled from the relatively simple fusion chamber, whereas MFE incorporates most of its complexity immediately around the fusion chamber. This provides IFE with potential advantages of (1) reducing development costs by allowing the chamber and driver modules to be developed and upgraded independently, and (2) reducing operational costs by delivering a higher availability to utilities. Higher availability is possible because the complexity within the fusion chamber is minimized, and repairs to systems outside the neutron shield can be made more rapidly, in some cases even while the driver continues to operate. Second, indirect-drive IFE (heavy-ions, or pulsed power) offers the potential for lifetime fusion chambers with renewable liquid coolants facing the targets, instead of solid, vacuum-tight walls that could be damaged by heat and radiation. Protected in this way, all of the chamber structural materials would be lifetime components. Their minimal residual radioactivity would mean that, at the end of the fusion plant's life, most of the materials could be recycled, rather than requiring deep underground disposal. As mentioned, laser-driven IFE is generally considered to be incompatible with thick-liquid walls, because of the risk of droplets or condensation on the final optics and subsequent damage to the optics.

Scientific understanding of inertial-confinement fusion is being significantly advanced by the current ignition campaign on the National Ignition Facility (NIF) at Lawrence Livermore National Laboratory in the United States, and the similar and soon to be completed Laser MegaJoule (LMJ) facility in France. The NIF is likely to be the first laboratory device to realize fusion energy gain. Although the primary missions of both the NIF and the LMJ are defense related, an important spin-off benefit is to show that inertial-fusion energy is feasible. To capitalize on this demonstration requires a parallel effort in developing suitable and cost-effective drivers and chambers for IFE.

9.3.5 Inertial Confinement—Concept Improvements

Recent concepts could dramatically reduce the driver energy for lower COE. These are fast ignition (Tabak et al. 1994) and shock ignition (Betti et al. 2007). With conventional IFE, the capsule is heated by a high degree of compression to ignite at the center, analogous to ignition in a diesel engine. With fast ignition, the capsule is compressed to moderately high density, but low temperature. A portion of the capsule edge is then heated to ignition by a very short energy pulse, with duration measured in picoseconds (millionth of a millionth of a second) analogous to spark-plug ignition in the internal-combustion engine. The hot spot heats the surrounding region, and the fusion burn propagates through the fuel. This concept can give gains (the ratio of fusion energy out to driver energy in) a factor of 5–10 higher than conventional IFE, because relaxing the compression ratio allows a lower energy, and therefore less expensive driver, resulting in a lower COE. However, several difficulties must be overcome first: the driver must bore a hole through the plasma surrounding the compressed capsule, which now has a diameter of order 100 μm, then deliver the fast-ignition pulse through the hole and focused to a very small size to hit the compressed capsule, at precisely the right time, with tolerances of less than 100 picoseconds (100×10^{-12} s), and finally it must heat the region it hits.

Shock ignition looks similar to conventional IFE ignition, except that the driver power is suddenly increased (spiked) when the capsule is reaching peak compression. This launches a shock wave that ignites a small region (similar to fast ignition), but this region is at the center of the capsule.

In addition to ignition concepts, proponents of each driver concept are studying ways to improve performance and reduce costs. Frequently, these start with innovative target designs because the requirements for igniting each target design set the requirements for the driver.

Fast, or shock, ignition in IFE and the several new MFE concepts mentioned earlier are examples of innovations that may significantly increase the attractiveness and lower the cost of electricity from a fusion power plant. Even 60 years into the fusion program, new ideas continue to emerge; and some old, previously rejected ideas become attractive with new technology.

Exercises

1. Pulsed versus steady state: Compute mass of coolant to keep temperature fluctuations to 1% for a thermal power of 3 GW and pulse rates of 10, 1, and 0.1 Hz. For estimating purposes, assume the specific heat is 1,000,000 cal/m^3°C, which is a reasonable approximation for many liquids, and the units conversion factor 1 cal = 4.186 J.
2. Based on deuterium being 1/6000 fraction of hydrogen in water, and the DT fusion reactions yielding 17.6 MeV plus another 5 MeV from generating T from Li, and the energy released from burning gasoline of 30 MJ/L, compute the energy equivalent of 1 L of water in terms of liters of gasoline. The unit conversion factor is 1 MeV = 1.6×10^{-13} J.

Question for Discussion

Does combining fusion with fission provide a desirable product with some of the best features of each, or is it the worst of both worlds? List the advantages and disadvantages of fusion–fission hybrids and the assumptions that you made to arrive at your answer.

References

Betti, R., C. D. Zhou, K. S. Anderson, L. J. Perkins, W. Theobald, and A. A. Solodov. "Shock ignition of thermonuclear fuel with high areal density." *Physical Review Letters* 98, 155001 (2007).

Chapman, B. E., et al. "Improved-confinement plasmas at high temperature and high beta in the MST RFP." *Nuclear Fusion* 49, 104020 (2009).

Chen, F. F. *Introduction to plasma physics and controlled fusion: plasma physics, ed. 2*, Springer (1984); for a discussion of fusion energy, see Chapter 9 of the Plenum Press 1974 Ed. 1. New York: Springer Science+Business Media.

Cuneo, M., et al. (2010). http://fire.pppl.gov/fpa10_PP_ICF_Cuneo.pdf. Accessed October 2012.

Delene, J. G., J. Sheffield, K. A. Williams, R. L. Reid, and S. Hadley. "An assessment of the economics of future electric power generation options and the implications for fusion." *Fusion Technology* 39, 228 (2001).

Department of Energy (DOE) Archives, http://science.energy.gov/fes/fesac/reports/reports-archive/. Accessed October 2012.

El-Guebaly, L., V. Massaut, K. Tobita, and L. Cadwallader. (2008). "Recent trend in managing fusion radwaste: recycling and clearance, avoiding disposal," http://fti.neep.wisc.edu/aries/TALKS/lae_radwaste0308.pdf. Accessed October 2012.

Fetter, S., and F.N. von Hippel. "Is U.S. reprocessing worth the risk?" *Arms Control Today*, Vol. 35, 6 (Sept. 2005).

Fetter, S., E. T. Cheng, and F. M. Mann. "Long-term radioactive waste from fusion reactors." *Fusion Engineering and Design* 13, 239 (1990).

Fischer, U., A. Möslang, and A. A. Ivanov. "Assessment of the gas dynamic trap mirror facility as intense neutron source for fusion material test irradiations." *Fusion Engineering and Design* 48, 307 (2000).

Goodin, D. T., et al. "Developing a commercial production process for 500,000 targets per day: A key challenge for inertial fusion energy." *Physics of Plasmas* 13, 056305 (2006).

Frosch, R. A. "Industrial ecology: A philosophical introduction." *Proceedings of the National. Academy of Sciences USA* 89, 800–803 (1992).

Henestrosa, E., B. G. Logan, and L. J. Perkins. "Quasispherical fuel compression and fast ignition in a heavy-ion-driven X-target with one-sided illumination." *Physics of Plasmas* 18, 032702 (2011).

Hudson, B., et al. "Energy confinement and magnetic field generation in the SSPX spheromak." *Physics of Plasmas* 15, 056112 (2008).

IFMIF neutron source for developing and testing materials for fusion power plants, http://www.frascati.enea.it/ifmif/. Accessed October 2012.

ITER, at http://www.iter.org/. Accessed October 2012.

Ivanov, A. A., et al. "Results of recent experiments on GDT device after upgrade of heating neutral beams." *Fusion Science and Technology* 57, 320 (2010).

Jarboe, T. R. "Review of spheromak research." *Plasma Physics and Controlled Fusion* 36, 945 (1994).

Kishimoto, H., S. Ishida, M. Kiduchi, and H. Ninomiya. "Advanced tokamak research on JT-60." *Nuclear Fusion* 45, 986 (2005).

Kruer, W. L. "*The physics of laser plasma interactions*," Boulder, CO: Westview Press (2003).

Lindl, J. D. "*Inertial confinement fusion*," New York: AIP Press (1998).

Logan, B. G., R. W. Moir, and M. A. Hoffman. "Requirements for low-cost electricity and hydrogen fuel production from multiunit inertial fusion energy plants with a shared driver and target factory." *Fusion Technology* 28, 1674 (1995).

MIT Fusion-Fission Research Workshop. 9/30-10/2/2009, http://web.mit.edu/fusion-fission/. Accessed October 2012.

MIT Issues http://hardware.slashdot.org/story/12/04/11/0435231/mit-fusion-researchers-answer-your-questions? utm_source = rss1.0mainlinkanon&utm_medium = feed. Accessed October 2012.

Moir, R. W., and T. D. Rognlien. "Axisymmetric tandem mirror magnetic fusion energy power plant with thick liquid walls." *Fusion Science and Technology* 52, 408 (2007).

Moir, R. W. "Liquid first walls for magnetic fusion energy configurations." *Nuclear Fusion* 37, 557 (1997).

Najmabadi, F, et al. "The ARIES-AT advanced tokamak fusion power plant with advanced technology." *Fusion Engineering and Design* 80, 3 (2006).

Peng, M., and R. Stambaugh. (2012). "Concepts for tokamak-based Fusion Nuclear Science Facility," http://vlt.ornl.gov/research_highlights.shtml. Accessed October 2012.

Peterson, P. F. "Design methods for thick-liquid protection of inertial fusion chambers." *Fusion Technology* 39, 702 (2001).

Petzoldt, R. W. "IFE target injection and tracking experiment." *Fusion Technology* 34, 831 (1998).

Petzoldt, R. W., E. Valmianski, L. Carlson, and P. Huynh. "Target injection placement accuracy improvement with electrostatic steering." *Fusion Science and Technology* 52, 459 (2007).

ReNew Report. "Research needs for magnetic fusion energy sciences," (2009) http://burningplasma.org/web/ReNeW/ReNeW.report.web2.pdf. Accessed October 2012.

Ryutov, D. D., H. L. Berk, B. I. Cohen, A. W. Molvik, and T. C. Simonen. "Magneto-hydrodynamically stable axisymmetric mirrors." *Physics of Plasmas* 18, 092301 (2011).

Sagaro, A.,Y. Igitkhanov, F. Najmabadi. "Review of stellarator/heliotron design issues towards MFE DEMO." *Fusion Engineering and Design* 85, 1336 (2010).

Sheffield, J. "The physics of magnetic fusion reactors." *Reviews of Modern Physics* 66, 1015 (1994).

Siemon, R. E., I. R. Lindemuth, and K. F. Schoenberg. "Why MTF is a low cost path to fusion." *Comments Plasma Physics Controlled Fusion* 18, 363 (1999).

Stambaugh, R. D. http://www.fusion.ucla.edu/FNST/FNST-18-20-Aug-2009/Presentations/Stambaugh-VG_FDF_UCLA_09.pdf. Accessed October 2012.

Steinhauer, L. C. "Review of field-reversed configurations." *Physics of Plasmas* 18, 070501 (2011).

Tabak, M., et al. "Ignition and high gain with ultrapowerful lasers." *Physics of Plasmas* 1, 1626 (1994).

Watkins, M. L., and JET Team. "Physics of high performance JET plasmas in DT." *Nuclear Fusion* 39, 1227 (1999).

Yu, S. S., et al. "An updated point design for heavy ion fusion." *Fusion Science and Technology* 44, 266 (2003).
Zinkle, S. J. and N. M. Ghoniem. "Operating temperature windows for fusion reactor structural materials." *Fusion Engineering & Design* 51–52, 55 (2000).
Zinkle, S. J. "Fusion materials science: Overview of challenges and recent progress." *Physics of Plasmas* 12, 058101 (2005).

10

Energy from Space for Sustainable Commercial Power for Earth

David R. Criswell

CONTENTS

10.1 Overview

In 2008, earth's commercial power systems provided approximately 2250 watts of thermal power per person (2250 W_t/person) to its 6.67 billon people. This totals to approximately 15 terawatts of thermal power (15 $TW_t = 15 \times 10^{12}$ $W_t = 2250$ W_t/person $\times 6.67 \times 10^9$ people). However, they need the equivalent of 6000–7500 W_t/person of sustainable power for sustainable prosperity. So, by 2050, 10 billion people will need approximately 75 TW_t of thermal power. Existing power systems are massive and move so much mass as mining wastes, coal, oil, natural gas, biomass, CO_2, ash and spent radionuclides, water, air, and other forms that they disrupt and contaminate the biosphere locally and globally and consume 10%–15% of gross world product (GWP) ($40 trillion GWP by year 2000). Both the International Panel on Climate Change and the commercially oriented World Energy Council repeatedly challenge world policy makers to enable a new sustainable global commercial power system that provides abundant and affordable electric power. Now, 3 W_t generate approximately 1 W_e of electric power. The global power system provides an average of 300 W_e of electric power per person (300 W_e/person). By 2050, almost all power will be delivered electrically, and 20 to 30 TW_e can enable a prosperous world of 10 billion people (20 TW_e = 2000 W_e/person × 10 × 10^9 people).

The sun dependably illuminates our moon with 13,000 TW_s of solar power or approximately 650 times the 20 TW_e needed by a prosperous earth. The lunar solar power (LSP) system uses solar-powered bases built on the moon to collect ≤1% of this sustainable power to dependably supply receivers on earth with safe microwave power beams (≤230 W_e/m^2 or ≤20% the intensity of noon sunlight). The 20 TW_e is output on earth from approximately 100,000 km^2 of receivers and then into electric power grids all about earth without significant movement of mass within the biosphere. The LSP system can pay for its own growth after an investment of approximately \$500 billion (\$500 × 10^9). LSP start-up, over 15 years, costs less than 1 year of the 2009 U.S. Department of Defense budget.

Since 1980, Japan and western Europe have produced approximately \$42 trillion of gross domestic product (GDP) from 1 terawatt year of electric energy (1 TWe-y). By 2050, a 20-TWe LSP system can enable a sustainable \$840 trillion GWP or >\$84,000 GWP/person. Additional clean LSP electric energy can be used to extract all industrial carbon dioxide from earth's atmosphere. A >\$10 trillion gross lunar product (GLP) is possible within the twenty-first century.

10.2 Global Wealth and Electric Energy

Both the International Panel on Climate Change and the commercially oriented World Energy Council have repeatedly challenged world policy makers to enable a new sustainable and affordable global commercial power system within the first part of this century [1, 2, 3]. However, they and most government, private, and nonprofit organizations tend to extrapolate the growth of commercial power from the capabilities of existing systems. They do not provide a systematic method by which to estimate the minimum commercial power needed to sustainably support earth's growing human population independent of our biosphere and then use those projections to identify and implement the required power systems.

Why is commercial electric power so important? Electricity, as opposed to chaotic thermal energy, is potentially the most effective source of useful work in final application. This is because an electric force vector, technology permitting, can be directed precisely, from the macrolevel to the atomic level, to deliver useful work and, thereby, clean new wealth and sustainable economic activity. Since commercial electric power was introduced in the 1880s, an increasing fraction of primary thermal fuels (wood, coal, natural gas) and natural mechanical power (hydro, winds, tides, etc.) has been converted into electric power. In 2006, the averaged world electric power output was 1.3 TW_e from fossil fuel; 0.3 TW_e from fission; 0.33 TW_e from hydroelectric; and only 0.04 TW_e from geothermal, terrestrial solar, wind, wood, and waste [4].

From 1980 to 2002, western Europe and Japan output approximately $4.80 GDP per kilowatt-electric hour of consumed electric energy ($4.80 GDP/$kW_e$-h or $42 trillion GDP/$TW_e$-y). The United States and the world averaged approximately $2.95 GDP/$kW_e$-h and the developing nations only $1.56 GDP/$kW_e$-h [5]. This is why rapidly developing nations such as China consider increasing production of electric energy and expanding their electric infrastructure among their highest national priorities.

Approximately 75 TW_t is required to output 20–30 TW_e [6—Chapters 2, 9]. By 2002, approximately 38% of all primary thermal energy in the United States was converted to electricity. The corresponding global trend implies a 100% electric world by the middle of this century. In 2000, approximately $40 trillion GWP/year was achieved. By providing the world 20 TW_e of sustainable electric power and enabling the demonstrated western European level of productivity, approximately $840 trillion GWP/year could be achieved.

10.3 Global Power Challenges

In 1976, H. E. Goeller and A. M. Weinberg provided a systematic method for estimating the commercial power needed by human civilization to live sustainably. They required the commercial power to be from a carbon-free source and adequate to

- Obtain all major nonrenewable material resources from their average crustal abundances and the ocean.
- Recycle cleanly all goods and water.
- Support sustainably all agricultural, industrial, residential, and transportation activities.

10.3.1 Economic Independence from the Biosphere

Goeller and Weinberg estimated that approximately 7500 watts of commercial thermal power per capita (= 7.5 kW_t/person) would be required. Ten billion people (10×10^9) would require 75 TW_t. A United States–style economy would require twice as much power per person [7 (see page 688, column 1)].

Electricity is the primary product of fossil and nuclear thermal power stations. A thermal power plant releases approximately two-thirds of its input power as waste heat to the biosphere. Releasing the waste heat from a 75-TW_t

fleet of coal or nuclear plants presents major engineering and environmental challenges. Using river or ocean water on a once-through basis to remove the waste heat requires approximately 3×10^{13} t/year. This approaches the 3.8×10^{13} t/year flow of all the world's rivers. Even the quantity of makeup water for use in cooling towers, at approximately 1.3×10^{12} t/year, is enormous [34].

10.3.2 Biosphere-Dependent Fossil Fuel Power Systems

Goeller and Weinberg were strongly motivated to seek a carbon-free source of power knowing that a 75-TW_t fleet of coal-fired power stations would consume on the order of 1×10^{11} t/year of coal. However, they did not place that number in a context directly relevant to earth's biosphere.

The burning of fossil fuel actually yields useful energy by "mining" the net energy from the chemically free oxygen molecules of our modern atmosphere. Earth's chlorophyll-based plants generate approximately 100 TW_t-y of net new combustible dry mass each year equally from the oceans (consumed quickly at sea) and from land (primarily wood). The plants use solar power and photosynthesis to convert water and CO_2 into new plant life and also release approximately 3×10^{11} t/year of new oxygen molecules [6 (chapter 9), 8]. A 75-TW_t fleet of coal-fired power stations would consume over 70% of new oxygen in direct competition with all animal life. The fossil carbon power fleet would also output 3×10^{11} t/year of CO_2 or 15 times the output of all industrial CO_2 in 2004 [9].

Kerogen, a brown-colored sediment, and sedimentary rocks are the source of oil and natural gas and include coal, oil shale, methane hydrates, and thousands of other types of carbon-rich remains of earlier biosphere/oxygen cycles. Geologists and mining engineers are increasingly adept at obtaining useful fuels from progressively lower grades of approximately 15×10^{15} t of kerogenic carbon [10]. Since there is only 1×10^{15} t of oxygen in our modern atmosphere, it is technically conceivable that an exponentially growing fossil fuel system could mine most of our atmospheric oxygen. Earth is a closed spacecraft. Its passengers must become aware of its biosphere's operational limits.

10.3.3 Nuclear Fission Reactors

Dr. Weinberg estimated that 15,000 5-GW_t reactors could supply 75 TW_t. Assuming a reactor lifetime of 30 years, 10 new reactors would be built every week as 10 old reactors are retired. For perspective, today, 439 commercial reactors collectively output approximately 1 TW_t.

A 75-TW_t fleet of light-water reactors would ingest approximately 4,200,000 t/year of natural uranium and thus reduce fuel mass flow through the biosphere by a factor of 84,000 compared to coal. However, proven continental reserves of uranium would last less than 4 years. Uranium (99.28% ^{238}U and 0.72% ^{235}U) extracted from seawater would last approximately 1000 years. Extraction of uranium with 100% efficiency would require the processing of 1.3×10^{15} t/year of seawater or 35 times the flow rate of all the rivers on earth [11]. The entire ocean would be filtered once over that thousand years. Ocean currents are driven by approximately 3 TW of mechanical power supplied by sun–moon tides, surface winds, surface solar heating, and high-latitude surface cooling. The pumping power to extract the uranium would exceed the natural mechanical power now driving the oceans. The vertical and horizontal thermal profiles of the oceans and ocean currents would be drastically changed [6 (chapter 9)].

A light-water reactor burns approximately 1% of its uranium fuel. In principle, a breeder reactor can burn approximately 99% of its uranium and thorium fuels and maintain a 75-TW_t economy for many thousands of years. Fission breeding reduces fuel and mining flows on land and sea by a factor of approximately 100, but it still requires moving matter on scales comparable to the natural matter flows of the biosphere. The breeder system also maintains an enormous inventory of plutonium that can be converted into nuclear weapons [6 (chapters 7, 11)].

10.4 Lunar Solar Power System

A new sustainable commercial electric power generation system that does not damage earth's biosphere must meet the following requirements:

1. Eliminate fuel gathering and transport and burning, ash disposal, oxygen consumption, CO_2 production, and their costs.
2. Eliminate terrestrial power plants and ancillary facilities and their costs.
3. Provide abundant electric energy that is independent of the biosphere and is affordable.

An LSP system, built on the moon from common lunar materials, can meet these three requirements [6 (chapters 9, 12 through 23)].

The sun dependably illuminates the moon with 13,000 TW_s of solar power. The LSP system shown in Figure 10.1 consists of power bases, located on the east and west limbs of the moon as seen from the earth (note the spot on the left limb of the moon). The power bases convert sunlight into electricity and then into low-intensity beams of microwaves. The beams illuminate the power receivers (rectennas) on earth that output commercial power to local and regional electric power grids. The lunar power bases need to collect less than 1% of the solar power that the moon intercepts to dependably provide rectennas on earth with more than 20 TW_e.

The power beams, as shown in Figure 10.2, are projected either directly from the moon to rectennas that can view the moon or by redirector satellites in orbit about earth to rectennas that cannot view the moon directly. The microwaves, approximately 10–13 cm in wavelength or approximately 3 to 2 GHz in frequency, dependably pass through all atmospheric conditions and

FIGURE 10.1
The lunar solar power system viewed from earth with a power beam from the moon feeding microwave power to a power receiver (rectenna).

provide electric power without the movement of mass through the biosphere, without power plants, and with power that is independent of the biosphere.

The rectennas are safely illuminated at a maximum of approximately 230 W_e/m^2 or approximately 20% of noon sunlight. They output approximately 200 W_e/m^2 and approximately 30 W_t/m^2 of waste heat. A rectenna can be approximately 90% transparent to sunlight. Areas under a rectenna can be whitened and can reflect, averaged over a year, more solar power to space than what the rectenna receives as microwaves. Thus, rectennas can be thermally neutral. Only 100,000 km^2 of rectennas would be required to dependably output 20 TW_e of virtually massless pure electricity that does not contaminate the biosphere or contribute directly or indirectly to global warming.

Rectennas would occupy much less area per unit power output than all other renewable systems. Stand-alone terrestrial solar and wind installations provide only 2–20 W_e/m^2 of delivered power because they are hostage to the biosphere and therefore require vast power storage and transmission systems. These global terrestrial systems have the potential to create economic, political, environmental, and even military conflicts. A globally distributed array of 20-TW_e terrestrial solar panels that feed global transmission lines and

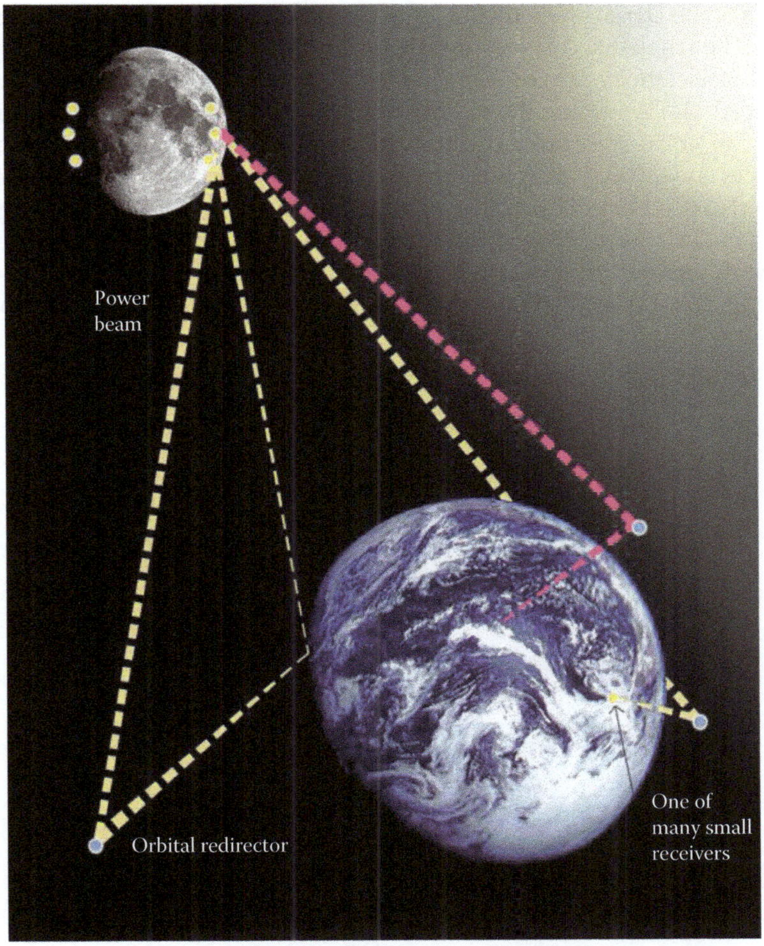

FIGURE 10.2
Sun–moon–earth lunar solar power system.

power storage facilities (45 days backup) would occupy over 2,000,000 km^2 compared with 100,000 km^2 for LSP rectennas [6 (chapter 9), 14, 19].

Also, rectennas can be placed over productive agricultural lands and grasslands, deserts, shallow fisheries, contaminated areas, and industrial parks. Humans and other life forms can safely enter a beam and adsorb the equivalent of less than 0.4% of noontime sunlight. However, almost all life forms will be excluded from the beam by a perimeter fence about each rectenna. Microwave power will be blocked from the area under the rectenna by the rectenna's elements, and additional electrical screening can be added. The microwaves scattered by a 20-TW$_e$ LSP system into the general environment can be less intense than the light of a full moon, which is far less than that associated with wireless phones.

In 2008, the U.S. average retail cost of electricity ranged from \$0.056 to \$0.29/ kW_e-h. The wholesale cost was approximately 50% of retail (EIA Electricity U.S. Data). LSP electricity can cost less than \$0.001/kW_e-h wholesale when LSP capacity exceeds 1 TW_e. Using the electricity from its own rectennas, any region or nation can sustainably produce almost all its essential goods from local material resources and through recycling. The local rectennas can also power services. No nation will need to be dependent on other nations for fuel, clean water, agricultural chemicals, or other commodity items critical to its sustainable prosperity.

Power bases on the moon will be constructed from local lunar materials (24, 25, 26, 27). Figure 10.3 illustrates a few demonstration power plots located

FIGURE 10.3
Lunar solar power demonstration base: multiple power plots (arrays of solar converters—#1, microwave transmitter—#2, and microwave reflector—#3, set of mobile factories (#4) and assembly units (#5), and habitat/manufacturing facility (#6).

within one of the LSP power bases. Mobile (Figure 10.3: #4, #5) and fixed (#6) factories, shops, and habitats are transported from earth to moon. These fixed and mobile factories then make the LSP power bases from local lunar materials. The factories produce hundreds to thousands of times their own mass in products. Production on the moon ultimately averages down the cost of transportation from earth to moon to a reasonable cost per unit of power-base component.

After transport from earth to moon of the initial mobile and fixed factories and habitats, lunar industrial materials can also be used on the moon to manufacture a large fraction of the additional production units. This process is called "bootstrapping." Using bootstrapping, detailed systems engineering models reveal that 1 kg of materials from earth enables the delivery of approximately 140,000,000 kW_e-h of electric energy to earth. By comparison, natural uranium burned once in a light-water reactor delivers approximately 80,000 kW_e-h. Burning 1 kg of oil with the oxygen of earth's atmosphere provides only approximately 12 kW_e-h.

10.5 Terrestrial Global Power Systems' Mass Effectiveness

As shown in Table 10.1, conventional solar, hydroelectric, fossil, and nuclear power systems not only process enormous flows of mass through the biosphere but also are themselves extremely massive per unit of delivered electric energy (tons per terawatt year of electric energy) and are relatively short lived. Therefore, they are expensive to construct and maintain, environmentally intrusive, and output expensive electric energy [6 (chapter 9), 14, 19, 28, 29].

The U.S. Department of Energy examined a stand-alone terrestrial solar power installation that consisted of solar cells of 16.5% efficiency operating under 18-to-1 solar concentrators (Table 10.1: 1). They assumed 2 hours of battery power storage each day. The total mass of the photovoltaic (PV) arrays, batteries, supporting structure, and foundations was calculated. The system required approximately 2×10^8 t of equipment to output 1 TW_e-y of energy. This system takes over 20 years to pay back the energy needed to construct it. As noted earlier, stand-alone terrestrial solar and wind power systems must be scaled up in area and electrical capacity, provided with global power lines and months-long power storage, and very carefully managed to output dependable power. Even then, large volcanoes, dust storms, and other events within the biosphere or political sphere can disrupt the power for an indefinite period of time.

Washington's Grand Coulee Dam has a mass of approximately 20,000,000 t and an average output of 0.0023 TW_e (Table 10.1: 2). Over the next 100 years, it could output 0.23 TW_e-y. This implies approximately 9×10^8 t

TABLE 10.1

Large Power Systems—Lifetimes, Mass Efficiency (t/TW_e-y), Electric Energy Payback Time (Year)

	Power System	System Lifetime (year)	t/TW_e-y	Electric Energy Payback (year)
1.	Terrestrial PV (16%, 18 × sun, 2-hour batteries)~	30	190,000,000	>20
2.	Hydroelectric (G. Coulee, 2.3 GW_e avg.)	100	91,600,000	0.7
3.	Light-water reactor	30	40,100,000	1.3
4.	Coal plant with mining and unit trains	30	10,000,000	1.3
5.	LSP earth rectennas (concrete supports)	30 (buildup) + 50	1,500,000	0.33
6.	LSP earth rectennas (electronics and wiring)	30 (buildup) + 50	15,000	0.02
7.	LSP (1980s tech., moon, and LLO components)	30 (buildup) + 50	1,200,000	99% direct solar
8.	LSP (1980s tech.) (equipment from earth)	30 (buildup) + 50	300	0.1

of equipment to output approximately 1 TW_e-y. Upriver silting, political objections, and major floods and seismic events limit a dam's useful life. Global hydroelectric resources are projected to supply ≤1 TW_e of averaged power [30].

Nuclear-fission light-water reactors have massive foundations, containment vessels, and facilities and are only 50% more mass efficient than Grand Coulee (Table 10.1: 3). They pay back their construction and maintenance energy in 1.3 years. But, like coal-fired plants, they consume a nonrenewable resource—uranium. The considerable mass of short- and long-term nuclear waste storage facilities is not included.

A coal-fired power plant, including the mining equipment and unit trains, requires only 25% of the operating mass of a light-water reactor (Table 10.1: 4). However, it consumes half the mass of Grand Coulee every year in coal for the same averaged power output. A 20-TW_e world would consume approximately 6600 Grand Coulees of coal per year. This includes the additional coal burned for mining, transport, and ash management, but not for CO_2 sequestration.

The LSP system rectennas on earth use ≤15% of the mass on earth of a comparable coal plant [5]. The rectennas (Figure 10.1) receive power as an inward flow of massless photons. The electrical and electronic components add only approximately 15,000 t/TW_e-y [6 (chapter 9)]. Better rectenna design can greatly reduce the mass of support structures.

The LSP system differs from all terrestrial power systems in its effects on earth. Reservoirs behind hydroelectric dams, farming for biomass, and large solar arrays change earth's reflectance, usually decreasing it and increasing the waste solar power that flows into the biosphere. A 20-TW_e collection of wind farms will have to capture approximately 10% of the wind power that flows at low altitudes over the continents. They can potentially modify global wind patterns. Coal and other fossil fuel plants emit CO_2. The CO_2 from fossil fuels added since the start of the Industrial Revolution converts an additional 1% of the 174,000 TW_s of solar power that earth intercepts into extra heat (~170 TW_t), which then recirculates within the biosphere, heats it, and contributes to global warming [6 (chapters 1, 2, 9)].

LSP system capacity can be increased at a faster rate to provide an extra 13 TW_e of clean, massless electricity to earth. The additional clean electric energy, approximately 650 TW_e-y, can be used to remove the 1900 Gt of industrially produced CO_2 in the oceans and the atmosphere and to convert it into inert carbonate rock.

10.6 Returns from Lunar Solar Power Investment

The LSP components built on the airless and mechanically quiet moon can be very thin and therefore have very low mass per unit of collected solar energy. Concentrated sunlight is the primary energy source to make glasses for the photovoltaics (Figure 10.3: #1), microwave reflectors (#2), microwave generators (#3), and other components shown in Figure 10.3 from the lunar soils. Solar electric energy is a small fraction of the input energy. Solar electric power is produced from the simple arrays placed directly on the lunar surface (Figure 10.3: #1). The solar power is collected in each of the many small power plots distributed within a larger power base. The thousands of individual subbeams generated from the power plots are then combined in free space above the power base into massless microwave power beams directed toward earth. Thus, the power components on the moon have only 0.6% of the mass per unit of energy delivered to earth of a terrestrial PV array with only 2 hours of battery power storage (Table 10.1: 7 and 1).

Less than 300 t of consumables and machinery are required to build the components to send back to earth 1 TW_e-y of clean energy (Table 10.1: 8). This highly efficient use of mass is a key to the projected low cost of solar electric energy from the moon and the short time to pay back the energy invested to install a unit of power capacity. The tons per terawatt year of electric energy of the lunar components can be reduced by another factor of 10 in the transition from 1980s to 2020s operating and production technology [16].

Restricting the initial lunar and space expenditures to $0.001/kW$_e$-h of returned energy implies that up to $140,000/kg can be expended on lunar demonstrations and space operations. The demonstration LSP system, with its industrial-scale manufacture of low-cost commodity goods, such as structures and life-support chemicals, and new lunar services, especially electric power, will enable safe living on the moon.

A 20-TW$_e$ LSP system will require major investments in machines, supplies, and people on the moon and for the transportation systems between earth and moon. The LSP demonstration can be deployed with operationally robust derivatives of the Space Transportation System, new expendable launch vehicles, and lunar facilities and systems derived from the operational International Space Station. The LSP system production phase will provide the motivation, operational scale, and financial resources to significantly reduce transportation costs. Table 10.2 [6] gives the approximate life-cycle cost of an LSP system that delivers, without the need for power storage, 20 TW$_e$ of load following power to earth and 1500 TW$_e$-y of energy between 2015 and 2100. The construction and maintenance of rectennas on earth is 87% of the life-cycle cost. This estimate assumes the 1980s levels of operating and production technologies on the moon. All labor and initial production machinery is brought from earth. The financing of the construction phase at 3%/year real interest for 30 years, approximately $34 trillion, is the largest single expense.

If 90% of the lunar manufacturing system is made from lunar materials and the rectennas on earth employ screen-like reflectors to concentrate the incoming microwaves, the cost drops by a factor of 10 (Table 10.2: 7). Again, the single major expense is interest, $3.2 trillion (Table 10.2: 7 and 8), on funds borrowed to conduct the construction phase. However, most of the LSP and rectenna construction can be a pay-as-you-go process. Rectennas can begin to efficiently output power after their diameter exceeds a few hundred meters.

An LSP demonstration project can likely break even by the time it sells approximately 0.5 TW$_e$-y of energy on earth at $0.1/kW$_e$-h. Total cost to this point is projected to be approximately $0.5 trillion measured in 2003 dollars (Table 10.2: 9). After breakeven, lunar construction could be financed from energy sales. Private and government bonds can be sold and the funds used to accelerate the growth of power bases and rectennas, thereby accelerating the growth of GWP. The projected unit costs of LSP electricity, averaged over 1500 TW$_e$-y, are $0.005/kW$_e$-h (Table 10.2: 6), $0.0005/kW$_e$-h (Table 10.2: 7), and $0.0003/kW$_e$-h (Table 10.2: 8). Several versions of the LSP system have been analyzed over a range of 0.1 to 10 times the reference industrial productivity on the moon, and all are profitable [16, 19].

Consider the implications of the LSP system for China. A prosperous China of 1.3 billion people requires at least 2600 GW$_e$. This would enable a sustainable $110 trillion GDP/year or $84,000 GDP/y-person by midcentury. The 15,000 km^2 of rectennas required would be 0.16% of the land area of China and 0.015% of its land, lakes, and estuaries. For scale, the Three Gorges Dam is projected to output an annually averaged power of approximately

TABLE 10.2

Major Cash Flows (T$/y = $1 × 10^12/year), Energy Cost (–T$), Positive Liquidity (+T$), and Summed Gross World Product (+T$)

	Economic Activity or Value	T$/y	T$
1.	GWP (@ $2000 in 2002, 6.03 billion people)	+40.3	
2.	Sum GWP (@ $6840/y-person, 2000–2100; 10 billion in 2050)		+6050
3.	Coal-fueled system cost (1500 TW_e-y and 3%/year for 30 years, $2000)		−1700
4.	Terrestrial solar power cost (*1500 TW_e-y; 1 day of thermal storage)		−1400
5.	Terrestrial PV cost (*; 45 days × 6.6 TW_e storage output) or GEO solar Power satellites from earth or moon (*; 40 TW_e for load following)		−10,000
6.	LSP (ref. 1980s) cost (1500 TW_e-y; rectennas 87%)		−72
7.	LSP (bootstrapped) cost (1500 TW_e-y; reflector rectennas 77%)		−6.9
8.	Engineering cost of #7 (1500 TW_e-y; @ 0%/year interest)		−3.7
9.	LSP demo breakeven (2003 dollars) (@ $0.1/$kW_e$-h		−0.5
10.	E&P (global oil and natural gas @ 2003) for ~1.6 TW_e equiv.	−0.2	
11.	U.S. corporate liquidity (2003)		+4.7
12.	Annual profits selling 20 TW_e (@ 1¢/kW_e-h)	+1.6	
13.	GWP (10 billion people @ 2050 with 20 TW_e)	+830	
14.	GWP (@ 2100 with 20 TW_e)	+1200	
15.	Sum of GWP with LSP (2000–2100: 1500 TW_e-y of LSP energy)		+66,000
16.	Gross lunar product (GLP) {0.3% energy @ 40 years, 5%/year growth}	+11	
17.	GLP-funded R&D (2050) {3.3% of GLP}	+0.36	

Note: This table was created by scaling the engineering model of LSP to the delivery of 1,500 TWe-y of net electric energy to receivers on Earth. This is indicated by "*1500 TWe-y". The next two "*" indicate this electric output without using the entire phrase.

11.41 GW_e. Its reservoir surface area will be approximately 1045 km² [35]. An equal area of rectennas would output approximately 209 GW_e of load following power. The use of the Three Gorges Dam can be optimized for water management and transportation. An LSP-powered China would be prosperous and would not have to mine coal or compete with other nations for fuel and mineral resources.

Implementing the LSP system will generate the following major resources on the moon, in cis-lunar space, and on earth:

- >5000 t of general lunar manufacturing systems at 10 years
- >500,000 t of lunar production machinery at 15 years
- Large-scale transport between earth and moon
- Thousands of people on the moon and many thousands more teleworkers on earth
- Multiple bases along and across the limbs of the moon as seen from earth
- Megawatts to terawatts of electric and beamed power at 5 years to 2050 (earth to cis-lunar space and moon, moon to cis-lunar space and earth)

Approximately a decade into the full-scale installation of the LSP system, a small fraction of the production facilities and people can be redirected to grow a generalized economy using local lunar materials. Assume that after the demonstration phase is complete the capacity of the LSP system is expanded by 0.3% and this extra capacity increases at 5%/year. The extra energy is directed to generating net new wealth on the moon and in cis-lunar space. Approximately 0.01 TW_e would be available to generate approximately \$0.4 trillion/year of new gross lunar product (GLP) 5 years into LSP construction. By year 40, when earth is provided 20 TW_e, the lunar economy receives 0.26 TW_e and achieves approximately \$11 trillion/year GLP (Table 10.2: 16). After year 40, most of the LSP production facilities can be redirected to economic development of the moon and human development of our solar system.

10.7 Twenty-First Century Power Tools

The power tools of the twentieth century were power stations (hydroelectric, coal, natural gas, oil, nuclear fission), local and long-distance electric transmission systems, electric motors, electric trains, and the thousands of other electric devices and electronic systems that have been used to enable exponential economic growth since 1880 [13].

The new power tools for the twenty-first century are as follows:

- Sun—the primary power source
- LSP bases
- Large-aperture radar facilities within the LSP bases

- Power beams
- Beam redirectors/rectennas in space and on earth and moon
- Beam-powered transports (ion and others)
- Rationalized industrial production facilities on earth, moon, and other moonlike bodies

Given adequate funding and focus, within 5–8 years microwave power beams can be sent from stations on earth to recycling ion-drive tugs that carry cargo from low orbits about earth to low orbits about moon [31]. High-tonnage cis-lunar transport cost can be reduced to the order of tens of dollars per kilogram. Low-cost commercial-scale power can be beamed to lunar bases to immediately enable the industrial-scale operations appropriate to the rapid growth of the LSP system.

Industrial-scale lunar power and lunar development can enable the following additional benefits:

- Provide approximately 300,000 km^2 of radio telescope collection area (LSP reflectors front and back areas).
- Locate, track, and characterize all comets and asteroids and directly deflect those sufficiently large to threaten earth.
- Enable radio, optical, x-ray, and γ-ray telescope networks distributed about the moon to continuously monitor our earth, sun, the solar system, and the universe.
- Extract all industrial carbon dioxide from earth's atmosphere by 2100.

10.8 Moving Forward to 2050

By 2050, sustainable, clean, dependable, and abundant electric power can enable the steady increase of sustainable wealth on earth by a factor of $10 to approximately $84,000 GWP/person-y. Energy, like agriculture has done, can decrease from approximately 15% of GWP today to approximately 1% of GWP and thus liberate considerable global economic activity for other uses. Over the twenty-first century, the LSP system can increase cumulative GWP, after energy expenditures, by approximately $60,000 trillion (Table 10.2: 15, 8, and 2). The LSP system will provide humanity a significant increase in sustainable profits and wealth with literally unlimited opportunities for growth [32, 33].

Development of the moon and industrial-scale exploration of the solar system requires the development of known lunar resources and a skilled population on the moon, in cis-lunar space, and on earth. These newly useful resources will generate growing off-earth profits and net new wealth that index humankind's growing independence from the ancient, limited resources of its birthplace.

Acknowledgments

It is a pleasure to acknowledge the comments on early drafts by Prof. Robert G. Watts of Tulane University (Department of Mechanical Engineering, retired), New Orleans, Louisiana, and by my clear-minded, determined editor Paula R. Criswell.

Questions for Discussion

The text and references in this chapter provide data and hints to answering the following questions. Some answers are based on solutions to previous questions.

Energy Resources and Power Flows

1. What is the temperature of the inside of our sun? What is the temperature of the universe?
2. Physical power (watts) moves as a flow of energy (joules) per unit time from a store of high-quality energy to an energy sink of lower quality. What is energy quality? What is high-quality energy? What is low-quality energy? Give examples.
3. Create a table that lists in the first column the energy resources within our solar system organized in descending order of energy quality. In the second column, list the total available energy (joules) of each resource. In the third column, list the average power flow from that resource (watts).
4. Create a table that lists in the first column the natural energy resources within the biosphere of earth. List them in descending order of energy quality. In the second column, list the total available energy (joules) of the resource. In the third column, list the average power flow from that resource (watts).

Human Scale

5. Consider preagricultural, agricultural, early coal–iron industrial, and modern electric societies. Discuss which natural and commercial

energy resources supported the major flow of commercial power within the biosphere and the human per capita power usage during these four periods.

6. What are the minimum power requirements of a preindustrial human being? List the sources of that power, the power level, and energy consumption over a typical lifetime about the globe.
7. What is the average power consumption of a person in a nation of the Organisation for Economic Co-operation and Development (OECD)? List the sources of that power, the power level, and the energy consumption over a typical lifetime.
8. In 1976, H. E. Goeller and A. M. Weinberg estimated the carbon-free per capita thermal power needed in western Europe and Japan to maintain their flows of physical goods and services. Update those estimates to today's western Europe, Japan, and the United States. Extrapolate these estimates for everyone in the world to have that level of prosperity.

Commercial Power and Wealth

9. Describe the relation between commercial power and human-enabled wealth for agricultural and early industrial societies. The work of A. Maddison (OECD, deceased) provides a useful starting point. Consider both the per capita and the total wealth of various regions.
10. End-use power is slowly converting from thermal to electric power. What is the economic output of a unit of thermal end-use energy in the OECD nations since 1980? What is the economic output of a unit of end-use electric energy in the OECD nations since 1980? What roles do thermal and electric power play in the recent rapid economic growth of China?
11. The combustion of ancient coal and hydrocarbons with modern atmospheric oxygen now generates over 60% of commercial electric power. What is the total life-cycle mass of all machinery and equipment required to deliver 1 GW_e of electric power to regional consumers for 40 years? Model a coal-fueled system supplied by Wyoming mines, coal trains to the U.S. East Coast, and subject to current environmental requirements. Estimate the average tons of machinery per gigawatt electric year of end-use electric power. Repeat the calculations for the total mass of fuel, oxygen, cooling water, and other materials consumed over the construction, operation, and decommissioning of the power system.
12. How much wealth was generated in the region over the life cycle of the coal-fired power system? Compare the wealth to the life-cycle cost of the coal-fired power system.

13. Serious students could use their methodology to examine existing natural gas, conventional fission, hydroelectric, biomass, wind, and terrestrial solar systems as both stand-alone and combined systems. Do electric energy costs increase or decrease as diverse systems are combined? What are the major variables?

Global Powers: Commercial and Natural

14. Humans are steadily increasing their consumption of earth's natural resources to provide commercial power. Suppose a world of 10 billion people achieves the present-day per capita power consumption in the United States. Compare the annual flows of consumables required by that world power system to the natural flows within the biosphere. List the major interactions between commercial and natural mass flows and power flows.
15. Within the core of the sun, hydrogen is converted by the process of nuclear fusion into a smaller mass of helium. The lost mass (∂m) is converted into energy ($E = \partial m \times c^2$), where c is the speed of light. How much mass is converted by fusion each second within the sun to illuminate earth, illuminate the moon, or provide 20 TW_e of commercial electric power to earth?
16. List the conventional and proposed power-generating technologies that, when operated on earth, can output high-quality power at a lower life-cycle cost than the sun. Consider operating the conventional and proposed technologies at great distances from the sun. How far from (or near) the sun must they be located so as to be less expensive than concentrating sunlight through mirrors or lens onto PV cells? Include waste heat rejection systems.

Lunar Commercial Power and Industries

17. The light of full moon provides sufficient visible illumination on a clear night for walking about. What are the maximum watts per square meter of the full moon on earth where the moon is directly overhead? What is the total power delivered to earth by the full moon? What was the "Lunar Society," and what is its relation to the Industrial Revolution?
18. What is the total mass of microwave photons required over 1 year from the moon to deliver 20 TW of power? Compare this mass of microwave photons to the mass of carbon, hydrocarbons, and oxygen molecules required to provide 20 TW_e of electric power using conventional fossil fuel power plants. What is the mass of trees, phytoplankton, and other photosynthetic plants required to provide the oxygen? What is the mass of nutrients and water required to support their growth?

19. Fields of antennas on earth receive microwave photons from LSP bases and rectify the photons into electric power (i.e., rectennas). Approximately 90% of the incoming microwave power is converted into commercial electricity. Approximately 10% of the input power appears as waste heat above the rectennas. Compare the flows of waste heat from rectennas and the electricity used finally to that from other power systems (fossil fueled, biomass fueled, nuclear fueled, fields of PV and thermal solar plants, etc.). Describe the systems and resources required to process and release that waste heat. Note the potential effects on the environment.
20. Beaming of power is an extrapolation of radar and radio astronomy technologies established during World War II and the Cold War to higher total power through larger apertures. Electrical engineers are encouraged to reexamine the fundamentals of radar and consider the limits of power beaming at the commercial-level terrestrial and lunar facilities and facilities in orbit about the earth and the moon?

Body Politic

21. President D. Eisenhower signed into law the National System of Interstate and Defense Highways (1956) and the National Aeronautics and Space Act (1958). Calculate the federal funds expended on both programs since their inception. List the major economic benefits and costs presented by both programs to U.S. citizens.
22. What are the benefits to U.S. citizens and others that can be provided by an abundant source of electric power? How can citizens, through their collectively government and public and private enterprises, establish profitable industrial and commercial power operations on the moon and within the cis-lunar space?
23. Is "less is more" true for global power systems?

References

1. Climate Change (2007). Synthesis Report, 73 pp, (especially note Section 6) http://www.ipcc.ch/pdf/assessment-report/ar4/syr/ar4_syr.pdf. Accessed October 29, 2012.
2. World Energy Council (2000, page 2). Energy for Tomorrow's World – Acting Now!, 175 pp, Atalink Projects Ltd., London.
3. WEC Statement (2009). Building the new World Energy Order, 4 pp, http://www.worldenergy.org/documents/wec_statement_2009.pdf. Accessed October 29, 2012.
4. Energy Information Agency, Table 8.9 Electricity End Use, 1949–2008, (2008). http://www.eia.doe.gov/emeu/aer/txt/ptb0809.html. Accessed October 29, 2012.

5. Criswell, D. R. (2009). Lunar Solar Power for the Human Race to Sustainable Prosperity, 31 pp. (manuscript).
6. Watts, R. G. (editor) (2002). Innovative Energy Strategies to CO2 Stabilization, Cambridge University Press. http://www.cambridge.org/uk/catalogue/catalogue.asp?isbn=9780521807258. Accessed October 29, 2012.
7. Goeller, H. E., and A. M. Weinberg (1976). The age of substitutability. *Science*, Vol. 191, 683–689.
8. http://en.wikipedia.org/wiki/Oxygen_cycle. Accessed October 29, 2012.
9. http://en.wikipedia.org/wiki/File:Global_Carbon_Emission_by_Type_to_Y2004.png. Accessed October 29, 2012.
10. Falkowsi, P., R. J. Scholes, E. Boyle, J. Canadell, D. Canadeld, J. Elser, N. Gruber, K. Hibbard, P. Hogberg, S. Linder, F. T. Mackenzie, B. Moore III, T. Pedersen, Y. Rosenthal, S. Seitzinger, V. Smetacek, and W. Steffen (2002, October 13). The global carbon cycle: A test of our knowledge of earth as a system. *Science*, Vol. 290, 293 pp, (Table 1).
11. Hoffert, M. I., K. Caldeira, G. Benford, D. R. Criswell, C. Green, H. Herzog, A. K. Jain, H. S. Kheshgi, K. S. Lackner, J. S. Lewis, H. D. Lightfoot, W. Manheimer, J. C. Mankins, M. E. Mauel, L. J. Perkins, M. E. Schlesinger, T. Volk, and T. M. L. Wigley (2002, November 1). Advanced technology paths to global climate stability: Energy for a greenhouse planet. *Science*, Vol. 298, 981–987.
12. Criswell, D. R. (2004). Lunar-solar power system, In Encyclopedia of Energy (Cutler J. Cleveland, Editor-in-Chief), Vol. 3 (Gl–Ma), 677–689, Boston: Elsevier Academic Press.
13. Criswell, D. R. (2003, December/2004, January). Lunar Solar Power: Reaching for the Moon? IEEE POTENTIALS, 20–25.
14. Criswell, D. R. (2002). Characteristics of commercial power systems to support a prosperous global economy. *Acta Astronautica*, Vol. 51, No. 1–9, 173–179.
15. Criswell, D. R. (2002, April/May). Solar power via the Moon, *The Industrial Physicist*, 12–15. See letters to the editor and Reprise in June/July, August/September, and October/November issues or at http://www.tipmagazine.com. Accessed October 29, 2012.
16. Criswell, D. R. (2001, October). Lunar Solar Power System: Industrial Research, Development, and Demonstration, [Discussion Sessions; Division 1: World Energy Market Challenges; 1.2.2 Hydroelectricity, Nuclear Energy and New Renewables] 17 pp, 18th World Energy Congress, Buenos Aires, Argentina (available from author).
17. Criswell, D. R. (2000). Lunar solar power system: Review of the technology base of an operational LSP system. *Acta Astronautica*, Vol. 46, No. 8, 531–540, Elsevier Sciences Ltd.
18. Waldron, R. D., and D. R. Criswell (1998). Costs of space power and rectennas on Earth, IAF-98-R.4.03, 5 pp, International Astronautical Congress.
19. Criswell, D. R., and R. G. Russell (1996). Data envelopment anaylsis of space and terrestrially-based large scale commercial power systems for Earth: A prototype analysis of their relative economic advantages. *Solar Energy*, Vol. 56, No. 1, 119–131, Elsevier Science Ltd.
20. Waldron, R. D., and D. R. Criswell (1993, September). Summary of characteristics of beamed power radiation patterns and side lobe residual power levels for Lunar Power System. *Acta Astronautica*, Vol. 29, No. 10/11, 765–769. Pergamon Press Ltd.

21. Criswell D. R., and R. D. Waldron (1993). International lunar base and lunar-based power system to supply Earth with electric power. *Acta Astronautica*, Vol. 29, No. 6, 469–480. Pergamon Press Ltd. Presented at 42nd Congress of the International Astronautical Federation, October 5–11, 1991 (Montreal, Canada). 15 pp. Paper #IAA-91–699.
22. Criswell, D. R., and R. D. Waldron (1990, August). Lunar System to Supply Solar Electric Power to Earth, Proceedings of the 25th Intersociety Energy Conversion Engineering Conf., Vol. 1, #900279, Aerospace Space Power Systems, 61–71, Reno, NV.
23. Criswell, D. R., and R. D. Waldron (1991, April). Results of Analyses of a Lunar-based Power System to Supply Earth with 20,000 GW of Electric Power, Proceedings of SPS 91 Power from Space: the Second International Symposium, page a3.6, 186–193. Paris, France.
24. Waldron, R. D., and D. R. Criswell (1980). Materials processing in space, In *Space Industrialization* (Ed. B. O'Leary), Vol. I, Ch. 5, 98–130, Boca Raton, FL: CRC Press.
25. Criswell, D. R., and R. D. Waldron (1980). Lunar utilization, In *Space Industrialization* (Ed. B. O'Leary), Vol. II, Ch. 1, 1–53, Boca Raton, FL: CRC Press.
26. Waldron, R. D., and D. R. Criswell (1983). Concept of the lunar solar power system. *Space Solar Power Rev.*, Vol. 5, 53–75.
27. Waldron, R. D., T. E. Erstfeld, and D. R. Criswell (1979, February 12). The role of chemical engineering in space manufacturing. *Chem Engg.*, 81–94, McGraw-Hill, New York, NY [Note table VI]. Translated into Chinese. 1979 American Industrial Report, (September) Issue No. 37: 26–43. (in #63.)
28. Cirillo, R. R., B. S. Cho, M. R. Monarch, and E. P. Levine (1980, April). *Comparative Analysis of Net Energy Balance of Satellite Power Systems and Other Energy Systems*, DOE/ER-0056, Argonne National Laboratory, Lamont, IL, 136 pp.
29. Criswell, D. R. (1991, August 27–30). Terrestrial and space power systems: Life-cycle energy considerations, SPS91: POWER FROM SPACE, A1.2, 71–81, SEE & ISF, Paris/Gif-sur-Yvette.
30. Hydropower and the World's Energy Future (2000). http://www.ieahydro.org/reports/Hydrofut.pdf. Accessed October 29, 2012.
31. Brown, W. C. (1992, July). *A Transportronic Solution to the Problem of Interorbital Transportation*, NASA CR-191152, National Technical Information Services, Springfield, VA, 168 pp.
32. Criswell, D. R. (1985). Solar System Industrialization: Implications for Interstellar Migration. In *Interstellar Migration and the Human Experience* (Eds. B. R. Finney and E. M. Jones), 50–87. 354 pp, Univ. Calif Press, Berkeley, CA. Noted in June 1986 "Book Reviews," by P. Morrison. Scientific American. 28–29.
33. Criswell, D. R. (1985). Cis-lunar industrialization and higher human options. *Space Solar Power Rev.*, Vol. 5, 5–38.
34. Wikipedia (2012, March). Cooling tower. http://en.wikipedia.org/wiki/Cooling_tower_system.
35. Wikipedia (2013, January 19) Three Gorges Dam. http://en.wikipedia.org/wiki/Three_Gorges_Dam.

11

Adapting to Climate Change

Donald J. Wuebbles

CONTENTS

11.1 Introduction

Given the earlier discussion on the importance of climate change and the potential for significant changes to the earth's climate over the twenty-first century and beyond, this chapter is aimed at discussing the societal basis for adaptation to climate change. Adaptation is about responding to the risks associated with climate vulnerability and building resilience in response to these risks. While it is hoped that the worst of the impacts will be avoided through mitigation actions, adaptation is essential to reduce the potential effects on society and ecosystems from unavoidable changes in the climate system. At the same time, and discussed further below, it is important to recognize that mitigation and adaptation policies can often be synergistic.

Climate adaptation can result either as a response to an existing impact or as anticipatory of a recognized risk (see Smit et al. 2000, for further discussion on these concepts). As an example, moving people and goods from a flooded area is reactive adaptation, while policies aimed at land-use planning because of a potential increasing risk for flooding would be anticipatory adaptation. In many cases, anticipatory adaptation is likely to have much less economic and well-being effects on society. The National Research Council (NRC) (2010) Panel on Adapting to the Impacts of Climate Change concluded that anticipatory climate change adaptation is a highly desirable risk management approach for the United States.

Decision makers need to draw on the best scientific understanding of climate change and societal vulnerabilities, and then carefully consider the likely efficacy and broader implications of different adaptation strategies (Frumhoff et al. 2007). Policies that aim to reduce the population, infrastructure, and economic activity in coastal floodplains, for example, can minimize the negative impacts on local businesses, facilitate relocation to higher ground, provide adequate compensation where necessary, avoid additional environmental damage, and rehabilitate threatened habitats. In addition, adaptation planning must be flexible—the understanding of future climate change can only be given in terms of a range because of various uncertainties, including the climate sensitivity because of feedbacks in the climate system and how future emissions related to human activities will change.

This chapter draws heavily on the little amount of available literature that exists about adaptation, including several reports coauthored by the author (e.g., Frumhoff et al. 2007, 2008; Hayhoe et al. 2008) resulting from regional climate assessments and the special National Research Council Report on Adapting to the Impacts of Climate Change (NRC 2010). One of the key conclusions, therefore, is that there is a lack of sufficient research on adaptation approaches, including what works and what does not. At this time, climate change adaptation has not been a research priority in the United States or other areas of the world (NRC 2010).

11.2 Adaptation Will Be Necessary

Human society and natural systems have always adjusted to climate variability, but the changes in climate occurring now and projected for this century are outside the range of human experience and, in any case, are occurring too rapidly for nature to readily respond. The prospect that the climate system, human society, and ecosystems may experience significant transitions to new states renders our previous experience an incomplete guide for future adaptation (NRC 2010).

The amount and rate of future changes in climate depend largely on the current and future emissions of carbon dioxide and other radiatively important gases and particles that are affecting the earth's climate, and the choices humanity makes to reduce those emissions. However, the slowness of the world's nations to respond toward reducing emissions or even slowing the increase in emissions makes it likely that significant changes in climate will occur over the future decades and perhaps over this century beyond even if mitigation eventually becomes a global priority.

The unavoided changes in climate are likely to result in impacts on human society and ecosystems because of local and regional vulnerabilities to climatic conditions. To reduce the effect of these changes, there is much that humanity can do toward adapting to the societal, economic, and ecological stresses that will accompany climate change. The planning for, and implementation of, adaptation policies and actions is therefore an important complement to the need to reduce human-related emissions to avoid the worst of the impacts from climate change. The sooner this planning can begin, the better policy makers and resource managers can prepare for the unavoidable consequences of climate change.

Climate change in the twenty-first century will create additional stresses on people and their environment and place growing demands on the ability to anticipate, prepare for, and respond to climate changes. For example, in coastal areas, even modest sea level rise, when combined with storms similar in frequency and intensity to those experienced today, can inflict heavy damage on the region's coastal infrastructure. Likewise, extreme heat threatens the health of an aging population. It is therefore not the region's capacity to adapt but its ultimate actions that will determine the severity of global warming's impact.

A delay in preparing for anticipated changes—or a continued reliance on infrastructure and emergency response plans based on historical experience rather than projected conditions—will increase the exposure to climate risks. The adaptation strategies most relevant to (and feasible for) any specific community or economic sector must be assessed on a case-by-case basis that addresses various technological, policy, financial, social, ecological, and ethical considerations. Any one of these factors may impose important constraints on the community's ability to adapt.

There are many components to getting to adaptation. The simple representation in Figure 11.1 shows how the understanding of the local/regional climate changes, the identified vulnerabilities, the assessed risks, adaptation strategies, and other factors can all interact with the concerns about various sectors. As an example, consider the possibility of reduced water supply in the western United States as the climate continues to change (as implied by existing climate modeling studies) because of reduced snowfall and snowpack in the mountains combined, especially in the southwest, with reduced rainfall. These changes interact with the current vulnerabilities in this region to droughts and the many demands for water supply. Options for adaptation to the prospect of severe water shortages might include improving efficiency in water use, reducing the need in competing purposes such as power plant cooling, finding ways to reduce evaporation from reservoirs, using groundwater smartly, reconsidering current water rights laws, developing new

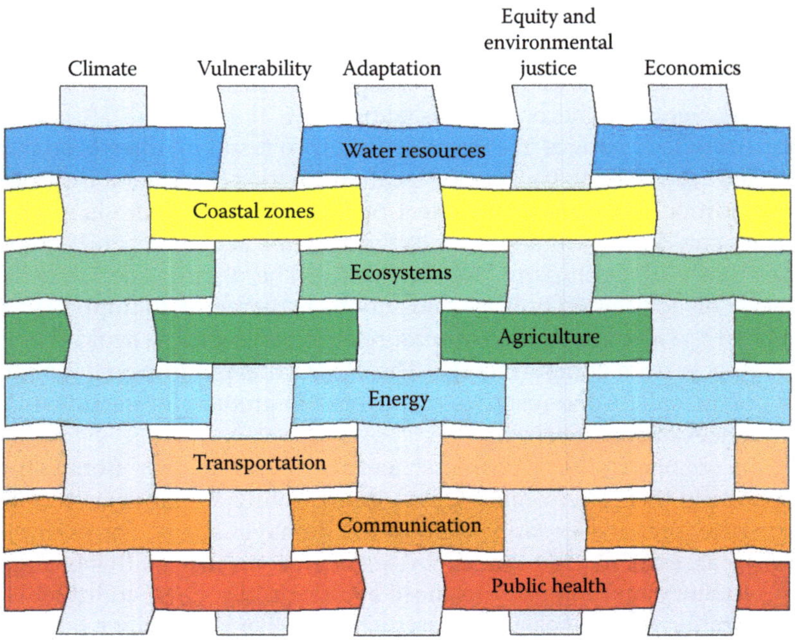

FIGURE 11.1
A simple cartoon of the components and complex interactions affecting adaptation across different sectors. (From New York State Climate Action Council. 2010. *Climate Action Plan Interim Report*, http://www.dec.ny.gov/docs/administration_pdf/irpart1.pdf.)

means for water transfers between basins, and the use of desalination. Thus, while it may be difficult to precisely know the impacts that will occur from climate change, adaptation can offer a way of preparing for and minimizing risks to the sectors that may be affected by these impacts. At the same time, there may be unintended and unanticipated consequences from some of these, and these must be explored.

11.3 Economics: The Costs of Inaction

The distinction between reactive adaptation and anticipatory adaptation can be very important to policy development because of the differences in motivations for these two types of adaptation. Anticipatory adaptation uses knowledge and resources that exist today to prevent potential future risks or to take advantage of likely climate changes. However, reactive adaptation uses existing capabilities and resources to deal with events at the time they occur. Practically speaking, political policy decisions are often easier to make after a crisis. However, the cost of preventive or anticipatory actions has historically often been much lower than the cost of reactive actions (Hallegatte et al. 2011).

Taking action to prepare for the likely consequences of climate change, while not cost-free, can prove to be less expensive than the economic damage that would result from doing nothing (and then being forced to take reactive adaptation measures). Less affluent people and communities, even in relatively wealthy regions, will be among the hardest hit by global warming in part because they can least afford to prepare for, or cope with, the impacts (such as extreme heat) once they occur (Frumhoff et al. 2007). Similarly, small or geographically isolated businesses may have fewer resources and available options for coping with climate change. And, as this report makes clear, some economically important species such as cod and lobster as well as other species of great intrinsic value such as the Bicknell's thrush in the U.S. Northeast will soon cross climate-related thresholds beyond which they will lose critical habitat or other conditions necessary for their continued survival.

For any given region, it is a moral obligation to focus attention on the plight of vulnerable communities, sectors, and ecosystems, and to increase their resilience to climate change—an outcome essential to the economic and ecological sustainability of the region as a whole. Decision makers, however, need a better understanding of the factors that contribute to climate vulnerability and the full suite of options that are available to reduce that vulnerability. Well-designed social or resource management policies, for instance, could substantially enhance the region's ability to adapt.

Therefore, moving swiftly to reduce vulnerability is also smart economics. For example, governments, businesses, and communities that plan ahead will be positioned to take advantage of the possible benefits of climate change. For example, communities that are modernizing their water and sewer infrastructure could protect their investments by incorporating near-term projections of more extreme rainfall into their plans, thus taking into account future changes in climate.

Some countries are likely to face major issues in the twenty-first century that may require drastic actions. For example, some island communities, such as the Maldives, are likely to be extremely vulnerable to sea level rise. The very existence of the community may require moving to different locations. Coastal regions of many countries are also likely to face major issues requiring long-term planning. Bangladesh is thought to be particularly vulnerable. The disappearance of glaciers may also affect water access in some countries, again requiring advanced planning if the community is going to prosper.

11.4 Getting Awareness of Potential Vulnerabilities

Climate change is already affecting the economies, lifestyles, and traditions of people across the world, and the societal and ecosystem vulnerabilities to a changing climate are expected to grow much more substantially in the decades to come (USGCRP 2000, 2009; IPCC 2007a). Assessments of the potential vulnerabilities under a range of possible futures are an important means to understand the possible vulnerabilities and how those vulnerabilities may be affected under different levels of climate change.

The higher-emissions scenarios described in Chapter 2 do not represent a ceiling for the changes the world may experience. At the same time, the lower-emissions scenario discussed there does not represent a floor. As an example, the lower-emissions scenario in Chapter 2 describes a world in which atmospheric concentrations of CO_2 rise from approximately 390 parts per million (ppm) today to approximately 550 ppm by the end of the century—in contrast to 940 ppm under the highest-emissions scenario. Substantially different impacts would be expected depending on the pathway we actually follow. However, many lines of evidence indicate that even lower emissions—and thus less severe impacts—are well within humanity's reach if we are ready to make the decisions that need to be made toward developing and applying the technology to reduce emissions of the gases and particles driving the changes in climate.

Various analyses (e.g., IPCC 2001a, 2007a) have evaluated the technical and economic potential for stabilizing atmospheric concentrations of heat-trapping gases at or below the CO_2 equivalent of 450 ppm. These studies suggest that achieving such a target would require the United States and other industrialized nations to reduce emissions by roughly 80% below 2000 levels by mid-century, along with substantial reductions by developing countries. This would be very difficult to achieve, and we currently show no sign that humanity is ready to tackle such a goal. Even if this goal were achieved, there would continue to be significant changes in climate and a number of resulting impacts requiring adaptation.

11.5 Building Resilience while Reducing Vulnerability

Institutions—not just physical organizations but the regulations, rules, and norms that guide behavior—can also influence the access to information and therefore the ability to use that information in decision-making. Environmental protection legislation, antidiscrimination policies, market regulations, and common expectations about socially acceptable behavior are all examples of such institutions. Well-functioning institutions can provide stability in an otherwise volatile environment, but when the environment changes fundamentally or rapidly—as is expected during this century due to climate change, especially if the higher-emissions scenario prevails—institutions can fail to serve their intended functions and hinder our ability to adapt. For example, in heavily developed and densely populated coastal zones, options for managing the mounting risks related to global warming are constrained by past development and land-use patterns, coastal laws and regulations, and the expectations that past policies have fostered among property owners (NOAA 2010). Coastal managers face an increasingly dynamic shoreline that will require regulations based on projected risks rather than historic risks.

Institutional mechanisms such as "risk spreading"—accomplished chiefly through insurance—have the potential to ease the risks of climate change. As an example, insurance is likely to play an important role in helping coastal residents and businesses improve their ability to cope with and recover from coastal storms (Frumhoff et al. 2007). Public education about the changing exposure to climate risks along the northeast's coastline will also need to be linked to strict enforcement of building codes and land-use regulations, and perhaps mandatory insurance coverage (especially for lower-income individuals).

11.6 Adaptation Planning

The ability to cope with and adapt to change is highly variable across populations, economic sectors, and even geographic regions. Climate change thus has the potential to aggravate resource constraints, social inequalities, and even public health disparities among different communities. For example, in urban populations, it is the elderly, the very young, and the poor that will likely suffer most from the stress of extreme heat.

In many instances, people's vulnerability to climate change is related to limited and climate-sensitive career options. Maine fishermen and women who can no longer make a living off fish such as cod, for example, have switched to lobster fishing, but if this fishery should suffer from lobster shell disease or a decline in the survival and growth of young lobsters, few income-generating alternatives will be readily available (Frumhoff et al. 2007).

It must be noted that climate change also presents opportunities to enhance economic development, human well-being, and social and environmental equity. In the countryside, small family farms—often vulnerable to economic trends and weather extremes—may be less capable of responding to the demands of a changing climate than industrial farms with greater resources, but the fact that many small farms are highly diversified in terms of the crops and livestock they produce could position them to take advantage of changing opportunities. In cities, efforts to build emergency response systems and support networks (e.g., buddy systems among neighbors during heat waves) may yield dividends in the form of new and stronger community ties.

Technological solutions for reducing climate vulnerability include seawalls for protection against rising seas, crop varieties better adapted to warmer conditions and longer growing seasons, and air conditioning to cope with summer heat waves. Some technologies may provide only temporary relief or none at all, or may have unintended side effects. Snow-dependent winter recreation provides a good example of the costs and limitations of using technology to cope with climate change: snowmaking equipment may provide an economically and technically feasible means of helping the alpine skiing industry cope with warmer winters (albeit with water and energy costs and certain ecological implications), but it is not a feasible solution for snowmobiling (Frumhoff et al. 2007).

Even for industries where technological solutions are available, managers must weigh the cost and timing of the transition to new technologies during what may be decades of quixotic weather. The commercial timber industry, for example, may adapt to warmer winters by switching to equipment that allows harvesting on unfrozen ground, but industry managers would need a clear climate signal to make this costly change.

Technological solutions can also seduce us into believing that a problem can be solved at the operational level, without requiring deeper systemic changes. Before embracing such solutions, therefore, policy makers must carefully weigh the environmental and social consequences by asking a set of questions starting with: who can *afford* to adopt the technology and who cannot? For example, are small farms as able as large farms to install costly air conditioning equipment that would maintain dairy (and poultry) production in a warmer climate?

It can be a struggle to integrate climate issues into daily work. Experience with major change and enhancement initiatives throughout the world has shown that successful implementation of adaptation requires more than good technical ideas (Parzen 2008). Governments and businesses need to focus a portion of their energies and resources on enhancing capacity as an organization to initiate and deliver climate change adaptation measures successfully. An organizational planning framework for a climate change adaptation program might include the following steps:

- Embed climate change adaptation considerations into planning processes.
- Breakdown existing "silo" mentalities and foster cooperation.
- Be capable of responding to climate change knowledge that is incomplete and dynamic.

A collaborative process may be most important, as climate-related risks touch many departments and cover many aspects of life. The approach that the City of Chicago has taken, for example, is to create a coordinating team of senior leadership, supported by interdepartmental and organizational working groups.

11.7 Using Chicago as an Example of Adaptation in Action

In 2007, the City of Chicago launched a comprehensive, multistakeholder planning process to tackle the very real impact of climate change on Chicago's environment and economy. The resulting assessment and the plans for adaptation are available both in journal publications (see Wuebbles et al. 2010, which introduces the other papers in the special issue in the *Journal of Great Lakes Research*) and in reports (e.g., Hayhoe et al. 2008) available at http://www.chicagoclimateaction.org. One of the key features of the process was the decision to rely on rigorous analysis to formulate policy decisions. Research on climate impacts (and costs of action and inaction) had a profound impact on the Chicago Climate Action Plan that was launched in September 2008. The analysis helped Chicago decision makers to understand the scale and scope of the problem. It provided a means to engage stakeholders in discussion about useful responses. It built the case for aggressive action, provided a means to prioritize actions, and produced tools for engaging the public in climate action.

The analyses in this assessment (Hayhoe et al. 2008, 2010a,b) to evaluate potential vulnerabilities and climate change impacts at the regional and city scales, which itself involved extensive interaction with city leaders toward determining key vulnerabilities, helped decision makers to understand the benefits of early action to address climate change, and provided a starting point for engagement with affected stakeholders about how to respond to climate change. Other analyses (Coffee et al. 2010) used economic risk analysis and interviews with numerous experts from 18 city departments to determine the extent to which each department's operations, assets, personnel, and services would be physically and operationally affected by projected climate changes. The research revealed that almost every department would be impacted by climate change and how. It revealed that the cost under the high-emissions scenario could be more than three times higher than the $700 million cost projected under the low-emissions scenario. Consequences of climate changes were rated on the overall severity of impact on health and safety and/or on economic impact. A list was then compiled of best practice adaptation tactics to address the highest risks and prioritized the list based on adaptation and mitigation benefits, costs, and catalytic potential.

An inventory of potential Chicago adaptation tactics was developed for reducing vulnerability to four key events: extreme heat, extreme precipitation, damage to infrastructure, and degradation of ecosystems. The inventory included tactics "best practice" adaptation measures planned in cities around the world that could be applicable to Chicago. For example, New Zealand offered a best practice for modifying planning processes to account for potential impacts of climate changes. New Zealand has issued guidelines for local authorities that include specific questions to be asked when drawing up individual plans, including whether the risk management analysis takes into account changes because of climate impacts and whether the plan includes a specific commitment to keep up-to-date with changing understanding of climate change implications.

As outlined in the Chicago Climate Action Plan (also see Coffee et al. 2010), the original list of potential adaptation actions was narrowed down to several dozen based on expected benefits and costs, time horizon, and barriers to implementation. Benefit types included life safety, human health, prevention of significant infrastructure damage, preservation of ecosystem health, prevention of major economic disruption, maintaining quality of life, uninterrupted city services, prevention of minor economic impact, maintenance of revenue, uninterrupted tourism, and reduced costs. These analyses categorized as "Must Do/Early Action" high net benefit adaptation tactics designed to prevent impacts with a short-term time horizon and with few impediments to implementation. "Must Do" Actions were high net benefit, but had potential impediments. It categorized as "Investigate Further" those tactics that addressed longer-term impacts, but had strong benefit-to-cost ratios. Finally, it categorized tactics that could have value in the long term, but were high cost as tactics to "Watch." Also identified were numerous "No Regret" options that could deliver benefits greater than their costs, regardless of the extent of future climate change. For example, the City of Chicago adopted the "No Regrets" tactic of updating its extreme weather operations plan using climate change projections.

The adaptation tactics for reducing vulnerability to extreme heat events that the City of Chicago adopted included examples from most of the categories. The City of Chicago began to focus existing tree planting by the Park District and Bureau of Forestry at locations where trees could reduce the urban heat island effect, which was a "Must Do" tactic. Another "Must Do" tactic the city pursued was to protect air quality in a higher-temperature environment by initiating a process to amend the air ordinance. An "Investigate Further" tactic that the city has made part of its plan is to develop thermal environment maps. Many of the tactics to address both extreme heat and extreme precipitation align with the City of Chicago's long-standing commitment to green development, including green urban design, green infrastructure for stormwater management, and greening for urban heat island reduction. This history has made it make climate change adaptation part of business as usual in Chicago.

Several examples of specific adaptation strategies and tactics from the Chicago Climate Action Plan are now presented to demonstrate how adaptation is being implemented.

11.7.1 Adaptation Strategy 1: Reduce Vulnerability to Extreme Heat Events

The 1995 heat wave was one of the most devastating in the history of Chicago with several hundred deaths and extensive hospitalizations associated with that week-long event. The climate projection analyses by Hayhoe et al. (2008, 2010a) suggest that such an unusual event could become commonplace by the end of the century (see low- and high-emission scenarios in Figure 11.2). While Chicago already has a sophisticated emergency response planning and coordination system in place, the city found it needed to be better prepared for the likelihood of an increasing number of heat waves.

Several different approaches have been put into place. The first is to be better prepared for heat waves by neighborhood checks, increasing the number of "cooling centers" where people can stay during heat waves, getting more air conditioners into neighborhoods with high-risk populations. The second tactic is to manage the regions where urban heat is a major issue by planting additional trees and adding to the number of "green" roofs. The third tactic is to manage concerns about urban ozone pollution under a warmer climate.

As a first step, the city evaluated medium- and high-risk areas of the city for heat effects. Landsat data were useful in helping determine the warmest areas (Figure 11.3). A series of potential adaptation actions were defined by (1) identifying "hotspots" within the city, including buildings and parking lots, and then determining decisions toward temperature reduction, energy savings, and air quality improvement; (2) optimizing routine and special tree planting to take into account and offset the urban heat island effect

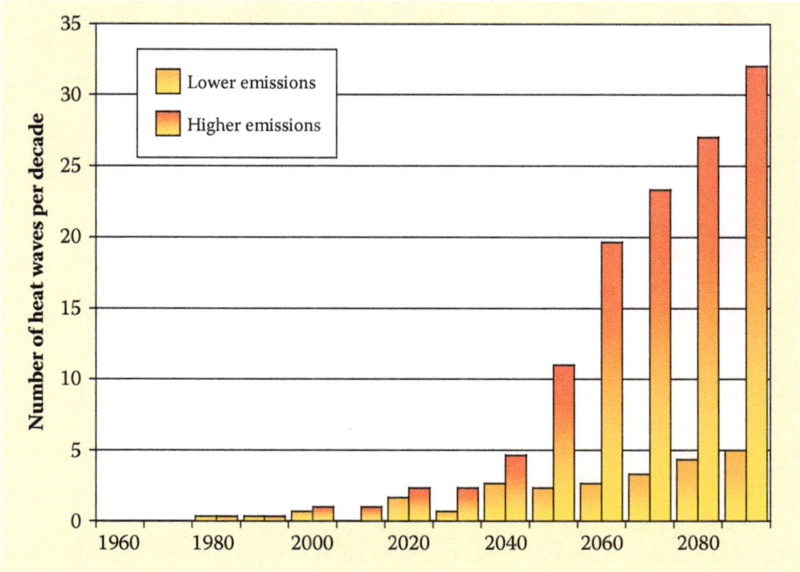

FIGURE 11.2
Number of 1995-like heat waves observed (1970–1999) and projected to occur over the twenty-first century as simulated by climate models based on the analyses by Hayhoe et al. (2008) using high (A1FI) and low (B1) SRES scenarios (Nakicenovic et al. 2000) for human-related emissions as used in the IPCC (2001a, 2007a) assessments.

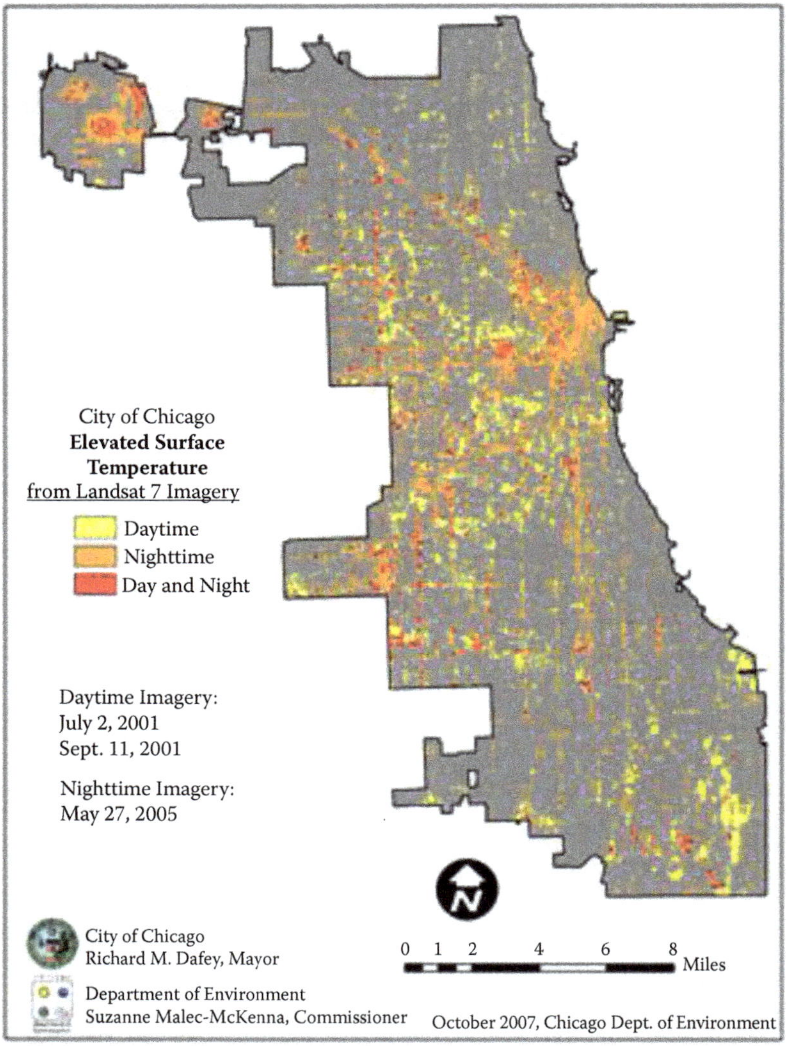

FIGURE 11.3
Using satellite data, the City of Chicago evaluated heat-related health issues, with the map showing those areas that are rated as high-risk and requiring immediate attention. This map also shows areas of the city subject to the "heat-island" effect. (Courtesy of the Chicago Department of Environment.)

(Figure 11.4 shows the downtown areas of the Chicago Marathon where a special effort was put into adding more trees to cool the area); and (3) further increasing the extent of "green" roofs (Figure 11.5), an urban response that the City of Chicago is already well known for being a leader.

Higher temperatures will exacerbate urban smog. Meeting the ozone standard is an important goal in addressing this challenge. Potential adaptation

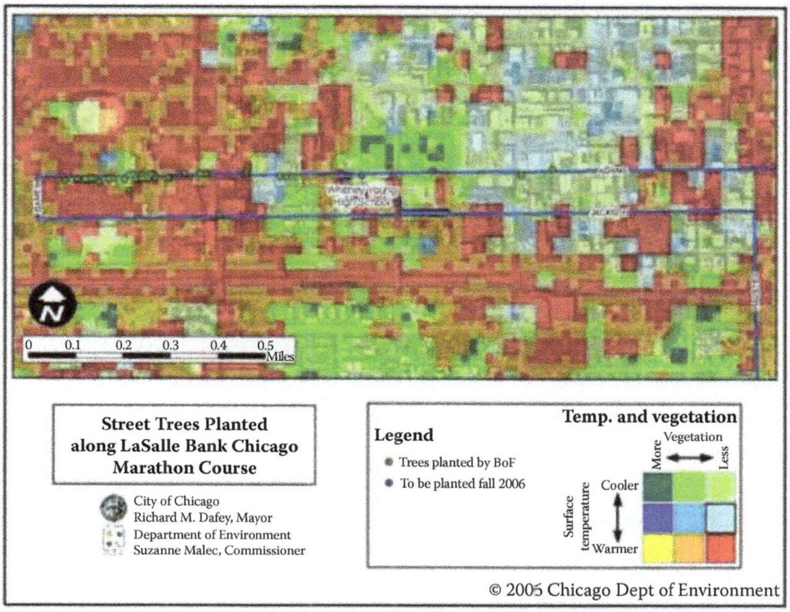

FIGURE 11.4
Analysis of the existing tree plantings along the course of the Chicago Marathon at the time of the Chicago climate assessment. Since 2008, there have been extensive new plantings of trees in this region of downtown Chicago. (Courtesy of the Chicago Department of Environment.)

FIGURE 11.5
The City of Chicago is well known for its use of green roofs to help cool the downtown area.

actions for municipalities, businesses, and other organizations in Chicago and the Chicago metropolitan area were defined, including the following steps to avoid worse air quality under a higher-temperature regime:

- Support public transit discount days, especially during the ozone season. Provide public transit, tax-free, flexible spending account benefits to employees.
- Share information with employees and residents about ways they can reduce ozone precursor emissions through energy conservation, lawn maintenance, and public transit.
- Encourage less engine idling for trucks, buses, locomotives, and marine vehicles.
- Use natural landscaping to reduce emissions from mowers and maintenance equipment.

11.7.2 Adaptation Strategy 2: Reduce Vulnerability to Extreme Precipitation Events

When we discussed potential vulnerabilities of the city to climate change, one of the first issues raised by city officials was the concern associated with high-precipitation events, especially when the city receives more than 2.5 inches of rain in a single day. In those events, the stormwater runoff can combine with sewer overflows, creating a major mess for the city, often leading to opening of the gates to dump this excess water and sewage into Lake Michigan. In addition, a major concern in some parts of the city was excessive water in the basements of houses and apartments.

As seen by our analyses (Hayhoe et al. 2008, 2010a) in Figure 11.6, the intensity of heavy precipitation events in Chicago is expected to continue to increase over this century, with the largest increases corresponding to the heavier events. This change means that the occurrence of such heavy precipitation events would arise from about once every four years to as much as once every other year by the late twenty-first century. Adaptation may thus be required to counteract the enhancement in flooding and stormwater contamination.

Several measures have been implemented by Chicago to address basement flooding, including:

- Installation of "rain blocker" devices that detain stormwater in the street rather than overloading the sewer system
- Encouraging the disconnection of downspouts, where appropriate, which also slows the rate at which water enters the system
- Recently adopting a new stormwater ordinance that requires large new developments to manage the first half inch of rain entirely on site

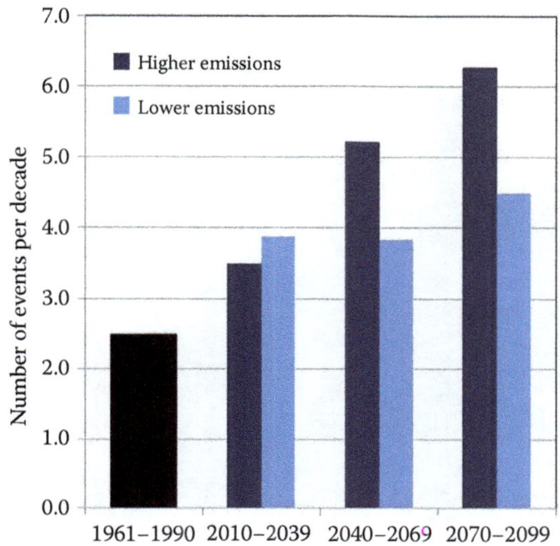

FIGURE 11.6
Projection changes in the frequency of heavy precipitation events (more than 2.5 in. per day) in the City of Chicago. (From Hayhoe et al., Report prepared for the City of Chicago, 2008, available at http://www.chicagoclimateaction.org/pages/research___reports/48.php; Hayhoe et al., *J. Great Lakes Res.*, 36, 7–21, 2010. With permission.)

For the larger-scale watershed issues, the first step was to incorporate climate projections into stormwater planning. As a result, a number of distributed solution possibilities either are under consideration or are being implemented. These include the following:

- Allowing permeable paving in parking lots
- Residential rear yard permeability—requiring rear yard open space to remain permeable
- Allowing use of perforated plastic pipes for underground drainage as part of subsurface stormwater management
- Developing green roof specifications that would absorb some of the excess water
- Creating and restoring wetlands (stormwater parks) to help manage stormwater entering the river
- Evaluating the capabilities and limitations of the current sewer system (see Figure 11.7)

The City of Chicago has 39 specific adaptation tactics in various stages of implementation, including:

- Reduce urban heat island effect through strategic planning
- Implement needs assessment to evaluate drainage infrastructure

FIGURE 11.7
The sewer model for Chicago can simulate future surface flooding impacts, helping to prioritize current resource use.

- Prepare Chicago drainage solutions strategy and commence water conservation strategy
- Develop energy resource management plan
- Analyze materials and methods for roadway and rail infrastructure
- Create and implement Chicago urban forest management plan

Each adaptation working group has been developing action plans that include primary actors, timelines, budgets, and performance measures that the city can tracked.

Recommendations were made not only for specific actions but also for an ongoing process to engage city decision makers in adaptation planning. Following this advice, the City of Chicago built the infrastructure it needs to continue to adapt as new climate research becomes available and to track its performance in both adaptation and mitigation. The City of Chicago's cabinet-level Green Steering Committee of departments and sister agencies, led by the city's chief environmental officer, formed five multidepartmental working groups to develop adaptation action plans: (1) extreme heat; (2) extreme precipitation events; (3) building, equipment, and infrastructure vulnerabilities; (4) ecosystem degradation; and (5) leadership, planning, and communication.

The success of climate action planning is dependent on tracking and reporting on performance over time. The City of Chicago's Continuous Improvement through Performance Measurement initiative for the Chicago Climate Action Plan includes periodic measurement of progress on each mitigation and adaptation tactic in the plan. However, it also includes broad quality of life indicators that capture adaptive capacity. The city is in the process of finalizing what indicators it will track. Some examples of likely measures are days of good air quality, morbidity due to elevated surface temperature, decreased urban heat island effects, increased permeable land area, percent of land covered by tree canopy, gallons of water usage per capita, and number of swim bans.

Chicago's climate impacts and adaptation analyses were of high value to decision makers during the development of the Chicago Climate Impact Plan, and continue to be so during implementation. Departments that gave input throughout the research process continue to be engaged in a continuous improvement process. The analyses helped to catalyze Chicago's long-term process of mitigation and adaptation. It also helped the city to set priorities for action and hone in on win-wins that address both mitigation and adaptation, as shown in Figure 11.8. This figure also indicates there are many situations where adaptation measures also contribute to reducing emissions of the radiatively important gases and particles.

The experiences and adaptation measures from Chicago can serve as useful guides for other cities and communities. While it was extremely useful to have impact analysis specific for Chicago, this is not essential for all cities. Cities increasingly can take advantage of existing research on climate impacts for their region, such as the ongoing work for the National Climate Assessment.

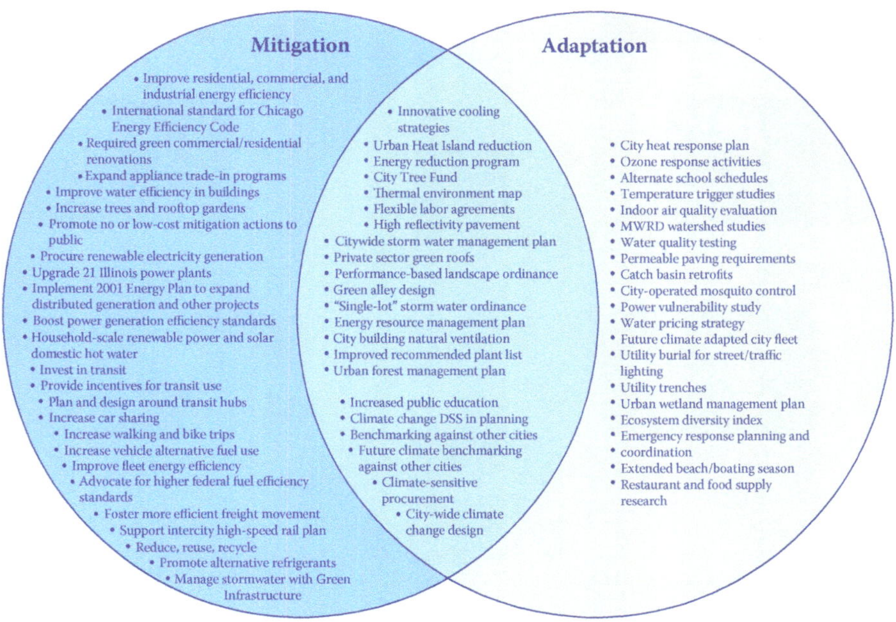

FIGURE 11.8
The nexus of mitigation and adaptation measures developed for the implementation of the Chicago Climate Action Plan.

11.8 A Methodology for Adaptation Strategies

Several studies (Frumhoff et al. 2007, 2008; NRC 2010) have considered general strategies for adaptation planning that are combined in the discussion below. Other studies, such as ClimAID, the Integrated Assessment for Effective Climate Adaptation Strategies in New York State, have adopted a methodology that works for their needs, as depicted in Figure 11.9.

It should be recognized that the various strategies with which countries, states, business sectors, and communities can prepare for climate change must be considered on a case-by-case basis. Each constituency is unique in the challenges it faces and its ability to adapt. However, the following principles (based primarily on Frumhoff et al. [2007] and Parzen [2008]) can help set priorities.

FIGURE 11.9

Methodology used in adaptation development in New York State. (From Rosenzweig, C. et al., eds, *Responding to Climate Change in New York State: The ClimAID Integrated Assessment for Effective Climate Change Adaptation in New York State,* New York State Energy Research and Development Authority, Albany, NY, 2011.)

- *Assess and monitor the changing environment.* Decision makers and resource managers must keep informed about the specific consequences of climate change for their region and areas of oversight. In particular, improved monitoring of both the climate and the condition of natural systems can give decision makers clearer signals about the need for action and more time to formulate appropriate adaptation strategies.
- *Track indicators of vulnerability and adaptation.* Monitoring both the progress of specific adaptation strategies and the social factors that limit a community's ability to adapt can enable decision makers to modify adaptation strategies and improve outcomes.
- *Build adaptive capacity.* Planning for climate change adaptation implies changing the human system as well as changing the "bricks and mortar" infrastructure.
- *Take the long view.* Decisions with long-term implications (e.g., investments in infrastructure and capital-intensive equipment, irreversible land-use choices) must be considered in the context of climate projections.
- *Embed climate change considerations into planning processes.* Planning processes identify future organizational operational, equipment, or infrastructure needs involving procuring new goods, services, and/or products or building new infrastructure that must function under a new set of climate conditions. To manage risk, planning processes should be modified to account for potential impacts of climate changes.
- *Consider the most vulnerable first.* Climate-sensitive species, ecosystems, economic sectors, communities, and populations that are already heavily stressed for nonclimatic reasons should be given high priority in policy and management decisions.
- *Look for Win-Wins.* Adaptation measures will almost always have multiple benefits, which may include reduced energy costs, improved aesthetics, reduced air and water pollution, and so on. These benefits can and should be considered and evaluated to convey the overall benefit of these measures.
- *Take incremental steps that maximize future options.* If possible, take incremental steps rather than large actions to keep options open to adapt in the future. Avoid making decisions that make it more difficult to manage future climate risks. Phased projects can help to avoid costly decisions. Also distributed infrastructure can be more flexible in responding to change than investment in large centralized systems.
- *Be dynamic, aware, nimble, and flexible.* The paradox of process planning is the intermixed integration of past, present, and future. We plan for the future, do so in the present, and use data from the past. Specifically, the planning process uses historical data to measure the

capability of current projects to achieve future goals. To adapt to climate change, municipalities and businesses need to incorporate new data and reassess decisions continually. The successful adaptation to climate change requires a learning organization, one that adapts to changing environmental factors.

- *Build on and strengthen social networks.* Connections and relationships between trusted individuals and organizations are an asset for adaptation at the community level and within business sectors. Strong leaders can inspire organizations in times of difficult change, and well-connected and well-informed individuals can disseminate information that may be critical for effective adaptation.
- *Put regional assets to work.* The United States and many other regions of the world have an enormous wealth of scientific and technological expertise in its universities and businesses that can be harnessed to improve our understanding of adaptation opportunities and challenges.
- *Improve public communication.* Regular, effective communication with and engagement of the public on climate change helps build our capacity to adapt.
- *Act swiftly to reduce emissions.* Strong, immediate action to reduce emissions can slow climate change, limit its consequences, and give our society and ecosystems a better chance to successfully adapt to those changes we cannot avoid.

11.9 Conclusions and Future Directions

Climate change represents an enormous challenge, but humans have the capacity to respond to avoid the worst of the changes and to adapt to those that cannot be avoided. Because the climate is already changing because of human activities and some amount of additional change is inevitable, adapting to higher temperatures and to trends in severe is an essential (and complementary) strategy to reduce emissions. For each adaptation measure considered, policy makers and managers must carefully assess the potential barriers, costs, and unintended social and environmental consequences.

Throughout the world, communities, industries, and individuals have, over time, developed ways to deal with their region's climate. This proven adaptability suggests that humanity has the resources and experience to cope with climate change (at the same time, we need to consider how to help ecosystems adapt). However, the rapid rate and widespread impacts of the expected climate changes necessitate immediate and sustained action by policy makers and resource managers—together with the engagement of scientific and technological expertise—will be needed to avoid the most dangerous consequences of the changing climate. As concluded in NRC (2010), the process of adapting to likely impacts of climate change poses a daunting challenge and the stakes are high. We need to take this seriously not only for ourselves but also for the sakes of our children and grandchildren.

Questions for Discussion

1. What are the some of the possible vulnerabilities to climate change close to where you live? In your country? What are possible approaches to adaptation that could be used to address those vulnerabilities?
2. Are the poorer developing countries likely to face more concerns about vulnerabilities from climate change? What will be some of their challenges toward adaptation?
3. What are some environmental consequences of massive desalination from the Pacific Ocean to provide freshwater for the U.S. west coast.
4. What would be the effects on the U.S. (or your country's) coastlines if sea level rises by 1 or 2 m (3 to 6 ft) over this century? What issues would develop requiring major adaptation?

References

Coffee, J. E., J. Parzen, M. Wagstaff, and R. S. Lewis, 2010. Preparing for a changing climate: The Chicago Climate action's Plan's adaptation strategy. *J. Great Lakes Res.*, 36, 115–117.

Frumhoff, P. C., J. J. McCarthy, J. M. Melillo, S. C. Moser, and D. J. Wuebbles, 2007. *Confronting Climate Change in the U.S. Northeast: Science, Impacts, and Solutions.* Synthesis report of the Northeast Climate Impacts Assessment (NECIA). Union of Concerned Scientists (UCS), Cambridge, MA.

Frumhoff, P. C., J. J. McCarthy, J. M. Melillo, S. C. Moser, D. J. Wuebbles, C. Wake, and E. Spanger-Siegfried, 2008. An integrated climate change assessment for the Northeast United States. *Mitig. Adapt. Strat. Global Change*, 13, 419–423.

Hallegatte, S., F. Lecocq, and C. de Perthuis, 2011. *Designing Climate Change Adaptation Policies: An Economic Framework.* Policy Research Working Paper 5568, The World Bank, Washington, DC.

Hayhoe, K., D. Wuebbles, and the Climate Science Team, 2008. Climate Change and Chicago: Projections and Potential Impacts. Report prepared for the city of Chicago, available at http://www.chicagoclimateaction.org/pages/research___reports/48.php. Accessed February 2012.

Hayhoe, K., J. VanDorn, D. J. Wuebbles, K. A. Cherkauer, and S. Vavrus, 2010a. Regional climate change projections for Chicago and the great lakes. *J. Great Lakes Res.*, 36, 7–21.

Hayhoe, K., M. Robson, J. Rogula, M. Aufhammer, N. Miller, J. VanDorn, and D. Wuebbles, 2010b. An integrated framework for quantifying and valuing climate change impacts on urban energy and infrastructure: A Chicago case study. *J. Great Lakes Res.*, 36, 94–105.

Intergovernmental Panel on Climate Change (IPCC), 2001a. *Climate Change 2001: The Scientific Basis.* Houghton, J. T., Y. Ding, D. J. Griggs, M. Noguer, P. J. van der Linden, X. Dai, K. Maskell, and C. A. Johnson (eds.), Cambridge University Press, Cambridge, UK.

Intergovernmental Panel on Climate Change (IPCC), 2001b. *Climate Change 2001: Mitigation. Contribution of Working Group III to the Third Assessment Report of the Intergovernmental Panel on Climate Change (IPCC).* Metz, B., O. Davidson, R. Swart, and J. Pan (eds.), Cambridge University Press, Cambridge, UK, 700 pp.

Intergovernmental Panel on Climate Change (IPCC), 2007a. *Climate Change 2007: The Physical Science Basis. Contribution of Working Group I to the Fourth Assessment Report of the Intergovernmental Panel on Climate Change (IPCC).* Cambridge University Press, Cambridge, United Kingdom and New York, NY, 996 pp.

Intergovernmental Panel on Climate Change (IPCC), 2007b. *Climate Change 2007: Impacts, Adaptation and Vulnerability. Contribution of Working Group II to the Fourth Assessment Report of the Intergovernmental Panel on Climate Change.* Parry, M. L., O. F. Canziani, J. P. Palutikof, P. J. van der Linden, C. E. Hanson (eds.), Cambridge University Press, Cambridge, UK and New York, 976 pp.

Intergovernmental Panel on Climate Change (IPCC), 2007c. *Climate Change 2007: Mitigation. Contribution of Working Group III to the Fourth Assessment Report of the Intergovernmental Panel on Climate Change,* Metz, B., O. Davidson, P. Bosch, R. Dave and L. Meyer (eds.), Cambridge University Press, Cambridge, UK, 851 pp.

Nakicenovic, N., O. Davidson, G. Davis, A. Grubler, T. Kram, E. L. L. Rovere, B. Metz, T. Morita, W. Pepper, H. Pitcher, A. Sankovski, P. Shukla, R. Swart, R. Watson, and Z. Dadi, 2000. *Emissions Scenarios: A Special Report of Working Group III of the Intergovernmental Panel on Climate Change.* Cambridge University Press, Cambridge, United Kingdom, and New York, NY.

National Research Council, 2010. *America's Climate Choices: Adapting to the Impacts of Climate Change.* National Academies, Washington, DC.

NOAA Office of Ocean and Coastal Resource Management, 2010. Adapting to Climate Change: A Planning Guide for State Coastal Managers. Available from NOAA at http://coastalmanagement.noaa.gov/climate/adaptation. html. Accessed February 2012.

Parzen, J., 2008. Chicago Area Climate Change Quick Guide: Adapting to the Physical Imapcts of Climate Change. City of Chicago. Available at http://www.chicagoclimateaction.org. Accessed February 2012.

Rosenzweig, C., W. Solecki, A. DeGaetano, M. O'Grady, S. Hassol, and P. Grabhorn (eds.) 2011. *Responding to Climate Change in New York State: The ClimAID Integrated Assessment for Effective Climate Change Adaptation in New York State.* Synthesis Report. New York State Energy Research and Development Authority, Albany, NY.

Smit, B., I. Burton, R. J. T. Klein, and J. Wandel, 2000. An anatomy of adaptation to climate change and variability. *Climatic Change*, 45, 223–251.

U.S. Global Change Research Program (USGCRP), 2009. *Global Climate Change Impacts on the United States.* Cambridge University Press, New York, NY.

Wuebbles, D. J., K. Hayhoe, and J. Parzen, 2010. Introduction: Assessing the effects of climate change on Chicago and the great lakes. *J. Great Lakes Res.*, 36, 1–6.

12

Climate Engineering: Impact Reducer or Suffering Inducer?

Michael C. MacCracken

CONTENTS

12.1 Introduction

Detection and attribution studies make clear that the growth of energy use and food production during the twentieth century has vaulted human activities into the leading cause of recent changes in the global climate (IPCC 2007a). In addition, there is also suggestive evidence that changes in land cover and agricultural practices going back several thousand years altered the atmospheric concentrations of carbon dioxide (CO_2) and methane (CH_4), initiating continental to hemispheric-scale changes in climate well before the industrialization of society (Ruddiman 2001, 2007). The disturbing consequence of these findings is that continuing reliance on the present approaches for supplying energy and other commodities needed to sustain and enhance the standard of living and welfare of people around the world will cause even more significant changes in climate over coming centuries (IPCC 2007a). Indeed, substantial consequences are expected to result even if aggressive near-term efforts are made to limit emissions from combustion of coal, petroleum, and natural gas.

The question considered in this chapter is whether the knowledge that has been gained about the functioning of the earth system offers the potential for implementing alternative approaches capable of counterbalancing or moderating some or all of the unintended changes in global climate being caused by large-scale human activities. This chapter, which is an update of Flannery et al. (1997), describes the approaches that have been proposed to geoengineer the climate—that is, to take deliberate actions aimed at ensuring that global reliance on fossil fuels as a source of energy will not cause "dangerous anthropogenic interference" with the global climate, as the 1992 United Nations Framework Convention on Climate Change set as its objective.

Section 12.2 summarizes the environmental and societal situation, options, and the urgency of the challenge facing the public and government leaders.* As this section indicates, despite remaining scientific uncertainties, the world community has taken so long to address the climate change issue that very significant environmental and societal disruption appears highly likely, even with aggressive limitations in emissions. Because of the failure to act to limit emissions over the last 50 years, projections of changes during the twenty-first century make clear that climate change will not and cannot soon be stopped. Although it may seem presumptuous, risky, and contentious to consider having the world community take on the role of master of the global climate system, geoengineering is starting to receive serious attention as a possible complementary policy to mitigation and adaptation because of the seemingly significant limits of existing approaches in preventing what appear to be unacceptable changes in climate in the decades ahead (Crutzen 2006).

Section 12.3 explores the options for directly reducing the causes of global climate change, that is, for reducing the rising atmospheric concentrations of greenhouse gases. Approaches range from enhancing the natural CO_2 uptake processes by land and ocean biota to direct air capture (DAC) using industrial techniques followed by geological sequestration. Together, these approaches are referred to as carbon dioxide removal (CDR) technologies.

Because the identified means for CDR have the potential for only slowly moderating the pace of projected climate change, Section 12.4 explores the options for faster action. Such actions would require taking steps to alter the global energy balance, with most options focused on reducing the earth system's uptake of solar energy by an amount roughly equivalent to the amount of energy that the rising concentrations of CO_2 and other greenhouse gases are trapping. Such approaches are generally referred to as solar radiation management (SRM) or, more appropriately, global SRM.

Seeking to reduce the scope of assuming responsibility for the global energy balance, Section 12.5 explores proposals for actions that focus on moderating the most significant environmental and societal consequences of climate change: an example being seeking to slow the rapidly accelerating pace of climate change in the Arctic. Such focused approaches might appropriately be thought of as proactive adaptation, or even geo-adaptation—that is, seeking to moderate or offset the actual impacts of ongoing changes in the global climate.

Establishing that there is a technical potential for intentionally moderating or counterbalancing human-induced climate change may well be less of a challenge than moving from capability to implementation. Section 12.6 briefly explores some of the policy implications of potentially incorporating climate engineering approaches into a comprehensive global response to climate change. Were technical capabilities to be demonstrated (and some argue there may be reasons to avoid undertaking the research needed to do this), moving to implementation would introduce many more challenges,

* More detailed discussions of these issues are presented in earlier chapters, particularly 3–5 and 11, where more extensive references are provided.

including issues of international governance, recalcitrant nations, winners versus losers in space and time, international and intergenerational equity and ethics, relative obligations and responsibilities, the potential for abuses, the effects of considering and implementing a geoengineering policy on the pursuit of other policy options, climate change impacts that cannot be countered, and more. Even though the intent of geoengineering is to try to maintain climatic conditions as near to present (or slightly earlier) conditions as possible, intentional intervention may be an indication of hubris that poses too difficult a decision for society and governments without significant objection.

12.2 The Climate Change Predicament

Since the start of the Industrial Revolution in the mid-eighteenth century, the atmospheric CO_2 concentration has risen from about 278 to 392 ppm (~40%), and the concentration is rising more rapidly now than at any earlier time. The rate of rise is consistent with roughly half of each year's fossil fuel emissions adding to the long-term elevation of the atmospheric CO_2 concentration, such that, numerically, the annual ppm increase is roughly one-quarter of the amount of the CO_2 emissions expressed in gigatonnes of C (billions of metric tons).* Consistent with this rule, global fossil fuel emissions are currently ~8.5 GtC/year and the annual rate of increase is ~2–2.5 ppm/year (Canadell et al. 2007; Le Quéré et al. 2009). Analysis of the global carbon cycle indicates that elevation of the atmospheric CO_2 concentration will persist for a very long time, even as individual molecules exchange with the ocean and biosphere on time scales of roughly 5 years (Archer 2005; Bala et al. 2005; Montenegro et al. 2007).

The atmospheric CH_4 concentration is also substantially higher than its preindustrial level, with human activities having caused its increase from about 750 ppb to over 1750 ppb (~130%). Major anthropogenic sources driving this increase include agriculture (e.g., rice production; pig, cattle, and poultry raising), leaks from coal mines and natural gas pipelines, sewage and waste disposal practices, and more. For reasons not yet fully understood, the concentration of CH_4, which had been rising rapidly through the 1980s, leveled out starting in the mid-1990s, meaning that the increased emissions have been matched by an increase in the loss rate, which mainly consists of chemical oxidation in the atmosphere. Over the last few years, however,

* This approximation to the airborne fraction will be drawn upon later. Note that while scientists express emissions in billions of metric tons of carbon (GtC) in order to keep track of each carbon atom and to work with manageable numbers, international negotiators express emissions in units of millions of metric tons of carbon dioxide ($MMTCO_2$), which yields numbers that are larger by a factor of 3670.

the concentration of CH_4 has again started to increase, apparently mainly a result of increasing emissions from low- and high-latitude pools of soil carbon (e.g., in tropical rainforests and thawing permafrost).

In addition to the warming influences of the increasing concentrations of CO_2 and CH_4, the rising concentrations of other greenhouse gases and emissions of black carbon are also contributing to global warming, while the oxidation of emissions of sulfur dioxide (SO_2) from coal combustion is leading to a widespread sulfate layer that exerts a cooling influence on the global climate. Collectively, detection–attribution studies summarized by Hegerl and Zwiers (2007) attribute most of the 0.8°C warming since the late nineteenth century to anthropogenic factors. Modeling studies suggest that the ultimate warming, were the present level of forcing to persist indefinitely, would be about twice its present amount.

With fossil fuels supplying roughly 80% of the energy for the seven billion people now alive and with per capita emissions and incomes of most of these people well below the per capita levels in the developed nations, energy and economic projections are that emissions of CO_2 will rise substantially through the twenty-first century (IPCC 2007c; Moss et al. 2007). Conservatively, assuming an average population over the twenty-first century of 8 billion and annual per capita emissions of 2.5 metric tons of C, which is roughly the present per capita emission levels of Europeans and about half the level of those living in North America, global emissions over the century would amount to ~2000 GtC. Assuming that the past airborne fraction provides a reliable approximation, this level of emissions would increase the CO_2 concentration by ~500 ppm this century, raising the concentration to ~900 ppm by 2100. With emissions then continuing into the twenty-second century, the concentration would be headed to well above 1000 ppm, which would lead to ultimate global warming reaching 5°C–10°C.

The environmental and societal impacts of such a warming would be enormous (IPCC 2007b; UNEP 2009). The present global average temperature is about 6°C higher than during the Last Glacial Maximum, a time when sea level was about 120 m lower than at present because of the water tied up in thick ice sheets covering much of North America and Europe. The present temperature is also about 6°C cooler than during the Cretaceous, the era of the dinosaurs more than 65,000,000 years ago during which palm trees grew near the Arctic Circle. At that time, the absence of ice sheets would be equivalent to sea level being ~70 m higher than at present. Such large changes in climate and sea level, which are being driven far more rapidly than in the past, can be expected to be very disruptive to biodiversity, to ecosystems, generally, and to society (IPCC 2007b).

In that the 0.8°C increase is already causing loss of mass from the Greenland and Antarctic ice sheets, retreat of sea ice, thawing of permafrost, an acceleration in the rate of sea level rise, shifts in ranges of species, and much more; the risk of severe consequences (summarized in Schellnhuber et al., 2006) has led international leaders to agree in the 2009 Copenhagen

Accord that it would be dangerous to society and the environment for the global average temperature to exceed 2°C above its preindustrial level.

With so much important environmental change occurring at 0.8°C, a growing number of scientists are suggesting that avoiding a high likelihood of significant environmental and societal consequences may well require the increase in global average temperature be no more than 0.5°C above preindustrial, a level that has already been exceeded (e.g., see Hansen et al. 2008). IPCC's Fourth Working Group 2 assessment (IPCC 2007b) describes the impacts to be expected at various levels of global average warming and the potential for building resilience to and adapting to various amounts of climate change. The alternative to being able to adapt, either proactively or reactively, is suffering, which could range from inconvenience, to relocation, to extinction, depending on the intensity and pace of climate change and the adaptive capacity of a species, ecosystem, or community, for example. A possible metaphor for the situation is a car with sluggish brakes accelerating downhill on a curvy expressway with poorly defined exits that each represent an irreversible impact once the exit is passed. Further downhill are an unknown number of sharp curves, each without a guardrail to protect falling off a cliff that represents an unforeseen, nonlinear threat with the potential for impacts that would cause severe societal disruption.

Studies by Meinshausen (2006) suggest that, to have at least a 50% chance of the equilibrium warming not exceeding 2°C, the CO_2 concentration cannot be allowed to exceed ~450 ppm. Using the airborne fraction relationship described earlier, limiting the concentration to 450 ppm would require that global fossil fuel emissions for the rest of this century total no more than ~240 GtC (i.e., [450–390 ppm] times 4 GtC/ppm). With present fossil fuel emissions being about 8.5 GtC/year, the allowable emissions limit would, at this rate, be reached in 30 years, requiring emissions to be cut to near zero beyond that time. Alternatively, the allowable emissions could be decreased linearly to zero emissions over ~60 years. Somewhat greater amounts would be allowed if emissions of other greenhouse gasses (GHG) and black soot were reduced significantly over the next couple of decades, as suggested in the recent UNEP assessment (UNEP and WMO 2011; Shindell et al. 2012), but the decline in the cooling influence of the sulfate loading as coal use declines would tend to cancel out some of this beneficial additional effort. More problematically, if indications are correct that 2°C is too high a level to ensure avoidance of dangerous anthropogenic interference with the climate system (Hansen et al. 2008), then the reductions in CO_2 emissions must be even sharper and more rapid, leading to the need for removal processes to be increased to exceed emissions in order to pull the atmospheric CO_2 concentration back toward preindustrial levels (IPCC 2007c).

Although peaking at an atmospheric concentration of ~450 ppm or less may be desirable from climate change and impact perspectives, the slow pace of international negotiations gives little reason to expect that sufficient action will be taken to accomplish this. Indeed, even limiting the concentration to a doubling of the preindustrial concentration seems likely to prove very challenging (Hoffert et al. 2002). Using the airborne fraction relationship, stabilizing at a CO_2 concentration of 560 ppm, which would likely be associated with an equilibrium warming of up to ~4°C–5°C, would allow CO_2 emissions of no more than ~680 GtC (i.e., [560–390] ppm times 4 GtC/ppm). Accomplishing this would require that annual average global emissions during the twenty-first century be no more than annual average global emissions during the first decade of the century, even though population and average standard of living are both headed up, and that annual average global emissions thereafter to be near zero. With current global emissions already 30% above the target annual average emissions level, keeping emissions over this century to less than ~680 GtC would likely require developed nations to soon be on a path to near-zero CO_2 emissions. Because it will be difficult for developing nations to forgo use of fossil fuels to alleviate the poverty of their peoples until the developed nations demonstrate that countries can prosper with such low CO_2 emissions, this too may be more aggressive than can be accomplished.

Although there are those that suggest that limiting the CO_2 concentration to 560 ppm will not be economically feasible, the environmental and societal consequences of not doing so are likely to be very significant. As is evident from studies of the earth system behavior during glacial cycling, there is a strong natural carbon feedback process that could be activated by rising temperatures. In particular, the ice-core record appears to indicate that the warming of ocean waters caused by orbital shifts leads to an increase in the atmospheric CO_2 concentration that adds to the warming influence as part of a positive feedback process (Shakun et al. 2012). While climate change deniers often cite this correlation as evidence that models and traditional scientific understanding are mistaken, the paleoclimatic record is really suggesting that future global warming is likely to excite a strong, positive carbon cycle feedback, amplifying the rise in the CO_2 concentration being caused by human activities.

Estimates of the amount of carbon tied up in the permafrost and in ocean sediments are very large, and there are increasing indications that warming could soon induce significant emissions of carbon, either in the form of CO_2 or in the form of CH_4, depending on whether local conditions would lead to oxidation of CH_4 as part of the release process. With the warming influence of equal masses of CH_4 being far greater than that of CO_2 over decades to a century,* even a relatively small, warming-induced release of CH_4 from this reservoir could significantly amplify the pace of warming. An important concern based on study of the very disruptive climatic warming that occurred

* On a per unit mass basis, the warming influence of a methane release is ~22 times as large as the same mass of carbon dioxide integrated over 100 years and ~75 times as large integrated over 20 years. See the discussion on global warming potential in chapter 2 of IPCC's Fourth Assessment Report (Forster and Ramaswamy 2007).

about 35 million years ago (the Paleocene–Eocene thermal maximum) is that global warming could lead to a large CH_4 release that would, as part of a strong positive feedback, lead to further rapid and substantial warming. Even if anthropogenic emissions of CH_4 can be limited, the risk of substantial warming due to a natural methane hydrate/clathrate feedback will remain.

With unfettered use of fossil fuels (particularly coal and oil shale) likely to lead to unprecedented warming and even restricted use of fossil fuels likely to lead to at least some dangerously disruptive impacts that cannot be adequately ameliorated through adaptation, the remaining option, other than simply suffering, appears to be deliberate intervention to try to limit the amount of climate change and its impacts. Although often referred to as "geoengineering," using this term has the potential to generate public confusion because the term can also be applied to modern agriculture (which now raises enough food to feed 30–40 times as many people as nature on its own was able to supply) and to water resource management (which can now sustain vast populations even in quite arid conditions). In addition, roughly 50 years ago, a few prominent scientists proposed that the global climate be altered to enhance global economic development, for example, by warming the Arctic, an act of hubris that glossed over many important uncertainties and ethical concerns (e.g., see Fleming 2004, 2007). Because the intent of current proposals is to limit climate change, so the exact opposite of those earlier notions, using *climate engineering* to describe the effort to counterbalance fossil fuel-induced climate change seems an important differentiation.

In the current context, the goal of climate engineering is to limit or offset changes in climate and resulting impacts, seeking to keep climatic conditions (and sea level and ocean acidification, if possible) near those of the mid-to-late twentieth century to which natural ecosystems, societal infrastructure, and economic development have generally been adapted. Although the term *climate remediation* has also been suggested as an appropriate descriptor for this effort (BPC 2011), the envisioned approaches do not really fully counterbalance all of the CO_2-induced impacts; instead, what is being proposed is to trade one set of risks for another, with the intent being to minimize, in some integrated way, the overall, or at least the most disruptive, impacts on the environment and society.

Sections 12.3 through 12.5 describe the possible approaches from a technical perspective. They explore the actions that might possibly be taken in an attempt to return the climate to conditions characterizing the late twentieth century while also reducing the overall risk and economic impacts of human activities. Section 12.6 then addresses issues that have arisen in considering the possible inclusion of climate engineering as a policy option, focusing on the question of whether it might be sensible to actually consider research into and eventually deployment of any of the proposed approaches as a way to reduce overall risk and what the policy, governance, and equity complications might be (albeit from the likely restricted perspective of a physical scientist and engineer).

12.3 Options for Extending Mitigation with Carbon Dioxide Removal

Mitigation involves reducing the release of greenhouse gases into the atmosphere. For CO_2 emissions from energy production, this can be accomplished by, for example, reducing demand for energy derived from combustion of coal, petroleum, and natural gas (Chapter 6); increasing efficiency of fossil fuel generating facilities; shifting the mix of fossil fuel-derived energy toward natural gas, which has the potential to lead to reduced greenhouse gas emissions per unit of energy generated, depending on associated leakage of CH_4 from production; substitution of energy derived from solar, wind, biomass, tidal/current, hydroelectric, and other renewable energy technologies (Chapter 7) (IPCC 2011); increasing use of nuclear and eventually fusion energy technologies (Chapters 8 and 9) or even energy from space (Chapter 10); or capturing fossil fuel-generated CO_2 emissions before they are released to the atmosphere and sequestering them underground or in deep ocean sediments where they will not later be released to the atmosphere (covered in this chapter). For greenhouse gas emissions from agriculture and changes in land cover and land use, mitigation can include reducing deforestation and plowing, altering the diets of sheep and cattle or capturing their emissions, and much more.

With roughly 80% of the world's energy being derived from fossil fuels and with fossil fuels being seen as the dominant energy source to lift many of the world's poor out of poverty, this emissions sector has received both particular attention and contention. While progress is being made in limiting deforestation, international negotiations have made little progress in even slowing the growth of global CO_2 emissions from the energy sector. There is, thus, very little reason to expect that the rise in the CO_2 concentration, which is what is required to stabilize the climate, can be ended in less than several decades, and perhaps much longer. As a result, the CO_2 concentration is currently projected to reach ~500 ppm by the mid twenty-first century and much higher by 2100 (IPCC 2000).

To the extent that traditional mitigation cannot adequately limit emissions of CO_2 and other greenhouse gases, the question becomes whether other human-managed approaches might be available to limit the increase in or even bring back down the elevated atmospheric CO_2 concentration.* Approaches to "engineering" the climate in this way are generally grouped into processes or technologies for CDR. Such CDR approaches are of two general types: those that augment natural biological processes that pull CO_2 from the atmosphere and store the carbon in vegetation, soils, or the ocean, and those that create new paths for CO_2 to be captured

* The primary focus is on CO_2 due to the long persistence of the perturbation to its natural atmospheric concentration and so the long-term perturbation to radiative warming influence.

and stored away (e.g., scrubbing CO_2 chemically from the atmosphere and injecting it into deep aquifers or ocean sediments where it will not be released for at least many millennia). This latter industrial approach is sometimes referred to as DAC, although, of course, the natural processes are also directly capturing and removing the CO_2 from the atmosphere as well.

There are several critical challenges for all of the CDR approaches. First, the atmospheric concentration of CO_2 is diluted by a factor of a few hundred compared to its concentration in a power plant plume, making it much more energy intensive (either for natural or industrial processes) to extract and store CO_2 once it has been emitted into the atmosphere. Second, the amount of material that must be handled can become very substantial, being several times the mass of the carbon that is captured because oxidation of the carbon to derive energy adds the mass of two oxygen atoms, and chemical combination with other atoms (e.g., calcium) may be needed to convert the CO_2 from a gas (or compressed liquid) to a storable solid. As a result, a meaningful CDR program would very likely require an industrial effort at least comparable with the ongoing effort to extract, process, and distribute carbon-based fuels, thereby imposing a cost comparable to the cost of the fuel itself. For this reason, most approaches to CDR would require substantial funding, likely via imposing a cost on the carbon being released. It is not yet clear whether addition of such a cost would favor adoption of CDR or a shift away from use of fossil fuels to derive energy, although the declining costs for renewable energy resulting from technological and engineering advances likely favor the latter, at least over the long-term.

Even if CDR were economically favored at the local scale to offset continued use of coal, for example, the extent of its influence is likely to be limited. Basically, for this type of approach to contribute to reducing the atmospheric CO_2 concentration, more CO_2 would have to be removed than is being emitted. For processes based on augmenting natural CO_2 uptake, the problem is having enough room (e.g., space for expanded forest cover) for the carbon to be stored. For industrial capture of CO_2, the cost and magnitude of the DAC effort needed may be the limiting factors (e.g., see Socolow et al. 2011; Keith 2011). Given these limits, it appears unlikely that CDR will play a significant role in reducing the changes in climate that are projected for the next several decades. Despite this, research on such approaches is likely justified because, once CO_2 emissions are substantially cut, CDR approaches could become useful in offsetting the emission of CO_2 where replacement of fossil fuels may be particularly difficult (e.g., aircraft fuels) and to moderate the rate of ocean acidification by pulling the atmospheric CO_2 concentration back toward its mid-twentieth century value, which would help to pull CO_2 out of the upper ocean.

Sections 12.3.1 through 12.3.5 briefly discuss the possibilities for CDR. Keith (2001) provides a framework for considering the possible approaches, and a recent overview is provided by IPCC (2005). Section 12.3.6 then highlights issues relating to the potential for and possible approaches to reducing the concentrations of radiatively active gases and aerosols other than CO_2.

12.3.1 Expanding the Terrestrial Biosphere

The aboveground terrestrial biosphere (so trees, grasses, etc.) totals ~500 GtC (Prentice 2001). Expanding or intensifying the extent of forest cover is, thus, at least conceptually, a way to store additional carbon (Kauppi and Sedjo 2001). However, increasing the uptake of the aboveground biosphere by 1 GtC/year for a century, which would account for only ~10% of present anthropogenic emissions, would require an expansion of the forest carbon reservoir by over 15%, which would be a very large change. For this reason, this approach alone would not solve the CO_2 emission problem. In addition, modifying the carbon uptake and release by existing ecosystems seems unlikely to be easy. For example, the stratospheric aerosols resulting from major volcanic eruptions reduce net solar radiation to the surface by order of 1% and convert about a quarter of the incoming direct solar radiation to diffuse radiation that can penetrate deeper into forest canopies. This radiation can temporarily enhance understory growth (Gu et al. 2003), unfortunately, however, by only ~1 GtC/year for a few years.

Similar discouraging results arise from an analysis of the global carbon cycle. The seasonal variation of the CO_2 concentration at the Mauna Loa Observatory makes clear that there is a seasonal variation in forest uptake of carbon, with the net biospheric uptake from spring to fall in the Northern Hemisphere exceeding the decay by ~7–8 GtC, with the reverse occurring from fall to spring. Net global deforestation is presently causing release of ~1 GtC/year to the atmosphere, with losses of ~1.5 GtC/year due primarily to harvesting and burning of low latitude forests (e.g., in the Amazon and Indonesia) exceeding carbon uptake of ~0.5 GtC/year by mid-latitude forests that are recovering from previous clear-cutting. Given the pressure for increasing use of land for food production and housing and the increasing demand for food and fiber to meet the needs of a growing and more prosperous population, even ending the existing net biospheric contribution to the rising CO_2 concentration, which is normally categorized as mitigation, will be a challenge.

Going further and inducing net carbon uptake is likely to be even more difficult. There are two fundamental challenges to promoting afforestation (i.e., the foresting of previously unforested lands): identifying sufficient suitable land areas where it would make economic and environmental sense to convert to forest cover, and providing the water, nutrients, and other resources needed to ensure that the growing of trees can be sustained and will make a sustained difference. If increased forest cover were the only approach to be used to offset current carbon emissions, Ornstein et al. (2009) estimated that it would require starting up and sustaining a forest roughly the size of Australia (or the Sahara). They calculated that, using solar energy to power desalination and water and nutrient distribution systems, continental-sized reforestation would cost less than carbon capture and storage of the CO_2 emitted in the exhaust plumes of coal-fired power plants, but this does not mean that their approach does not face other limitations. For example, to avoid saturating the uptake potential and not return the trapped CO_2 to the atmosphere, the trees would have to be harvested every few decades and their trunks buried or stored (e.g., in the deep ocean, as first suggested by Marchetti [1975]). In addition, their model calculations indicated that the presence of the forest would, by modifying the albedo, surface friction, and soil moisture, alter the region's weather. In addition, they did not continue their analysis to consider the broader environmental and cultural impacts of such a transition or changes in the risk of fire, biodiversity, land use, and more.

Overall, while there is some potential to increase carbon uptake and storage above ground and consideration of this approach does provide a sense of the overwhelming impact of the fossil fuel emissions on the earth's carbon balance, analyses of proposed approaches for land-based CDR suggest that it would be very difficult to offset a significant fraction of the existing level of CO_2 emissions.

12.3.2 Increasing Carbon Stored in Terrestrial Soils

The amount of carbon stored in the earth's soils is roughly 2–3 times larger than the amount of carbon stored in living biomass (Prentice 2001). While much of the carbon in roots rapidly decays and is returned to the atmosphere, a small fraction is transformed into forms that are quite long-lasting in the soils. The oxidation of carbon already in soils can be reduced by greater use of no-till agriculture and, with sufficient nutrients, planting of particular crops, or even perhaps, in the future, genetic modification of particular plant species.

A more direct proposal for increasing soil carbon is to simply add carbon to the soils. The leading approach for doing this is to convert waste biomass to charcoal (often called "biochar") and to mix this biochar into the soils in the process of tilling the soils (Biofuelwatch 2011). Conversion to biochar ties up carbon for an extended time because, like charcoal, it has a very long lifetime in the soils. An added advantage is that increasing soil carbon tends to increase the water holding capacity of the soils, lengthening the time that increased temperature and evaporation due to climate change will take to induce soil moisture stress.

Estimates of the potential for biochar vary (Biofuelwatch 2011). With the net greening of the biosphere from spring to fall in the Northern Hemisphere amounting to ~7–8 GtC, sequestering 1 GtC/year would require converting about 15% of this amount to charcoal and burying it. Going beyond that amount would require harvesting mature trees and replacing them with faster growing new vegetation, so basically resorting to tree farming.

Although the process appears quite straightforward, incompletely oxidized carbon is a fuel, and so this process is essentially burying incompletely burned biomass that, at least conceptually, could instead be used to generate energy that would reduce the need for fossil fuel-derived energy. Kreutz et al. (2008) seek to overcome the problem of burying potential energy by using available biomass in combination with fossil fuels, the extra energy being used to cover the energy needed to sequester all of the generated CO_2. Comprehensive carbon, climate, and sustainability footprints are, thus, needed to evaluate the net costs, benefits, and potential for this type of approach.

12.3.3 Increasing Chemical Uptake by Rocks and Minerals

Bypassing the biological step has also been proposed. Weathering of carbonate and silicate rocks is one natural way that CO_2 is taken out of the atmosphere, converted into a solid, eroded by rain (which is slightly acidic due to the dissolved CO_2), carried into the ocean, and eventually deposited in ocean sediments, leading, over very ling times, to formations such as the white cliffs of Dover (e.g., Rau et al. 2007). In that most rocks at the surface have already been exposed to the atmosphere, accelerating this process requires exposing unweathered rock surfaces to the air (e.g., see Keleman and Matter 2008). Another proposed approach for doing this would be to mine fresh rock and disperse it over agricultural land areas; unfortunately, the amount of material that would need to be processed to have a significant effect, especially with CO_2 emissions continuing to rise, would necessitate very large mining operations and only be slow to have an effect. Because of the economic and energy costs involved, this approach seems unlikely to be practical until there is a substantial price on carbon and overall global emissions have dropped enough to make such efforts meaningful. For the foreseeable future, at least, such efforts seem likely to be far more costly than efficiency improvements and converting away from fossil fuel to non–carbon based energy systems. A report from the Royal Society (2009) provides a recent summary and evaluation.

12.3.4 Increasing Ocean Uptake of Carbon

Nature is constantly trying to reduce gradients. As a result of temperature and CO_2 gradients and the natural circulation of the ocean, the atmosphere and ocean exchange (i.e., transfer in each direction) of carbon totals ~90 GtC/year. The release of CO_2 to the atmosphere is largest in low latitudes where super-saturated cold waters rise, warm, and release CO_2 (just as occurs from a warming bottle of carbonated soda), and the largest uptake occurs in high latitudes where ocean waters cool, take up additional CO_2, and then sink, carrying CO_2 back into the deep ocean. Because the rising waters are cold and supersaturated in CO_2, the natural overturning of the ocean brings almost 10 GtC more to the upper ocean each year than descending currents take back to the deep ocean. This net upward transport has historically been balanced by a net conversion of dissolved CO_2 into dissolved organic carbon (DOC) and by transfer of calcium carbonate to the intermediate and deep ocean via sinking fecal pellets and skeletons (the so-called "biological pump"). Of the carbon carried to the deep ocean by the biological pump, only 10%–15% of the additional carbon ends up adding to the sediments. The rest is dissolved as it sinks and so is available to be returned to the upper ocean over periods of decades to centuries as the ocean overturns.

The rising atmospheric CO_2 concentration caused by fossil fuel emissions alters the CO_2 gradients and, thus, the fluxes. Each year, an amount equal to about one quarter of each year's CO_2 emissions (so presently about 2 GtC/year) is forced by the altered air–sea CO_2 gradient into the wave-mixed upper ocean. The higher loading in this layer increases the carbon loading of sinking waters without altering the CO_2 loading of upwelling waters, leading to a net transfer to the deep ocean of about 80% of the additional carbon taken up each year from the atmosphere (Caldeira et al. 2005).

Because the transfer of carbon from the oceans to the sediments is only ~4% of each year's human emissions, however, the carbon loadings of all the active reservoirs (i.e., the atmosphere, living biosphere, soils, and upper and deep ocean) end up increasing and will stay significantly elevated for centuries to millennia. It is this strong tendency for the elevated carbon loading of these reservoirs to persist for very long periods that makes the climate change and ocean acidification impacts of significance on not just human, but geological timescales.

Martin (1990) was the first to suggest that the rate of transfer of carbon from the upper to the deep ocean could be increased. He had noted that there were regions of the ocean where nutrients were not getting quickly utilized by the marine biota, as was normally the case. He hypothesized that, in these generally remote areas of the Southern Hemisphere, this was a result of a lack of the micronutrient iron, which is normally added to the ocean by dust blown off large continental areas. He proposed that adding iron would create blooms of biological activity and that additional carbon would get carried to the deep ocean and sediments, either directly by sinking or by being taken up by fish and then sinking as fecal pellets or skeletons. The net effect would thus be to pull carbon out of the atmosphere and increase transfer to the deep ocean and sediments.

Roughly a dozen field experiments have been carried out to see if this would indeed be the case (Strong et al. 2009). Although all of the experiments led to phytoplankton blooms, an increased flux of carbon to the deep ocean was not convincingly demonstrated. Remaining questions about this approach include the following: Would the nutrients have been used later after drifting into regions where iron would be naturally deposited, so all that is really happening is displacing the uptake of carbon? Would the carbon that sinks actually make it to the ocean bottom or would it mostly dissolve on the way down and, in that the fertilization is occurring in or near an upwelling region (i.e., how the nutrients got to the surface in the first place), be carried back to the upper ocean's mixed layer within a decade or two, where the now supersaturated amounts of CO_2 would be released back to the atmosphere? What near and distant effects would taking such actions have on the marine biosphere? How much uptake of carbon could be stimulated in this manner?

Only for the last question is there a sense of the answer. Estimates are that the upper end potential for this approach is likely ~1 GtC/year. This limit is a result of the limited areas in which upwelling occurs in remote areas of the ocean where there is little dust deposition. With global CO_2 emissions from fossil fuel combustion and agriculture currently ~10 GtC/year and growing, iron fertilization could, thus, at best, offset the growth in emissions by several years, an effect so small that, even if the transfer to the deep ocean worked perfectly, the question arises if it would be worth putting the marine biosphere at potential, but uncertain, risk.* Although limited scientific research continues, primarily to better understand the cycling of nutrients, there seems little likelihood that iron fertilization, at least as a means to limit the rise in the global CO_2 concentration, would play a significant role until conventional approaches have been used to reduce global CO_2 emissions to far below current levels and there is a demonstrated need to do even more to moderate ocean acidification.

Lovelock and Rapley (2007) have proposed another approach for augmenting the ocean's biospheric uptake of CO_2.† While bringing the deeper water up would likely enhance overall marine productivity and perhaps increase carbon transfer to the deep ocean, bringing the waters up would also bring up CO_2 that would be given off as the waters warmed. The enhanced fish production and/or growing of biomass for marine biofuels would, however, also be beneficial, and these two outcomes might be used to limit the need of farmers and ranchers for fossil fuel-derived energy, so the complete carbon, climate, and sustainability footprint remains uncertain.‡ Whether this approach might directly make a noticeable net contribution to carbon transfer to the deep ocean remains to be demonstrated.

* There have been a few proposals to undertake iron fertilization to generate credits for carbon sequestration, but the significant uncertainties and the unquantified risk to the marine biosphere, much less the commercial use of the global ocean commons, have led to a ban on such plans under the 1972 U.N. Convention on the Prevention of Marine Pollution by Dumping of Wastes and Other Matter (commonly referred to as the London Convention).

† Such pumps can be quite simple, basically consisting of a flexible tube extending about 1 km deep into the ocean, held up at the top by a float and at the bottom having a check valve that lets the cold water in as the structure moves down in the troughs of waves and that closes as the float is carried upward to the crest of waves. Such pumps have, to date, been demonstrated to work for hours to days, but have generally not been strong enough to last for extended periods in realistic ocean environments.

‡ Such outcomes were also once touted as beneficial side effects of ocean thermal energy conversion (OTEC) systems, which were designed to use the temperature gradient between warm upper-ocean waters and cool deeper waters to produce energy. For such systems, the deeper waters would be brought to the surface, where they would be warmed in the generation of energy, and then would be released near the surface, providing nutrients to support marine life, but also releasing their excess CO_2 to the atmosphere. Whether this release of CO_2 would be balanced by biological or other processes carrying carbon to the deep ocean was not typically considered.

Distinct from attempting to increase biological uptake, suggestions have also been made about how to increase oceanic uptake, generally in ways that would also help to reduce ocean acidification. For example, House et al. (2007) suggested an approach that imitated natural silicate weathering by electrochemically removing HCl from the ocean in a way that would increase CO_2 uptake and, over the very long term, increase deposition of calcium carbonate. Although implemented at a power plant to have access to the elevated CO_2 concentration in the exhaust plume, Rau and Caldeira (1999) proposed that limestone (which would need to be mined) and water could be used to create bicarbonate that could then be discharged into the ocean, helping to buffer the drop in pH caused by the increased CO_2 uptake caused by the higher CO_2 concentration (also see Rau 2008; Langer et al. 2009). The Royal Society (2009) reviewed a wider range more of proposed approaches. What is especially problematic, however, is the large magnitude of the intervention that would be needed to have a significant and timely influence, making such approaches very expensive unless CO_2 emissions are very substantially reduced and there is a substantial price put on carbon.

12.3.5 Scrubbing CO_2 from the Atmosphere*

Just as photosynthesis and formation of marine skeletons are chemical processes that can extract CO_2 from the environment, there are a number of industrial processes that can do this. For example, cooling air down to the sublimation point for CO_2 (195 K) leads to formation of dry ice (or this can be done under pressure to generate CO_2 as a liquid). Rather than do this, present DAC approaches seek to bring CO_2 into contact with various liquids that precipitate the CO_2 out as part of a solid material. The challenge investigators are facing is to do this efficiently, cost effectively, and in ways that can handle large amounts of carbon (Keith et al. 2010).

A number of groups are working on this (e.g., Keith et al. 2006; Stolaroff et al. 2008; Keith 2008, 2009; Krevor and Lackner 2011). Each seeks to expose large amounts of air in a clever way (so avoiding use of an energy-intensive fan) to a liquid catalyst that would rapidly bind to the CO_2, even at its relatively low atmospheric concentration (e.g., Mahmoudkhani and Keith 2009). The now-bound CO_2 and catalyst would then be transferred into a separator where a change in a condition such as temperature or humidity would cause the CO_2 molecule to be given off in a highly concentrated form. The next step would be, for example, to compress the captured CO_2 and then transfer it to an isolated storage location such as a deep underground aquifer, a depleted oil field, the floor of the deep ocean (where the CO_2 would tend to remain,

* Note that extracting CO_2 from the exhaust stream of a power plant, for example, by bubbling the CO_2 through nutrient rich water in a greenhouse environment and growing biomass that is later used for generating energy, is considered mitigation rather than CDR.

although perhaps held in some sort of flexible container), or well into ocean sediments (Goldberg et al. 2008). Powering such a process with renewables or nuclear would be required to keep the carbon footprint low. While not inconceivable, there is considerable debate about whether carbon scrubbing could be done at a cost that is low enough to allow continuing use of fossil fuels (Socolow et al. 2011; Keith 2011).

At present, research on this approach is generally being done as part of projects aimed at providing CO_2 for use, for example, in enhanced recovery of oil. If a cost-effective process can be developed, however, that simply involves injection of the CO_2, then a climate change policy that puts a value on carbon might make it cost-effective for sequestering CO_2. With ocean acidification likely to become a more and more serious impact, scrubbing CO_2 from the atmosphere may become essential, so continued research seems worthwhile, even if economically viable approaches are not yet evident.

12.3.6 Moderating the Warming Influence of Non-CO_2 Greenhouse Gases and Aerosols

Carbon dioxide has been the major focus of climate engineering proposals because of the long persistence of the elevated CO_2 concentration. However, there are a number of other gases and aerosols that are adding to the warming influence of human activities for which considering the potential for interventions may make sense. For example, the lifetimes of chlorofluorocarbons and perfluorocarbons tend to be centuries to millennia, so their warming influence will persist for very long times. While the Montreal Protocol and subsequent amendments have limited halocarbon emissions, laser dissociation, although perhaps energy intensive, may make sense to shorten their atmospheric lifetime, especially if renewable energy was to be used to drive the lasers. Because halocarbons are well mixed in the atmosphere, the lasers could be located, for example, in subtropical deserts where solar radiation is high, and by reflecting the laser signal back and forth between mountaintop mirrors, a relatively long pathlength could be created to raise the likelihood of the beam striking a sufficient number of molecules. Nitrous oxide (N_2O) also has a relatively long lifetime; in addition to efforts used to improve the efficiency of fertilizer application, Richardson et al. (2009) propose an enzymic regulation approach. Although every contribution helps, lowering the concentrations of these greenhouse gases will only be of significance if a full range of other approaches is also being taken.

For greenhouse gases that have relatively short atmospheric lifetimes (e.g., CH_4 has a lifetime of about a decade), limiting emissions is likely the most cost-effective approach. Relatively inexpensive approaches can be used, for example, to limit emissions from refineries and pipelines, sewage plants, waste disposal sites, and even from some agricultural sources (e.g., from cattle-feeding lots, where the CH_4 concentration is high enough that conventional extraction techniques allow it to be captured and used as fuel). Because the global warming potential of CH_4 is much higher relative to CO_2, it may, however, make sense to develop climate engineering approaches to limit the increased emission of CH_4 from such natural reservoirs as thawing permafrost and warming ocean sediments (Stolaroff et al. 2011). The preferred approach would be to extract the CH_4 for use as a fuel before it wafts off into the atmosphere. Where that is not possible, approaches that have been suggested generally involve converting the CH_4 to CO_2 to reduce the overall warming influence. This could be done, for example, by either igniting or capturing and combusting CH_4 coming off concentrated sources, as simply as by firing artillery at the emission locations or as involved as laying out blankets and channeling the CH_4 to separation devices.

For dark aerosols such as black carbon, the atmospheric lifetime is of order a week or two, so it is likely much more cost effective to limit source emissions than to try to recapture and remove black carbon once it has been released to the atmosphere. Because black carbon landing on snow amplifies its warming influence, particular attention could be devoted to the emissions from forest fires and other sources in high latitudes, especially during the sunlit season. Whether this would be at all cost-effective is simply not clear.

12.3.7 Summary on the Potential for Carbon Dioxide Removal

Although there are some opportunities to enhance the uptake of CO_2 by natural processes (thus, essentially, relying on solar energy to power the uptake), their potential is relatively limited given the current high level of CO_2 emissions from fossil fuel combustion. Nature helps out on its own to some extent because CO_2 is a plant nutrient and some types of plants grow better with a higher CO_2 concentration if sufficient water and nutrients are available. Allowing and promoting greater growth of forests is generally beneficial, although the additional use of land generally competes with land demand for human occupation and for agriculture. Converting agricultural waste to charcoal and creating biochar that would be mixed in with agricultural soils, would be beneficial, but at least in some areas it may well be more efficient to use the biomass to make electricity (e.g., using with coal and then capturing and sequestering the CO_2) or biofuels.

Even if global emissions can be brought down, however, CDR only has the potential to slowly change the CO_2 concentration, so that the moderating effect on the climate would be very slow to materialize, even though likely quite expensive. In the long-term, the most important potential role for CDR may be to contribute to reducing ocean acidification, especially in sensitive areas (e.g., in coral reefs). However, with the maximum potential CDR capacity likely being less than a few gigatonnes of C per year, its influence is likely to be small until fossil fuel emissions are sharply reduced rather than increasing.

12.4 Options for Counterbalancing Global Warming with Solar Radiation Management

The second major approach to climate engineering is to counterbalance the increased long wave radiation being absorbed and re-radiated downward by the elevated concentrations of greenhouse gases and because of the solar radiation being absorbed by dark-colored aerosols. While it would be optimal to directly offset the additional trapping of long wave radiation, this would require reducing either the heat-trapping effects of greenhouse gases, and reducing the extent of high-level clouds so that more thermal radiation from lower part in the atmosphere reaches space and less thermal radiation is emitted downward from the upper troposphere, or increasing the emissivity and/or decreasing the absorptivity of the surface and clouds. The potential for reducing the concentrations of greenhouse gases was covered in Section 12.3 and will not be repeated here; the reduction of high-level cloud cover will, however, be covered later. With respect to emissivity changes, both the surface and clouds are generally radiatively black (i.e., have an absorptivity and emissivity near one), and so adjustments would not lead to additional thermal radiation being emitted to space.

With so little opportunity to increase outgoing thermal radiation, the primary opportunity for altering the global energy balance is reducing the absorption of incoming solar energy. Employing the metaphor of trying not to get too warm when lying in bed in a very cold room, if one cannot remove the additional blanket added by the greenhouse gases, then reducing the amount of energy shining on the blanket is the remaining alternative. Because the objective of this category of climate engineering is focused primarily on adjusting the amount of incoming solar radiation, the group of proposed approaches is generally lumped under the designation "solar radiation management" (SRM).

Ideally, the reduction in absorption of solar energy in space and time would exactly offset the additional trapping of infrared (IR) radiation, the expectation being that, because energy is what matters, independent of source, this would return the climate to conditions prevailing before human-induced modification of the concentrations of greenhouse gases. Obviously, however, achieving this is not going to be possible. First, the greenhouse gas-induced alteration of the IR radiation occurs both day and night and varies due to the particular weather situation; the reduction in solar radiation just cannot be matched to this variability. Second, the amount of incoming solar radiation at the top of the atmosphere varies by latitude and cycles through the seasons; even reducing solar radiation to match the daily or monthly average change in IR trapping at each location would be complex and difficult. If indeed the alterations in solar radiation have to be so finely tuned, SRM would seem likely to be so impractical that there would be virtually no chance that it could be technically implemented, much less become a viable policy option.

Interestingly, however, there are indications that such fine-scale alteration of the incoming solar radiation is not needed. For example, early weather forecast models showed some success out to a few days without even including treatment of IR or solar radiation, indicating that the heat content of the atmosphere provides some thermal inertia. Tests with climate models have looked at longer time intervals, comparing the outcome of model simulations when perturbed by either a doubling of the CO_2 concentration, which leads to a relatively smooth trapping of IR radiation by latitude and season, or by an increase in solar output by ~1.8%, which adds about the same amount of energy on an annual basis, but with a very strong seasonal and latitudinal pattern.* Interestingly, when the two perturbations are compared, the resulting changes in latitudinal and seasonal patterns of temperature are comparable (e.g., Manabe and Wetherald 1980; Hansen et al. 1984), and Govindasamy and Caldeira (2000) also found comparable changes in precipitation. The similar spatial and seasonal patterns of the results apparently occur because of the large inertial effects of the oceans and the redistributional effects of the atmosphere and oceans.

Based on model results such as these, the IPCC assessments use the annual, global-average change in net solar minus IR radiation at the tropopause as the metric for comparing the potential climatic effects of different radiative forcings. The total forcing is, thus, the sum of the annual global average of the changes in radiative forcing at the tropopause caused by natural and human-caused changes, even though the forcings caused by each factor may have quite different latitudinal and seasonal distributions. That this is an appropriate metric is somewhat surprising because paleoclimatic evidence suggests that changes in the seasonal and latitudinal distributions of incoming solar radiation at the top of the atmosphere caused by cyclic changes in the orbit of the earth about the sun, as first hypothesized by Milankovitch (1941), were the driving force for glacial cycling over the last few million years (Mysak 2008). That the orbital changes can exert such a strong influence may result from the longer times for their effects to alter the climate and the different atmospheric compositions and geographic characteristics during the Pleistocene (roughly the last 1–2 million years) as compared to earlier epochs when orbital variations did not result in glacial cycling. These results suggest that close monitoring and, quite possibly, iteration of the imposed

* A doubling of the CO_2 concentration leads to an increased trapping of IR radiation, measured at the tropopause because of the close thermal coupling of the surface-troposphere system, of about 3.7 W/m^2. Solar irradiance is about 1360 W/m^2; after dividing by 4 to account for the sphericity of the earth, multiplying by 0.97 to account for absorption of solar ultraviolet radiation by stratospheric ozone, and then multiplying by 0.7 so as not to count the 30% of solar radiation reflected mainly by clouds, a 1.8% increase in irradiance would lead to an increase in energy absorbed by the surface–troposphere system of about 4.1 W/m^2. The remaining difference is mainly a result of how adjustments in the IR radiation balance are accounted for and the rough estimate of the planetary albedo.

changes in solar radiation would be needed to ensure that an SRM-based approach is neither over- or under-compensating for the warming influence of the increases in greenhouse gas concentrations.

Accepting the results of the model simulation that the CO_2-induced warming influence could, in principle, be offset by reductions in absorbed solar radiation, a number of possible approaches for accomplishing this have been proposed (see reviews, for example, by MacCracken [1991], NAS [1992], Leemans et al. [1995], Flannery et al. [1997], Keith [2000], Schneider [2001], Lane et al. [2007], and Schneider [2008]). In their broadest conception, the proposed approaches include reducing the amount of solar radiation reaching the top of the atmosphere, imitating the cooling effect of volcanic aerosols by injecting aerosols into the stratosphere (or above*), increasing the reflectivity of clouds and/or of the tropospheric aerosol layer, and increasing the reflectivities of the land and/or ocean surfaces. Sections 12.4.1 to 12.4.5 describe the range of approaches that have been suggested.

12.4.1 Proposed Approaches for Reducing the Amount of Solar Radiation Reaching the Earth

If all forcings can be treated equally, as is done by the IPCC, the most straightforward approach to counterbalance the increased trapping of energy by the rising concentrations of greenhouse gases would be a dimming of the sun. Historical records suggest that past decreases in solar radiation (e.g., as indicated by sunspot records) have led to cooling, so dimming of the sun would presumably do likewise. Because other options discussed below would require injecting substances into the atmosphere, a key advantage of taking action in space would be that it would neither pollute the atmosphere nor be complicated by atmospheric physics. The key problem, however, would be that the deployment would have to be done in outer space, and this requirement tends to significantly raise the cost, create navigational challenges, and require an extended time for implementation. Several approaches have been suggested, but all seem to have serious shortcomings.

Low-earth orbit is the minimum height that deflectors would need to be located so that they would not be slowed and lost by interaction with the earth's atmosphere. In the simplest configuration, to counterbalance about half of the warming influence of a CO_2 doubling, the mirrors would, at any given time, need to cover about 1% of the earth. If one imagines the mirrors as thin sheets of, for example, aluminum foil (so equivalent to the very reflective Echo balloon satellites of the early 1960s), and one imagines each being structurally held in place as a square of dimension *10 km by 10 km* and with no overlap as they orbit, there would need to be over 50,000 reflecting foils

* For example, Keith (2010) proposes lofting specially designed aerosols into the mesosphere.

in orbit at the same time,* each moving at nearly 30,000 km/hr (NAS 1992). Although only 1% of the sky would be obstructed, each overpass would create a 1–2 second eclipse to an observer at the surface. While going to smaller mirrors would shorten the period of the eclipse, this would sharply increase the required number of orbiting mirrors and, thus, further exacerbate the problem of making sure they stayed in orbit without crashing into each other. In addition, the spherical geometry of the earth and complexities of orbital paths make it unlikely that a set of orbits could be found that would approximate an equal diminution of solar radiation over the earth for all season and latitudes, the more likely result would seem to be a greater reduction in high latitudes that would alter the latitudinal distribution and gradients of temperature.

In an attempt to get around the need to manage so many mirrors, there have been suggestions that lofting small absorbing or reflecting particles could be used to create an optical shield. A critical problem with this approach, however, is that the solar wind would tend to push small particles out of orbit, eventually leading to their burn up in the earth's atmosphere. A key desired advantage of blocking solar radiation in space would be that the reflectors would be expected to have a longer lifetime than similar materials lofted into the atmosphere, the rapid removal of such particles would be an important disadvantage of this approach. An additional disadvantage would be that the particles would likely scatter at least as much radiation as they reflect, so there would be a whitening (or diffusing) effect on incoming solar radiation.

To overcome the navigation, mass, and reflector lifetime problems, Early (1989) proposed placing a deflector shield at the earth's first Lagrange (L-1) point, which is the spot between the earth and the sun about 1.6 million kilometers above the earth where there would be equal gravitational pull from each. While this equal pull would create a metastable location, active efforts would be needed to keep any small force from pushing the deflector away from this location. Early calculated that to block 1% of incident solar radiation, the shield would need to have a diameter of about 1800 km and that active measures would be needed to prevent the solar wind from pushing the shield away from the L-1 location. Early went further, calculating that the most efficient and cost-effective way to create and deploy such a shield would be to manufacture and loft it from the moon. The potential benefit would be that active management of the shield's configuration and orientation could modulate incoming solar radiation, thus, providing an active control option.

* In low earth orbit, only half this number would be exposed to and blocking sunlight at any given time. Interestingly, it is possible to orbit satellites in near-earth orbit while keeping them always exposed to sunlight by having the orbit be over the sunrise–sunset boundary, and some proposals for solar satellites as an energy source take advantage of this (e.g., Hoffert and Potter 1997).

Concerned about the delay and cost that would be imposed by the requirement to build a lunar base to deploy such a shield, Angel (2006) proposed creating a composite sunshade at the L-1 point by lofting a very large number of small deflectors directly from the earth's surface. When fully deployed, Angel's proposed sunshade (or parasol) would consist of roughly 10 trillion autonomous flyers, each being a very thin sheet having a diameter of about a meter and a thickness of about 0.5–1 μm. Each of the flyers would have an onboard control mechanism capable of maintaining location, orientation, and separation by "sailing" in the solar wind. Angel proposed that the flyers be launched directly from the earth's surface to the L-1 point using an electromagnetic launcher that would be fired 20 million times over coming decades, each launch carrying 800,000 flyers in a stack weighing 1000 kg. While the counterbalancing effects of the reflected radiation could be incrementally increased to the level needed, the estimated costs of constructing and then operating the needed launch facility are so high that, even though it addresses shortcomings of some other approaches, the solar parasol approach is not considered a practical alternative, at least in the next few decades, unless the climate warms very rapidly.

12.4.2 Proposed Approaches for Increasing the Reflectivity of the Stratosphere

Augmenting the lower stratospheric sulfate loading has been proposed as a counterbalancing technology since the 1960s (Budyko 1974). Such proposals were based on the relatively rapid and significant cooling (in most, but not all seasons and locations) that has been observed following major volcanic eruptions, most of which loft many megatons of sulfur (generally as SO_2) to roughly 20 km or above, where it is converted to sulfate aerosols over the following few months. The attraction of injecting the aerosols into the stratosphere, as compared to the troposphere where coal-fired power plants loft the SO_2 that forms the tropospheric sulfate/haze layer, is that the global-average lifetime of stratospheric aerosols is typically one to a few years, depending on the altitude of injection, as compared to 1–2 weeks for tropospheric aerosols. To first order this would mean that, for the same optical effect, only 1%–2% as much material would need to be injected into the stratosphere as into the troposphere. The longer lifetime would also mean that a relatively uniform aerosol layer could be created with far fewer injection locations.

Quite a number of analyses and simulations have been done to estimate the climatic effects of volcanically created veils (e.g., Oman et al. 2005; Kravitz and Robock 2011). The model simulations project, on average, global cooling of order a degree Celsius lasting for a year or two. Model simulations and observations also suggest that some land areas might warm during the winter (when solar radiation is not a major source of the atmosphere's energy at mid to high latitudes) and that precipitation over land could be suppressed in summer because of the reduced energy available to drive the hydrological cycle (Bala et al. 2008). A critical question, however, is the potential applicability of findings from the effects of individual eruptions for evaluation of proposals for deployment of approaches that would lead to sustained increases in stratospheric aerosol loading (Trenberth and Dai 2007).

Because aviation fuels had a few percent sulfur content to serve as a lubricant, the first major assessment of human-induced augmentation of the stratospheric sulfate loading was carried out in the early 1970s as part of the evaluation of the potential climatic consequences of a proposed fleet of supersonic aircraft. In addition to the finding that the nitrogen oxides in the exhaust would tend to reduce stratospheric ozone, the evaluation found that a cooling would indeed be expected from the increased sulfate loading in the stratosphere. Since those early studies, an increasing number of studies have been done, both schematically by reducing the solar constant (Govindasamy and Caldeira 2000) and more comprehensively starting with injections of SO_2 and even treating the formation and transport of sulfate aerosols (Robock et al. 2008; Rasch et al. 2008a, 2008b).

These results indicate that the surface–troposphere system will cool and that a significant fraction of the warming and changes in precipitation caused by the rising concentrations of CO_2 and other greenhouse gases could be offset on both a regional and seasonal basis. While the offset would not be perfect, the remaining changes would generally tend to be within the levels of present variability, so potentially within the range of plausible adaptation efforts. Even these relatively schematic studies, however, have identified a number of important limits and qualifications (Robock 2008; Brovkin et al. 2009). The major challenges include the following:

- Although relatively low-cost stratospheric injection of SO_2 appears to be possible using aircraft or even hoses held aloft by balloons, ensuring that there is efficient SO_2 conversion to sulfate particles of the optimal size for reflecting radiation while achieving few-year lifetimes may well require adjustments of the proposed approaches. Analyses also indicate that as more and more SO_2 is injected, the particles that are created are larger, have shorter stratospheric lifetimes, and cause reduced radiative effects, thus, indicating that there is apparently a maximum aerosol loading, and therefore radiative offset, that can be achieved (e.g., see English et al. 2012). Although it may be that an alternative chemical form of injected sulfur could be used, that alternative injection and dispersion approaches could be developed (Pierce et al. 2010), or that an alternative material could be used (e.g., Chang and Shih 1991), it still appears that there is a limit to the amount of offset that could be achieved, reinforcing the view that climate engineering is a potential complementary rather than alternative approach to mitigation.

- Sulfate aerosols, even optimally sized, tend to be relatively inefficient at back-scattering radiation. While this increases the amount of aerosol needed to reflect a given amount of incoming solar radiation back to space, the most important radiative effect is that the aerosols tend to forward-scatter roughly four times as much radiation as they backscatter (Murphy 2009; Kravitz et al. 2012). The forward-scattered radiation tends to whiten the sky during the day while also leading to more colorful sunrises and sunsets. Although perhaps aesthetically acceptable, one of the major negatives is that a reduction of, for example, 20% of the direct radiation reaching the surface would reduce the power that could be derived from energy technologies that concentrate the solar beam using mirrors (e.g., thermal solar technologies).
- Observations following major volcanic eruptions suggest that significant ozone depletion can occur when sulfate particles are present in the wintertime polar stratosphere as air temperatures decrease to ~200 K or below. While the intent would be to refrain from augmenting the sulfate layer in this region (Crutzen 2006), the natural flow in the stratosphere is from low to polar latitudes, and so sulfate particles created at low latitudes would get carried by the natural flow into the polar stratosphere and then eventually into the troposphere, where they would be removed from the atmosphere by precipitation. With the rising CO_2 concentration tending to cool the stratosphere and slow the multi-decadal recovery of stratospheric ozone concentrations from the emissions of CFCs and other halocarbons, calculations indicate that augmenting the natural sink mechanism for ozone would extend and even amplify an environmental threat now being slowly alleviated (Tilmes et al. 2008). Whether nonuniform seasonal, latitudinal, and/or altitudinal injection patterns offer the potential for limiting the potential impacts on the ozone layer is yet to be investigated.

Present research efforts on the potential for augmentation of the stratospheric aerosol loading to counterbalance greenhouse gas–induced climate change are focused mainly on increasingly detailed model simulations. A model intercomparison project (GeoMIP) has been initiated by nearly 20 modeling groups to evaluate the levels of agreement and difference among model simulations with respect to the formation of a sulfate layer, regional changes in temperature and precipitation, and consequent impacts on natural and societal systems (Kravitz et al. 2011). Roughly speaking, annual injection of 5–10 TgS/year would be needed to offset global warming to date; this amount compares to annual near-surface anthropogenic emissions of roughly 70–90 TgS/year. With the injected amount being well below current global emissions, the injected material would be unlikely to pose a significant threat to ecological systems or human health (Kravitz, et al. 2009).

A second research thrust is focused on exploring potential technical approaches for injecting suitable materials with an optimal distribution and in a cost-effective manner (McClellan et al. 2011). Because of concerns about the potential impacts from field research into, much less actual creation of, a stratospheric aerosol approach, consideration is also needed concerning the strengths and weaknesses of various climate engineering governance structures (Rayner et al. 2009; Asilomar Scientific Organizing Committee 2010), including the potential for and ramifications of an individual country (or even an individual) proceeding with such an approach, either on a research or implementation basis, without first having international agreement (Victor et al. 2009).*

12.4.3 Proposed Approaches for Increasing the Reflectivity of the Troposphere

Clouds in the troposphere reflect about 25% of incoming solar radiation back to space.† Theoretically, increasing the amount of solar radiation reflected to 26.8% would be enough to approximately counterbalance the increased trapping of infrared radiation caused by a doubling of the CO_2 concentration. Were appropriate policies being put in place, an even smaller increase would suffice to offset the increase in forcing occurring over the next several decades while emission reductions are cut enough to stop the ongoing increase in radiative forcing.

That human activities can increase tropospheric reflectivity is made clear by the present effects of SO_2 emissions that come mainly from coal-fired power plants (Penner 2001). Until the middle third of the twentieth century, SO_2 associated with the burning of coal was mainly emitted near the surface, where wet and dry deposition processes involving filtering by vegetation and absorption by dew were generally effective in removing the SO_2 within a day or two, often damaging vegetation in the process. To reduce the resulting near-surface air pollution and health impacts, the exhaust plumes from power plants were then increasingly routed through tall stacks, leading to dispersion of the pollutants above the surface boundary layer and into the lower troposphere.‡ The higher injection level reduced rates of removal, especially over oceans where inversions limited air mixing toward the surface where removal would be rapid. As a result, there was more time for oxidation of gaseous SO_2 into particulate sulfate, creating whitish haze layers

* In addition, see information based on the workshop entitled "Geoengineering: Workshop on Unilateral Planetary Scale Geoengineering" held under the sponsorship of the Council of Foreign Relations. Materials are accessible at http://www.cfr.org/projects/world/geoengineering-workshop-on-unilateral-planetary-scale-geoengineering/pr1364.

† The surface reflects back to space about 5% of the incoming top-of-the-atmosphere solar radiation, bringing the total global albedo up to about 30%.

‡ At the same time, associated particulate pollution, which tended to be dark-colored and, therefore, tended to absorb solar radiation, was usually filtered out.

that extended for thousands of kilometers downwind of industrial areas. Even with the decreasing overall emissions that resulted as residential and industrial use of coal declined, the clear-sky reflection of sunlight as a result of SO_2 emissions increased due to the jump in the atmospheric lifetime of sulfur compounds from 1–2 days to 1–2 weeks.

In addition, because sulfate particles serve as cloud condensation nuclei (CCN), the reflectivity of clouds was increased, particularly over ocean areas where stratus clouds tend to cover large areas. This brightening occurs because, with the higher sulfate loading, the available water vapor is spread across many more small droplets, and clouds made up of small droplets are more reflective than clouds made up of relatively few larger droplets (Twomey 1977). That this is the case is most obviously shown in satellite photos of what appear to be contrails that exhaust-spewing ships often create when passing below low-lying marine stratus cloud decks, particularly in areas of relatively clean air where the exhausted particles can significantly enhance the natural CCN, and therefore cloud droplet, concentration.

In listing the human-induced alterations to the global fluxes of radiation, the most recent IPCC assessment estimated direct, clear-sky forcing resulting from the current atmospheric sulfate layer to be -0.4 ± 0.2 W/m^2 and the indirect, cloudy-sky forcing to be -0.7 (range -0.3 to -1.8) W/m^2 (Forster and Ramaswamy 2007). While associated human-induced emissions of black carbon, organics, and other substances, mainly from coal-fired power plants and biomass fires, lead to some counterbalancing increases in solar absorption by dark-colored aerosols, augmenting the tropospheric brightening effects of the existing sulfate loading (or otherwise increasing the CCN concentration in clouds) represents a potential approach to climate engineering that is already, although inadvertently, contributing to moderating global warming.

Recognizing that there are vast areas of marine stratus clouds that are not currently being brightened by upstream emissions of sulfate-generating SO_2, Latham (1990) suggested that the warming influence from the rising CO_2 concentration could be largely counterbalanced by increasing the cloud-level CCN loading. To accomplish this, it was proposed that fleets of wind-powered trimarans* roughly the size of clipper ships be operated under marine stratus decks, each ship spraying upward a fine mist of seawater that, upon evaporation, would enhance the prevailing CCN loading by adding appropriately sized sea-salt particles that would be carried up to cloud level by the natural wind-driven mixing (Latham et al. 2008; Wang et al. 2011). According to their

* To keep costs down (especially of the manpower to man the ship and set sails), the power would be generated using Flettner rotors, which are basically vertically oriented, finned pipes that, essentially, take the place of the masts (Salter et al. 2008). These pipes are rotated in order to generate different net air speeds on opposite sides. By Bernoulli's principle, which also explains the lift generated by an airplane's wing, a thrust is generated that is dependent on the wind speed and speed of rotation of the vertical pipes. The thrust generated would then be used both to move the boat and to generate the power to spray seawater out through the top of the masts.

estimates, of order a score of such trimarans, so well within ship-building capacities, would need to be deployed each year to offset the warming influence of a year's CO_2 emissions.

Global model simulations indicate that increasing the brightness of marine stratus decks around the world would lead to a cooling influence that could offset a significant share of the warming from a doubling of the CO_2 concentration (Salter and Latham 2007). Boats powered by Flettner rotors were built first in Germany in the 1930s, and experimental versions have been built recently (Salter et al. 2008). Modern implementations could use electronic navigation systems to eliminate the need for a crew, and so could be deployed relatively inexpensively. An alternative approach that has been proposed is to place CCN generators on existing freighters where power would be available to power the needed pumps. While this approach could reduce injection costs, only making injections along shipping routes would seem unlikely to be optimal in terms of timing, location, and maximizing the global effect by emitting the CCN in relatively pristine areas.

Satellite photos of ship tracks (and limited field experiments) indicate that adding CCN can indeed brighten clouds, in some cases leading to brightened cloud decks that appear to persist for a day or more (Schreier et al. 2006). In addition to the need for information about the persistence of the modified cloud albedo, other issues to be addressed include developing the capability to actually generate the CCN so that they do not quickly coalesce (although this is appearing near to being solved) and determining how best to optimize the injection process so that overseeding does not lead to the cloud water precipitating out, which would lead to loss of the cloud and additional absorption of solar radiation rather than its increased reflection. In addition, analysis is needed to determine the extent and significance of unintended side effects, such as modifying the weather (and ecosystems) of nearby land areas, affecting precipitation patterns and amounts, and altering light levels for marine biota.

An alternative approach to tropospheric brightening would be to increase the climatic influence of aerosols that are injected above the boundary layer, just as is the case for SO_2 emissions emitted from tall stacks (MacCracken 2009). This might be particularly important to consider because efforts to limit use of coal are also likely to reduce the emissions of SO_2 from energy generation, thus, reducing the existing cooling influence of sulfate aerosols and exposing the world to an additional warming influence just as it is trying to switch away from use of fossil fuels. Through most of the twentieth century, SO_2 emissions were concentrated in eastern North America and greater Europe. With emissions occurring year-round, wintertime emissions likely had a very small climatic effect due to the reduced solar radiation during that season, but at the same time had a relatively large ecological effect as deposition onto snow built up and acidified lakes and streams during springtime snowmelt. With emissions of Atlantic basin nations now sharply reduced, the centroid of SO_2 emissions has switched to southern and eastern Asia, where the emissions create thick polluted layers which so concentrate the sulfate loadings that, in addition to amplifying health impacts, the per unit mass reflective effect is likely reduced to well below its potential maximum.

Separating the injection of SO_2 from the generation of electric power offers a number of opportunities for increasing the global cooling influence of injections of SO_2 into the troposphere above the boundary layer. First, relocating emissions to lower latitudes where the solar radiation is greater through the year would increase the amount of energy reflected. Similarly, locating emissions in a way that would spread the sulfate out over larger areas would increase the per unit mass effect of the injected material. Having the emissions be only SO_2 rather than the mix of pollutants typically coming out of power plants would also increase the per unit mass effect. Third, locating the sulfate layer out over the dark ocean creates the maximum albedo contrast while also moving the sulfate away from centers of population. Together, these factors suggest that increasing the sulfate loading over the very large, but mainly remote, areas of the low latitude Pacific and Indian Oceans could lead to much greater reflection of solar radiation per unit of sulfur injected and much reduced ecological and health impacts.

For the sulfate to persist in the atmosphere for 1–2 weeks, the injections would need to be into the atmospheric layer above the boundary layer. If such a lifetime could be achieved, it would be considerably longer than would be achieved by injecting CCN into the boundary layer, especially if emissions are made only during times of favorable wind directions and weather.

With respect to other potential consequences, sulfate deposition into the ocean would be unlikely to be harmful, especially because overall emissions could likely be reduced due to the greater per unit mass efficiency. While at least some of the injected sulfate would likely get carried by the winds to surrounding land masses where deposition might lead to some adverse impacts, the more appropriate question to be investigated is how such impacts would compare to the benefits of reduced global warming. If unintended consequences do turn out to be significant and/or if the supply of sulfur becomes restricted, there would always be the option of injecting alternative substances (e.g., sea salt) into the lower troposphere above the boundary layer to induce similar types of impacts.

12.4.4 Proposed Approaches for Reducing the Infrared Opacity of the Troposphere

Although not strictly SRM, increasing long wave emissions from the troposphere (and, thus, from the surface–troposphere system) has also been proposed and is considered here because the approach involves modifying the planetary energy balance by altering cloud effects. While low-level clouds have their major effect on absorption of solar radiation, high-level clouds have their primary effect on emission of long wave (i.e., infrared, or IR) radiation. This is the case because, even though they are cold with respect to typical tropospheric temperatures, they are very warm compared to the cold of outer space. Thus, downward-directed IR radiation from the upper troposphere is greatly increased by the presence of high-level cirrus clouds. These clouds are so thin that most of the solar radiation passes through, but so opaque at IR wavelengths that most of the upward-directed IR radiation is absorbed, with about half of it radiated back toward the surface. The upper tropospheric water vapor that, under certain conditions, makes up the cirrus clouds, also plays a significant role in limiting outward-directed IR radiation and warming the surface. Because of the critical role of upper tropospheric water vapor, drying this region, has the potential to limit warming.

Recognizing this possibility, Mitchell and Finnegan (2009) suggested that it might be possible to reduce the IR-limiting effect of cirrus clouds by creating larger ice particles, thus, increasing their fall speed. Increasing the loss of water from the upper troposphere would both reduce the IR-trapping effects of cirrus directly and water vapor indirectly. One proposed approach for doing this is to inject low concentrations of ice nuclei that would take up water vapor at lower levels of super-saturation than natural ice nuclei. As a result, water vapor would be attracted to fewer particles, and each ice particle would become larger and, therefore, have a higher fall velocity, tending to reduce water vapor loading and cirrus clouds in the upper troposphere.

A range of traditional materials could potentially be used, including, for example, sulfuric acid, ammonium sulfate, or ammonium nitrate; more exotic options might include bismuth trioxide or silver iodide. Injection of such compounds at altitude could be by aircraft flying routes that are fairly widely separated. This is the case because the needed loading is quite low and winds would fairly quickly spread the injected ice nuclei widely. On the other hand, the lifetime of the injected materials at altitude would be 1–2 weeks, so emissions would need to be done on a continuing basis over large areas, even though the mass of injected materials is relatively low. Cost-effective possibilities for the needed injections would include commercial aircraft and drones.

Global model simulations reported by Lohmann et al. (2008) and Mitchell et al. (2008) indicate that a reduction in the allowed cirrus-cloud super-saturation from about 140% to 130% would lead to larger ice nuclei that would fall out more rapidly. If this could be accomplished over the entire earth, this would reduce the IR trapping by enough to offset a significant fraction of the increase in trapping due to a doubled CO_2 concentration. With cirrus spread widely over the earth, there would be large regions where reduction of cirrus trapping could be practiced, perhaps enabling some fine tuning of the geographic pattern of injections to moderate possible impacts of the increasing greenhouse gas concentrations on the atmospheric circulation. And because the approach would be reducing the amount of energy trapped in the upper troposphere rather than SRM approaches that reduce energy at the surface, the hydrological cycle would be less likely to be adversely impacted. A comprehensive comparison to other approaches has not yet, however, been conducted.

12.4.5 Proposed Approaches for Increasing the Reflectivity of the Surface

Proposals to increase the surface albedo as a means to reduce the amount of absorbed solar radiation go back to the mid-twentieth century (e.g., PSAC 1965), but most approaches seem to have a very limited potential. Roughly 50% of incoming surface radiation makes it through the clouds, ozone, and water vapor to be absorbed at the surface, with an additional 5% being reflected from the surface back out to space as a result of a surface albedo of a bit over 10%. To increase the amount of solar radiation striking the surface and reaching back out to space enough to offset a doubling of the CO_2 concentration, the albedo over 20% of the area of the earth (roughly 2/3 of the land area or 1/3 of the ocean area) would have to be increased by about 20%. One suggestion has been to lighten the surface of urban areas (Akbari et al. 2009); the potential coverage is quite limited, however, and whitening buildings is likely most appropriate when seeking to reduce energy use in buildings and to moderate the intensity of urban heat islands. There have also been proposals to genetically modify the reflectivity of trees (e.g., Ridgwell et al. 2009), although demand for land for agriculture and living space and the need for trees to absorb significant solar radiation to continue their growth likely makes this impractical. In addition, climate engineering with living things would seem much more difficult to stop or reverse if problems arose—basically, it is much more difficult to call back genetic changes as compared to stopping injection of sulfate or ice nuclei, which natural processes would quickly remove.

It has also been suggested that reflectors could be deployed over land areas to increase surface reflectivity. The darkest land areas where the effect could be greatest, however, are heavily used for forests and agriculture, so further brightening relatively reflective, often quite arid, areas is the remaining option. Gaskill and Reese (2003), for example, propose brightening deserts by covering them with a more reflective material, although this would likely be costly to maintain and clean, and the spatial pattern of the distributed pattern of the perturbation might well adversely impact local to regional weather.

Over the oceans, two different approaches to increase the surface albedo, both based on emulating natural processes, have been suggested. One approach has been to imitate ship wakes and near shore wave crests by injecting very small bubbles into the water. Specifically, Seitz (2011) has proposed that it might be possible to inject microbubbles into the ocean, essentially creating an in-water cloud that would increase the surface albedo. Key questions include how long the bubbles last and how difficult it would be to inject enough bubbles to make a difference. The second approach would be to increase the near-surface biological loading of algae or other organisms. The brightening efforts of such increases have been evident from satellites. The key questions on this approach are on how it might be controlled, whether there might be adequate nutrients, and how serious the unintended consequences might be. It has also been suggested that one might produce, for example, billions of pieces of Styrofoam to brighten the ocean. Basically, however, this would require floating enough material to cover a continental sized region, and the likely adverse impacts on the marine biosphere might well be significant.

Related to changing the albedo of the surface to change its temperature, there have been proposals to alter the flow of the ocean to accomplish this. For example, Johnson (1997) proposed to dam the Strait of Gibraltar, and other proposals have included damming the Bering Strait, etc. While there would certainly be climatic responses, likely mainly regional, it is not likely that such modifications would be enough to change the global climate.

12.4.6 Summary

The significant advantages of global SRM as compared to CDR are the rapidity of the cooling that could be induced, its relatively low cost as compared to both CDR and most approaches to mitigation, and the potential to have an effect within months to years as compared to over many decades. Critical limitations of global SRM as compared to CDR are that SRM will not limit the increasing acidification of the ocean, that it would need to be continued indefinitely, and that it would require global agreement that might be hard to gain due to there being both beneficial and detrimental impacts of global intervention.

Among the approaches to SRM, each proposed approach has strengths and weaknesses. While limiting the amount of solar radiation absorbed by the earth could, at least conceptually, be scaled up to any required reduction in solar radiation, creating a shield in space would be very expensive, time consuming, and raise complex international governance and economic issues. Building up the stratospheric sulfate layer is estimated to be relatively inexpensive to accomplish, but there are limits to how much greenhouse gas warming can be offset. The governance issues are likely to be quite challenging, and potential impacts on stratospheric ozone and light transmission through the atmosphere could be quite important.

Because of the shorter lifetime of materials injected into the troposphere, increasing tropospheric brightness requires more material than stratospheric approaches. As compared to stratospheric approaches, tropospheric sulfate approaches would likely raise stickier governance issues and could have more severe unintended consequences. The types of impacts, however, are much more analogous to effects with which there is experience in its impacts and costs are also likely to be quite low. Overall, approaches involving changing surface albedo of the land or oceans as a means to limit global warming are very hard to scale up to adequate magnitude and their regional application may well affect regional weather. Reducing cirrus cloud cover in order to increase transmission of IR radiation to space appears plausible if ice nuclei dispersion could be done globally by commercial aircraft at reasonable cost, but, as for the other approaches, moving forward would likely raise complex governance issues.

Given the many complications and challenges, it might well be that the most appropriate application of approaches changing the surface albedo would be to address severe impacts in limited regions, as explained in Section 12.5.

12.5 Options for Focusing Climate Engineering Technologies on Moderating Specific Impacts

With CDR likely to be costly and only able to slowly limit global warming and global SRM likely to lead to at least some important unintended consequences and to require international agreement that could be achieved only after very significant impacts are affecting most countries of the world (which, unfortunately, might mean they are essentially irreversible), MacCracken (2009, 2012) suggested consideration of an interim approach that would use selected approaches to moderate specific, often regional, phenomena that are now or are very likely to soon be causing particularly detrimental impacts. Examples of changes that are likely to lead to adverse impacts include warming of the Arctic (including permafrost thawing, leading to release of large amounts of CO_2 and/or CH_4), deterioration of ice sheets (and so associated sea level rise), drought-inducing shifts of storm track, intensification of tropical cyclones and other storms, and, as already discussed in Section 12.4, reduction of the existing sulfate offset as global CO_2 emissions are reduced. Broadening the concept, it might also be possible to use CDR approaches to limit the effects of ocean acidification on, for example, specific coral reefs and other sensitive marine systems.

There would seem to be a number of potential advantages to at least initially considering intentional human intervention on a focused, regional scale rather than on the global scale. While an intervention anywhere would lead to a response everywhere as the atmosphere and oceans spread out a localized deficit in the global energy balance, the counterbalancing effect could likely be made to be largest in an intended target area while being within the range of normal variability over much of the world. This was pretty clearly the case, for example, during the middle third of the twentieth century when SO_2 emissions were concentrated in eastern North America and greater Europe. Focusing attention on moderating a particular impact would also provide a more specific objective for the potential application of the climate engineering technology as compared to the rather amorphous objective of moderating overall global climate change, much as a scalpel would seem to be more useful than a blunt hammer for dealing with a localized injury. In this regard, having a specific objective, and so metric for determining progress, might well also be more understandable to the public, and so contribute to moderating public and governance concerns. Considering how a few possible objectives might be pursued helps to illustrate what might be possible.

12.5.1 Potential for Moderating Arctic Warming

The Arctic is warming at a rate 2–3 times the global average (ACIA 2004; Winton 2005); excluding the warming in the Arctic, overall global warming would be 10%–20% less. At the same time, sea ice retreat is accelerating and ocean waters are warming (apparently contributing to increased emission of CH_4 from coastal sediments), mountain glaciers and ice sheets are losing mass (increasingly contributing to sea level rise), and permafrosted areas are thawing (risking significant emission of the large amounts of carbon stored in high latitude soils, either as CO_2 or, even worse from a climatic perspective, as CH_4 [see Ping et al. 2008]). The warming in high latitudes is amplified because the melting, thinning, and areal shrinkage of sea ice is reducing the surface albedo and allowing more absorption of solar radiation during the sunlit months. Proceeding into winter, this additional heat must escape so that sea ice can form and thicken. This ocean heat reaches the atmosphere directly or, more slowly, by conduction through thin sea ice, keeping the surface temperature near freezing.

Basically, sea ice cover needs to be nearly continuous and 1–1.5 m thick so that the upward transmission of infrared radiation can chill the sea–ice surface to about –40°C, creating the strong near-surface inversion that in the past characterized Arctic wintertime conditions. While the warming of the Arctic troposphere is not much different than warming of the troposphere at lower latitudes, the particular effects of thinning of the sea ice and lowering surface albedo have led to the surface temperature change being especially large as overall global warming proceeds. In addition, because of the cold temperature, little of the trapped energy is devoted to evaporation, so that, in contrast to low latitudes, most energy goes into melting the ice and thereby contributes to amplified warming. With Arctic warming already exceeding 2°C, there is essentially no chance that reductions in CO_2 emissions can prevent much further warming from occurring, which will lead to the loss of the Arctic climate that has existed for many millennia. Climate engineering appears to be the only way to moderate, or even reverse, this very significant change.

Although suggested earlier, Caldeira and Wood (2008) were the first to carry out model calculations to explore the potential for specifically limiting Arctic warming. Starting from warming associated with a doubled CO_2 concentration, they arbitrarily reduced top-of-the-atmosphere (TOA) solar radiation over variously sized high-latitude bands.* These reductions approximately returned Arctic temperatures and sea ice extent to conditions calculated for their CO_2 baseline case. Interestingly, however, the elevated rate of high latitude precipitation brought on by global warming persisted, presumably because the solar reductions were at high latitudes and so did not reduce the increased energy that is intensifying the hydrologic cycle at low and mid-latitudes. With the Arctic cooled, the increased precipitation would presumably fall as snow, thus, contributing to rebuilding, or at least limiting the loss of mass from, mountain glaciers and ice sheets (Irvine et al. 2009).

Closer examination of the results indicates that cooling the Arctic with respect to the rest of the planet tends to draw heat out of mid- and lower latitudes, thus, helping moderate global warming (MacCracken et al. 2012). However, this also leads to a slight southward shift of the Intertropical Convergence Zone (ITCZ), giving the Northern Hemisphere atmospheric circulation greater access to the solar heat absorbed in tropical oceans. Because this ITCZ shift would also shift the associated precipitation belts, MacCracken et al. (2012) have conducted simulations that would comparably reduce the amount of solar radiation for each pole's summer season. The results indicate that the ITCZ shift would be reduced and that polar cooling would contributed to some moderation of the warming over much of the earth. In terms of relative effectiveness, they found that the reductions in solar absorption in high latitudes are generally 2–3 times as effective in contributing to cooling as equivalent global reductions in solar radiation.

There are several possibilities for reducing polar radiation. Robock et al. (2008) considered the potential for cooling the Arctic by injecting SO_2 into the lower polar stratosphere, with subsequent conversion to sulfate. The result was indeed a cooling of the Arctic, but some of the sulfate spread to mid-latitudes and reduced the intensity of the Asian monsoon, perhaps in part because the SO_2 injection was continued all year long rather than limited to only the periods that would contribute to lower stratospheric sulfate over the polar region during its sunlit season. Other possible approaches to altering the polar energy balance might include increasing the sulfate loading of the Arctic summer troposphere (MacCracken 2009),† increasing the tropospheric CCN concentration in order to brighten clouds over the Arctic and perhaps over the areas where ocean currents are carrying heat into the Arctic (Salter 2012), and using bubbles to brighten the ocean surface, each of which would contribute to reducing solar absorption, and injecting ice nuclei (Mitchell and

* The specific cases they studied involved reducing incoming solar radiation by either 10% north of 61°N or 25% north of 71°N.

† An interesting unanswered question is whether the cleaning up of springtime Arctic haze because of reductions in SO_2 emissions from sources in North America, Europe, and Russia might have, by letting more solar radiation reach the surface of the snow-covered sea ice, contributed to the greater observed rate of sea ice retreat than what is calculated in global climate change simulations.

Finnegan 2009), which would allow increased infrared cooling (an option that could be applied year-round). None of these tropospheric or surface-based systems would be likely to contribute to depletion of stratospheric ozone, and none of these approaches would increase the ratio of diffuse to total solar radiation over low- and mid-latitude regions that might come to rely on direct solar energy-production technologies.

12.5.2 Potential for Limiting Ice Sheet Deterioration

While mountain glaciers and ice caps hold about 1 m of sea level, the Greenhouse ice sheet contains about 7 m and the Antarctic ice sheet about 61 m of sea level equivalent (Church and Gregory 2001).* The rise during the twentieth century was roughly 0.2 m, and the IPCC (Meehl and Stocker 2007) projected that the rise during the twenty-first century from thermal expansion and melting mountain glaciers and ice caps will be about double this amount. Because realistic ice sheet models are only just being constructed, the IPCC's twenty-first century estimate essentially leaves off the contribution from the ice sheets, which paleoclimatic analyses suggest could contribute to a meter or more of sea level rise per century (e.g., see Jevrejeva et al. 2010; Vermer and Rahmstorf 2010) and, once initiated, could continue for many centuries. With summertime melting now occurring over much of the Greenland ice sheet and a lowering of the ice sheet elevation resulting in warmer temperatures and more melting, it may well be that a transition to long-term melting is occurring that mitigation alone cannot prevent or reverse. Some suggest that evidence of such a transition (e.g., a significantly increased rate of sea level rise) might be the basis for undertaking global climate engineering, although by the time that occurs it may be too late to stop the transition without a truly massive intervention.

With such a serious and significant risk, seeking out other options would seem to be important (Moore et al. 2010). Generally cooling the Arctic and Antarctic, as discussed earlier, would be helpful, especially because the cooling would be expected to lead to increased snowfall and general cooling of ocean and air temperatures. However, the increasing loss of ice mass from the ice sheets is primarily from increased loss rather than diminished deposition. In particular, warmed ocean waters (i.e., perhaps up to 2°C) are increasingly weakening the glacial streams at their ocean terminus, so in fjords along the coast of Greenland and under the ice shelves reaching out from Antarctic ice streams. The question then becomes whether climate engineering technologies, especially brightening of the oceans with micro-bubbles or overlying clouds with additional CCN, could be used to cool the

* During the Last Glacial Maximum about 20,000 years ago when global average temperature is estimated to have been about 6°C below present, sea level was down about 120 m as a result of the ice piled on land. This is the most definitive evidence that climate change can lead to very substantial changes in sea level.

waters that are moving toward the termini of the ice streams taking water temperatures back toward the freezing point of sea water (i.e., –1.8°C), so that they would not contribute so significantly to accelerating the flow of glacial streams into the ocean.

Other proposed approaches for limiting access of the relatively warm water to the glacial streams pushing out through a few of the fjords reaching into the interior of Greenland might include building a weir in the fjord so that the meltwater can flow out but the denser, warmer salt water cannot flow in at depth; impeding the escape of icebergs from the calving glacial terminus; and vertically mixing the waters in the entrance to the fjord to reduce the temperature of the waters at depth. Increased attention to researching such approaches may be merited because it is now recognized that the weight of the ice sheet has depressed much of the interior of Greenland to below sea level, which means that heat can be transferred to the ice much more rapidly by contact with ocean water than by transfer from air to the upper ice surface. As has been recognized for a long time for the West Antarctic ice sheet (Joughin and Alley 2011), ice sheets grounded below sea level have the potential to deteriorate much more rapidly than ice sheets resting above sea level, as is the case for most of the East Antarctic ice sheet. The ice shelves holding back the flow of Antarctic glacial streams are also being diminished by relatively warm ocean waters, and so cooling of the Southern Ocean may also prove beneficial for moderating sea level rise.

12.5.3 Potential for Nudging Storm Tracks

As the world has been warming, the atmospheric circulation has been changing. Basically, the Hadley Cell has been intensifying, with greater upward motion leading to more ITCZ precipitation and a poleward expansion of the dry subtropics. As the mid-latitude jet and storm tracks have shifted poleward, there is an increasing likelihood of persistent drought in, for example, southern Australia and southwestern North America. With relatively large populations in both regions, there is increasing pressure on water resources and rising social tensions as water management is becoming more and more restrictive.

Reversing the poleward shift of the mid-latitude jets is likely only possible with very large-scale intervention, such as might be attempted by increasing the stratospheric aerosol loading, brightening marine stratus cloud decks, or globally decreasing the amount of cirrus cloud cover around the world. The GeoMIP set of modeling experiments should allow evaluation of the extent that this may be possible. Whether polar-only reductions would be adequate to shift the jets is not yet clear, but worth investigating.

There might also be an alternative approach that would focus on reducing the long-term drought stress in particularly hard hit areas. For example, off the west coasts of both Australia and southwestern North America, there are storm tracks that show considerable year-to-year variability that is, at least in part, a result of at most few-degree gradients in sea surface temperature (SST) over areas of the ocean that are, during the summer season, covered by marine stratus cloud decks; for example, Namias (1977, 1978) describes the relationships of southwestern North America. An interesting, researchable question is whether seasonal cloud brightening could be used, at least during some years with particularly suitable SST conditions, to create an SST gradient that would increase the likelihood that at least some storms would generate precipitation in areas that are experiencing drying. Pretty clearly, if recent shifts continue, the optimism of suggesting that the affected regions are experiencing drought will have to recognize that the climate has changed and the adjacent subtropical desert climates are really expanding—after all, we do not say that the Sahara is experiencing a 6000-year drought.

12.5.4 Potential for Moderating the Intensity of Tropical Cyclones

In high latitudes, the energy trapped by greenhouse gases goes mostly into warming, and there is much attention to the high-latitude amplification of temperature change. In low latitudes, most of the energy goes into evaporating moisture, basically limiting the warming by evaporative cooling; the consequence, however, is that with more water vapor in the atmosphere, precipitation intensity increases, creating different types of adverse weather. In particular, although considerable uncertainty remains, recent trends and model simulations give a hint that, while the total global number of tropical cyclones (in the Atlantic called hurricanes, in the Pacific called typhoons) may go down, the fraction that are intense may increase, leading to a net increase in severe tropical cyclones, or least in the number of days that the tropical cyclones that do occur are in a high intensity stage (Christensen and Hewitson 2007). Physically, the warmer ocean temperatures have the potential to provide more water vapor and energy to the storms that do form. On the other hand, by trapping infrared energy in the upper troposphere, the higher CO_2 concentration tends to stabilize the atmosphere so that it takes larger energy release, and so more power, to sustain the storm. And more power means higher winds, larger storms, and, consequently, much more damage because damage goes up roughly with the *ninth* power of the peak wind speed (Nordhaus 2010).

Well-developed hurricanes process tremendous amounts of energy. Based on typical amounts of precipitation, the processed energy can be equivalent to explosion of a megaton-sized nuclear explosion every several seconds. Although only a small percentage of the energy is dissipated through surface friction (and so damage), this is still a very large amount. Initial efforts to use cloud seeding to alter the patterns of heat release in an attempt to reduce the intensity of a particular hurricane go back to the mid-twentieth century, but the research was stopped due to the difficulty of detecting and attributing changes in such variable storm systems to the imposed perturbation and to the fear of financial liability if an affected storm did substantial damage, even if the storm had not been significantly altered (Fleming 2004, 2010; Mooney 2007).

With hurricane models substantially improved, although presently calculating changes in track considerably better than changes in intensity, renewed consideration is being given to the potential for moderating hurricane intensity, especially because intense storms can generate very strong storm surges that can do tremendous damage. For example, Bowers et al. (2009) filed a patent application for an approach that would use one or more vessels or barges placed in the predicted path of a hurricane to replace warm surface waters with deeper and cooler waters using a wave-powered pumping method. Given the large expanse of very warm open water in a body such as the Gulf of Mexico and the limitations in model forecasts, the logistics and effectiveness would seem to be problematic. Other approaches to modify individual storms are also being considered, now that they can be tested in model simulations.

Perhaps more practical would be seeking to limit the seasonal warming of water bodies that provide the energy to sustain a hurricane's intensity as it approaches land, or perhaps in its early stages so a hurricane is less likely to form (MacCracken 2012). For example, might brightening clouds over the Gulf of Mexico or brightening the surface by injection of microbubbles from spring to fall be able to reduce the peak late summer temperatures that power the intensification of storms approaching the coastline? Might brightening of clouds or increasing surface albedo over the Caribbean Sea and the neighboring western Atlantic Ocean as well as the Gulf of Mexico reduce the peak intensity of hurricanes over the whole region? In that hurricanes are an important way for the atmosphere to transport energy from low to higher latitudes, might reducing the transport in some areas lead to more transport in other areas? In that much of the damage and injury comes from the heavy precipitation and flooding, would reducing peak wind speeds tend to reduce damages or cause a storm to be slower moving so total precipitation and damage would go up? While there appears to be potential to moderate intense hurricanes, virtually no research has been done.

12.5.5 Potential for Limiting the Effects of Ocean Warming on Ocean Reefs

In addition to stresses caused by harvesting of fish, pollution, and increased tourism, reef systems face several stresses related to on-going reliance on fossil fuels. These include acidification of ocean waters (Royal Society 2005), ocean warming, and, potentially, more severe storms and sea level rise. Reducing acidification would require continuing efforts to add buffering materials to ocean waters that come into contact with reef systems, the intent being to imitate the effects of natural weathering of exposed rock by rainfall, but on a greatly accelerated time scale (e.g., see Royal Society 2009). The scale of such an endeavor would likely be enormous.

Reducing coral bleaching caused by ocean warming, however, might be more feasible. Both regional cloud brightening by sea salt injection and sea surface brightening by microbubble injection might be options for limiting solar absorption and warming, at least in some situations during periods of peak temperature. While such shading might be useful, the impacts of ongoing ocean acidification are expected to become so significant by 2050 that very few, if any, regions of the ocean will have a pH that is conducive to growth of most types of coral, and so it might well be that actions to limit the warming would, in effect, only slightly delay the inevitable.

12.5.6 Summary

Even though only very limited analysis has been carried out, there do appear to be possibilities for counterbalancing or moderating at least a few of the most significant effects and impacts of human-induced climate change. What is also clear, however, is that the potential for such efforts is limited to perturbations that are relatively limited. There is simply no way that very large changes to the system caused by changes in the global climate can be offset by localized offsets; the smoothing tendencies of the global atmosphere and oceans are very difficult to impede.

This suggests that such regional approaches may have the potential to play a role in moderating peak and potentially irreversible impacts if there is an international strategy aimed at limiting overall global climate change and returning climate toward present or preindustrial conditions. In that these approaches are, thus, mainly limiting impacts rather than pulling back on the factors causing the climate to change, they might appropriately be considered an extension of approaches for adapting to the changes in global climate (so geo-adaptation) rather than geo- or climate engineering. Nonetheless, it might well be that they could serve a very useful purpose, especially because moving forward with them may present fewer challenges in getting permission to proceed with field testing, demonstration, and even implementation than seeking to geoengineer the global climate.

12.6 Implications of Incorporating Climate Engineering into a Comprehensive Response Strategy

Table 12.1, with axes indicating increasing societal significance and increasing long-term global implications, provides a matrix that encompasses the range of possible responses to the challenge of human-induced climate change. The most direct way to limit climate change, indicated in the upper left of the table, is to phase down emission of greenhouse gases into the atmosphere—that is, to stop causing the problem. Indeed, this was essentially the only

TABLE 12.1

Matrix of Options for Responding to Human-Induced Global Warming

	Increasing Commitment to Long-Term Global Implications→		
Approaches to Dealing with Climate Change	**Control and Response at Local to National Levels**	**Control and Response Needed over Regional Domains**	**Control and Response Needed at Global Scale**
Approaches to mitigation	Conservation, efficiency, alternative energy technologies, carbon capture and storage, reduced deforestation, and so on	Increasing uptake by natural sinks of carbon, including by afforestation, ocean fertilization, biochar burial, and so on	Scrubbing of CO_2 from the atmosphere with permanent burial and sequestration
Approaches to adaptation	Adjusting, sheltering, system strengthening, proactively reducing vulnerability, increasing resilience	Regional interventions to reduce the intensity of specific impacts (e.g., from hurricanes, the warming Arctic, shifting storm tracks)	Taking control of the planetary climate by management of incoming solar radiation; ocean acidification through bicarbonate addition
Degrees of suffering	Inconvenience, temporary relocation and sheltering, protection of existing infrastructure, modest costs of accommodation	Significant and repeated disruption, repeated repair of infrastructure, gradual relocation of communities and infrastructure	Significant loss/extinction of species and changes in ecosystems, widespread coastal retreat

Moving left to right increases the commitment to international action and maintaining actions over longer periods. Moving top to bottom increases the involvement of society in acting and dealing with climate change.

approach being considered in negotiations preceding international agreement on the United Nations Framework Convention on Climate Change in 1992 (U.N. 1994, p. 9). Article 2 of this Convention agreed that

> The ultimate objective of this Convention and any related legal instruments that the Conference of the Parties may adopt is to achieve, in accordance with the relevant provisions of the Convention, stabilization of greenhouse gas concentrations in the atmosphere at a level that would prevent dangerous anthropogenic interference with the climate

> system. Such a level should be achieved within a time frame sufficient to allow ecosystems to adapt naturally to climate change, to ensure that food production is not threatened and to enable economic development to proceed in a sustainable manner.

In that forests, for example, mature on a time scale of centuries, the first qualification in the second sentence would seem to mean that only quite slow climate change should be allowed, meaning emissions can grow only slowly. Similarly, in that projected changes in climate are likely to cause significant shifts in where important crops can be grown, restrictions on emissions would seem likely to need to be quite strict. On the other hand, with combustion of fossil fuels providing roughly 80% of the global energy that supports improving the standard of living of 7 billion people, slowing the growth in emissions is proving problematic, much less cutting CO_2 emissions by roughly 90%, which would be what is required to stabilize its atmospheric concentration (Matthews and Caldeira 2009).

While stabilizing the CO_2 concentration would be a major achievement, this would not mean that there would not be further climate change. Because the ocean heat capacity causes present warming to be less than would be expected at equilibrium, because cutting CO_2 emissions by 90% would likely reduce the offsetting cooling influence provided by SO_2 emissions by an equivalent amount, and because present and ongoing warming is likely triggering climate feedbacks that further amplify warming, there are built-in pressures favoring further warming. Quite clearly, while mitigation is essential to limit ultimate warming, the near-term challenge of reducing poverty in the developing nations and economic and political constraints in the industrialized nations have so far meant that the rate of emissions is still increasing, meaning that both the concentration of CO_2 and the rate of warming are also still increasing.

Assuming that no special policies were adopted to limit use of fossil fuels, IPCC (2000) projected a range of emissions trajectories out to 2100 and beyond. The high end projected a 350% increase in year-2000 emissions by 2100; the middle scenario projected roughly a doubling of emissions; and the low, very optimistic and very green scenario projected average twenty-first century emissions would be up about 50%. A bit over a decade after the scenarios were published, annual emissions are up over 40% and there is little sign that global emissions are near to peaking. Model projections are that these scenarios would raise global average temperature to 2.5°C–4°C above preindustrial temperatures, with further increase in global average temperature over ensuing centuries (up to levels of roughly 3°C–8°C by 2300). Keeping the rise in global average temperature below about 2°C, which is the level international leaders have at least tentatively accepted as likely to trigger "dangerous" impacts, is, thus, clearly going to be a very difficult challenge. Roughly speaking, annual emissions of CO_2 and other long-lived species over the twenty-first century would need to average no more than their year-2000 level (and then go to virtually zero after 2100), deforestation would need to be reversed, and the emissions of black carbon, CH_4, and other relatively short-lived species would need to be cut significantly. There is little sign at present that the policies to make this happen are near to adoption, and the longer nations wait, the sharper will need to be the reductions.

With the inadequacy so far of mitigation, adaptation is necessarily emerging as an essential component of the overall response strategy, the intent being to ensure that the changes in climate and environment that do occur are bearable and do not do great harm. In concept, proactive adaptation should be able to make society less vulnerable to the types of projected impacts, and perhaps some steps can be taken to help limit ecosystem damage. As for mitigation, however, the ability of adaptation to moderate impacts seems likely to be limited. In that the global average temperature during the Last Glacial Maximum (~20,000 year ago) was about 6°C colder than present and during the Cretaceous (>65 million years ago, the age of dinosaurs) was only about 6°C above the present average, projected changes due to reliance on fossil fuels are very large from both geological and historical perspectives, and are very rapidly, limiting the potential for adaptation.

For the United States, the projected changes in weather and climate by 2100 are likely to cause, for example, significant shifts in land cover. Roughly, the climate, and thus, perhaps the land cover, of the northern tier of states is likely to become like that of the central tier, and the climate of the central tier of states like that of the southern tier of states. As important as the projected changes are the steps that occur during the transition. Generally what happens is that the existing land cover is stressed and dies, with fire often accelerating the process. In contrast to the rapidity of the deterioration process, the growth of replacement species that are better adapted to the new climate occurs relatively slowly—or at least would if the climate settled at a new equilibrium rather than kept changing, so starting the cycle over and over again. Assuming there are adequate water resources, farmers, collectively, can likely sustain large-scale food production and become more resilient to the ongoing changes by becoming better prepared to raise a wider range of crops; for individual farmers, the challenge could well be more difficult because of the need to link their changing crop selections with new market pathways.

More generally, dealing with the increasing stresses likely to impact water resources may be the most serious challenge over much of the interior of the country as the snow line recedes, evaporation increases, and demand for water for agriculture, industry, energy generation, communities, and the stream and river environment all increase. Greater efficiency and water pricing can be helpful, but persistent drought can change landscapes, allow more frequent wildfires, and force further dependence on nonrenewable groundwater resources. While moving water from water-rich to water-poor areas has been done in the past, this is no longer as flexible an option because recent usage patterns have tightened water availability in virtually all locations.

Along coastlines there is the challenge of rising sea level, which is projected to be 1 ± 0.5 m by 2100, plus more intense storms, higher storm surges, coastal erosion, and rising temperatures. In some low lying regions, including the Mississippi River Delta and Chesapeake Bay, the effective rise could be even more because these lands are slowly sinking due to the region's geology and/or other human activities. While levees can, in many regions, delay near-term inundation, rising sea level is pushing salt water into coastal aquifers, such as in Florida, which is underlain by limestone, salt water will be able to push far inland. Even more problematic is that, over the next several centuries, sea level will continue to rise as warming penetrates into the deep ocean and ice sheets continue to lose mass. At the peak of the last glacial, sea level was down about 120 m, so about 20 m per degree change in global average temperature.

During the Pliocene and Cretaceous when temperatures were several degrees higher, there was very little ice at either pole, so sea level was, in effect, 60–70 m higher, so perhaps 10–20 m per degree increase in global average temperature. During the retreat of the ice sheets, sea level rose, on average, about 1 m per century for 120 centuries. The rate of warming is now faster than during that period—it may, thus, be that unconstrained warming could initiate a rate of sea level rise of a meter per century or more for several millennia. The uncertainty in sea level rise is primarily *when* sea level rise could reach 1, 3, 5, even 10 m above present, not so much whether it will happen, especially if emissions are not aggressively reduced, and quite likely reversed (Wigley 2005). Adaptation is clearly not going to be easy—and suffering and displacement are the only alternatives unless aggressive actions are undertaken.

It is the increasingly serious situation that the world faces and the lack of international progress on rapidly reducing greenhouse gas emissions that prompted Wigley (2006) and Crutzen (2006) to lift the issue of geoengineering (referred to here as climate engineering) from the dark background to serious consideration. What their papers, and many papers by many authors since, make clear is that nonaggressive mitigation and adaptation will not be adequate to prevent a quite high likelihood that most nations will experience very serious, even "dangerous," environmental and societal consequences from the rising greenhouse gas concentrations.

Table 12.1 is set up such that the components of climate engineering are considered as extensions of mitigation and adaptation. Traditional mitigation encompasses cutting back emissions by reducing the use of fossil fuels to generate energy. Intervening via CDR approaches to increase the transfer of CO_2 from the atmosphere to the terrestrial biosphere and oceans presents a set of options that imitates how the natural system has previously responded to disruptions, and thus, seems a relatively low-risk response option; unfortunately, it has relatively limited capacity. Very greatly expanding industrial scrubbing of CO_2 from the atmosphere and then permanently sequestering it in geological formations or ocean sediments represents the ultimate mitigation option for it is really building the capability to exert intentional human control over atmospheric composition.

Adaptation can also be subdivided into three increasingly aggressive policy approaches. The least prescriptive includes variously preparing for and adjusting to the changes in climate that are occurring. This approach is what is most widely discussed, and ranges from, for example, extending access to air-conditioning to building new levees and reservoirs to updating the choice of crops grown in particular locations as the climate changes. The second possible level of adaptation is intended to encompass proactive actions that might be taken to limit or moderate regional impacts. Examples might include actions to moderate the intensity of tropical cyclones, limit Arctic warming, or alleviate drought. Such actions would not deal with the global problem, being instead a limited way of dealing with particularly severe impacts. The most aggressive response option, again dealing with the consequences of the increase in the CO_2 concentration rather than the cause, would be to take control of the global climate by altering the fluxes of solar radiation or the climate system elements, like clouds, that influence the infrared radiation balance. Dealing with solar radiation allows a much more rapid influence than does changing the CO_2 concentration, and so would put humans in charge of the global climate on a decade to decade basis.

The third general response option, and the path the world is generally on, is to suffer, experiencing damage and dislocation. There seems to be an impression that for most people this will only involve acceptable inconvenience—a level of damage easily repaired, short-term evacuation in the face of hurricanes, fires, severe heat waves, and more. The projected consequences for ongoing climate change are, however, much more severe. The next degree of suffering would encompass degrees of damage that would require rebuilding infrastructure and buildings, with some relocation of those in vulnerable communities (such as indigenous peoples who live on barrier islands, those displaced by storm surges and severe flooding, etc.)—basically where staying in place would require more and more cost and protection. Beyond this level are situations that require permanent, large-scale relocations or that lead to extinctions, such as seem likely to be required when sea level rise exceeds a meter or more. Examples likely include the Everglades and the Mississippi River Delta region, including New Orleans. For such cases, the habitat and environment that have served people and the natural world just disappear—there is no longer a place to go back to.

12.6.1 Policy Implications of CDR

As described in Section 12.2, there are some approaches that would be able to make a difference if global CO_2 emissions were to be brought down significantly; without such reduction, all that CDR, even aggressively implemented, would accomplish would be to slightly slow climate change over the second half of the twenty-first century. While at least some of the CDR approaches could be undertaken on a national level without international approval and would not seem to raise significant ethical issues, at least some of the approaches would require land, competing with other pressing needs. Actions aimed at increasing oceanic uptake of CO_2, assuming the CO_2 would really be sequestered for an extended period, are already being brought under the auspices of an existing international framework, on the one hand, providing a designated forum for discussion and, on the other hand, making clear that international agreement will be needed to proceed. And this is occurring for a general category of climate engineering that would only have the potential to have a relatively small effect spread over many decades.

Whether there is a cost-effective approach for scrubbing and sequestering CO_2 from the atmosphere is a subject of current research. At present, suitable technologies have not been developed for scrubbing CO_2 out of power plant exhaust plumes where the concentration is roughly 10% (100,000 ppm), much less for doing so from the free atmosphere where the concentration is roughly 400 ppm. It would seem that building a capacity to scrub enough CO_2 from the atmosphere to make a significant difference in the global carbon cycle would likely require an industrial infrastructure comparable in size to the present infrastructure for providing fossil fuel energy. At least until global emissions are brought down by a factor of 5 to 10, it is not clear that this level of effort would be feasible. However, once global emissions are brought down to of order 1 GtC/yr, such an approach might become very important in order to counterbalance ongoing use of fossil fuels in critical applications. In addition, CO_2 removal could be used to help pull the atmospheric CO_2 concentration down to lower levels than prevail today, helping to reduce global average temperature and limit ocean acidification. Because of such possible applications, long-term research on potential technologies seems justified, especially if it might also help determine how best to limit emissions from fossil-fueled power plants.

12.6.2 Policy Implications of SRM

With respect to SRM, it would seem most plausible to place the initial focus on moderating specific, generally regional, climate impacts such as Arctic warming, shifts in storm tracks, intensification of tropical cyclones, and accelerating loss of mass from the Greenland and Antarctic ice sheets. While intervening in the climate in regionally focused ways certainly introduces issues of governance, ethics, liability, and more, the relative risk evaluation of global warming with and without climate engineering can likely be more specifically engaged at the regional than at the global scale. Certainly, the degree to which a regionally focused intervention might affect the global environment and community of nations needs consideration, but it would likely be most efficient if the first evaluation were to be undertaken by those likely to be most directly affected. At least for some of the potential regional interventions, such as moderating Arctic warming and slowing ice sheet loss, the major global consequences would seem to be generally beneficial, while for others, such as nudging storm tracks to alleviate drought, the global scale consequences might well be relatively limited compared to effects around the world of global-scale oscillations and natural variability. Given the limits of traditional mitigation and adaptation, research on, and even testing of, possible options for regional climate engineering would seem to merit near-term consideration (Boyd 2008), although there are many questions to be considered (Robock 2008).

Although the potential for regionally focused climate engineering has received relatively little attention, there has been considerable discussion, even if only limited research to date, regarding potential approaches for global climate engineering. Most attention has focused on the potential for augmenting the stratospheric sulfate layer or brightening marine stratus clouds. While model simulations indicate that such approaches have the potential, with some uncertainty, to moderate roughly as much warming as might be induced by a doubling of the CO_2 concentration (so, a few degrees Celsius), a number of policy-related issues have arisen that have the potential to greatly complicate consideration of such an approach. First, some in the policy community are under the misconception that global climate engineering could serve as a substitute or alternative for mitigation, and even adaptation—basically a Plan B. Such views represent a serious misunderstanding of what could possibly be accomplished by global climate engineering. For example, climate engineering would have no influence on ocean acidification, the amount of warming that could possibly be offset is limited, and there are likely to be some serious unintended side effects. At best, global geoengineering might be a complement to mitigation and adaptation, but not a substitute.

Even conceptualized as a complement to mitigation and adaptation, the relatively low expense of the leading SRM approaches raises the issue of whether this might lead to an over-reliance on SRM and less aggressive efforts to limit emissions, essentially seeming to provide more time for the transition to a non–fossil fuel based global energy system. If all that is included in the cost-benefit analysis that is done is to include the costs of the energy transition and the costs of climate engineering, slowing mitigation efforts might well appear to be a cost-effective step. However, if the near- and long-term costs of the impacts of climate change are also included in the analysis, and a discount rate is used that is appropriate to considering the irreversibility of many of the impacts (e.g., see Stern 2007; although also see Nordhaus 2007), it seems likely that an aggressive mitigation with climate engineering as a possible complement would be the preferred approach.

Even with agreement that mitigation must be actively pursued and that climate engineering is a possible complement that would be used only to counterbalance the warming that is going to be causing (or already is causing) unacceptable impacts, there is the question of who would get to make decisions about the extent, character, and safety of a climate engineering intervention, what sort of governance structure might be needed, and what issues they would face? With greenhouse gas–induced warming causing differential climatic impacts around the world, mostly detrimental but some beneficial, and global climate engineering causing a rough rather than exact counterbalancing, complex discussions would be needed to agree on the most appropriate and inclusive international governing structure and on how much counterbalancing there would need to be, how those adversely affected by the climate engineering technology and by remaining climate change might be compensated, how adverse impacts would be convincingly proven, who would pay what to whom, and more. The discussion would likely also be complicated by a prevailing perception that all aspects of the climate (including severe tropical cyclones, heat waves, etc.) would now be under human control—that is, the calamities that in the past have been viewed as unfortunate natural variations would now be seen as a failure of the control system and rationale for demanding compensation. How to address potential legal and liability aspects is yet to be addressed.

Out of frustration with the difficulty of working through the process of setting up a global governance structure and driven by frustration with intensifying regional climate change, it has been suggested that one or a small group of nations, or even an individual, might, given the relatively low cost, unilaterally move ahead with climate engineering. While it is likely that such actions could be detected, the question then becomes what actions would be available to limit any adverse impacts. It would be conceivable for other nations to emit additional greenhouse gases, but this would really be self-defeating, so the response would likely need to involve economic and political sanctions, further polarizing the international negotiations. Personally, it seems to me relatively unlikely that, because of its diffuse influence, a particular nation would choose global climate engineering to address a particularly severe climatic impact that it was experiencing, even if this were economically feasible. What seems more likely to me would be the possibility that one or a few nations would seek to pursue a regionally focused climate intervention, especially if the impact of concern were of a magnitude that it would seriously be affecting major national capabilities for growing food, supplying water, generating energy, or protecting public health.

Even if the governance concerns can be addressed, an important aspect of climate engineering is that, to maintain a particular offset climate, the intervention must be continued for as long as the elevated greenhouse gas concentration being offset persists. For CO_2 and the other long-lived species, this could be centuries, or even longer, depending especially on how long very high CO_2 emissions continue and how much their effects need to be offset. This raises both practical and ethical issues. First, governance and implementation that functions effectively for centuries on the global scale would be unprecedented because there are few, if any, institutions, much less governments, that adopt policies and implement programs that have time horizons beyond decades—religious organizations and universities do, but hardly seem to fit the characteristics of an implementing organization.

At least as important, taking actions at present that would impose obligations for continuing action by future generations would seem to raise considerations of ethics and equity, especially in the case of global-scale climate engineering, because failing to continue the effort would lead to a relatively rapid readjustment toward the higher equilibrium temperatures that are being offset.

12.7 Summary

The world community has gotten itself into quite a predicament. Recognition that fossil fuel emissions could increase the atmospheric CO_2 concentration and warm the climate goes back to Arrhenius in 1896 to high-level attention. The first official scientific report bringing this issue (and the possible role of geoengineering) was delivered to President Johnson in 1965 by the President's Science Advisory Committee (PSAC 1965). The first international reports on this issue go back to the early 1970s, and there were contemporaneous reports from national academies of science. The U.S. Department of Energy published a comprehensive assessment on the carbon and climate change science in 1985 (DOE 1985a, 1985b, 1985c, 1985d), the same year that an international meeting organized by WMO/UNEP/ICSU (1985) gave notice to international policy makers that they could no longer assume that the future climate would be like the past. From 1990 to 2007, the Intergovernmental Panel on Climate Change (1990, 1995, 2001, 2007a), drawing upon the international scientific community, has published assessment reports that have been *unanimously* endorsed by the U.N. community of nations covering climate change science, impacts, and options for taking action. All that has resulted on the policy front is a convention with an aspirational objective. This convention, however, has led to not much more than an agreement among leading nations that they are committed to keeping the increase in global average temperature over preindustrial to less than 2°C; unfortunately, there is no plan on how to do this and it is very unlikely that this can be achieved. In that an increase of only 0.8°C has led to significant reductions in Arctic Sea ice extent and thickness, accelerated loss of mass from the Greenland and Antarctic ice sheets, an increased rate of sea level rise and permafrost thawing, increasing stress on and shifts in range of many types of flora and fauna, and much more, allowing the global average temperature to have gone up by more than about 0.5°C has likely significantly increased the risk of extensive damage and disruption.

With fossil-fueled energy still providing roughly 80% of the world's energy and halting climate change requiring an 80%–90% (or more) replacement with non-fossil sources of energy, it will clearly take decades more to halt and then reduce the ongoing increases in global average temperature and, especially, in sea level and stress on water resources. Greatly increased efforts to adapt to and improve societal resilience to climate change and other stresses will be helpful, but there are significant limits to the damage and disruption that can be alleviated. In this predicament, it is likely going to become increasingly attractive to seek out whatever other options may exist—and the only other option seems to be climate engineering. Pulling additional CO_2 out of the atmosphere by natural approaches such as afforestation and enhancing ocean uptake have limited capacity, industrial removal is likely to be expensive relative to switching energy generation away from fossil fuels, and all such efforts are likely to take many decades to have an effect, especially as global emissions continue to increase. Intervening in the global energy balance by reducing solar absorption or enhancing infrared loss may indeed be feasible and not unduly expensive, but such approaches introduce very complex issues of hubris, governance, and the ethics and equity of passing along the decade-to-multicentury obligation of maintaining the effort to future generations.

Despite the complications of climate engineering, the very limited actions being taken to limit emissions, especially in the context of the very difficult situation that exists, may well make climate engineering too attractive to resist, despite the governance and intergenerational implications and the likely unintended consequences (Barrett 2008). Right now, however, interest in the potential application of climate engineering seems to far exceed the limited knowledge and research that has been conducted on its physical, environmental, and societal aspects. In my view, it is past time that the international research program on climate engineering be expanded, with the intent being to, at the very least, determine if climate engineering might be a credible option for complementing mitigation and adaptation.

Acknowledgments

The author is grateful to Professor Robert Watts for the invitation to prepare this chapter and for his forbearance in its delayed submission. The views expressed are those of the author and not necessarily of the Climate Institute with which he is affiliated or other organizations with which he is associated.

Questions for Students

1. Biochar (i.e., carbonized agricultural waste) is proposed as a means to sequester carbon below ground. What are the factors that you would consider in determining the relative advantages and disadvantages

of using agricultural waste to make and then bury biochar versus using the agricultural waste to displace use of fossil fuels?

2. To significantly reduce the problem of forward scattering by sulfate aerosols injected into the stratosphere, Dr. Edward Teller (Teller et al. 1997, 2002) proposed instead using particles or balloons dimpled with corner reflectors that would backscatter incident radiation. Assuming hydrogen-filled Mylar balloons of 1-m diameter floating freely in the stratosphere, how many balloons would be needed to counterbalance the radiative trapping of a CO_2 doubling? What would be the complicating factors that would need to be considered in a detailed analysis, and what, if anything, might be done to address them?
3. Salter and Latham (2007) and Salter et al. (2008) have proposed using trimarans powered by Flettner rotors that generate power by spinning the masts and relying on the Bernoulli principle. What is the relationship between energy generated and the rate of rotation of the masts?
4. During spring, bright, snow-covered sea ice reflects 60%–80% of incident solar radiation. During the summer, surface melting causes the surface albedo to drop, so underlying ocean and surface temperatures are nearly the same. In the fall, as surface temperature drops due to reduced sunlight, sea ice reduces the flow of heat from the ocean to the atmosphere. Assuming that all the heat lost to the atmosphere through the ice goes into creation of ice at the air-sea interface, thus, thickening the ice, what are the factors to consider when calculating how using an icebreaker to break through the ice would contribute to thickening of the ice? What factors would you use in a calculation of the thickness of the ice for icebreakers to break through to maximize the rate of sea ice thickening over the ice area?

References

Akbari, H., S. Menon, and A. Rosenfeld, 2009. Global cooling: Increasing world-wide urban albedos to offset CO_2. *Climatic Change* **94**, 275–286, doi:10.1007/s10584-008-9515-9.

Angel, R., 2006. Feasibility of cooling the earth with a cloud of small spacecraft near the inner Lagrange point (L1). *Proceedings of the National Academy of Sciences* **103**, 17184–17189.

Archer, D., 2005. Fate of fossil fuel CO_2 in geologic time. *Journal of Geophysical Research* **110**, C09S05, doi:10.1029/2004JC002625.

Arctic Climate Impact Assessment (ACIA), 2004. *Impacts of a Warming Arctic: Arctic Climate Impact Assessment*, Cambridge University Press, 140 pp.

Asilomar Scientific Organizing Committee (M. MacCracken [chair], S. Barrett, R. Barry, P. Crutzen, S. Hamburg, R. Lampitt, D. Liverman, T. Lovejoy, G. McBean, E. Parson, S. Seidel, J. Shepherd, R. Somerville, and T. M. L. Wigley), 2010. *The Asilomar Conference Recommendations on Principles for Research into Climate Engineering Technologies: Conference Report*, Climate Institute, Washington DC, 37 pp.

Bala, G., K. Caldeira, A. Mirin, M. Wickett, and C. Delire, 2005. Multicentury changes to the global climate and carbon cycle: Results from a coupled climate and carbon cycle model. *Journal of Climate* **18**, 4531–4544.

Bala, G., P. B. Duffy, and K. E. Taylor, 2008. Impact of geoengineering schemes on the global hydrological cycle. *Proceedings of the National Academy of Sciences* **105**, 7664–7669.

Barrett, S., 2008. The incredible economics of geoengineering. *Environmental Resource Economics* **39**, 45–54.

Biofuelwatch, 2011. Biochar: A Critical Review of Science and Policy (http://www.biofuelwatch.org.uk/).

Bipartisan Policy Center (BPC), 2011. *Geoengineering: A National Strategic Plan for Research on the Potential Effectiveness, Feasibility, and Consequences of Climate Remediation Technologies*. Task Force on Climate Remediation Research, Bipartisan Policy Center, Washington, DC, 36 pp.

Bowers, J. A., K. Caldeira, G. Kenneth, A. K. Chan, W. H. Gates, III, R. A, Hyde, M. Y. Ishikawa, J. Kare, J. Latham, N. P. Myhrvold, S. H. Salter, C. T. Tegreene, W. H. Wattenburg, J. R. Wood, and L. Lowell, 2009. Water alteration structure applications and methods, United States Patent Application: 0090173386.

Boyd, P. W., 2008. Ranking geo-engineering schemes. *Nature Geoscience* **1**, 722–724.

Brovkin, V., V. Petoukhov, M. Claussen, E. Bauer, D. Archer, and C. Jaeger, 2009. Geoengineering climate by stratospheric sulfur injections: Earth system vulnerability to technological failure. *Climatic Change* **92**, 243–259, doi:10.1007/s10584-008-9490-1.

Budyko, M. I., 1974. The method of climate modification. *Meteorology and Hydrology* **2**, 91–97 (in Russian).

Caldeira, K., and L. Wood, 2008. Global and Arctic climate engineering: numerical model studies. *Transactions of the Royal Society A* **366**, 4039–3056, doi:10.1098/rsta.2008.0132.

Caldeira, K., M. Akai, P. Brewer, B. Chen, P. Haugan, T. Iwama, P. Johnston, H. Kheshgi, Q. Li, T. Ohsumi, H. Poertner, C. Sabine, Y. Shirayama, and J. Thomson, 2005. Ocean storage, pp. 277–317 (Chapter 6) in *IPCC Special Report on Carbon Dioxide Capture and Storage*, prepared by Working Group III of the Intergovernmental Panel on Climate Change, Metz, B., O. Davidson, H. C. de Coninck, M. Loos, and L. A. Meyer (eds.), Cambridge University Press, Cambridge, UK, and New York, 442 pp.

Canadell, J. G., C. Le Quéré, M. R. Raupach, C. B. Field, E. T. Buitenhuis, P. Ciais, T. J. Conway, N. P. Gillett, R. A. Houghton, and G. Marland, 2007. Contributions to accelerating atmospheric CO_2 growth from economic activity, carbon intensity, and efficiency of natural sinks. *Proceedings of the National Academy of Sciences* **104**, 18866–18870.

Chang, D. B., and I-Fu Shih, 1991. Stratospheric Welsbach Seeding for Reduction of Global Warming, U.S. Patent 5,003,186, issued March 26, 1991 (also refers to U.S. Patents 4,755,673 on Selective Thermal Radiators, issued July 5, 1988, and 3,222,675 on Means for Positioning a Plurality of Elements in Orbit about a Celestial Body, issued December 7, 1965).

Christensen, J. H., and B. Hewitson, 2007. Regional climate projections, pp. 847940 (Chapter 11) in *Climate Change 2007: The Physical Science Basis*, Contribution of Working Group I to the Fourth Assessment Report of the Intergovernmental Panel on Climate Change, S. Solomon, D. Qin, M. Manning, et al. (eds.), Cambridge University Press, Cambridge, UK, and New York, 996 pp.

Church, J. A., and J. M. Gregory, 2001. Changes in Sea Level, pp. 639–693 (Chapter 11) in *Climate Change 2001: The Scientific Basis*, Contribution of Working Group I to the Third Assessment Report of the Intergovernmental Panel on Climate Change, J. Houghton, Y. Ding, D. J. Griggs, M. Nouger, P. J. van der Linden, X. Dai, K. Maskell, and C. A. Johnson (eds.), Cambridge University Press, Cambridge, UK, and New York, 881 pp.

Crutzen, P. J., 2006. Albedo enhancement by stratospheric sulfate injections: A contribution to resolve a policy dilemma? *Climatic Change* **77**, 211–219.

Department of Energy (DOE), 1985a. *Atmospheric Carbon Dioxide and the Global Carbon Cycle*, J. R. Trabalka, ed., DOE/ER-0239, U.S. Department of Energy, Washington DC, 316 pp.

Department of Energy (DOE), 1985b. *Detecting the Climatic Effects of Increasing Carbon Dioxide*, M. C. MacCracken, and F. M. Luther, eds., DOE/ER-0235, U.S. Department of Energy, Washington DC, 198 pp.

Department of Energy (DOE), 1985c. *Projecting the Climatic Effects of Increasing Carbon Dioxide*, M. C. MacCracken, and F. M. Luther, eds., DOE/ER-0237, U.S. Department of Energy, Washington DC, 381 pp.

Department of Energy (DOE), 1985d. *Direct Effects of Increasing Carbon Dioxide on Vegetation*, Boyd R. Strain, and J. D. Cure, eds., DOE/ER-0238, U.S. Department of Energy, Washington DC, 286 pp.

Early, J. T., 1989. Space-based solar screen to offset the greenhouse effect. *Journal of the British Interplanetary Society* **42**, 567–569.

English, J. M., O. B. Toon, and M. J. Mills, 2012. Microphysical simulations of sulfur burdens from stratospheric sulfur geoengineering. *Atmospheric Chemistry and Physics Discussions* **12**, 2517–2558. doi:10.5194/acpd-12-2517-2012.

Flannery, B. P., G. Marland, W. Broecker, H. Ishatani, H. Keshgi, H. Komiyama, M. MacCracken, N. Rosenberg, M. Steinberg, and T. Wigley, 1997. Geoengineering climate, pp. 379–427 (Chapter 8) in *The Engineering Response to Climate Change*, Robert G. Watts (ed.), Lewis Publishers, Boca Raton FL, 492 pp.

Fleming, J. R., 2004. Fixing the weather and climate: Military and civilian schemes for cloud seeding and climate engineering, pp. 175–200 (Chapter 9) in *The Technological Fix: How People Use Technology to Create and Solve Problems*, L. Rosner (ed.), Routledge, New York.

Fleming, J. R., 2007. The climate engineers, *Wilson Quarterly* (spring), 46–60.

Fleming, J. R., 2010. *Fixing the Sky: The Checkered History of Weather and Climate Control*, Columbia University Press, New York, 344 pp.

Forster, P., and V. Ramaswamy, 2007. Changes in atmospheric constituents and radiative forcing, pp. 129–234 (Chapter 2) in *Climate Change 2007: The Physical Science Basis*, Contribution of Working Group I to the Fourth Assessment Report of the Intergovernmental Panel on Climate Change, S. Solomon, D. Qin, M. Manning, et al. (eds.), Cambridge University Press, Cambridge, UK, and New York, 996 pp.

Gaskill, A., Jr., and C. E. Reese, 2003. Global Warming Mitigation by Reduction of Outgoing Longwave Radiation Through Large-Scale Surface Albedo Enhancement of Deserts Using White Plastic Polyethylene Film—the Global Albedo Enhancement Project (GAEP) Research Plan, Report from Environmental Reference Materials, Inc., Research Triangle Park, NC, 90 pp.

Goldberg, D. S., T. Takahashi, and A. L. Slagle, 2008. Carbon dioxide sequestration in deep-sea basalt. *Proceedings of the National Academy of Sciences* **105**, 9920–9925.

Govindasamy, B., and K. Caldeira, 2000. Geoengineering earth's radiative balance to mitigate CO_2-induced climatic change. *Geophysical Research Letters* **27**, 2141–2144.

Gu, L., Baldocchi, D. D., Wofsy, S. C., Munger, J. W., Michalsky, J. J., Urbanski, S. P., and T. A. Boden, 2003. Response of a deciduous forest to the Mount Pinatubo eruption: Enhanced photosynthesis. *Science* **299**, 2035–2038.

Hansen, J., A. Lacis, D. Rind, G. Russell, P. Stone, I. Fung, R. Ruedy, and J. Lerner, 1984. Climate sensitivity: Analysis of feedback mechanisms, pp. 130–163 in *Climate Processes and Climate Sensitivity*, AGU Geophysical Monograph 29, Maurice Ewing Vol. 5. J. E. Hansen and T. Takahashi (eds.), American Geophysical Union, Washington DC.

Hansen, J., M. Sato, P. Kharecha, D. Beerling, R. Berner, V. Masson-Delmotte, M. Pagani, M. Raymo, D. L. Royer, and J. C. Zachos, 2008. Target atmospheric CO_2: Where should humanity aim? *The Open Atmospheric Science Journal* **2**, 217–231, doi:10.2174/1874282300802010217.

Hegerl, G. C., and F. W. Zwiers, 2007. Understanding and attributing climate change, pp. 663–745 (Chapter 9) in *Climate Change 2007: The Physical Science Basis*, Contribution of Working Group I to the Fourth Assessment Report of the Intergovernmental Panel on Climate Change, S. Solomon, D. Qin, M. Manning, et al. (eds.), Cambridge University Press, Cambridge, UK, and New York, 996 pp.

Hoffert, M. I., and S. D. Potter, 1997. Beam it down: How the new satellites can power the world, in *Solar Power Satellites: A Space Energy System for Earth*, P. E. Glaser, F. P. Davidson and K. I. Csigi (eds.), John Wiley, New York, 694 pp. (download from http://www.spacefuture.com/pr/archive/beam_it_down_how_the_new_satellites_can_power_the_world.shtml).

Hoffert, M. I., et al., 2002. Advanced technology paths to global climate stability: Energy for a greenhouse planet. *Science* **298**, 981–987.

House, K. Z., C. H. House, D. P. Schrag, and M. J. Aziz, 2007. Electrochemical acceleration of chemical weathering as an energetically feasible approach to mitigating anthropogenic climate change. *Environmental Science and Technology* **41**, 8464–8470, doi:10.1021/es0701816.

Intergovernmental Panel on Climate Change (IPCC), 1990. *The IPCC Scientific Assessment*, J. T. Houghton, G. J. Jenkins, and J. J. Ephraums, eds., Cambridge University Press, Cambridge, UK, 365 pp.

Intergovernmental Panel on Climate Change (IPCC), 1995. *Climate Change 1995: The Science of Climate Change*, J. T. Houghton et al., eds., Cambridge University Press, Cambridge, UK, 572 pp.

Intergovernmental Panel on Climate Change (IPCC), 2000. *Special Report on Emissions Scenarios* (SRES), N. Nakicenovic, et al., eds., Cambridge University Press, Cambridge, UK, and New York, 599 pp.

Intergovernmental Panel on Climate Change (IPCC), 2005. *IPCC Special Report on Carbon Dioxide Capture and Storage,* Prepared by Working Group III of the Intergovernmental Panel on Climate Change, Metz, B., O. Davidson, H. C. de Coninck, M. Loos, and L. A. Meyer (eds.), Cambridge University Press, Cambridge, UK, and New York, 442 pp.

Intergovernmental Panel on Climate Change (IPCC), 2007a. *Climate Change 2007: The Physical Science Basis,* Contribution of Working Group I to the Fourth Assessment Report of the Intergovernmental Panel on Climate Change, S. Solomon, D. Qin, M. Manning, et al. (eds.), Cambridge University Press, Cambridge, UK, and New York, 996 pp.

Intergovernmental Panel on Climate Change (IPCC), 2007b. *Climate Change 2007: Impacts, Adaptation and Vulnerability,* Contribution of Working Group II to the Fourth Assessment Report of the Intergovernmental Panel on Climate Change, M. Parry, O. Canziani, J. Palutikof, P. van der Linden, and C. Hanson, et al. (eds.), Cambridge University Press, Cambridge, UK, and New York, 976 pp.

Intergovernmental Panel on Climate Change (IPCC), 2007c. *Climate Change 2007: Mitigation,* Contribution of Working Group III to the Fourth Assessment Report of the Intergovernmental Panel on Climate Change, B. Metz, O. Davidson, P. Bosch, R. Dave, and L. Meyer (eds.), Cambridge University Press, Cambridge, UK, and New York, 851 pp.

Intergovernmental Panel on Climate Change (IPCC), 2011. *Renewable Energy Sources and Climate Change Mitigation,* O. Edenhofer, R. Pichs-Madruga, Y. Sokona, K. Seyboth, P. Matschoss, S. Kadner, T. Zwickel, P. Eickemeier, G. Hansen, S. Schloemer, and C. von Stechow (Eds.), Cambridge University Press, Cambridge, UK, and New York, 1075 pp.

Irvine, P. J., D. J. Lunt, E. J. Stone, and A. Ridgwell, 2009. The fate of the Greenland Ice Sheet in a geoengineered, high CO_2 world. *Environmental Research Letters* **4**, 045109, doi:10.1088/1748-9326/4/4/045109.

Jevrejeva, S., J. C. Moore, and A. Grinsted, 2010. How will sea level respond to changes in natural and anthropogenic forcings by 2100? *Geophysical Research Letters* **37**, L07703. doi:101029/2010GL042947.

Johnson, R. G., 1997. Climate control requires a dam at the Strait of Gibraltar. *EOS* **78**, 277, 280–281.

Joughin, I., and R. Alley, 2011. Stability of the West Antarctic ice sheet in a warming world. *Nature Geoscience* **4**, 506–513, doi:10.1038/ngeo1194.

Kauppi, P., and R. Sedjo, 2001. Technological and economic potential of options to enhance, maintain, and manage biological carbon reservoirs and geo-engineering, pp. 301–343 (Chapter 4) in *Climate Change 2001: Mitigation,* Contribution of Working Group III to the Third Assessment Report of the Intergovernmental Panel on Climate Change, B. Metz O. Davidson, R. Swart, and J. Pan (eds.), Cambridge University Press, Cambridge, UK, 752 pp.

Keith, D. W., 2000. Geoengineering the climate: History and prospect. *Annual Review of Energy and the Environment* **25**, 245–284.

Keith, D. W., 2001. Sinks, energy crops, and land use: Coherent climate policy demands an integrated analysis of biomass. *Climatic Change* **49**, 1–10.

Keith, D. W., 2008. CO_2 capture from air, presentation at the MIT Carbon Sequestration Forum, Cambridge, MA (downloadable from http://www.ucalgary.ca/~keith/Misc/AC%20talk%20MIT%20Sept%202008.pdf).

Keith, D. W., 2009. Why capture CO_2 from the atmosphere. *Science* **325**, 1654–1655.
Keith, D. W., 2010. Photophoretic levitation of engineered aerosols for geoengineering. *Proceedings of the National Academy of Sciences* **107**, 16428–16431.
Keith, D. W., 2011. Comments on the APS report on Direct Air Capture, posted at http://www.carbonengineering.com/wp-content/uploads/2011/05/CE_APS_DAC_Comments.pdf.
Keith, D. W., M. Ha-Duong, and J. K. Stolaroff, 2006. Climate strategy with CO_2 capture from the air. *Climatic Change* **74**, 17–45.
Keith, D. W., K. Heidel, and R. Cherry, 2010. Capturing CO_2 from the atmosphere: Rationale and process design considerations, p.107–126 in *Geo-Engineering Climate Change,* B. Launder, and M. Thompson, (ed.), Cambridge University Press, Cambridge, UK.
Kelemen, P. B., and J. Matter, 2008. in situ carbonation of peridotite for CO_2 storage. *Proceedings of the National Academy of Sciences*, published online, doi:10.1073/pnas.0805794105.
Kravitz, B., and A. Robock, 2011. The climate effects of high latitude volcanic eruptions: The role of the time of year. *Journal of Geophysical Research* **116**, D01105, doi:10.1029/2010JD014448.
Kravitz, B., A. Robock, L. Oman, G. Stenchikov, and A. B. Marquardt, 2009. Sulfuric acid deposition from stratospheric geoengineering with sulfate aerosols. *Journal of Geophysical Research* **114**, D14109, doi:10.1029/2009JD011918.
Kravitz, B., A. Robock, O. Boucher, H. Schmidt, K. Taylor, G. Stenchikov, and M. Schulz, 2011. The Geoengineering Model Intercomparison Project (GeoMIP). *Atmospheric Science Letters* **12**, 162–167, doi:10.1002/asl.316.
Kravitz, B., D. G. MacMartin, and K. Caldeira, 2012. Geoengineering: Whiter skies? *Geophysical Research Letters* **39**, L11801.
Kreutz, T. G., E. D. Larson, G. Liu, and R. H. Williams, 2008. *Fischer-Tropsch Fuels from Coal and Biomass,* Report from the Princeton Environmental Institute, Princeton University, Princeton NJ, 86 pp.
Krevor, S. C. M., and K. S. Lackner, 2011. Enhancing serpentine dissolution kinetics for mineral carbon dioxide sequestration. *International Journal of Greenhouse Gas Control* **5**, 1073–1080, doi:10.1016/j.ijggc.2011.01.006.
Lane, L., K. Caldeira, R. Chatfield, and S. Langhoff (eds.), 2007. Workshop Report on Managing Solar Radiation, NASA Ames Research Center, 18–19 November 2006, NASA/CP2007-214558, 31 pp.
Langer, W. H., C. A. San Juan, G. H. Rau, and K. Caldeira, 2009. Accelerated weathering of limestone for CO_2 mitigation: opportunities for the stone and cement industries, *Mining Engineering*, pp. 27–32 in Special Editorial Supplement: 2009 SME/CMA Annual Meeting Official Show Guide.
Latham, J., 1990. Control of global warming? *Nature* **347**, 339–340.
Latham, J., P. J. Rasch, C. C. Chen, L. Kettles, A. Gadian, A. Gettelman, H. Morrison, K. Bower, and T. W. Choularton, 2008. Global temperature-stabilization via controlled albedo enhancement of low-level maritime clouds. *Philosophical Transactions of the Royal Society A* **366**, 3969–3987, doi:10.1098/rsta.2008.0137.
Leemans, R., S. Agrawala, J. A. Edmonds, M. C. MacCracken, R. Moss, and P.S. Ramakrishnan, 1995. Mitigation: Cross-Sectoral and Other Issues, pp. 799–819 (Chapter 25) in *Climate Change 1995: Impacts, Adaptations and Mitigation of Climate Change: Scientific-Technical Analyses*, Contribution of Working Group II to the Second Assessment Report of the Intergovernmental Panel on Climate Change,

R. T. Watson, M. C. Zinyowera, and R. H. Moss, (eds.), Cambridge University Press, Cambridge, UK, 879 pp.

Le Quéré, C., M. R. Raupach, J. G. Canadell, G. Marland, et al., 2009. Trends in the sources and sinks of carbondioxide. *Nature Geoscience* **2**, doi:10.1038/ngeo689.

Lohmann, U., P. Spichtinger, S. Jess, T. Peter, and H. Smit, 2008. Cirrus cloud formation and ice supersaturated regions in a global climate model. *Environmental Research Letters* **3**, 045022.

Lovelock, J. E., and C. G. Rapley, 2007. Ocean pipes could help the earth to cure itself. *Nature* **449**, 403, doi:10.1038/449403a.

MacCracken, M. C., 1991. *Geoengineering the climate,* Lawrence Livermore National Laboratory Report UCRL-JC-108014, Livermore CA.

MacCracken, M. C., 2009. On the possible use of geoengineering to moderate specific climate change impacts. *Environmental Research Letters* **4**, 045107, doi:10.1088/1748-9326/4/4/045107.

MacCracken, M. C., 2012. Potential Applications of Climate Engineering Technologies to Moderation of Critical Climate Change Impacts, pp. 55–56 in *Meeting Report of the IPCC Expert Meeting on Geoengineering* held 20–22 June 2011 in Lima, Peru Intergovernmental Panel on Climate Change, Geneva, Switzerland, 108 pp.

MacCracken, M. C., H-J. Shin, K. Caldeira, and G. A. Ban-Weiss, 2012. Climate response to imposed solar radiation reductions in high latitudes, in preparation.

Mahmoudkhani, M., and D. W. Keith, 2009. Low-energy sodium hydroxide recovery for CO_2 capture from atmospheric air—Thermodynamic analysis. *International Journal of Greenhouse Gas Control* **3**, 376–384.

Manabe, S., and R. T. Wetherald, 1980. On the distribution of climate change resulting from an increase in the CO_2 content of the atmosphere. *Journal of the Atmospheric Sciences* **37**, 99–118.

Marchetti, C., 1975. *On Geoengineering the CO_2 Problem,* International Institute for Applied Systems Analysis, Laxenburg, Austria.

Martin, J. H., 1990. Glacial-interglacial CO_2 change: The iron hypothesis. *Paleoceanography* **5**, 1–13.

Matthews, H. D., and K. Caldeira, 2009. Stabilizing climate requires near-zero emissions. *Geophysical Research Letters* **35**, doi:10.1029/2007GL032388.

McClellan, J., J. Sisco, B. Suarez, G. Keogh, 2011. *Geoengineering Cost Analysis,* Aurora Flight Sciences Corporation, Cambridge, MA, 87 pp.

Meehl, G. A., and T. F. Stocker, 2007. Global climate projections, pp. 747–845 (Chapter 10) in *Climate Change 2007: The Physical Science Basis,* Contribution of Working Group I to the Fourth Assessment Report of the Intergovernmental Panel on Climate Change, S. Solomon, D. Qin, M. Manning, et al. (eds.), Cambridge University Press, Cambridge, UK, and New York, 996 pp.

Meinshausen, M., 2006. What does a 2°C target mean for greenhouse gas concentrations? A brief analysis based on multi-gas emission pathways and several climate sensitivity uncertainty estimates, pp. 265–279 in *Avoiding Dangerous Climate,* J. H. Schellnhuber, H.J., et al., (eds.), Cambridge University Press, Cambridge, UK, and New York, 392 pp.

Milankovitch, M., 1941. *Kanon der Erdbestrahlung und seine Anwendung auf das Eiszeitproblem,* Serbian Academy of Science, Belgrade, 633 pp.

Mitchell, D. L., and W. Finnegan, 2009. Modification of cirrus clouds to reduce global warming. *Environmental Research Letters* **4**, 045102, doi:10.1088/1748-9326/4/4/045102.

Mitchell, D. L., P. Rasch, D. Ivanova, G. McFarquhar, and T. Nousiainen, 2008. Impact of small ice crystal assumptions on ice sedimentation rates in cirrus clouds and GCM simulations. *Geophysical Research Letters* **35**, L09806.

Montenegro, A., V. Brovkin, M. Eby, D. Archer, and A. J. Weaver, 2007. Long term fate of anthropogenic carbon. *Geophysical Research Letters* **34**, L19707, doi:10.1029/2007GL030905.

Mooney, C., 2007. *Storm World: Hurricanes, Politics and the Battle Over Global Warming*, Houghton Mifflin Harcourt, Boston MA, 400 pp.

Moore, J. C., S. Jevrejeva, and A. Grinsted, 2010. Efficacy of geoengineering to limit 21st century sea-level rise. *Proceedings of the National Academy of Sciences* **107**, 15699–15703.

Moss, R., et al., 2007. *Towards New Scenarios for Analysis of Emissions, Climate Change, Impacts, and Response Strategies*, IPCC Expert Meeting Report, Intergovernmental Panel on Climate Change, Geneva, 132 pp.

Murphy, D. M., 2009. Effect of stratospheric aerosols on direct sunlight and implications for concentrating solar power. *Environmental Science and Technology* **43**, 2784–2786, doi:10.1021/es802206b.

Mysak, L. A., 2008. Glacial inceptions: Past and future. *Atmosphere-Ocean* **46**, 317–341, doi:10.3137/ao.460303.

Namias, J., 1977. Forecasting climate fluctuations: The winter of 1976–77. *Science* **196**, 1386–1387.

Namias, J., 1978. Multiple causes of the North American abnormal winter 1976–77. *Monthly Weather Review* **106**, 279–295.

National Academy of Sciences (NAS), 1992. *Policy Implications of Greenhouse Warming: Mitigation, Adaptation, and the Science Base*, National Academy Press, Washington DC.

Nordhaus, W. D., 2007. Critical assumptions in the Stern review on climate change. *Science* **317**, 201–202.

Nordhaus, W. D., 2010. The economics of hurricanes and implications of global warming. *Climate Change Economics* **1**, 1–20, doi:10.1142/S2010007810000054.

Oman, L., A. Robock, G. Stenchikov, G. A. Schmidt, and R. Ruedy, 2005. Climatic response to high-latitude volcanic eruptions. *Journal of Geophysical Research* **110**, D13103, doi:10.1029/2004JD005487.

Ornstein, L., Aleinov, I., and D. Rind, 2009. Irrigated afforestation of the Sahara and Australian outback to end global warming. *Climatic Change* **97**, 409–437.

Penner, J. D., 2001. Aerosols, their direct and indirect effects, pp. 289–348 (Chapter 5) in *Climate Change 2001: The Scientific Basis*, Contribution of Working Group I to the Third Assessment Report of the Intergovernmental Panel on Climate Change, J. Houghton, Y. Ding, D. J. Griggs, M. Nouger, P. J. van der Linden, X. Dai, K. Maskell, and C. A. Johnson (eds.), Cambridge University Press, Cambridge, UK, and New York, 881 pp.

Pierce, J. R., D. K. Weisenstein, P Heckendorn, T. Peter and D. W. Keith, 2010. Efficient formation of stratospheric aerosol for geoengineering by emission of condensible vapor from aircraft. *Geophysical Research Letters* **37**, L18805, doi:10.1029/2010GL043975.

Ping, C-L, G. J. Michaelson, M. T. Jorgenson, J. M. Kimble, H. Epstein, V. E. Romanovsky, and D. A. Walker, 2008. High stocks of soil organic carbon in the-North American Arctic Region. *Nature Geoscience*, published online: 24 August 2008, doi:10.1038/ngeo284.

Prentice, I. C., 2001. The Carbon Cycle and Atmospheric Carbon Dioxide, pp. 183–237 (Chapter 3) in *Climate Change 2001: Impacts, Adaptation and Vulnerability*, Contribution of Working Group II to the Third Assessment Report of the Intergovernmental Panel on Climate Change, J. J. McCarthy, O. F. Canziani, N. A. Leary, D. J. Dokken, and K. S. White, (eds.), Cambridge University Press, Cambridge, UK, 1032 pp.

President's Science Advisory Committee (PSAC), November 1965. Appendix Y4: Atmospheric Carbon Dioxide, pp. 111–133 in *Restoring the Quality of our Environment, Report of the Environmental Pollution Panel*, The White House, Washington DC.

Rasch, P. J., P. J. Crutzen, and D. B. Coleman, 2008a. Exploring the geoengineering of climate using stratospheric sulfate aerosols: The role of particle size. *Geophysical Research Letters* **35**, L02809, doi:10.1029/2007GL032179.

Rasch, P. J., S. Tilmes, R. P. Turco, A. Robock, L. Oman, C.-C. Chen, G. L. Stenchikov, and R. R. Garcia, 2008b. An overview of geoengineering of climate using stratospheric sulfate aerosols. *Philosophical Transactions of the Royal Society A*, **366**, 4007–4037, doi:10.1098/rsta.2008.0131.

Rau, G. H., 2008. Electrochemical splitting of calcium carbonate to increase solution alkalinity: Implications for mitigation of carbon dioxide and ocean acidity. *Environmental Science and Technology* **42** (23), 8935–8940, doi:10.1021/es800366q.

Rau, G. H., and K. Caldeira, 1999. Enhanced carbonate dissolution: A means of sequestering waste CO_2 as ocean bicarbonate. *Energy Conversion and Management* **40**, 1803–1813.

Rau, G. H., K. G. Knauss, W. H. Langer, and K. Caldeira, 2007. Reducing energy-related CO_2 emissions using accelerated weathering of limestone. *Energy* **32**, 1471–1477.

Rayner, S., C. Redgwell, J. Savulescu, N. Pidgeon, and T. Kruger, 2009. Memorandum on draft principles for the conduct of geoengineering research, (the 'Oxford Principles') reproduced in House of Commons Science and Technology Committee, The Regulation of Geoengineering, Fifth Report of the Session 2009–10, Report together with formal minutes, oral and written evidence, HC 221, March 18, 2010 (see www.publications.parliament.uk/pa/cm200910/cmselect/cmsctech/221/221.pdf).

Richardson D., H. Felgate, N. Watmough, A. Thomson, and E. Baggs, 2009. Mitigating release of the potent greenhouse gas N_2O from the nitrogen cycle—could enzymic regulation hold the key? *Trends in Biotechnology* **27**, 388–397.

Ridgwell, A., J. S. Singarayer, A. M. Hetherington, and P. J. Valdes, 2009. Tackling regional climate change by leaf albedo bio-geoengineering. *Current Biology* **19**, 146–150.

Robock, A., 2008. 20 reasons why geoengineering may be a bad idea. *Bulletin of Atomic Scientists* **64**(2), 14–18, 59, doi:10.2968/064002006.

Robock, A., L. Oman, and G. Stenchikov, 2008. Regional climate responses to geoengineering with tropical and Arctic SO_2 injections. *Journal of Geophysical Research* **113**, D16101, doi:10.1029/2008JD010050.

Royal Society, 2005. *Ocean Acidification due to Increasing Atmospheric Carbon Dioxide Concentration*, Science Policy Section of The Royal Society, 68 pp.

Royal Society (prepared by J. Shepherd et al.), 2009. *Geoengineering the Climate: Science, Governance and Uncertainty*, Science Policy Centre of The Royal Society, 98 pp.

Ruddiman, W. F., 2001. *Earth's Climate: Past and Future*, W.H. Freeman and Company, New York 465 pp.

Ruddiman, W. F., 2007. The early anthropogenic hypothesis: challenges and responses. *Reviews of Geophysics* **45**, RG4001, doi:10.1029/2006RG000207.

Salter, S., 2012. Testimony to British MPs, reported in "Climate 'tech fixes' urged for Arctic methane," Richard Black, BBC News Science & Environment, March 17, 2012 (http://www.bbc.co.uk/news/science-environment-17400804).

Salter, S., and J. Latham, 2007. The reversal of global warming by the increase of the albedo of marine stratocumulus cloud, International Climate Change Conference, Hong Kong, China, May, 2007 (viewable at http://www.theengineer.co.uk/assets/getAsset.aspx?liAssetID=25650).

Salter S., G. Sortino, and J. Latham, 2008. Sea-going hardware for the cloud-albedo method of reversing global warming. *Philosophical Transactions of the Royal Society A* **366**, 3989–4006, doi:10.1098/rsta.2008.0136.

Schellnhuber, H. J., W. Cramer, N. Nakicenovic, T. Wigley, and G. Yohe (eds.), 2006. *Avoiding Dangerous Climate Change*, Cambridge University Press, Cambridge, UK, and New York, 392 pp.

Schneider, S. H., 2001. Earth systems: Engineering and management. *Nature* **409**, 417–421.

Schneider, S. H., 2008. Geoengineering: Could we or should we make it work? *Philosophical Transactions of the Royal Society A*, published online, doi:10.1098/rsta.2008.0145.

Schreier, M., A. A. Kokhanovsky, V. Eyring, L. Bugliaro, H. Mannstein, B. Mayer, H. Bovensmann, and J. P. Burrows, 2006. Impact of ship emissions on the microphysical, optical and radiative properties of marine stratus: a case study. *Atmospheric Chemistry and Physics Discussions* **6**, 1023–1071.

Seitz R., 2011. Bright water: hydrosols, water conservation and climate change. *Climatic Change* **105**, 365–381, doi:10.1007/s10584-010-9965-8.

Shakun, J. D., P. U. Clark, F. He, S. A. Marcott, A. C. Mix, Z. Liu, B. Otto-Bliesner, A. Schmittner, and E. Bard, 2012. Global warming preceded by increasing carbon dioxide concentrations during the last deglaciation. *Nature* **484**, 49–54.

Shindell, D., J. C. I. Kuylenstierna, E. Vignati, R. van Dingenen, M. Amann, Z. Klimont, S. C. Anenberg, N. Muller, G. Janssens-Maenhout, F. Raes, J. Schwartz, G. Faluvegi, L. Pozzoli, K. Kupiainen, L. Höglund-Isaksson, L. Emberson, D. Streets, V. Ramanathan, K. Hicks, N. T. Kim Oanh, G. Milly, M. Williams, V. Demkine, and David Fowler, 2012. Simultaneously Mitigating Near-Term Climate Change and Improving Human Health and Food Security. *Science* **13**, 183–189, doi:10.1126/science.1210026.

Socolow, R., et al., 2011. *Direct Air Capture of CO_2 with Chemicals: A Technology Assessment for the APS Panel on Public Affairs*, American Physical Society, Washington DC, 100 pp.

Stern, N., 2007. *The Economics of Climate Change: The Stern Review*, Cambridge University Press, Cambridge, UK, and New York, 712 pp.

Stolaroff, J. K., D. W. Keith, and G. V. Lowry, 2008. Carbon dioxide capture from atmospheric air using sodium hydroxide spray. *Environmental Science and Technology* **42**, 2728–2735.

Stolaroff, J. K., S. Bhattacharyya, C. A. Smith, W. L. Bourcier, P. J. Cameron-Smith, and R. D. Aines, 2011. Review of methane mitigation technologies with application to rapid release of methane from the Arctic. *Environmental Science and Technology* **46** (12), 6455–6469, doi:10.1021/es204686w.

Strong, A. L., J. J. Cullen, and S. W. Chisolm, 2009. Ocean fertilization: Science, policy and commerce. *Oceanography* **22**, 236–261.

Teller E., R. Hyde, and L. Wood, 2002. *Active Climate Stabilization: Practical Physics-Based Approaches to Prevention of Climate Change*, Lawrence Livermore National Laboratory, UCRL-JC-148012.

Teller, E., L. Wood, and R. Hyde, 1997. *Global Warming and Ice Ages: I. Prospects for Physics-based Modulation of Global Change*, UCRL-JC-128715, Lawrence Livermore National Laboratory, Livermore CA.

Tilmes, S., R. Müller, and R Salawitch, 2008. The sensitivity of polar ozone depletion to proposed geoengineering schemes. *Science* **320**, 1201–1204.

Trenberth, K. E., and A. Dai, 2007. Effects of Mount Pinatubo volcanic eruption on the hydrological cycle as an analog of geoengineering. *Geophysical Research Letters* **34**, L15702, doi:10.1029/2007GL030524.

Twomey, S., 1977. The influence of pollution on the shortwave albedo of clouds. *Journal of the Atmospheric Sciences* **34**, 1149–1152.

UNEP (prepared by C. P. McMullen and J. Jappour), 2009. *Climate Change Science Compendium 2009*, United Nations Environment Programme, Nairobi, Earthprint, 70 pp.

UNEP (United Nations Environment Programme), and WMO (World Meteorological Organization), 2011. *Integrated Assessment of Black Carbon and Tropospheric Ozone*, D. Shindell, Chair, UNEP, Nairobi, Kenya, 249 pp.

U.N. General Assembly, 1994. *United Nations Framework Convention on Climate Change*, resolution/adopted by the General Assembly, January 20, 1994, A/RES/48/189, available at: http://www.unhcr.org/refworld/docid/3b00f2770.html (accessed January 31, 2013).

Vermeer, M., and S. Rahmstorf, 2010. Global sea level linked to global temperature. *Proceedings of the National Academy of Sciences* **106**, 21527–21532, doi:101073/pnas.0907765106.

Victor, D. G., M. G. Morgan, J. Apt, J. Steinbruner, and K. Ricke, 2009. The geoengineering option, *Foreign Affairs* **88**(2), 64–76.

Wang H., P. J. Rasch, and G. Feingold, 2011. Manipulating marine stratocumulus cloud amount and albedo: A process-modeling study of aerosol-cloud-precipitation interactions in response to injection of cloud condensation nuclei. *Atmospheric Chemistry and Physics* **11**, 4237–4249, doi:10.5194/acp-11-4237-2011.

Wigley, T. M. L., 2005. The climate change commitment. *Science* **307**, 1766–1769.

Wigley, T. M. L., 2006. A combined mitigation/geoengineering approach to climate stabilization. *Science* **314**, 452–454, doi:10.1126/science.1131728.

Winton, M., 2005. Amplified Arctic climate change: What does surface albedo feedback have to do with it? *Geophysical Research Letters* **33**, L03701, doi:10.1029/2005GL025244.

World Meteorological Organization (WMO), United Nations Environment Programme (UNEP), and International Council of Science (formerly, International Council of Scientific Unions, ICSU), 1985. *Report of the International Conference on the Assessment of the Role of Carbon Dioxide and of Other Greenhouse Gases in Climate Variations and Associated Impacts*, World Meteorological Organization Publication #661, Geneva.

Index

J

K

L

M

T

U

V

W

For Product Safety Concerns and Information please contact our EU
representative GPSR@taylorandfrancis.com
Taylor & Francis Verlag GmbH, Kaufingerstraße 24, 80331 München, Germany

www.ingramcontent.com/pod-product-compliance
Ingram Content Group UK Ltd.
Pitfield, Milton Keynes, MK11 3LW, UK
UKHW022122190926
475702UK00013B/1586
* 9 7 8 1 1 3 8 0 7 4 1 1 8 *